Effects of Pollution on Fish

Effects of Pollution on Fish

Molecular Effects and Population Responses

Edited by

Andrew Lawrence

Department of Biological Sciences,
University of Hull, UK

and

Krystal Hemingway

Institute of Estuarine & Coastal Studies,
University of Hull, UK

Blackwell
Science

Editorial offices:
Blackwell Science Ltd, 9600 Garsington Road, Oxford OX4 2DQ, UK
Tel: +44 (0) 1865 776868
Iowa State Press, a Blackwell Publishing Company, 2121 State Avenue, Ames, Iowa 50014-8300, USA
Tel: +1 515 292 0140
Blackwell Science Asia Pty Ltd, 550 Swanston Street, Carlton, Victoria 3053, Australia
Tel: +61 (0)3 8359 1011

First published 2003

Library of Congress Cataloging-in-Publication Data

Lawrence, A. J. (Andrew J.)
Effects of pollution on fish : molecular effects and population responses / A.J. Lawrence, K.L. Hemingway.
p. cm.
Includes bibliographical references and index.
ISBN 0-632-06406-4 (hardback : alk. paper)
1. Fishes—Effect of pollution on. I. Hemingway, Krystal. II. Title.

SH174.L39 2003
571.9′517—dc21

2003005890

ISBN 0-632-06406-4

A catalogue record for this title is available from the British Library

Set in 10/13pt Times
by Graphicraft Limited, Hong Kong
Printed and bound in Great Britain using acid-free paper
by MPG Books Ltd, Bodmin, Cornwall

For further information on Blackwell Publishing, visit our website:
www.blackwellpublishing.com

Contents

List of Contributors

Augustine Arukwe Great Lakes Institute for Environmental Research (GLIER), University of Windsor, 401 Sunset Avenue, Windsor, Ontario, Canada N9B 3P4.

Miren P. Cajaraville Laboratory of Cell Biology and Histology, Department of Zoology and Animal Cell Dynamics, Science Faculty, University of the Basque Country, PO Box 644, E-48080 Bilbao, Basque Country, Spain.

Gary Carvalho Department of Biological Sciences, University of Hull, Cottingham Road, Hull, HU6 7RX, UK.

Kevin Crean Sunnydene, Blacktoft, Nr. Howden, ON14 7YN, UK.

Mike Elliott Institute of Estuarine & Coastal Studies (IECS), University of Hull, Cottingham Road, Hull, HU6 7RX, UK.

Stephen Feist Centre for Environment, Fisheries and Aquaculture Science (CEFAS), Weymouth Laboratory, Barrack Road, The Nothe, Weymouth, Dorset, DT4 8UB, UK.

Lars Förlin Department of Zoology, Zoophysiology, Göteborg University, Box 463, SE-40530, Göteborg, Sweden.

Anders Goksøyr Department of Molecular Biology, University of Bergen, HIB, Bergen, N-5020, Norway & Biosense Laboratories AS, Thormøhlensgate 55, N-5008 Bergen, Norway.

Lorenz Hauser School of Aquatic and Fishery Sciences, University of Washington, 1122 NE Boat Street, Box 355020, Seattle, WA 98195-5020, USA.

Krystal L. Hemingway Institute of Estuarine & Coastal Studies (IECS), University of Hull, Cottingham Road, Hull, HU6 7RX, UK.

Ketil Hylland Norwegian Institute for Water Research (NIVA), PO Box 173 Kjelsås, Brekkeveien 19, Oslo, N-0411, Norway.

Dagmar Krueger Arbeitsgemainschaft Forschungstauchender Biologen und Geowissenschaftler (ARFOBIG), Gaußstraße 38, 22765 Hamburg, Germany.

Carmen Lacambra UNEP-WCMC, 219 Huntingdon Road, Cambridge, CB3 0DL, UK.

Joakim Larsson Department of Physiology/Endocrinology, Sahlgrenska Academy, Göteborg University, Medicinaregatan, SE-40530 Göteborg, Sweden.

Andrew J. Lawrence Department of Biological Sciences, University of Hull, Cottingham Road, Hull, HU6 7RX, UK.

David Lowe Plymouth Marine Laboratory, Prospect Place, The Hoe, Plymouth, PL1 3DH, UK.

Peter Matthiessen Centre for Ecology and Hydrology, Ferry House, Far Sawrey, Ambleside, Cumbria, LA22 0LP, UK.

Mike N. Moore Plymouth Marine Laboratory, Prospect Place, The Hoe, Plymouth, PL1 3DH, UK.

Bente M. Nilsen Biosense Laboratories AS, Thormøhlesgt. 55, N-5008 Bergen, Norway.

Igor Olabarrieta Laboratory of Cell Biology and Histology, Department of Zoology and Animal Cell Dynamics, Science Faculty, University of the Basque Country, PO Box 644, E-48080 Bilbao, Basque Country, Spain.

Martin Sayer Scottish Association for Marine Science (SAMS), Dunstaffnage Marine Laboratory, Dunbeg, Oban, Argyll PA37 1QA, UK.

John Thain Centre for Environment, Fisheries and Aquaculture Science (CEFAS), Burnham Laboratory, Remembrance Avenue, Burnham-on-Crouch, Essex, CM0 8HA, UK.

Ralf Thiel German Oceanographic Museum, Katharinenberg 14/20, 18439 Stralsund, Germany.

John Wedderburn Coastal and Marine Biotechnologies Ltd, Tamar Science Park, 1 Davy Road, Derriford, Plymouth, Devon, PL6 8BX, UK.

Preface

In April 1999 an article appeared in *Fishing News* under the headline 'Dredging and pollution hit stocks far more than fishing'. The article reported on claims made by an environmental group that these two anthropogenic impacts are causing a far greater decline of fish stocks in the North Sea than 'supposed overfishing' and that if these were remedied 'you would be able to walk to Europe on the fish concentrated in the North Sea'. Whilst there is no clear evidence to support these claims, the article does raise a real issue of concern to EU policy makers, the general public, and the fishing community as a whole. This is the impact of pollution on commercial fish and fisheries and the health implications of eating contaminated fish products.

This book has resulted from the Commission of the European Communities, Agriculture and Fisheries (FAIR) specific Research and Technological Development programme, CT97 3827, *Impacts of Marine Xenobiotics on European Commercial Fish – Molecular Effects & Population Responses*. However, it does not necessarily reflect the Commission's views and in no way anticipates the Commission's future policy in this area.

This book has brought together experts from across Europe to examine the literature both on marine and freshwater fish and, where necessary, invertebrates and other model organisms, to produce a status report on pollution impacts and to construct a conceptual model to describe these impacts – from the subcellular and molecular level, through organism to population and community levels and subsequently to socio-economic implications.

The group of scientists involved in this book include individuals with expertise in each of the hierarchic levels of organisation on which pollution can impact. They encompass molecular geneticists, biochemists, physiologists, population and community biologists and fishery economics experts. Throughout the 2-year duration of the concerted action on which this volume is based, the group met on three occasions during which they worked on each of the thematic topics which form the basis to the chapters.

Chapter 1 introduces the subject and context of the volume. It also presents a conceptual model which was developed following the first meeting of the group in Oslo, Norway. The model presented in this chapter is a simplified version of that presented in the report to the European Commission. The model describes the way in which pollution may impact on a fishery and highlights the potential linkages between the various biological levels of organisation from molecular to community and economic. It is used to identify the direct links between the hierarchic levels of impact identified in the literature together with feedback or homeostatic mechanisms within the system.

The model additionally provides the framework around which the rest of the literature is presented in the book. Each of the following chapters were outlined at subsequent meetings of the group, first in Aveiro, Portugal, and later in Bilbao, Spain. The chapters represent linked technical themes which together describe the potential impact of pollution on a fishery. The chapters are designed both to describe the impact of pollution on the specific level of biological organisation and to highlight any linkages between this level and other, higher levels of complexity. The aim here was to confirm any pathways in which subcellular detection of pollution in the individual might lead to changes in population and community. A further important goal of the study was to identify and highlight any gaps in the literature that might help to direct future research in the field.

Chapter 2 reviews the forms of genetic damage that occur within the cell, either as a direct result of pollution perturbation or due to the production of genotoxic by-products of the detoxification process. Forms of damage include those caused by oxygen radicals, and the formation of adducts and mutations, together with direct effects on chromosomes. The chapter links genetic damage to examples of the consequence of this at higher levels of organisation from tumour formation, cell death, lesions, production of neoplasms, altered enzyme function and protein turnover rates. In addition, protection mechanisms are identified. This chapter is seen to have clear links with many of the other chapters within the book.

The links between molecular and cellular responses to pollution and the physiological response of individuals, including links to higher orders of organisation, are considered in Chapter 3. Principle components of this theme include the role of the lysosome and lysosome dysfunction related to altered rates of protein turnover and the energetic cost of altered gene expression. The chapter identifies links between these cellular events and organism effects including impacts on growth (including age/size trade-offs) and energy budget and scope for growth. Physiological effects examined included impacts on developmental processes such as osmoregulation, respiration and excretion, neuroendocrine and immune responses and impacts on reproduction.

Chapter 4 examines aspects of the physical health and immune system of fish in relation to pollution exposure and the links between these responses and those seen at higher and lower levels of organisation. Aspects of macrohealth considered include parasite load, presence of lesions and papillomas, spine and other deformities, anaemia, fin rot and fungal infection. These are examined in relation to cellular and molecular damage and the causative agents. Disorders are classified into pathological, physiological and developmental.

Chapter 5 considers one of the critical processes in the hierarchic chain of response through which impacts on the individual may be reflected in the population or natural homeostatic mechanisms override any pollution damage on the individual. Pollution impacts on reproduction and fecundity are linked directly to genetic damage in the gamete and impaired physiology through reallocation of energy. Additionally, the direct effects of endocrine disrupters on the reproductive process are considered. Evidence is reviewed on effects of pollutants on aspects of reproduction from egg size and viability to vitellogenic processes and fecundity.

Chapter 6 highlights the links between the individual response and population response to pollution. Concepts of community structure together with functioning and effects of pollutants on these, through transfers of effects from cellular to individual and then population and community, are considered in this chapter. The potential effects are analysed through the impacts on individual health in relation to condition and individual health in relation to production. Links between reproduction and population structure and survival to population yield are additionally reviewed. Production and yield are evaluated both in terms of quality and quantity of the population and quality of the individual.

The implications to population genetics and fitness as a result of pollution exposure, genetic damage, and cellular and molecular events are reviewed in Chapter 7. The theme reviews microevolutionary processes including mutation, selection, genetic drift and inbreeding. Sublethal responses and the potential for selection are examined, including fitness effects such as viability and fecundity. The literature on existing polymorphisms in pollutant metabolising genes is reviewed, together with the molecular basis of adaptation and the evolution of tolerance. The consequences of these adaptations in terms of reduced genetic heterogeneity and future fitness are considered.

Consequences of pollution impacts in relation to socio-economic effects are considered in Chapter 8. This incorporates the various bio-economic models currently employed by fisheries scientists. These models give information on fish quality and population, both of which are impacted upon by the effects of pollution. An important element of this chapter is not simply the impacts of pollution of fish and fish quality, but also the perceived impacts and the effects that this can have on individual fisheries and the overall resource. Direct links between lower order effects and impacts on the market are highlighted. Human health consequences are additionally considered.

Finally, Chapter 9 summarises the evidence presented in each of the previous chapters, with emphasis on the linkages identified between the hierarchic levels of biological organisation outlined in the conceptual model. The chapter highlights the areas in which recent advances have been made, as well as the aspects of the subject that require further study. In addition, it considers the limitations with current empirical approaches in quantifying some of the links in the hierarchic response from cell to population. The potential role of mathematical modelling in the field of ecotoxicology is briefly reviewed. In particular, recent developments and advances are outlined with regard to how models may be used to overcome some of the limitations with empirical study.

Acknowledgements

The editors and main authors would like to express their thanks to the many other scientists from a variety of countries who also contributed significantly to the knowledge and ideas within this book. In particular, Ionan Marigómez from the University of País Vasco, Bilbao, Basque Country, Spain for the conceptual model, and Victor Quintino, José Rebelo, Maria Ana Monteiro Santos and António Correia from the University of Aveiro, Portugal.

Although only the main authors are named on each chapter, the editors wish to note that everyone involved in the preparation of the book contributed significantly, both in the exchange and supply of information, to all chapters.

Chapter 1
Introduction and Conceptual Model

A.J. Lawrence and M. Elliott

1.1 Background

During the last decade of the twentieth century and the beginning of the twenty-first century the dependency of man on the earth's natural resources has become increasingly apparent. Fish and fisheries, worldwide, are one of the most important marine resources exploited by man. In addition to providing an extremely important source of protein on which many communities are reliant, they also create much needed employment in coastal areas around the world, including Europe. In 1999 total landings of fish, crustacea and molluscs in Europe were approximately 55 165 000 tonnes with a value of 64 510 million EUR and the fishing fleet alone employed 142 908 people (EUROSTAT, 2000).

The importance of the earth's natural resources and the need to protect them and use them sustainably was recognised by the nations of the world in Rio in 1992 during the United Nations Conference on Environment and Development (UNCED). This meeting led to the ratification of the 1993 UN Convention on Biological Diversity (CBD). The main objectives of the CBD are the conservation and sustainable use of biological diversity and the fair and equitable sharing of benefits arising from its utilisation. Biodiversity is defined here as the variability among living organisms from all sources including terrestrial, marine and other aquatic organisms and the ecological complexes of which they are a part, including diversity within species, between species and of ecosystems (Article 2, CBD).

The conference of the parties (COP) also identified marine and coastal biological diversity as an early priority. This was highlighted at COP 2 which adopted the Jakarta Mandate on Marine and Coastal Biological Diversity in 1995. The ministerial statement highlighted the urgent need for the COP to address the conservation and sustainable use of marine and coastal biodiversity and urged parties to initiate immediate action to implement decisions on this issue.

The urgent need to protect the marine environment and its resources recognises the impact that man is currently having on this environment. The Jakarta Mandate identifies the principle threats to marine biodiversity as being overexploitation of resources, habitat destruction, pollution and invasion by alien species. These threats do not have equal weighting. For example, in a study of local species extirpation in the Wadden Sea, Wolff (2000) found that the major threats to marine biodiversity were overexploitation and habitat loss and fragmentation. Marine pollution was responsible for the loss of two species whilst the introduction of exotic species had not caused any local extirpations.

The CBD recognises that a key factor in the conservation of biological diversity is its sustainable use. Unfortunately, fisheries throughout the world have a history of overexploitation. In Europe, according to the International Council for the Exploration of the Sea (ICES), two-thirds of North Atlantic commercial fish stocks are in serious decline. Spawning stock biomass (SSB) of North Sea, Irish Sea and Acto-Norwegian cod is currently at or near a historic low and 49% of the most important commercial fish stocks are considered to be outside safe biological limits (ICES, 1997). Reduction in SSB is reflected in low average recruitment during the 1980s and 1990s with up to 50% reduction in some stock and 25% reduction in cod, haddock and whiting (WWF, 2000).

Sea fish represent a natural and renewable resource. Healthy stocks can sustain a reasonable level of exploitation but for this they need a healthy marine environment. Unfortunately, the OSPAR Commission Quality Status Report (OSPAR, 2000) states that the marine environment is also at risk from hazardous substances. These include antifouling treatments, endocrine disrupters, radioactive substances, nutrient pollution and consequences of shipping activities including oil spills and ballast water discharges (WWF, 2001).

Whilst pollution may not be the single most important factor having impacts on European and global fisheries, it is clear that it may be having a significant impact on stocks already depleted by other factors, including overexploitation and habitat loss. Unfortunately, the impacts of pollution on already depleted fish stocks are not known but they are of increasing concern to EU policy makers, the general public and the fishing community. This was highlighted by the Select Committee of Science and Technology (1995) which noted that when areas of knowledge on pollution research are integrated, insufficient emphasis is placed on pollution issues within fisheries management. Policy makers within the EU need to understand and build pollution impacts on fish into fishery management under the European Union's Common Fisheries Policy (CFP). In December 2002, the conservation and management policy (EEC Regulation 3760/92) came to the end of a 20-year term and is currently under review. Consequently, it is essential that any pollution impacts on fisheries are now recognised, understood and built into any modified quota scheme under the Total Allowable Catch (TAC). These problems were highlighted at the Intermediate Ministerial Meeting on the Integration of Fisheries and Environmental Issues (Svelle *et al.*, 1997).

Whilst the European Union is mostly concerned with fishery management, the public are far more concerned with quality issues related to fish as well as human health-related aspects of eating contaminated fish products. Reports of significant numbers of fish being landed with spine deformities, skin lesions and cancers all add to the perception that pollution is having a dramatic impact on fish and that these may be passed on to the consumer. These perceptions, whether correct or not, may have significant ramifications on the economics of the fishery.

The third interest group with concerns related to pollution impacts is fishing communities themselves who are reliant on fishing as their economic base. As previously noted, with spawning stock biomass of several European cod populations at or near historic lows and with some stocks in danger of commercial collapse (WWF, 2000), the precise causes of the collapse and the interactions between these causes need to be determined and remedied so that a sustainable fishery can be developed. If these causes include pollution impact then the importance of this, relative to other causes, must be appreciated. The long-term survival of many of these communities will be dependent on this.

The concerns raised with each of these user groups are largely a result of work being undertaken by scientists in Europe and around the world looking at various aspects of pollution impact on fish and invertebrates from molecular responses to population and community changes. These studies may often be of a completely academic nature and simply advance scientific knowledge on the mechanism of impact and responses by organisms. However, the misinterpretation and communication of this information to the wider community may have led to unfounded concerns about the overall consequence of individual studies.

There are various levels of biological organisation on which pollution can impact. Anthropogenic effects may lead to severe consequences for populations or species occupying the area (Möller & Dieckwisch, 1991; Bernát *et al.*, 1994). However, consequences at the ecosystem level may display a long response time and when effects occur it may be too late to take countermeasures. Pollution exposure may also lead to decreased growth rates and increased infection but even these responses are preceded in time by effects at the molecular level (Blackstock, 1984; Boon *et al.*, 1992). An attempt has been made to combine impacts on systems at various levels under the concept of ecosystem health assessment and ecosystem pathology (Harding, 1992).

Extensive studies have been undertaken to examine and determine the impact of a wide range of xenobiotics on various individual aspects of fish and invertebrate biochemistry, physiology and population structure (Bayne *et al.*, 1988; Förlin *et al.*, 1995). These studies have been performed throughout Europe and worldwide to resolve various objectives. They may have been performed to determine the toxicity of a specific chemical or compound or to determine the potential hazard to individual species or ecosystems of the disposal of waste (Donkin & Widdows, 1986). They may have been developed to provide a rapid biomarker for pollution (Lawrence & Poulter, 1998; Goksøyr *et al.*, 1991; Depledge, 1994) or to act as bioassays of health, fitness and growth, under varied complex environmental parameters (Lawrence & Poulter, 2001). In some cases, commercial fish and shellfish have been used, whilst in others, species of ecological importance or that fulfil various monitoring criteria may have been chosen (Elliott *et al.*, 1988).

In the last two decades of the twentieth century attention was focused on subcellular responses to pollution. The need to detect and assess the impact of pollution, particularly low concentrations of increasingly complex mixtures of contaminants, on environmental quality has led to the development of molecular indicators of exposure to and effects of contaminants on aquatic organisms. Molecular indicators are often referred to as biomarkers but simply represent a subcellular response to exposure. Such diagnostic and prognostic early warning tests offer the potential of specificity, sensitivity and application to a wide range of organisms. Known biomarkers of early warning capacity include induction of metallothioneins, stress proteins, the cytochrome P450 enzyme system e.g. CYP1A, UDP glucuronosyl and glutathione transferases to detect exposure/effects of different metal and organic contaminants (Goksøyr & Förlin, 1992; Lawrence & Nicholson, 1998).

cDNA probes against fish CYP1A are now available and new and sensitive biomarkers include NADPH: quinone reductase (DT-diaphorase). Antioxidant defence systems include detection of glutathione reductase and GSH:GSSG. Additionally, glutathione peroxidases, lipid peroxidation and protein oxidation are of increasing interest. Enzymes such as superoxide dismutase and catalase are also now included in current studies (Cajaraville *et al.*, 1992).

More recently research has focused on direct damage to DNA caused by xenobiotics, and here mainly on chromosomal aberrations, micronucleus formation, DNA adducts (covalent attachments of a chemical to DNA) and strand breakage. The latter two of these responses have been used as relatively quick and sensitive biomarker assays for exposure to genotoxic compounds (e.g. Stein *et al.*, 1993, 1994; Theodorakis *et al.*, 1994; Shugart & Theodorakis, 1994).

This wealth of information appears to demonstrate an impact of pollution on a variety of organisms at a subcellular level. However, a direct link between effects at this molecular level and population/yield impacts is yet to be demonstrated in any species. Indeed, there are many stages in the hierarchy of response from molecular to population in which homeostatic mechanisms within an individual, population or community may act to absorb or nullify the response seen at the subcellular level. Consequently, the science is most precise, and there is least noise, at the lower levels of biological organisation. However, management is only willing to operate at the higher levels of organisation (population and community) and the link between these two levels needs to be established.

Despite this, individual laboratories around the world continue to develop ever more sensitive biomarkers for genetic/biochemical response. At an academic level this is extremely valuable in advancing the field of environmental ecotoxicology and how pollutants affect organisms. However, there may also be a problem in that these studies may also drive pollution legislation and clean-up, without ever showing a clear impact on the individual at higher levels of organisation (reproduction, fecundity, population/yield).

No study has currently attempted to link each of the response criteria (biochemical, cellular, physiological, reproduction, population/yield) to evaluate the ramifications of low level, sublethal effect to population and community structure and thence the socio-economic impact to communities exploiting the resource.

1.2 Aims and objectives

The aim of this book is to review and synthesise information and literature on the impacts of pollution at hierarchic levels of organisation in fish, and where necessary other invertebrates and model organisms, to produce a status report which:

- Identifies any mechanistic links between the hierarchic levels of biological organisation (genetic, subcellular, physiological, reproduction and fecundity, population and yield, socio-economic);
- Presents a conceptual model which can be used to illustrate the recognised and potential links between biological levels of organisation, and around which gaps in the current knowledge and research priorities might be identified;

- Assesses the present ability to quantify the links and cause/effect relationships between the various biological levels, to consider how near or far the science is from offering any predictive capability for fisheries management;
- Identifies or highlights any potential links in the biological system in which homeostatic mechanisms may have an ability to absorb any effects of change detected at lower levels of organisation.

Whilst the principle aim of the review is to identify and demonstrate the links between the various hierarchic levels of response within a species, it should be noted that for any study attempting to demonstrate the linkages to be truly rigorous, effects response should also be directly linked to body burden of pollutant.

1.3 Contaminant, environmental and life history stage factors

Before considering the conceptual model it should first be noted that the type of response to pollution elicited in an organism will depend on the type of contaminant and whether this is acting singularly or in combination with others. It will also depend on the interaction of the contaminant(s) with other environmental factors such as temperature, salinity and dissolved oxygen levels. Finally, it will depend on the stage of development and health of the organism as it comes into contact with the pollutant. Whilst it is beyond the scope of this book to consider these factors in detail, some consideration must be given to them because of the implications of these on the response of the organism to the pollution event.

1.3.1 Contaminants

Whilst it is beyond the scope of this book to detail the various types of contaminant that impact on an organism, it is necessary to give a basic classification of the major classes of pollutant. This is important because often the mechanism of impact or subcellular response elicited is specific to a particular type of contaminant, whether it is a single contaminant or, as is more likely, a mixture of contaminants, and finally the level of contamination, i.e. whether the impact is lethal or sublethal, acute or chronic. Briefly, contaminants may be divided into the following categories.

1.3.1.1 Halogenated hydrocarbons

Since the discovery of widespread distribution of chlorinated contaminants in aquatic organisms in the 1960s, there have been numerous reports on the bioaccumulation of halogenated hydrocarbons. This term spans a wide range of contaminants including: DDT and its metabolites, polychlorinated biphenyls (PCB), polychlorinated dibenzodioxins and dibenzofurans, hexaclorobenzene (HCB), octachlorostyrenes (OCS), toxaphene, chlordanes, dieldrin, hexachlorohexane (HCH, lindane), polybrominated diphenylethers (PBDE), polybrominated biphenyls (PBB), polychlorinated paraffins (CP) and polychlorinated naphthalenes (PCN). Halogenated organic contaminants are more or less resistant to degradation in biological systems and some of them, e.g. DDTs and PCBs, have been found in all biological samples studied. In addition to methylmercury, halogenated hydrocarbon contaminants predominantly contribute towards problems for the use of marine organisms as a food resource. Many of these chemicals are synthetic and thus the mechanisms evolved for dealing with them are poor or non-specific.

1.3.1.2 Non-halogenated hydrocarbons

Non-halogenated hydrocarbons can be divided into aromatic and non-aromatic. Hydrocarbons with non-aromatic groups are generally degraded quickly and form little risk to the environment. Aromatic hydrocarbons may accumulate in organisms with low metabolic activity towards planar substances, such as some bivalves, but are generally metabolised in fish. Some results indicate that fish species with high fat-content of non-metabolic tissues, e.g. eel, may accumulate polycyclic aromatic hydrocarbons (PAHs). Treatment of fish products, especially smoking, in general causes much higher levels of non-halogenated hydrocarbons in fish than environmental exposure to such substances.

1.3.1.3 Organometals

The single most serious incidence of human consumption of contaminated seafood, the 'Minimata' incident, was caused by an organometal. In this incident, methylmercury, produced by the methylation of industrially discharged mercury, was taken up by marine invertebrates and fish. Consumption of these products is thought to have been responsible for over 100 deaths and many cases of severe disability. In addition, organic forms of lead, tin, selenium, antimony and arsenic are found in the marine environment. With respect to metals (section 1.3.1.4), it is important to distinguish between organic metals or metalloids that may have metabolic roles, e.g. arsenobetaine in crustaceans and selenium-dependent enzymes, and those that have no known function, e.g. methylmercury, alkyl-lead and tributyltin. For reasons which are not entirely clear, methylmercury tends to accumulate in muscle, even in species with low-fat muscle tissue.

1.3.1.4 Non-organic metals

Metals can be divided into three principle groups: bulk metals, essential (trace) metals and non-essential (heavy) metals. Most metals do not form stable alkylated forms, but some (e.g. Cu, Hg) have high affinity for organic material and may be found associated with organic macromolecules in water and/or sediment. Hence, there is the need to consider the behaviour of the pollutant in the environment as well as in the organism. In any discussion concerning tissue metal levels it is vital to consider essential and non-essential metals separately. In addition, natural levels vary widely between species and taxonomic groups. Essential metals, i.e. elements which all living organisms need to exist, include Fe, Cu, Zn, Mn, Mo and Ni. Whereas the lack of one or more of these elements is not uncommon in terrestrial organisms, such deficiencies have not been reported for marine invertebrates or vertebrates. Non-essential metals, e.g. elements for which there is no known function, include Cd, Hg, Pb, Ag and Au. A typical difference in the accumulation of essential and non-essential elements is the longer biological half-life of the latter. The highest levels of both essential and non-essential metals are generally found in the liver.

1.3.2 Life-stage interactions

In addition to the type and level of contaminant and its behaviour in the environment (air, water, sediment and their interfaces), the response of the organism will also depend on its developmental stage and the interaction between it and its environment. For example, the consequences of any impact of genotoxic compounds is likely to be far more severe if the impact is on germ cells than if it is on adult somatic cells. Embryonic and larval stages of organisms are generally recognised as being far more sensitive to contaminants than adult stages given their often unprotected nature and smaller size (larger surface area to volume ratio). Even in adults, however, the response to pollution, seen at each level of organisation, will depend on whether the organism has been previously exposed to the pollutant or is from a genotypically adapted population. The organism's health, age, reproductive state and nutritional state will all affect its response to pollution load.

Furthermore, many organisms and particularly fish are mobile and operate in an open rather than closed system. Being mobile, they may potentially avoid discrete pollution incidents such as oil spills by moving around or away from the area. Alternatively, however, they may be exposed to a broader range of diffuse pollutants at various times and in various locations within their geographic range. Ideally, therefore, tracking an animal through its life cycle and geographic range may be important to determine cumulative toxic effects at different developmental stages.

1.3.3 Environmental factors

In addition to contaminants and stage specific responses, fish also have to respond to environmental change brought about by climatic conditions. Often, extremes of these natural factors can induce subcellular and physiological responses in fish similar to those caused by pollution. Environmental factors include: salinity, pH, temperature, current strength and oxygen levels. Impacts may include starvation, chemical stress from for example hypoxia, increased predation and altered population density (Bucke, 1993).

1.3.4 Summary

The combination of type and mixture of contaminant, its interaction with the environment and the life-stage of the organism being impacted, make it very difficult to tease out the specific impact of pollution on an organism over its lifetime when this impact is at a sub-lethal level. In addition, it makes the development of a conceptual model to illustrate this impact much more complicated. Indeed, to model all of these interactions together would require a complex multidimensional model. Whilst it is possible to do this, the model would become too complex to be readily interpretable visually. Consequently, a more simplistic schematic model is presented which allows the links between the hierarchic levels of biological response to be identified.

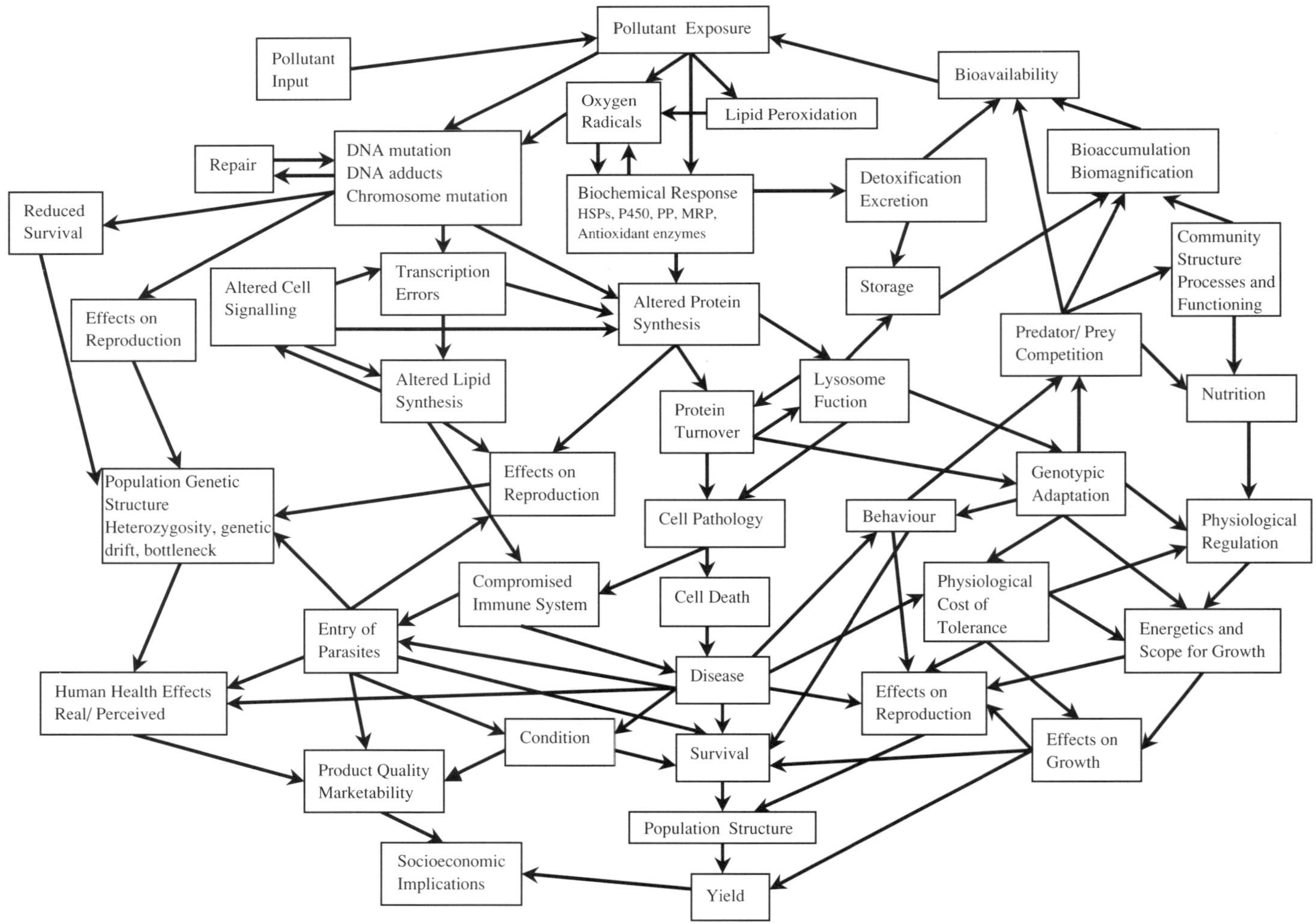

Fig. 1.1 Conceptual model.

1.4 Overview of the conceptual model

Figure 1.1 shows the conceptual model developed for this book. At its simplest, this model shows the possible mechanistic linkages between the various hierarchic levels of biological response to pollution from molecular to population and then socio-economic. However, in more detail, it could potentially provide the framework around which a mathematical model can be developed with predictive capability.

As already described, the focus of toxicological studies at a subcellular level has resulted in a good appreciation of how pollutants can affect DNA and proteins. To summarise, impacts of pollutants at a biochemical level can result in the induction of a variety of proteins and enzymes involved in xenobiotic detoxification, metabolism and excretion. The pathway induced depends on the species of contaminant but includes metallothioneins, stress proteins, CYP1A, UDP glucuronosyl and glutathione transferases and NADPH: quinone reductase (DT-diaphorase) (Fig. 1.1). Antioxidant defence systems include glutathione reductase and GSH:GSSG. In addition, glutathione peroxidases and lipid peroxidation and protein oxidation have been identified and enzymes such as superoxide dismutase and catalase have also been included in studies.

Direct damage to DNA caused by genotoxic xenobiotics and UV radiation, includes chromosomal aberrations such as micronucleus formation, DNA adducts (covalent attachments of a chemical to DNA) and strand breakage (Fig. 1.1). Although little is known about mechanistic links between DNA damage and effects on the individual and population, biomarkers like DNA strand breakage, chromosome aberration and DNA adducts have been correlated with mortality, malformations and fecundity. This is obviously, therefore, a direct route between subcellular damage and possible impacts on fecundity, larval survival and consequently population.

Many of the products of detoxification or the site of the detoxification process are the lysosomes (Fig. 1.1). Lysosomes are subcellular organelles bounded by a semi-permeable lipoprotein membrane that act optimally at an acid pH and are collectively capable of degrading all classes of macromolecules of endogenous (intracellular) and exogenous (extracellular) origin. Lysosomes appear to be ubiquitous in animal cells, with the notable exception of mammalian red blood cells, and their role includes sequestration of foreign compounds, the immune response and intracellular digestion as well as an involvement in reproduction, embryonic development and programmed cell death (apoptosis). In addition to the sequestration of pathogens, lysosomes also accumulate a diverse range of chemical contaminants from the environment, that are damaging to cells.

Chemical exposure can have a stabilising or destabilising effect on the lysosome membrane. In addition, it can activate or inhibit the acid hydrolases within the lysosome. In combination, membrane destabilisation and acid hydrolase activation leads to lysosomal damage and this has been described in a variety of finfish. Impacts on the lysosomal system, particularly in eggs and early life-stages of commercially important fish and molluscs, may have significant implications for higher levels of organisation. Additionally, effects on the lysosome rich liver and hepatopancreas of finfish and molluscs respectively have been reported and may be important because these tissues are central to numerous biological processes which become impaired.

However, pollutants may impact on the physiology and reproductive system of finfish in other ways. For example, the induction of detoxification systems required the diversion of energy away from other metabolic processes (Fig. 1.1). Sequestration of pollutants may, therefore, lead to reduced energy for growth and reproduction in exposed populations. The concept of scope for growth (SfG) was developed in fish but used more effectively with invertebrates. More recent approaches to examining the impact of xenobiotics on organism energetics include cellular energy allocation (CEA) and adenylate energy charge (AEC). Changes in energy balance related to rates of protein turnover have also been suggested as a potential mechanism which affects an individual's fitness in terms of its potential to survive contamination and to reproduce under these conditions.

In addition, adverse effects of environmental pollutants are known to interact with the endocrine system (endocrine disrupters) in fish. Endocrine disrupters can affect normal function in all organs that are regulated by hormones. Also, small disturbances in endocrine function especially during early life-stages lead to adverse and lasting effects. For example, xenoestrogens have the potential to affect sex differentiation in fish, and steroid hormone dysfunction may link to embryonic malformation.

Cellular, tissue and organ pathologies such as fin erosion, ulcers and tumours are often recorded as evidence of pollution impact although their appearance is not enough to attribute them to pollutants. However, the relationship between particularly the liver as a site of detoxification, lysosome activity and the occurrence of neoplastic lesions in this tissue does suggest that a mechanistic link should exist. More difficult to interpret are the consequences of these pathologies on the fitness of the fish in terms of reproductive output and fecundity.

Impacts on reproduction and fecundity may ultimately be expressed at the population level. However, this is one of the most difficult steps to demonstrate scientifically. Changes in population may be linked with many other environmental parameters including changes in species interaction within a community. There are studies that have demonstrated the loss of species or reduction in populations in relation to pollution gradients. Examples of these include the impact of bleach kraft mill effluent (BMKE) on fish communities in Sweden. However, whilst correlations between population change and pollution load may be demonstrated, this does not prove cause and effect. Furthermore, studies on the impact of pollution on populations are not consistent. Other cases have shown that a particular fish population may increase in environments exposed to pollutants and that this is due to reduced competition for resources resulting from the loss of a competitor.

Although spatial differences in pollutant effects are commonplace, relatively little attention has been paid to differences between populations in the responses to pollution. One major problem is that in most marine and estuarine areas, stressors work in combination. Anthropogenic impacts on the environment, other than pollution, may work additively, antagonistically or synergistically to affect a population. It is, therefore, difficult to separate the effects of pollution from these other anthropogenic factors.

Pollutants can, however, be expected to exert strong selection pressures on a population. If the pollution load exceeds the ability to survive of some of the individuals within a population, this will lead to an increase in the frequency of tolerant genotypes. Those individuals within the population that have the ability to survive and reproduce under the pollution stress express these genotypes. There is evidence for differential pollution tolerances between genotypes and for the predominance of such tolerant genotypes in field populations from exposed sites. However, these tolerant genotypes will result in a weaker population exhibiting effects such as increased mortality and reduced fecundity.

Selection caused by pollution, together with reduction in population sizes due to increased mortality, may also lead to a reduction in genetic variability of exposed populations which in turn has been shown to result in reduced fitness parameters like growth, fecundity and survival. Knowledge of such changes in population genetic structure is crucial for the assessment of the long-term effects of pollution, as populations may be able to adapt to certain pollutants but may lose genetic variability and fitness and, therefore, be more vulnerable to other stochastic events such as genetic drift, demographic stochasticity and environmental stochasticity.

There are two significant ways in which these combined effects of pollution on individuals and populations may have consequences on man. The first is through socio-economic impacts and the second is through human health implications. These, therefore, form the final links in the model (Fig. 1.1). Pollution impacts on fish species of economic importance may have two principle consequences. Firstly, if the impact results in a reduction of the fishery population, then there will be an equivalent reduction in economic benefit from the stock. Alternatively, however, there may not be a reduction in the quantity of the fish but in the quality of the product. These, together with any perceived human health-related aspects to impacted fish, could result in a collapse in that particular market with reduced prices for the product until such time as consumer confidence has been regained. The perceived effects of pollution may, therefore, be as important to the fishery as any actual pollution impact. Bioeconomic models are currently being developed which try to build in the impact of pollution on fisheries economics.

1.5 Conclusions

Based on this broad overview it is not difficult to construct a simplistic conceptual model from which evidence for the links between hierarchic levels of impact of pollution on fish can be examined. Whilst the model provides a useful framework around which the review can be developed, it is important to remember that there are many other factors, both biotic and abiotic, which affect an organism throughout its life. However, to incorporate these parameters into the conceptual model would make it too complicated to identify the potential mechanistic links between the levels of response.

This book presents a status report on the published relationships between each of the hierarchic levels of response to pollution from molecular to population and economic. It identifies and confirms links where possible, whether these are real or perceived, using literature on fish and, where necessary, invertebrates and other model organisms. The following chapters describe the impacts of pollution on and between each level of organisation in much more detail, highlighting links where applicable and homeostasis.

1.6 References

Bayne, B.L., K.R. Clarke & J.S. Gray (eds) (1988) *Biological Effects of Pollutants: results of a practical workshop*. Inter-Research, Amelinghausen, 278 pp.

Bernát, N., B. Köpcke, S. Yasseri, R. Thiel & K. Wolfstein (1994) Tidal variation in bacteria, phytoplankton, zooplankton, mysids, fish and suspended particulate matter in the turbidity zone of the Elbe Estuary: interrelationships and causes. *Netherlands Journal of Sea Research*, **28** (3–4), 467–476.

Blackstock, J. (1984) Biochemical metabolic regulatory responses of marine invertebrates to natural environmental change and marine pollution. *Oceanography and Marine Biology*, Annual Review, **22**, 263–313.

Boon, J.P., J.M. Everaarts, M.T.J. Hillebrand, M.L. Eggens, J. Pijnenburg & A. Goksøyr (1992) Changes in levels of hepatic biotransformation enzymes and haemoglobin levels in female plaice (*Pleuronectes platessa*) after oral administration of a technical polychlorinated biphenyl mixture (Clophen A40). *Science of the Total Environment*, **114**, 113–133.

Bucke, D. (1993) Aquatic pollution: effects on the health of fish and shellfish. *Parasitology*, **106**, S25–S37.

Cajaraville, M.P., J.A. Uranga & E. Angulo (1992) Comparative effects of the WAF of three oils on mussels. 3. Quantitative histochemistry of enzymes related to the detoxication metabolism. *Comparative Biochemistry and Physiology*, **103C**, 369–377.

Depledge, M. (1994) Genotypic toxicity: implications for individuals and populations. *Environmental Health Perspectives*, **102** (12), 101–104.

Donkin, P. & J. Widdows (1986) Scope for growth as a measure of environmental pollution and its interpretation using structure-activity relationships. *Chemistry and Industry*, **21**, 732–735.

Elliott, M., A.H. Griffiths & C.J.L. Taylor (1988) The role of fish studies in estuarine pollution assessment. *Journal of Fish Biology*, **33** (Suppl. A), 51–61.

EUROSTAT (2000) The Statistical Office of the European Union. *European Parliament Fact Sheet 4.2.5 Fisheries Policy*. In: www.europarl.eu.int/factsheets4_2_5fi.htm

Förlin, L., T. Andersson, L. Balk & Å. Larsson (1995) Biochemical and physiological effects of bleached pulp mill effluents in fish. *Ecotoxicol. Environmental Safety*, **30**, 164–170.

Goksøyr, A. & L. Förlin (1992) The cytochrome P450 system in fish, aquatic toxicology and environmental monitoring. *Aquatic Toxicology*, **22**, 287–311.

Goksøyr, A., T.S. Solberg & B. Serigstad (1991) Immunochemical detection of cytochrome-P4501A1 induction in cod larvae and juveniles exposed to a water-soluble fraction of North sea crude oil. *Marine Pollution Bulletin*, **22** (3), 122–127.

Harding, L.E. (1992) Measures of marine environmental quality. *Marine Pollution Bulletin*, **25** (1–4), 23–27.

ICES (1997) Report of the study group on the precautionary approach to fisheries management. ICES CM 1997/Assess: 7, Copenhagen.

Lawrence, A.J. & B. Nicholson (1998) The use of stress proteins in *Mytilus edulis* as indicators of chlorinated effluent pollution. *Water Science and Technology*, **38**, 253–261.

Lawrence, A.J. & C. Poulter (1998) Development of a sub-lethal pollution bioassay using the estuarine amphipod *Gammarus duebeni*. *Water Research*, **32**, 569–578.

Lawrence, A.J. & C. Poulter (2001) The impact of copper, PCP and benzo[a]pyrene on the reproduction of *Chaetogammarus marinus*. *Marine Ecology Progress Series*, **223**, 213–223.

Möller, H. & B. Dieckwisch (1991) Larval fish production in the tidal River Elbe 1985–1986. *Journal of Fish Biology*, **38**, 829–838.

OSPAR (2000) *Quality Status Report 2000*. OSPAR Commission for the Protection of the Marine Environment of the North-East Atlantic. OSPAR Commission, London, 108 pp.

Select Committee of Science and Technology (1995) *Second Report on Fish Stock Conservation and Management*, HL25, 25–1 and 50–1 (session 1994/5). The House of Lords Select Committee on Science and Technology, the UK Parliament. www.parliament.the-stationery-office.co.uk

Shugart, L. & C. Theodorakis (1994) Environmental genotoxicity: probing the underlying mechanisms. *Environmental Health Perspectives*, **102**, 13–18.

Stein, J.E., T.K. Collier, W.L. Reichert, E. Casillas, T. Hom & U. Varanasi (1993) Bioindicators of contaminant exposure and sublethal effects in benthic fish from Puget Sound, WA, USA. *Marine Environmental Research*, **35** (1–2), 95–100.

Stein, J.E., W.L. Reichert & U. Varanasi (1994) Molecular epizootiology: assessment of exposure to genotoxic compounds in teleosts. *Environmental Health Perspectives*, **102** (Suppl. 12), 19–23.

Svelle, M., H. Aarefjord, H.T. Heir & S. Overland (eds) (1997) *Assessment Report on Fisheries and Fisheries Related Species and Habitats Issues*. Intermediate Ministerial Meeting on the Integration of Fisheries and Environmental Issues, Ministry of Environment, Norway.

Theodorakis, C.W., S.J. Dsurney & L.R. Shugart (1994) Detection of genotoxic insult as DNA strand breaks in fish blood cells by agarose gel elecrophoresis. *Environmental Toxicology and Chemistry*, **13** (7), 1023–1031.

Wolff, W. (2000) Causes of extirpations in the Wadden Sea and estuarine areas in The Netherlands. *Conservation Biology*, **14**, 876–885.

WWF (2000) *The Common Fisheries Policy: Background and Review for 2002*. Vol. 42. Marine Update, WWF-UK.

WWF (2001) *Time for a different approach for the marine environment*. Vol. 45. Marine Update, WWF-UK.

Chapter 2

Genetic Damage and the Molecular/Cellular Response to Pollution

M.P. Cajaraville, L. Hauser, G. Carvalho, K. Hylland, I. Olabarrieta, A.J. Lawrence, D. Lowe and A. Goksøyr

2.1 Damage to DNA by oxygen radicals

2.1.1 Contaminants

A wide range of contaminants can give rise to an increased generation of free radicals, notably oxygen free radicals, also known as 'reactive oxygen species' (ROS) or 'reactive oxygen intermediates' (ROI). Elevation of ROS production with exposure to pollution can occur by several mechanisms. These include the uptake of redox cycling metals and organic xenobiotics, the metabolism of xenobiotics to redox cycling derivatives such as quinones and the induction of oxyradical generating enzymes (Livingstone *et al.*, 1989).

Redox cycling contaminants include some transition metals such as iron, copper and manganese, and organic xenobiotics including aromatic diols and quinones, nitroaromatics, aromatic hydroxylamines, and bipyridyls (Di Giulio, 1991). Other contaminants such as polycyclic aromatic hydrocarbons (PAHs) can be metabolised to redox cycling compounds (i.e. quinones) through the cytochrome P450 system, previously called the mixed function oxidase or MFO system. This is a universally distributed system involved in the metabolism of both endogenous compounds and xenobiotics (see Chapter 3). In addition, the biotransformation enzymes cytochrome P450, cytochrome P450 reductase and other flavoprotein reductases are considered to generate ROS as by-products (Livingstone *et al.*, 1989). Other enzymes that give rise to ROS formation are the peroxisomal oxidases, some of which are induced after exposure to peroxisome proliferating drugs and xenobiotics (Reddy & Mannaerts, 1994).

Peroxisome proliferators include a vast array of structurally unrelated compounds such as hypolipidemic drugs and other therapeutic drugs, phthalate ester plasticisers, steroids, pesticides, solvents and diverse industrial chemicals, food flavours, hydrocarbons (PAHs) and polychlorinated biphenyls (PCBs) (Beier & Fahimi, 1991; Bentley *et al.*, 1993; Fahimi & Cajaraville, 1995; Lake, 1995). One common feature of these compounds or their metabolic derivatives is a hydrophobic-lipophilic backbone with an acidic function, generally a carboxylic group (Fahimi & Cajaraville, 1995).

2.1.2 *Production mechanisms*

2.1.2.1 General aspects

Reactive oxygen species (ROS) or reactive oxygen intermediates (ROI) consist of the superoxide anion radical (O_2^-), hydrogen peroxide (H_2O_2) and the hydroxyl radical (.OH). Other biologically relevant ROS include singlet oxygen and alkoxyl radicals. Superoxide anions are generated by enzymatic or non-enzymatic univalent reduction of oxygen. In the endoplasmic reticulum or microsomes where the cytochrome P450 system resides, O_2^- can be generated by redox cycling or by other cytochrome P450 catalysed reactions. Following this, O_2^- can dismutate to H_2O_2 either spontaneously or in a reaction catalysed by the antioxidant enzyme superoxide dismutase. H_2O_2 can be further reduced to give .OH by different mechanisms involving catalytic redox cycling of metals, biological metal chelates or other oxyradicals (Halliwell & Gutteridge, 1986). Iron is required for the production of .OH from H_2O_2 via the Fenton reaction and the conversion of O_2^- to .OH via catalysis of the Haber-Weiss reaction (Halliwell & Gutteridge, 1986). Other transition metals such as copper and manganese can also catalyse the reaction.

Consequently, exposure to contaminants inducing the cytochrome P450 system leads to an increased generation of ROS in target cells (section 2.1.2.2). In addition, the metabolism of certain xenobiotics by microsomal enzymes can directly render free radical metabolites, which may themselves produce several deleterious effects.

Oxygen consumption and hence ROS production occurs by a multitude of oxidative processes in various cell compartments including the mitochondria, cytosol, endoplasmic reticulum, peroxisomes and lysosomes (Fig. 2.1). In specific cell types such as activated phagocytic cells, plasma membrane-bound NADPH oxidase is an important additional source of ROS (mainly O_2^-) during the oxidative burst (Adema *et al.*, 1991; Wientjes & Segal, 1995). During the latter process, lysosomal myeloperoxidase catalyses the formation of hypochlorite or HOCl, a potent oxidant acting on amines, amino acids, thiols, thioethers, nucleotides and haemoproteins (Sies & de Groot, 1992).

The mitochondrial electron transport system is a well-known source of ROS, as demonstrated both *in vivo* and *in vitro* (Loschen & Flohé, 1971; Nohl & Hegner, 1978). The mitochondrial enzymes involved in ROS production include NADH-coenzyme Q complex, succinate-coenzyme Q complex, and coenzyme QH2-cytochrome c reductases complex (Kehrer, 1993). The ROS primarily generated in these reactions seem to be the superoxide anion that may give rise to H_2O_2 after dismutation.

In mammalian liver, peroxisome respiration can account for 10–35% of total respiration (de Duve & Baudhuin, 1966). The various flavin oxidases present in peroxisomes reduce molecular oxygen to H_2O_2. As well as generating 34% of the H_2O_2 found in the cell, peroxisomes also produce reduced amounts of O_2^- by means of their xanthine oxidase, cytochrome b5, cytochrome P450 and 20 kDa membrane protein activities (Dhaunsi *et al.*, 1992; Zwacka *et al.*, 1994; Singh, 1997). This O_2^- can then give rise to the extremely reactive hydroxyl radical in the presence of transition metals. The peroxisomal generation of ROS is greatly increased during peroxisome proliferation, a process characterised by induction of several ROS-producing peroxisomal enzymes (section 2.1.2.3).

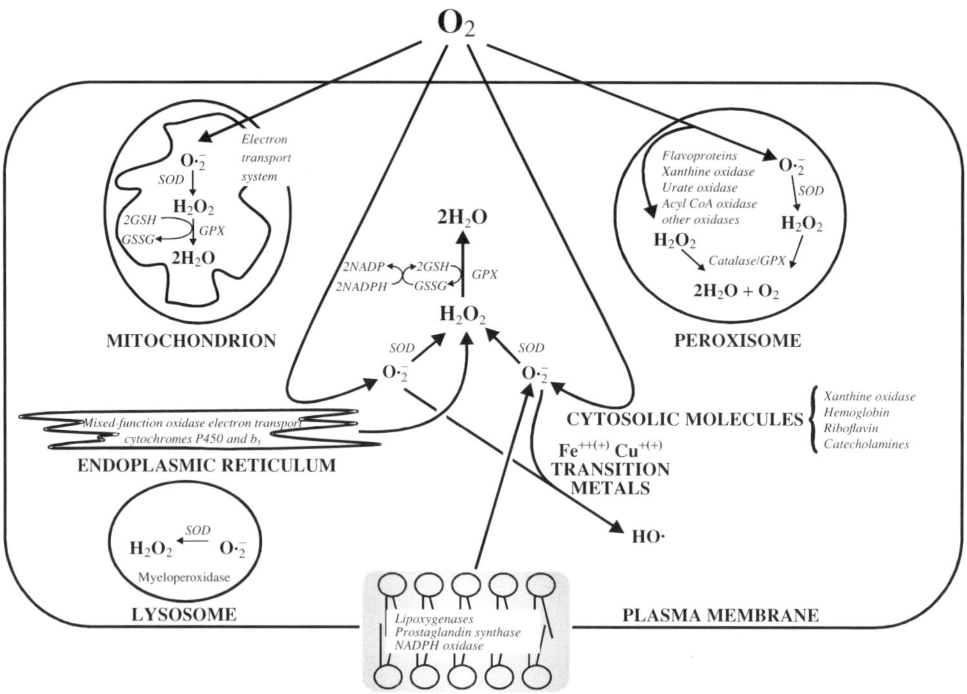

Fig. 2.1 Endogenous sources of ROS. Production of ROS occurs in different cell compartments including plasma membrane, mitochondria, cytosol, endoplasmic reticulum, peroxisomes and lysosomes. The importance of the various routes of ROS production varies from cell to cell, depending on the relative abundance of each cellular compartment, the physiological status of the cell and the extracellular environment of the cell. Antioxidant enzymes are found in several cell compartments and act as a primary defence by scavenging newly produced ROS (refer to Fig. 2.5).

There are reports indicating that oxyradicals (possibly superoxide but not hydrogen peroxide or hydroxyl radical) may be produced within the lysosomal compartment in association with the pynocytotic activity of molluscan digestive gland cells (Winston *et al.*, 1991). Thus, non-fluorescent dihydrorhodamine 123 was endocytosed by isolated digestive gland cells and oxidised to give fluorescent products presumably by superoxide radicals within lysosomes. Additionally, several studies have demonstrated that oxyradicals generated in the cytosol can cross the lysosomal membrane and cause damage to the lysosomal membrane in a number of cultured mammalian cell systems (Brunk & Cadenas, 1988; Brunk *et al.*, 1995; Roberg & Öllinger, 1998). According to these studies, hydrogen peroxide produced during oxidative stress may cross the lysosomal membrane. Inside the lysosome, the acidic pH and the occurrence of reducing compounds promotes iron reduction and Fenton reactions. This gives rise to hydroxyl radicals that can destabilise the lysosomal membrane through lipid peroxidation. This may cause leakage of lysosomal hydrolytic enzymes into the cytosol which can damage various cell organelles (section 2.1.3.6). In aquatic organisms, Winston *et al.* (1991) have found that ROS produced outside the lysosomal membrane cause a decrease in the stability of the lysosomal membrane in isolated mussel digestive gland cells.

2.1.2.2 Induction of cytochrome P450 system

The cytochrome P450-catalysed insertion of oxygen into a substrate is the culmination of a process that reduces molecular oxygen to a species equivalent to an oxygen atom, in terms of electron count and reactivity. Uncoupling of catalytic turnover from substrate oxidation can divert the consumption of reducing equivalents toward the production of superoxide, H_2O_2, or water rather than substrate-derived products (reviewed by Ortiz de Montellano, 1995).

Several studies have demonstrated that exposure to a variety of contaminants including PAHs and PCBs induces the activity of enzymes of the cytochrome P450 system in fish, particularly of the CYP1A subfamily (Goksøyr & Förlin, 1992; Stegeman & Hahn, 1994; Bucheli & Fent, 1995; Goksøyr, 1995; see also Chapter 3) and this could lead to increased formation of ROS.

In rodents xenobiotics causing peroxisome proliferation induce enzymes of the peroxisomal β-oxidation and also the microsomal cytochrome P450 4A family or CYP 4A family (4A1, 4A2, 4A3) involved in the Ω-oxidation of fatty acids (Bell *et al.*, 1992; Muerhoff *et al.*, 1992). Peroxisome proliferator response elements (PPREs) have been reported for rodent cytochrome P450 4A6 and P450 4A1 (Muerhoff *et al.*, 1992; Aldridge *et al.*, 1995; reviewed by Simpson, 1997). Intraperitoneal injection of peroxisome proliferators (the hypolipidemic drugs clofibrate or ciprofibrate) into bluegill (*Lepomis macrochirus*) and catfish (*Ictalurus punctatus*), causes induction of both CYP2M1 and CYP2K1 cytochrome P450 isozymes, known to be associated with lauric acid hydroxylase activity (Haasch, 1996; Haasch *et al.*, 1998). As in mammals, induction was sex-specific, the protein being more inducible in male bluegill liver and in male catfish kidney and possibly liver.

2.1.2.3 Peroxisome proliferation

Peroxisomes are membrane-bound cytoplasmic organelles appearing in most eukaryotic cells (Hruban & Rechcigl, 1969). Although the enzyme composition of peroxisomes is variable depending on the species, organ or cell type studied, all peroxisomes contain a variety of H_2O_2 producing oxidases and catalase, which degrades H_2O_2 (de Duve, 1965; Hruban & Rechcigl, 1969; Böck *et al.*, 1980; Fahimi & Cajaraville, 1995). Apart from their pivotal role in oxyradical metabolism, peroxisomes are involved in several aspects of lipid metabolism. These include β-oxidation of long chain and very long chain fatty acids, bile acid formation, biosynthesis of ether lipids, biosynthesis of cholesterol and dolichol, and catabolism of prostaglandins and leukotrienes (Reddy & Mannaerts, 1994; Singh, 1997).

One of the unique features of peroxisomes is their ability to undergo a massive proliferation, a phenomenon termed 'peroxisome proliferation' which is induced by a number of endogenous compounds and xenobiotics. Peroxisome proliferation consists of an increase in peroxisome number and fractional volume. This is usually accompanied by the induction of some peroxisomal enzyme activities, particularly those of the fatty acid β-oxidation system (Reddy & Mannaerts, 1994). Acyl-CoA oxidase (AOX), the enzyme catalysing the first reaction of the β-oxidation pathway, the multifunctional enzyme (enoyl-CoA hydratase/3-hydroxyacyl-CoA dehydrogenase/isomerase or PH) and thiolase (3-ketoacyl-CoA thiolase or PT) are most significantly elevated both at the protein (up to 30x) and mRNA level. In contrast, the activity of catalase is not induced or is only slightly elevated (up to 4x) (Reddy & Mannaerts, 1994; Fahimi & Cajaraville, 1995). Therefore, peroxisome proliferation is considered to be a potential source of oxidative stress for cells since ROS-generating enzymes are induced to a much higher extent than ROS-detoxifying catalase (Reddy & Lalwani, 1983; Nemali *et al.*, 1989). This can be accentuated further by the fact that the antioxidant enzymes superoxide dismutase and glutathione peroxidase are inhibited in situations of peroxisome proliferation (Ciriolo *et al.*, 1982; Awashi *et al.*, 1984; Furukawa *et al.*, 1985; Cattley *et al.*, 1987).

Most studies on peroxisome proliferation have been carried out in mammalian systems. However, evidence clearly indicates that induction of peroxisome proliferation also occurs in invertebrates (Cajaraville, 1991; Fahimi & Cajaraville, 1995; Cajaraville *et al.*, 1997; Cancio *et al.*, 1998) and fish (Yang *et al.*, 1990; Mather-Mihaich & Di Giulio, 1991; Scarano *et al.*, 1994; Pedrajas *et al.*, 1996; Ruyter *et al.*, 1997; Au *et al.*, 1999). The initial experiments in fish were performed using typical mammalian peroxisome proliferators such as the hypolipidemic drugs ciprofibrate and gemfibrozil. Intraperitoneal injection of both drugs induces hepatic peroxisome proliferation in rainbow trout (*Oncorhynchus mykiss*) as measured by increased levels of AOX, PH, catalase, polypeptide PPA-80 and increased liver to body weight ratios (Yang *et al.*, 1990; Scarano *et al.*, 1994). Ciprofibrate injection also causes a 2.3-fold increase of peroxisomal volume density but no significant difference in peroxisomal numerical density (Yang *et al.*, 1990). Comparable effects have been demonstrated in the Japanese medaka (*Oryzias latipes*) which when exposed to gemfibrozil, showed increases in peroxisomal AOX and PH (Scarano *et al.*, 1994). Similarly, in *in vitro* experiments a strong and dose-dependent induction of AOX and PH has been found in rainbow trout isolated hepatocytes exposed to clofibrate and ciprofibrate but not to gemfibrozil (Donohue *et al.*, 1993). Clofibrate and bezafibrate, administered to salmon (*Salmo salar* L.) hepatocytes in culture also resulted in an increased activity of AOX (Ruyter *et al.*, 1997). In contrast to this, Pretti *et al.* (1999) were unable to detect any induction of several marker enzymes of peroxisome proliferation including AOX in sea bass (*Dicentrarchus labrax*) injected with clofibrate.

Apart from hypolipidemic drugs, certain environmental pollutants including various pesticides, bleached kraft pulp and paper mill effluents (BKME) and PAHs appear to cause peroxisome proliferation in fish liver. For example, exposure of the European eel *A. anguilla* to the pesticide dinitro-*o*-cresol (DNOC), resulted in a stimulation of peroxisomal enzymes (catalase, allantoinase and urate oxidase) and a higher number of peroxisomes in liver (Braunbeck & Völkl, 1991). Combined exposure to the pesticides endosulphan and disulphoton also provoked a transient increase in the absolute volume occupied by peroxisomes in liver of rainbow trout (Arnold *et al.*, 1995). Peroxisome proliferation has also been reported in kidney proximal tubules of rainbow trout treated with atrazine and linuron (Oulmi *et al.*, 1995a,b). Injection of the herbicide dieldrin in *Sparus aurata* markedly induced the activity of AOX and protein concentration of the peroxisomal fraction (Pedrajas *et al.*, 1996). BKME has been shown to provoke increases in catalase, lauroyl CoA-oxidase and AOX in the channel catfish (*Ictalarus punctatus*) and induce an increase in the number of liver peroxisomes in *Cottus gobio* downstream of two paper mills (Mather-Mihaich & Di Giulio, 1991; Bucher *et al.*, 1992). Following intraperitoneal injection of the

PAH benzo[a]pyrene in the demersal fish *Solea ovata*, increases in the numerical densities of hepatic lipofuscin granules and peroxisomes occurred, and most interestingly, these parameters were significantly correlated with EROD activities (Au *et al.*, 1999).

In mammals the induction of peroxisomal proteins is mediated by a 'peroxisome proliferator-activated receptor' (PPAR) which belongs to the nuclear hormone receptor superfamily of transcription factors together with the oestrogen receptor, the retinoid receptors and thyroid hormone receptors (Issemann & Green, 1990; Cancio & Cajaraville, 2000). Of the different PPAR isoforms found, only PPARα, and more recently PPARγ, appears to be related to peroxisome proliferation events in mammals. This PPARα binds to a peroxisome proliferator binding protein involved in the translocation of the PPARα from the cytoplasm to the nucleus (Reddy & Mannaerts, 1994). The PPARα then forms a heterodimer with a retinoid-X-receptor (RXRα) prior to binding to peroxisome proliferator response elements (PPRE) on the genes of the peroxisomal β-oxidation enzymes (Fig. 2.2). Apart from these peroxisomal β-oxidation enzymes, several mitochondrial (i.e. carnitine acetyltransferase), microsomal (i.e. cytochrome P450 4A1) and cytosolic (i.e. epoxide hydrolase) enzymes are also induced by peroxisome proliferators in mammals, the majority of which are involved in lipid metabolism and transport (Cancio & Cajaraville, 2000).

Leaver *et al.* (1997) found in the plaice (*Pleuronectes platessa*) that the promoters of the genes of the glutathione-*S*-transferase enzyme contain sequence elements identical to PPRE in mammals. The same authors have also cloned a plaice PPAR gene which may be more closely related to PPARγ than to PPARα, β or δ (Leaver *et al.*, 1998). Ruyter *et al.* (1997) had previously cloned a salmon (*Salmo salar* L.) PPARγ gene which is induced by clofibrate and bezafibrate in cultured hepatocytes. The three subtypes of PPAR have been detected in several tissues of adult zebrafish (*Danio rerio*) using immunohistochemistry (Cajaraville *et al.*, 2002).

Clearly, additional work is required on peroxisome proliferation in fish and other aquatic organisms given its importance as a mechanism of ROS production and its possible association with ROS-induced DNA damage and initiation/promotion of liver neoplasia. Hypolipidemic drugs, certain pesticides, BKME and benzo[a]pyrene appear to induce peroxisome proliferation in fish. Future studies should in part, therefore, evaluate the possible peroxisome proliferating ability of other environmentally relevant xenobiotics known to act as peroxisome proliferators in mammals (i.e. PCBs, phthalate ester plasticisers, steroids). In addition, the possible species and gender-specific sensitivity to peroxisome proliferators and their toxic effects also requires attention (Cancio & Cajaraville, 2000; Cajaraville *et al.*, 2002).

The effects of natural variables such as water temperature, salinity, season, reproductive stage and feeding habits on fish peroxisomes also need to be determined. For example, high fat diets, cold adaptation, vitamin E deficiency, riboflavin deficiency, genetic obesity, diabetes and starvation are known to induce peroxisomal changes in rodents (Bentley *et al.*, 1993). It has also been reported that peroxisomal enzyme activities and peroxisomal structure vary depending on season and tidal level in marine bivalve molluscs (Ibabe, 1998; Cancio *et al.*, 1999). Studies with the fish *Mugil cephalus* indicate that there are differences in liver peroxisomes depending on the age of the animals as well as on the sampling season and site (Orbea *et al.*, 1998a, 1999). In the brown trout (*Salmo trutta*) seasonal differences have been found in peroxisomal volume and surface densities and size (Rocha *et al.*, 1999).

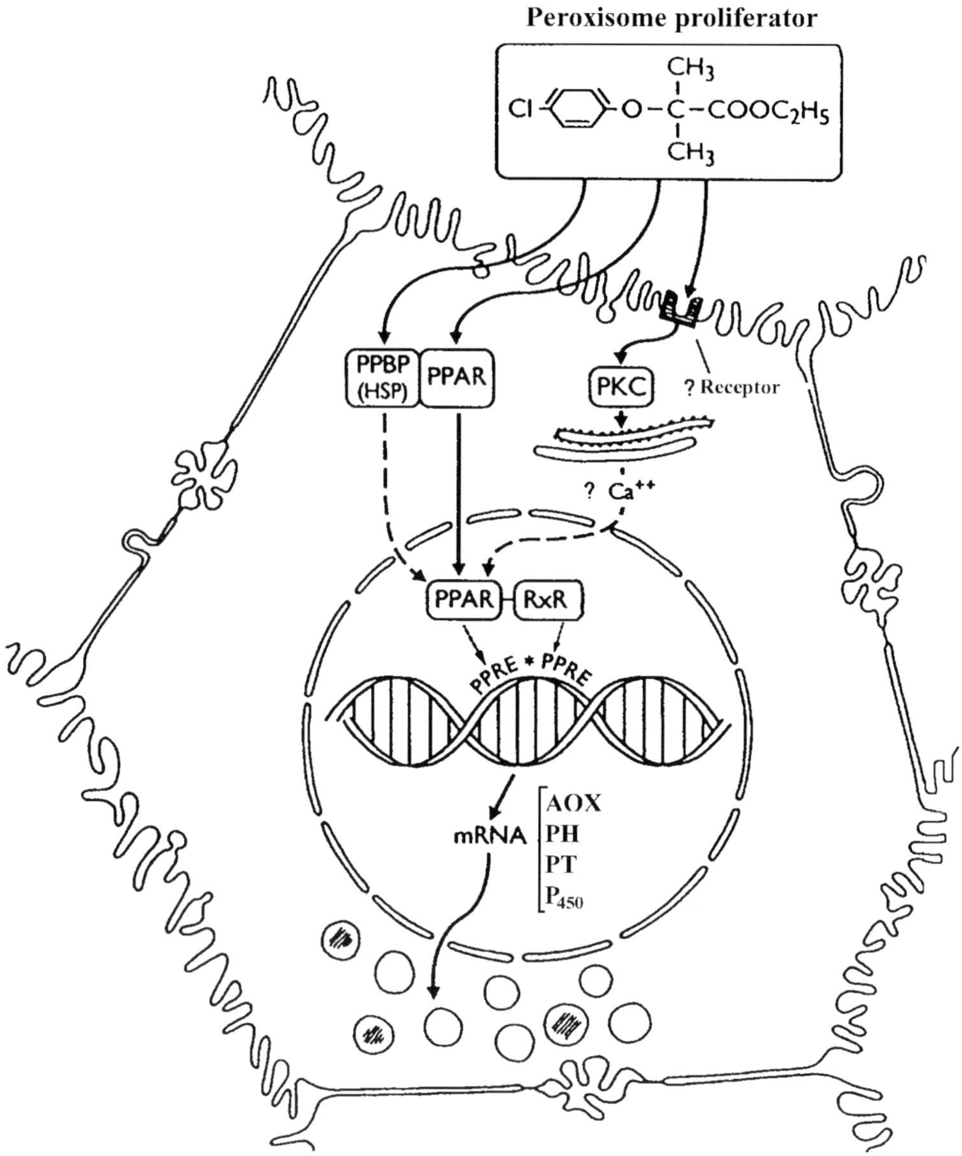

Fig. 2.2 Model of the induction of peroxisomal and other proteins by the typical peroxisome proliferator clofibrate in rat hepatocytes. This process is mediated by a 'peroxisome proliferator activated receptor' (PPAR) which forms a heterodimer with the retinoic acid receptor (RXR) and then binds to peoxisome proliferator response elements (PPRE) on the genes of peroxisomal β-oxidation enzymes (acyl-CoA oxidase, AOX; multifuntional enzyme, PH; thiolase, PT) and other genes and activates their transcription. A member of the heat shock protein (HSP) family, the peroxisome proliferator binding protein (PPBP), is involved in the translocation of PPAR from the cytoplasm to the nucleus. In addition to PPAR, the activation of a protein kinase C (PKC) type receptor and the elevation of cytosolic calcium have also been implicated. Modified from Fahimi & Cajaraville (1995).

The influence of various confounding factors on peroxisomes of aquatic organisms has been reviewed (Cajaraville *et al.*, 2002).

2.1.2.4 Markers of oxyradical production

The best markers of oxyradical production are those measuring directly the formation of radicals. However, the measurement of cytochrome P450 induction or peroxisome proliferation induction may also be used as indirect evidence of oxyradical production.

Induction of cytochromes P450 can be measured by enzymatic, immunochemical or molecular assays, using substrates, antibodies or probes, respectively, that specifically reflect the levels of a particular isozyme. For CYP1A induction, the 7-ethoxyresorufin O-deethylase (EROD) or aryl hydrocarbon hydroxylase (AHH) enzymatic assays have been shown to be specific. Alternatively, a number of studies have employed fish-specific CYP1A antibodies in immunochemical analyses such as ELISA, western blotting or immunohistochemistry (see review by Goksøyr & Husøy, 1998). The use of both an immunochemical technique and an enzyme assay gives additional quality control. This may be important in studies where samples may have been affected by inhibiting compounds, including high levels of pollutants such as PCBs, or by sample degradation during poor or difficult sampling conditions (Peters *et al.*, 1994; Collier *et al.*, 1995; Goksøyr & Husøy, 1998). These different methodological strategies have also been applied to the study of cytochromes P450 induced specifically upon peroxisome proliferation in fish (CYP2M1 and CYP2K1) associated with lauric acid hydroxylase activity (Haasch, 1996; Haasch *et al.*, 1998).

Induction of peroxisome proliferation can be studied using complementary morphological and biochemical approaches (Cajaraville *et al.*, 2002). In the former, peroxisomes are specifically stained by using enzyme histochemical methods for marker enzymes (the alkaline DAB method for catalase demonstration) or immunochemical methods. Then, the volume density, surface density, size and numerical density of peroxisomes are measured by means of quantitative microscopical methods such as stereology or image analysis (Beier & Fahimi, 1991; Fahimi & Cajaraville, 1995; Cajaraville *et al.*, 1997, 2002). In the biochemical approach, the induction of the peroxisomal β-oxidation system is quantified either by measuring the activities of the peroxisomal β-oxidation enzymes (Acyl-CoA oxidase, multifunctional enzyme and 3-ketoacyl-CoA thiolase) or the protein levels by immunoblotting or immunocytochemistry.

It is necessary to apply the morphological and biochemical approaches simultaneously because certain peroxisome proliferators such as the drug BM-15766 induce significant proliferation of peroxisomes without simultaneous induction of peroxisomal β-oxidation (Baumgart *et al.*, 1990). With rare exceptions (Yang *et al.*, 1990; Au *et al.*, 1999) the few studies reporting peroxisome proliferation in fish have used only biochemical enzyme measurements, complemented in some cases with qualitative estimates of peroxisome size or numbers. The application of quantitative methods to assess peroxisome size or numbers is of utmost importance in environmental studies to allow correlations to be determined between peroxisomal parameters and other biomarker measurements and environmental pollution levels.

2.1.3 Protection mechanisms

2.1.3.1 Induction of antioxidant enzymes

Animal cells possess a defence system for the detoxification of potentially harmful oxygen free radicals, most importantly the antioxidant enzymes catalase, Cu,Zn-superoxide dismutase (Cu,Zn-SOD), Mn-superoxide dismutase (Mn-SOD) and glutathione peroxidase (GPX) (Fig. 2.3). The antioxidant enzymes are localised in the sites of oxyradical generation in order to defend the cell from the deleterious effects of these highly reactive molecules. Catalase and GPX decompose H_2O_2. Catalase is located in the peroxisomes and is the most abundant of the peroxisome enzymes (50% in liver peroxisomes). Cu,Zn-SOD is mainly a cytosolic enzyme that converts O_2^- to H_2O_2. It has also been found in the matrix of peroxisomes in human hepatocytes and fibroblasts (Keller *et al.*, 1991; Crapo *et al.*, 1992) and in rat hepatocytes (Chang *et al.*, 1988; Dhaunsi *et al.*, 1992). The mainly mitochondrial Mn-SOD has also been located in peroxisomes, more specifically to the peroxisomal membrane, while GPX has been demonstrated to be partially a peroxisomal membrane enzyme in rat liver (Dhaunsi *et al.*, 1993; Singh *et al.*, 1994; Singh, 1997). In fish, Cu,Zn-SOD has been found in isolated peroxisomal fractions of the gilthead seabream (*Sparus aurata*) (Pedrajas *et al.*, 1996). More recently, in mullet (*Mugil cephalus*) hepatocytes, Cu,Zn-SOD and GPX, but not Mn-SOD, have been found to be localised within peroxisomes (Orbea *et al.*, 1998b, 2000). The peroxisome is, therefore, a cell organelle with active implication in

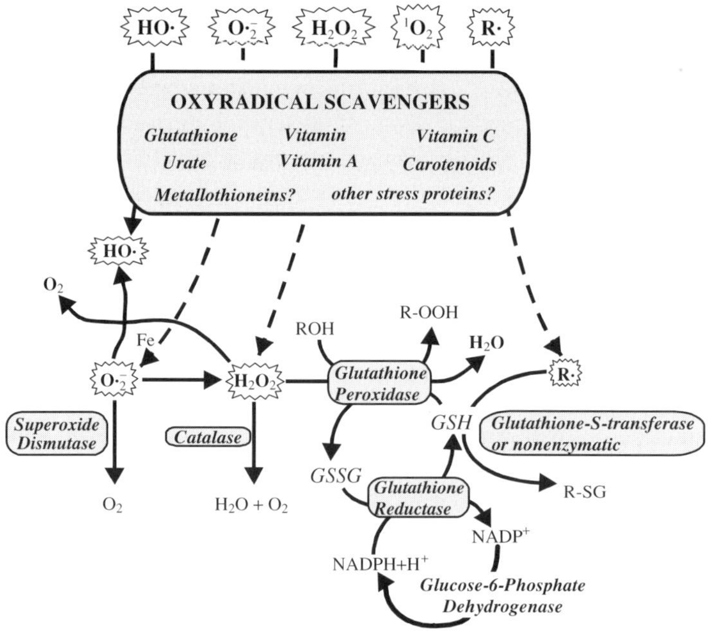

Fig. 2.3 Roles of the antioxidant enzymes catalase, superoxide dismutase (SOD), glutathione peroxidase (GPX) and other primary antioxidant defences in ROS detoxification. The involvement of other glutathione-requiring enzymes in these processes, i.e. glutathione-S-transferase (GST) and glutathione reductase (GRE) is also indicated. GSH, reduced glutathione; GSSG, glutathione disulphide.

oxyradical metabolism and homeostasis, since it is an important site for oxyradical production (Fig. 2.1) and for the activity of antioxidant enzymes (Figs 2.1 and 2.4).

Induction of antioxidant enzymes can be used as an exposure biomarker of pollutants acting through enhanced generation of ROS. These pollutants include some transition metals and other redox cycling compounds, xenobiotics metabolised through the cytochrome P450 system and peroxisome proliferators.

There is increasing laboratory and field-based evidence of antioxidant enzyme induction in fish treated with oxyradical-generating contaminants. Hepatic peroxisomal catalase is increased transiently in fish treated with the redox cycling herbicide paraquat (Di Giulio *et al.*, 1989). However, SOD activity is diminished in fish (*Sparus aurata*) injected with the same herbicide but increases in Cu-injected animals (Pedrajas *et al.*, 1995). Injection of the insecticide dieldrin in *Sparus aurata* induces the activity of catalase and SOD (Pedrajas *et al.*, 1996). Similarly, catalase activity has been induced in the livers of rainbow trout injected with ciprofibrate, *Ictalurus punctatus* exposed to BKME, *Anguilla anguilla* exposed to dinitro-*o*-cresol and *I. punctatus* and *Limanda limanda* exposed to PAHs (Yang *et al.*, 1990; Mather-Mihaich & Di Giulio, 1991; Braunbeck & Völkl, 1991; Di Giulio *et al.*, 1993; Livingstone *et al.*, 1993). However, the combined exposure to disulphoton and endosulphan had no effects on catalase activity in rainbow trout (Arnold *et al.*, 1995). In sea bass (*Dicentrarchus labrax*) and dab (*L. limanda*), intraperitoneal injection of 3-methylcholanthrene caused only slight and transient induction of catalase and SOD (Lemaire *et al.*, 1996). Injection of β-naphthoflavone in rainbow trout marginally and transiently induced SOD while catalase activity was reduced (Lemaire *et al.*, 1996).

In field studies, catalase and other antioxidant enzyme activities increased in the livers of grey mullets (*Mugil* sp.) sampled from an area polluted with metals, polyaromatic hydrocarbons, polychlorinated biphenyls and pesticides when compared with a reference area (Rodríguez-Ariza *et al.*, 1993). Similarly, catalase and SOD activities decrease away from coastal organic contamination in liver of dab (*L. limanda*), from the North Sea and, with some exceptions, in larvae of sardine (*Sardina pilchardus*), from the Biscay Gulf (Livingstone *et al.*, 1992; Peters *et al.*, 1994). In a field study carried out in the Mediterranean Sea with red mullet (*Mullus barbatus*), catalase activity was maximal in the site with highest levels of pollution and minimal in a reference site, but varying responses were found regarding SOD, GPX and DT-diaphorase (Burgeot *et al.*, 1996a). No changes in SOD activity were detected in chub (*Leuciscus cephalus*) from a polluted river when compared with those from reference river areas, although livers of polluted fish contained higher concentrations of transition metals, especially copper and iron, which are known redox cycling compounds (Lenártová *et al.*, 1997). In a study carried out in the Great Lakes, the activity of catalase was higher in lake trout (*Salvelinus namaycush*) from a reference lake when compared with those from a contaminated lake (Palace *et al.*, 1998).

Consequently, it appears that oxyradical-generating contaminant exposure is not always matched by antioxidant enzyme induction in fish. Furthermore, the interpretation of data on antioxidant enzyme activities is often obscured by the fact that several abiotic and biotic variables may also influence their activity. Environmental factors such as water temperature, salinity, season and feeding habits are known to exert changes on antioxidant enzyme activities in fish. Radi *et al.* (1987) have analysed the effect of the feeding habit on antioxidant enzyme activities. Seasonal, geographical and species-specific variations have been

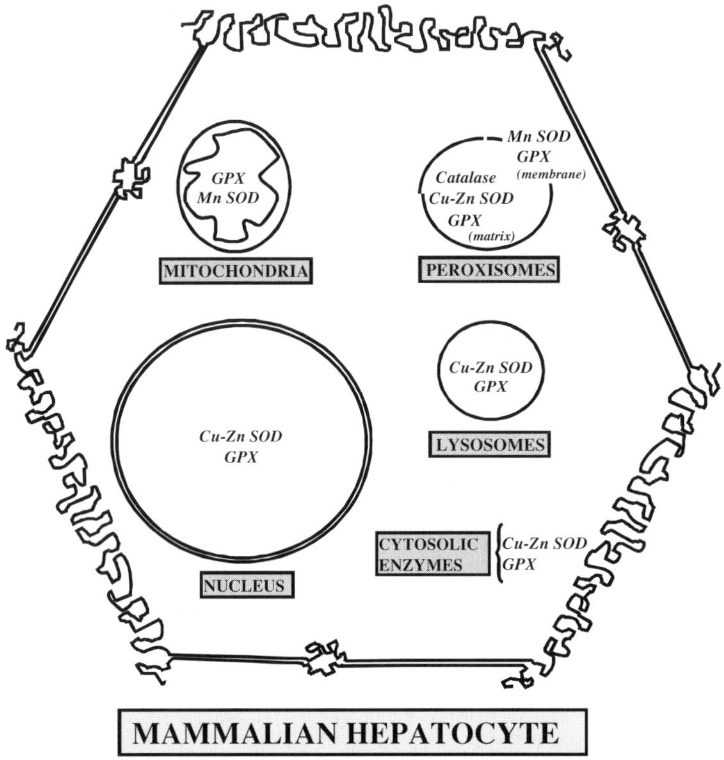

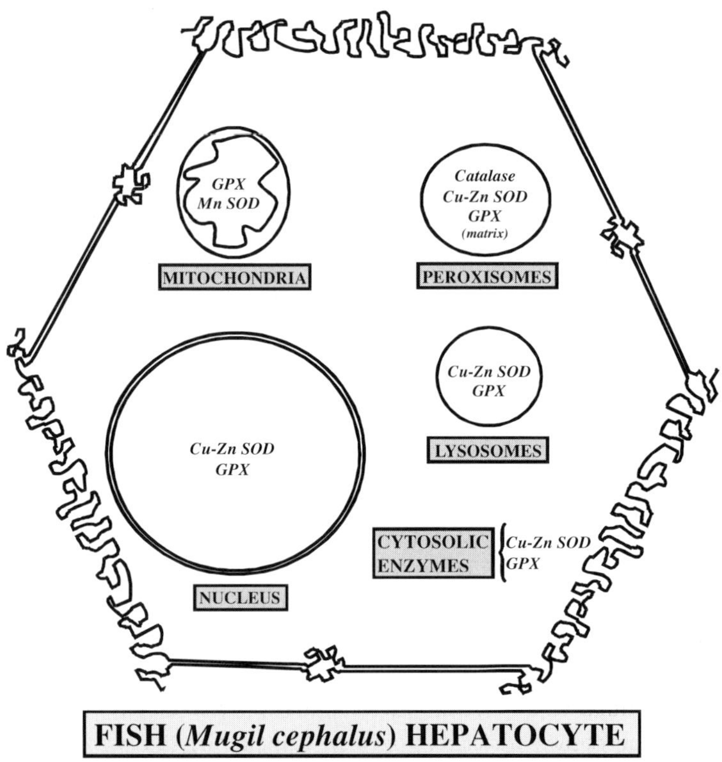

Fig. 2.4 Subcellular localisation of antioxidant enzymes in mammalian and fish hepatocytes. The fish model is based on immunocytochemical studies carried out with mullet (*Mugil cephalus*). Special emphasis has been placed on the role of hepatic peroxisomes as sites of ROS production and detoxification.

reported in the activities of catalase and SOD in three species of freshwater fish (Palace & Klaverkamp, 1993). In addition, catalase activity is decreased during cold adaptation in the hepatocytes of the golden ide (*Leuciscus idus melanotus*) (Braunbeck *et al.*, 1987).

2.1.3.2 Oxyradical scavengers

Whilst there are specific enzymes that catalyse reactions to eliminate superoxide anions and hydrogen peroxide, there is no enzyme activity specifically involved in the elimination of the highly reactive hydroxyl radicals (Brunk & Cadenas, 1988). However, hydroxyl radicals, together with other radicals generated via lipid peroxidation or via xenobiotic metabolism, can be neutralised through the interaction with small molecules known as ROS scavengers (Fig. 2.3). These can be either hydrophilic molecules (glutathione, vitamin C or ascorbate, urate) or lipophilic molecules (vitamin E or tocopherols, vitamin A or retinoids, carotenoids). These low molecular weight ROS scavengers have been shown to be elevated in some contaminant-induced oxidative stress situations (Andersson *et al.*, 1988).

However, in a study in the Great Lakes, trout from the reference Lake Superior had greater concentrations of tocopherol in liver and kidney than fish from the contaminated Lake Ontario. However, total alcohol and esterified concentrations of retinoids were higher in the kidneys of trout from Lake Ontario (Palace *et al.*, 1998). Hepatic and renal ascorbic acid concentrations were not different between the two populations.

2.1.3.3 Glutathione status

Organic xenobiotics usually undergo biotransformation via phase I (cytochrome P450 system) and phase II or conjugative reactions which produce derivatives that may be readily excreted in urine. Conjugation with reduced glutathione (GSH) is a very important route of detoxification of electrophilic xenobiotics and is catalysed by glutathione-*S*-transferases or GST, a multigene family of enzymes (James, 1987; George, 1994).

Reduced glutathione also acts as a ROS scavenger and protects cells against oxidative stress. Additionally, glutathione and other cellular thiols have been shown to protect cells from metal toxicity through their ability to sequester the metals. The importance of cellular glutathione status in metal toxicity has been underlined in an *in vitro* study using a continuous rainbow trout cell line (Maracine & Segner, 1998). Thus, the toxicity of Hg, Cu and Cd was significantly increased in GSH-depleted cells, whereas the toxicity of Zn, Ni and Pb was not altered. In a fish hepatoma cell line (*Poeciliopsis* hepatoma cells), Cd treatment led to significant increases of GSH, Zn treatment produced no change in GSH and treatment with H_2O_2 reduced cellular GSH concentration (Schlenk & Rice, 1998).

Usually, xenobiotics that cause induction of the cytochrome P450 system (such as PAHs or the model drug phenobarbital) also cause an elevation of GSH-requiring enzyme levels and reduction of cellular GSH concentrations (Peterson & Guengerich, 1988). Also, lipid peroxidation products reduce the levels of reduced glutathione, thus leading to alterations in the redox status of cell (Viarengo, 1989). Examples of contaminant-related induction of GST have been reported in rainbow trout exposed to disulphoton and endosulphan in combination (Arnold *et al.*, 1995).

However, no induction of GST has been found in sea bass (*Dicentrarchus labrax*) injected with the three model PAHs and typical inducers of the cytochrome P450 1A system: benzo[a]pyrene, 3-methylcholantrene and β-naphtoflavone (Lemaire *et al.*, 1992, 1996; Novi *et al.*, 1998). The lack of co-induction of CYP1A and GST indicates that in fish these two xenobiotic metabolising systems might be regulated independently at the transcriptional level.

In field studies, GPX and GST activities were higher in chub (*Leociscus cephalus*) from polluted areas than those from reference river areas, while GRE activity was not significantly changed (Lenártová *et al.*, 1997). In two freshwater fish species, gudgeon (*Gobio gobio)* and roach (*Rutilus arcasii*), animals from contaminated sites showed reduced GSH concentration and elevated GPX and GRE activities but a tendency to decrease GST activity (Almar *et al.*, 1998). However, decreases in GPX activity with contamination have also been reported. For instance, trout from the relatively uncontaminated Lake Superior had greater hepatic and renal activities of GPX than fish from the contaminated Lake Ontario (Palace *et al.*, 1998).

Peroxisome proliferators such as hypolipidemic drugs generally decrease GSH-requiring enzyme levels, i.e. GST and GPX levels, in mammals (Ciriolo *et al.*, 1982; Awashi *et al.*, 1984; Furukawa *et al.*, 1985; Cattley *et al.*, 1987). The inhibition of these enzymes has been linked with the increased generation of ROS caused by peroxisome proliferators (Furukawa *et al.*, 1985). A similar pattern of response has been demonstrated in fish. For example, the herbicide dieldrin caused induction of peroxisomal ROS-producing enzymes and ROS-mediated decreases in microsomal GST activity (Pedrajas *et al.*, 1995, 1996). Malathion injection also led to reduced GST activities in fish (Pedrajas *et al.*, 1995). However, it has been been shown that GST genes from plaice (*Pleuronectes platessa*) are up-regulated after administration of peroxisome proliferators, and the products of these genes appear to be efficient in the conjugation of some of the end products of lipid peroxidation (Leaver *et al.*, 1997; Leaver & George, 1998). Similarly, sea bass (*Dicentrarchus labrax*) injected with clofibrate showed significantly induced hepatic GST activity (Pretti *et al.*, 1999).

2.1.3.4 Induction of metallothioneins

Metallothioneins (MT) are cytosolic and nuclear proteins that are induced by and bind mono- and divalent metals such as Cu, Zn, Cd and Hg (Kägi, 1993). This peculiar low-molecular-weight protein consists of 20–30% cystein and few or no aromatic amino acids. All the cystein appears to be involved in metal-binding (Nielson *et al.*, 1985; Huang, 1993).

Metallothionein is present in the tissues of most vertebrates and some invertebrates. More than ten isoforms have been described for some mammalian species, whereas there is generally one or two isoforms in fish (Gedamu *et al.*, 1993; Olsson, 1993). The main function of MT has been thought to be its involvement in the regulation of intracellular Zn and/or Cu availability (Bremner, 1991a,b; Bremner, 1993). Other proposed functions include free radical scavenging, metal detoxification and its presence as part of the acute phase response (Marafante *et al.*, 1972; Marafante, 1976; Thornally & Vasak, 1985; Schroeder & Cousins, 1990).

There are few studies that directly address the function of MT in fish. The fact that metallothionein will bind nearly all Cd in fish liver cells was the major reason for the initial interest in the protein – as a detoxifying mechanism for Cd and possibly other metals (Hamilton

& Mehrle, 1986; Cosson *et al.*, 1991; George & Olsson, 1994). Some fish species have been found to have phenomenal levels of metals in their tissues, especially in liver. In some tropical fish species high natural metal (Zn) levels have been suggested to be associated with reproductive processes, although it is unclear why these particular species should require more Zn than other related species (Hogstrand *et al.*, 1996; Hogstrand & Haux, 1996). Similarly, the white perch (*Morone americana*), accumulates high Cu levels in the liver in an age-dependent fashion (Bunton *et al.*, 1987; Bunton & Frazier, 1989).

Knowledge of the regulation of MT in different fish species has increased (Zafarullah *et al.*, 1989; Kille *et al.*, 1993; Olsson *et al.*, 1995). Consequently, it appears that there are species-dependent differences in the structure of metal promoter regions (MREs), but there are also other factors that influence the sensitivity of fish species to metal exposure (Olsson & Kille, 1997).

In the liver of fish under nearly all conditions, the concentrations of Zn and Cu will supersede the concentrations of toxic metals such as Cd by orders of magnitude. The non-essential metals, e.g. Cd and Hg, have higher affinity for metallothionein than Zn and will thus preferentially be bound to MT, thereby liberating Zn which is thought to interact with cytosolic elements to induce MT synthesis. Considering the levels of Zn naturally present in the cell, it is somewhat surprising that such small increases induce MT synthesis. There is no doubt, however, that exposure to both essential and non-essential metals does cause induction of MT mRNA and MT protein in fish tissues (George & Olsson, 1994). Studies on haemoglobin-less Antarctic fish indicate that there are mechanisms for the expression of MT protein other than through Zn promotion. In these fish, MT is found at very low basal levels (hardly detectable), but protein synthesis is strongly induced following Cd exposure (Carginale *et al.*, 1998; Scudiero *et al.*, 1992, 1997).

The molecular properties of metallothionein have made it a candidate for being a free radical scavenger (Thornally & Vasak, 1985; Hidalgo *et al.*, 1988; Sato, 1991). Indeed, studies in mammalian systems have indicated that metallothionein provides protection against free-radical generating treatments, although there is some controversy concerning the mechanism (Min *et al.*, 1993). However, increased content of MT protein does appear to provide protection against DNA damage in mammalian systems (Abel & de Ruiter, 1989; Chubatsu & Meneghini, 1993; Cai & Cherian, 1996).

There is not overwhelming evidence for an involvement of MT in antioxidant processes in fish, but there are some indications. Olsson and co-workers have identified promoter-regions in fish MT genes (AP1) that indicate that synthesis of MT in fish could be induced by exposure to free radicals (Kling & Olsson, 1995; Olsson & Kling, 1995). Furthermore, an induction of MT was seen in fish cell cultures following exposure to hydrogen peroxide (Kling *et al.*, 1996). As yet, there is no knowledge of whether increased MT levels in fish tissues or cells confer protection against DNA damage.

2.1.3.5 Induction of stress proteins

'Stress proteins' is a general term used to describe any protein for which there is an increased synthesis in response to a stressor. In addition to heat-shock proteins (HSPs), this term also includes metallothionein, heme oxygenase and acute phase proteins. Only HSPs will be considered here.

Heat-shock proteins, or HSPs, are a heterogeneous collection of proteins that are induced by thermal shock, contaminant exposure and other stressors (Anderson, 1989; Lindquist & Craig, 1988). The HSPs are generally denoted by their apparent size in SDS-polyacrylamide electrophoresis and the most commonly used categories are: HSP100 (100 kDa), HSP90 (90 kDa), HSP70 (70–75 kDa), HSP60 (58–65 kDA) and low-molecular-weight HSPs (16–35 kDa). Much of the interest in HSPs has been triggered by the observation that these proteins are highly conserved between different animal phyla and appear to be present in all living organisms (Miller, 1989).

Heat-shock proteins have many different functions in cells. Whereas HSP70s appear to be 'chaperonins', predominantly involved in the handling of other proteins, HSP90s appear to be involved in the regulation of the synthesis of other proteins and ubiquitin in the removal of damaged proteins. There is also a related group of membrane-bound proteins of similar sizes, glucose regulated proteins (GRP), that also appear to be involved in the handling of proteins in the cell.

There are indications that increased synthesis of HSPs may confer protection against DNA damage in mammalian systems (Minisini *et al.*, 1994; Richards *et al.*, 1996; Kwak *et al.*, 1998) or protect against apoptosis (Samali & Cotter, 1996).

The similarity of HSPs between species has prompted studies using various aquatic organisms and antisera raised against human or rodent HSPs. These have mostly focused on invertebrates rather than fish (Sanders & Martin, 1993; Lawrence & Nicholson, 1998). There is thus limited knowledge concerning the presence and behaviour of HSPs in fish tissues (reviewed by Iwama *et al.*, 1998). As with mammalian HSPs, there appear to be both stressor inducible HSPs and HSPs that are constitutively expressed, but that are not inducible by heat stress or contaminants (Grøsvik & Goksøyr, 1996). A range of different HSPs have been described, mainly from cell line studies or studies with primary cell cultures. The major families of stress proteins, together with their location and function, are shown in Table 2.1. The current knowledge of whole-fish HSP responses to contaminants or other environmental stressors is not sufficient to indicate whether HSPs may provide protection against damage to DNA in fish tissues.

2.1.3.6 Lysosomal sequestration

Lysosomes are cytoplasmic organelles involved in several important cell functions. These include the digestion of both endogenous materials, such as cellular macromolecules and organelles, and exogenous materials internalised through endocytic and phagocytic processes.

Lysosomes are able to accumulate and sequester a wide range of both organic and inorganic chemical compounds (Moore, 1980). This protects the cell by isolating potentially toxic compounds within the membrane-bounded lysosome compartment. In addition, lysosomes are also involved in the sequestration of oxidatively damaged lipids, proteins and carbohydrates caused by xenobiotic induced cell injury. These sequestered macromolecules may be further degraded in the lysosome and the degradation products made available to the cell for reuse, or eliminated through exocytosis. For example, the end products of lipid peroxidation accumulate in lysosomes where they precipitate in the form of an insoluble and undegradable fluorescent pigment called lipofuscin (Sunderman, 1986; Sohal & Brunk, 1990).

Table 2.1 Major families of stress proteins together with their location and function (adapted from Parsell & Lindquist, 1993 and de Pomerai, 1996).

Protein family	Monomer size (KDa) and Eukaryotic location	Stress functions
HSP 100	80–110 KDa, Cytoplasm, nucleus	Extreme heat tolerance, ethanol tolerance, regulation of CLpP protease, disaggregation of protein complexes
HSP 90	82–96 KDa Cytoplasm, ER, nucleus	ATP dependent chaperone, folding and functional association with kinases, steroid receptors. Under normal conditions modulates many cellular activities by binding target proteins. Under stress conditions, synthesis increases and may redirect cellular metabolism to enhance tolerance. Specific mechanism not identified
HSP 70	67–76 KDa, Different members occupy different compartments: cytoplasm, nucleus, mitochondria, cholorplast, ER	ATP dependent molecular chaperone with ATPase activity, conveys unfolded proteins to various cell compartments, association with misfolded proteins to allow refolding where possible, breaking up of protein aggregates and vectoring badly damaged proteins for destruction by ubiquitination and proteolysis
HSP 60	58–65 KDa, Mitochondria	ATP dependent molecular chaperone. Major mitochondrial HSP 60 that acts to receive and correctly fold mitochondrial proteins imported from the cytoplasm. At least one other family member may act in a similar fashion in the ER lumen. Under normal conditions binds incompletely folded proteins and directs the folding peptide to the correct conformation. Chaperonin synthesis increases under adverse conditions
HSP 27	16–28 KDa, Cytoplasm, ER, nucleus	Various ATP independent chaperone functions, inhibition of actin polymerisation. Synthesis induced under adverse conditions. Little known regarding specific cellular functions
HSP 10	9–12 KDa, Mitochondria, cholorplasts	Stimulates hsp 60 functions
Ubiquitin	8 KDa, Cytoplasm, nucleus	Tags irreversibly denatured HSP 70 associated proteins for proteolytic degradation

However, the protective role of lysosomes can be reversed once the storage capacity of these organelles is overloaded. This could in turn lead to severe damage of the lysosomal membrane. Injury to lysosomes may also occur through direct damage to the lysosomal membrane by toxic compounds or by oxyradicals produced during metabolism of certain xenobiotics (Winston *et al.*, 1991). Due to the pivotal role of lysosomes in intracellular degradation of both exogenous and endogenous macromolecules, impairment of lysosomes could cause severe metabolic disorders and pathological alterations including preneoplastic and neoplastic liver lesions in fish (Köhler, 1991; Köhler *et al.*, 1992; Köhler & Pluta, 1995) (see also Chapter 4). Furthermore, the damage of lysosomal membranes results in the release of lysosomal acid hydrolases into the cytosol, as demonstrated by *in situ* enzyme cytochemistry at the ultrastructural level (Cancio *et al.*, 1995), and this could give rise to a cascade of alterations involving nearly all cell components and ultimately cell death.

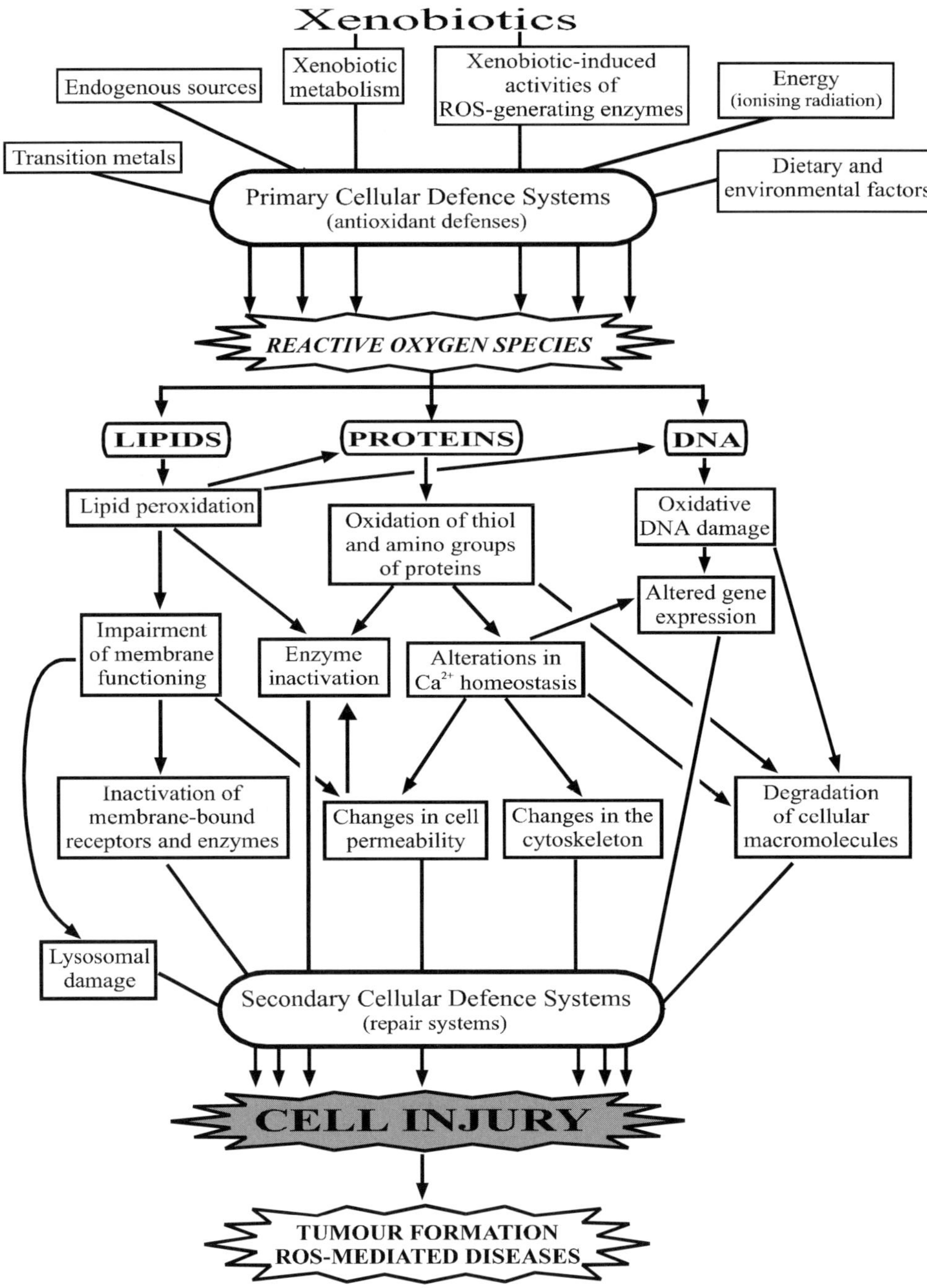

Fig. 2.5 Damage caused by reactive oxygen species. Reactive oxygen species or ROS are produced through different mechanisms in aerobic organisms. A major part of these are detoxified by primary cellular antioxidant defences and secondary repair systems. However, when ROS production mechanisms overwhelm cellular defences, ROS can readily interact with cellular macromolecules. The schematic diagram shows the main cellular targets of ROS-induced damage. Lipids, proteins and DNA are all known to be target molecules of ROS and their alterations can give rise to a cascade of events eventually leading to cell injury and dysfunction.

2.1.3.7 Markers of cell protection against oxyradicals

Enhanced activity/expression of antioxidant enzymes and increased concentrations of non-enzymatic ROS scavengers including glutathione, metallothioneins and possibly stress proteins all represent possible markers of cell protection against oxyradicals. Methods to determine antioxidant enzyme activities, glutathione and other related parameters have been reviewed by Lackner (1998). The concentration of metallothionein in tissues can be determined using immunochemical methods (RIA, ELISA) (Hogstrand & Haux, 1990; Hylland, 1999), electrochemical methods (differential pulse polarography (Olafson & Olsson, 1991), metal-replacement assays (Piotrowski *et al.*, 1973; Scheuhammer & Cherian, 1986; Bartsch *et al.*, 1990), chromatographic separation followed by metal or protein quantification (Carpenè & Vasak, 1989) or spectrophotometric methods (Viarengo *et al.*, 1997). To study induction kinetics, MT mRNA can be determined, most commonly by Northern blot or slot-blot. Heat-shock proteins (HSP) are highly conserved and antisera produced against mammalian HSP cross-react with peptides (presumably HSPs) in invertebrates (Lawrence & Nicholson, 1998) and fish. As there is some uncertainty as to the nature of the proteins actually being measured, HSP is most commonly determined semi-quantitatively using Western blot (electrophoretic separation followed by immunochemical identification and densitometric quantification) (Dunlap & Matsumura, 1997). Induction kinetics are generally studied using quantification of HSP mRNA (Abe *et al.*, 1995).

The extent of lysosomal accumulation and sequestration of ROS-producing xenobiotics may represent a sensitive index of cell protection against ROS. A simple autometallographical method coupled to image analysis has been applied in aquatic organisms to assess intralysosomal accumulation of metallic contaminants (Soto *et al.*, 1996; Soto & Marigómez, 1997). The intralysosomal accumulation of lipid peroxidation end products in the form of lipofuscin can be measured by using specific stains for lipofuscins (e.g. Schmorl reaction) and image analysis, as mentioned above (Moore, 1990; Krishnakumar *et al.*, 1995).

Since the accumulation of ROS-producing xenobiotics or ROS-damaged cellular macromolecules could severely injure the lysosomal membrane, assessment of the integrity of the lysosomal compartment appears necessary. This could be accomplished by measuring lysosomal enzyme activity (acid phosphatase, β-glucuronidase and other acid hydrolases), lysosomal membrane stability, and lysosomal structure (volume density, surface density, surface to volume ratio and numerical density) (see review by Cajaraville *et al.*, 1995). Other methods developed more recently include the measurement of lysosomal biomarker protein levels through the use of specific antibodies by quantitative immunoblotting or immunohistochemistry (Lekube *et al.*, 1998, 2000). The *in vitro* neutral red assay has also gained increased attention as a measure of endocytic-lysosomal function in molluscan isolated digestive cells and haemocytes (Lowe *et al.*, 1995; Cajaraville *et al.*, 1996; Robledo & Cajaraville, 1996) and in fish blood cells (Lowe *et al.*, 1992).

2.1.4 Damage

ROS can cause severe damage to cellular macromolecules through the oxidation of DNA, membrane lipids and proteins (Fig. 2.5). As cells possess efficient antioxidant systems to detoxify ROS (section 2.1.3), the extent of damage to cellular macromolecules will depend on the balance between ROS production and detoxification. In addition to antioxidant 'primary defences' that prevent ROS production, there is a group of 'secondary defences' that repair oxidatively damaged DNA, proteins and lipids (Davies, 1986; Kehrer, 1993). The latter include DNA repair mechanisms, considered in detail in sections 2.2.1.2 and 2.2.2.4, and a number of proteases and lipases that may degrade damaged proteins and oxidised fatty acids, respectively. Higher level consequences of ROS-induced oxidative damage may include tumour formation and other oxyradical-mediated diseases (sections 2.1.5 and 2.4) but these would greatly depend on several factors such as the species, organ or cell type considered.

2.1.4.1 Oxidative DNA damage

Free radicals and other ROS are very reactive molecules that can readily react with DNA and other cellular macromolecules. For example, it has been demonstrated that hydroxyl radicals damage DNA by converting guanine to 8-hydroxyguanine (Kasai & Nishimura, 1986). In addition, the products of oxyradical-induced lipid peroxidation are known to react with DNA (Comporti, 1985; Viarengo, 1989). Therefore, overproduction of ROS during xenobiotic metabolism or in situations of peroxisome proliferation can cause direct damage to DNA (Di Giulio *et al.*, 1993). Nishimoto *et al.* (1991) have reported oxidative DNA damage in English sole (*Parophrys vetulus*) exposed to nitrofurantoin. Furthermore, studies in the same fish species indicate that DNA lesions induced by oxidative injury are causally linked to tumourigenesis (Malins & Haimanot, 1991; see section 2.1.5.1).

An interesting model linking oxidative DNA damage to tumourigenesis is the peroxisome proliferation model. However, most of the data on peroxisome proliferators and carcinogenesis has been produced in mammals rather than fish. It has been shown, for example, that peroxisome proliferators induce hepatocarcinomas in rodents under chronic exposure. Since this class of xenobiotics is non-mutagenic and non-genotoxic, the neoplastic transformation of liver cells could be caused by the oxidative damage to DNA related to an imbalance between oxyradical producing processes and antioxidant defences (Reddy & Lalwani, 1983; Reddy & Rao, 1989). Indeed, the administration of some peroxisome proliferators has been shown to increase the levels of 8-hydroxy-deoxyguanosine in rat liver DNA (Takagi *et al.*, 1990).

2.1.4.2 Lipid peroxidation

Lipid peroxidation refers to the oxidative deterioration of polyunsaturated lipids and occurs in several steps (Gutteridge & Halliwell, 1990). During the initiation step, ROS react with polyunsaturated fatty acid chains resulting in the subtraction of a hydrogen atom and production of semi-stable lipid hydroperoxides. These molecules evolve rapidly into lipid radicals that, upon reaction with fatty acids, can cause impairment of cell membrane structure and function. As a result of this oxidative damage a complex mixture of molecules such as aldehydes is produced, which could further react with thiol and amino groups of proteins and, most importantly, with DNA. The end products of oxidative membrane damage can precipitate in the form of lipofuscin, an insoluble pigment which has been found to accumulate in cells of aquatic organisms treated with contaminants (Wolfe *et al.*, 1981; Aloj Totaro *et al.*, 1986; Pipe & Moore, 1986; Viarengo, 1989; Cajaraville *et al.*, 1990; Moore, 1990; Krishnakumar *et al.*, 1995). In fish, Aloj Totaro *et al.* (1986) have reported an increased formation of lipofuscins in the nervous tissue of *Torpedo marmorata* upon copper exposure. More recently, applying quantitative morphometric methods, Au *et al.* (1999) found increased lipofuscin granules in the hepatocytes of *Solea ovata* injected with benzo[a]pyrene. The increased number of lipofuscin granules were suggested to be linked to increased lipid peroxidation driven by ROS produced during redox cycling of benzo[a]pyrene quinone metabolites. The occurrence of lipid peroxidation in biological membranes causes impaired membrane functioning, decreased membrane fluidity, inactivation of membrane bound receptors and enzymes and increased non-specific permeability to ions such as calcium (Gutteridge & Halliwell, 1990).

It is widely accepted that pollutant-induced ROS can initiate or promote lipid peroxidation (Comporti, 1985; Brunk & Cadenas, 1988; Farber *et al.*, 1990; Gutteridge & Halliwell, 1990). For instance, heavy metal-induced lipid peroxidation has been reported extensively for transition metals, especially copper and iron, and to a lesser extent lead and zinc. Radi and Matkovics (1988) have reported increased lipid peroxidation in carp (*Cyprinus carpio*) exposed to copper and zinc and similar results have been found for gilthead seabream (*Sparus aurata*) injected with polar xenobiotics, copper and paraquat (Pedrajas *et al.*, 1995). This has also been reported in rainbow trout (*Oncorhynchus mykiss*) treated with endosulphan and disulphoton simultaneously (Arnold *et al.*, 1995). In addition, there is evidence that exposure to PAHs can result in lipid peroxidation in fish such as channel catfish (*Ictalurus punctatus*) and dab (*Limanda limanda*) (Di Giulio *et al.*, 1993; Livingstone *et al.*, 1993). The fish peroxisome proliferators dieldrin and clofibrate caused lipid peroxidation in *S. aurata*. This was detected by an increase in the levels of microsomal thiobarbituric acid reactive substances in the fish, whilst malondialdehyde content was not altered (Pedrajas *et al.*, 1998). In contrast to results obtained in laboratory experiments, field studies have shown unchanged lipid peroxidation levels in control and contaminated fish populations, this probably reflecting an adaptation to the chronic oxidising conditions in contaminated fish (Lenártová *et al.*, 1997).

2.1.4.3 Alterations in protein function

Another known consequence of enhanced ROS production is enzyme inactivation (Wolff *et al.*, 1986). For instance, ROS oxidise several membrane proteins such as sodium channels and Ca-ATPases thereby causing an increase in cytosolic concentrations of free Ca (Srivastava *et al.*, 1989). Oxidative stress also causes the release of Ca from mitochondrial and endoplasmic reticulum stores through the interaction of ROS with the thiol groups of proteins (Orrenius & Nicotera, 1987). The alterations in Ca homeostasis lead to multiple consequences for the cell, including the activation of non-lysosomal Ca-dependent proteases and lipases and changes in the organisation of the cytoskeleton, that can ultimately cause cell death according to some authors (Orrenius & Nicotera, 1987).

ROS can also cause the inactivation of numerous enzymes directly or through the indirect action of lipid peroxidation products (Comporti, 1985; Viarengo, 1989). Furthermore, oxidised proteins have been found to be more susceptible to proteolysis (Davies, 1986). In the liver of gilthead seabream injected with copper, dieldrin or malathion, the appearance of new oxidised forms of superoxide dismutase (SOD) has been reported as due to the increased production of ROS (Pedrajas *et al.*, 1995). New isoforms of Cu, Zn-SOD have also been detected in chub (*Leuciscus cephalus*) living in contaminated rivers when compared to fish from reference river areas (Lenártová *et al.*, 1997). In addition, the formation of new isoforms of SOD can be reproduced *in vitro* by incubation of liver cell-free extracts with malondialdehyde and by incubation of isolated pure SODs with malondialdehyde and 4-hydroxy-2-nonenal (Pedrajas *et al.*, 1998). The oxidation of haemoglobin to methemoglobin in fish erythrocytes has been reported in some studies (reviewed by Lackner, 1998). However, it has additionally been demonstrated that peroxisome proliferators inhibit the enzymes superoxide dismutase, glutathione peroxidase and glutathione-S-transferase (see below). Also, the peroxisomal β-oxidation system is very sensitive to H_2O_2 at least in rat kidney peroxisomes (Gulati *et al.*, 1993).

2.1.4.4 Markers of oxyradical-mediated cell injury

Markers of oxyradical-mediated cell injury consist of measurements of oxidative alterations in DNA, proteins and lipids. Oxidative DNA damage is conventionally measured as increased 8-hydroxydeoxyguanosine (8-OH-dG) (Lake, 1995). There are a number of methods available to measure oxidative damage to proteins including myoglobin oxidation, haemoglobin oxidation to form methemoglobin, inhibition of Ca-ATPase and others (Kehrer, 1993; Lackner, 1998). However, a major problem with these end-points is that they are not specific for ROS. For instance, measurement of methemoglobin formation cannot be used as a specific index of oxidative protein damage because some pollutants can specifically oxidise haemoglobin.

One of the most commonly measured end-points of oxidative damage to membrane lipids is lipid peroxidation. Lipid peroxidation can be detected using different markers such as malondialdehyde formation (measured generally using the thiobarbituric acid test), conjugated dienes, ethane/pentane ratios, fatty acid analyses etc. (see reviews by Gutteridge & Halliwell, 1990; Kehrer, 1993; Lackner, 1998). Lipofuscin accumulation can also reflect oxidative damage to lipids. Histochemical methods coupled to image analysis techniques are reliable means to measure lipofuscin accumulation in aquatic organisms (Moore, 1990; Krishnakumar *et al.*, 1995).

2.1.5 Consequences of damage

As a consequence of immediate oxidative damage exerted by ROS on cellular macromolecules including DNA, membrane lipids and proteins (Fig. 2.5), a cascade of reactions could be triggered leading to various dysfunctions at higher levels of biological organisation, i.e. tissues and organs, individuals, populations and ecosystems. At this stage only relationships between ROS-induced genetic damage and molecular/cellular/tissue effects are apparent while the effects at higher levels of biological organisation remain largely unexplored and are considered in section 2.4.

2.1.5.1 Tumour formation

There is a substantial body of literature linking ROS-induced DNA damage with tumourigenesis in a variety of experimental models. One of the best established models is the English sole (*Parophrys vetulus/Pleuronectes vetulus*) carcinogenesis model. Several studies have demonstrated a good correlation between the occurrence of liver tumours and related liver lesions in English sole and environmental or laboratory exposure to certain pollutants such as PAHs (Malins *et al.*, 1988; Myers *et al.*, 1990, 1998; Schiewe *et al.*, 1991). The PAH benzo[a]pyrene is also carcinogenic to rainbow trout (Hendricks *et al.*, 1985). Although indirect, there is evidence indicating that oxidative damage to DNA associated to environmental contaminant exposure is linked to tumourigenesis in English sole. Thus, Malins and Haimanot (1991) have found that concentrations of DNA modification caused by hydroxyl radicals are higher in apparently healthy fish from contaminated areas with respect to healthy uncontaminated fish, and in contaminated fish with hepatic tumours when compared to contaminated apparently healthy fish. An additional relevant factor in contaminant-induced tumourigenesis is DNA adduct formation and mutagenesis, as discussed in section 2.2.

Another well-studied model linking oxidative stress with tumour formation is the peroxisome proliferation model in responsive or sensitive species. Peroxisome proliferators are non-genotoxic and non-mutagenic compounds that do not bind covalently to DNA. However, chronic treatment of mice and rats with peroxisome proliferators leads to a higher liver tumour incidence (Reddy *et al.*, 1980; Reddy & Lalwani, 1983; Stott, 1988; Lake, 1995). Reddy and co-workers have proposed that the overproduction of H_2O_2 derived from increased activities of peroxisomal β-oxidation and microsomal Ω-oxidation enzymes could not be detoxified by catalase, whose activity is induced only slightly (Reddy *et al.*, 1980; Reddy & Lalwani, 1983; Reddy & Rao, 1989). Then, the excess H_2O_2 could diffuse outside peroxisomes and react, directly or after conversion into hydroxyl radicals, with cellular macromolecules including DNA. This would eventually cause DNA damage and ultimately tumour formation (Fig. 2.6). Additionally, the administration of peroxisome proliferators in responsive species leads to a reduction in the activity of antioxidant and GSH-requiring enzymes (SOD, GPX and GST) and in the amount of oxyradical scavengers (reduced glutathione and vitamin E), thus facilitating the process of oxyradical-mediated hepatocarcinogenesis (Cattley *et al.*, 1987; James & Ahokas, 1992; Demoz *et al.*, 1993; Grasso, 1993; Lake, 1995).

It has been demonstrated that the administration of peroxisome proliferators leads to an increased H_2O_2 formation (Demoz *et al.*, 1993; Lores Arnaiz *et al.*, 1995), which is followed by oxidative DNA damage (Takagi *et al.*, 1990) and an increased lipid peroxidation and lipofuscin deposition (Cattley *et al.*, 1987; Demoz *et al.*, 1993; Grasso, 1993). However, according to Demoz *et al.* (1993) and Lake (1995), the magnitude of effects described is not enough to account for the tumour promotion ability of peroxisome proliferators. Thus, James and Roberts (1995) have concluded that oxidative damage cannot cause the early stages of tumour formation and expansion although it could increase the number of cells initiated and these could be later on promoted by peroxisome proliferators.

In addition to oxidative stress induction, the hepatocarcinogenic activity of peroxisome proliferating agents in responsive species could be related to their effects as enhancers of cell replication or mitogenesis, as promoters of liver lesions, or as suppressors of cell

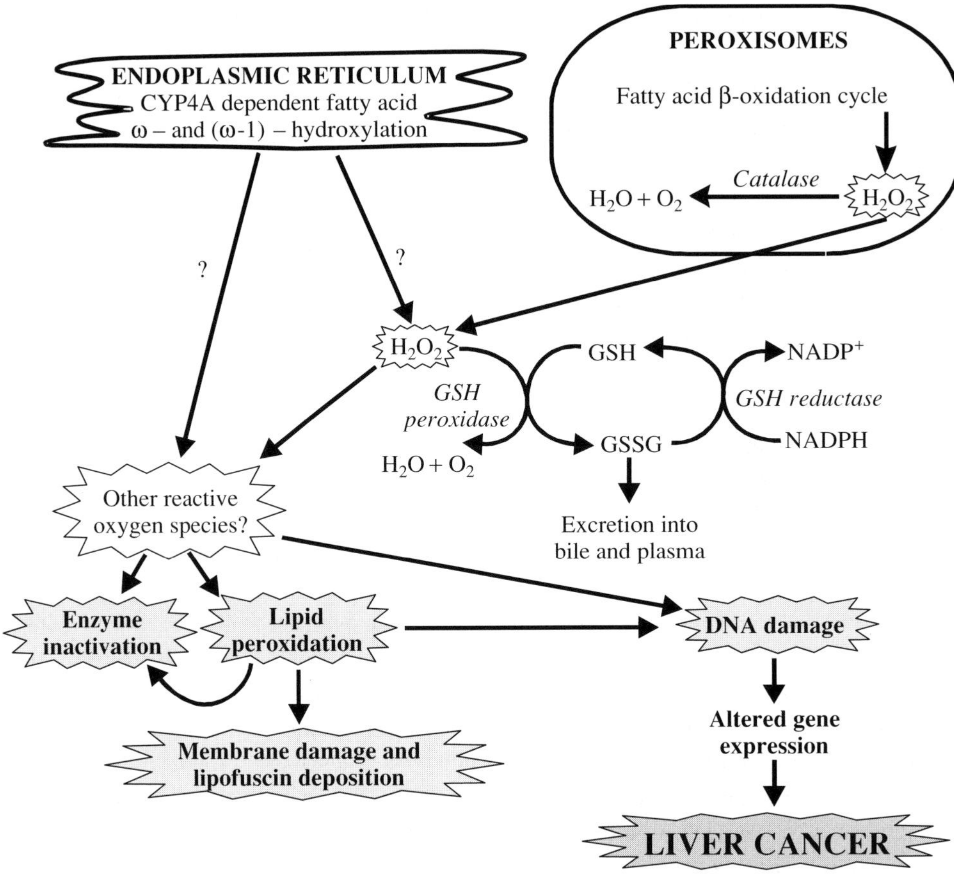

Fig. 2.6 Hypothetical model linking the overproduction of ROS under sustained peroxisome proliferation with liver tumour formation in sensitive species. The induction of hydrogen peroxide-producing enzymes in peroxisomes and endoplasmic reticulum leads to an increase in the cytosolic levels of ROS that would result in oxidative stress, cell injury and tumour formation. Modified from Lake (1995).

apoptosis (see reviews by Bentley *et al.*, 1993; Fahimi & Cajaraville, 1995; Lake, 1995). Therefore, it appears that a multifactorial etiology may be responsible for the hepatocarcinogenic effect of peroxisome proliferators in rodents and other sensitive animals. To the authors' knowledge there is no study addressing the possible hepatocarcinogenic effect of peroxisome proliferators in fish and thus this is an important gap to be filled in future studies. The literature concerning tumour appearance in fish is reviewed in Chapter 4.

2.1.5.2 Other oxyradical-mediated diseases

A role for free radicals has been proposed in the toxicity of numerous chemicals and in the pathogenesis of many diseases in humans or other mammalian organisms (see review by Kehrer, 1993). However, the direct link of free radicals with any specific toxicity or disease has been difficult to establish even in the better studied species and to the best of our knowledge there is no information on this subject in fish.

2.2 Direct damage to DNA by mutagenic chemicals and radiation

Direct damage to DNA is an increasingly important focus in ecotoxicology research for two reasons: firstly, because of the far-reaching effects of genotoxins on the health of an organism and the possible future implications if the germline is affected, and secondly, because extremely sensitive methods of detecting DNA damage have been developed, which allowed the development of early biomarkers for xenobiotic exposure (Groopman & Skipper, 1991; Stein *et al.*, 1994; Nestmann *et al.*, 1996). There are two principle mechanisms by which pollutants can cause direct damage to DNA. These are through the formation of adducts and via direct mutation.

2.2.1 Adducts

2.2.1.1 Contaminants and production mechanisms

DNA adducts arise from covalent binding of electrophilic xenobiotics to DNA, and are structures ranging in complexity from simple alkyl groups to large multi-ring residues (Harvey, 1995). There are numerous electrophilic chemicals which are capable of forming such structures, such as carbonium ions, nitronium ions, free radicals, diazonium ions, epoxides, aziridinium ions, episulfonium ions, strained lactones, sulfonates, halo ethers and enals (Williams & Weisburger, 1991). The formation of adducts is widely thought to trigger a cascade of biochemical changes leading to neoplasia and sometimes malignancy (Weinstein, 1988; Depledge, 1994), though some chemicals may exert genetic damage by mechanisms other than DNA binding (Hemminki, 1990). Because of their reactivity, many of the electrophilic chemicals causing DNA adducts are unstable and degrade rapidly (Harvey, 1995).

In addition to such chemicals directly binding to DNA, a large number of chemically inert compounds may be converted into metabolites with electrophilic properties, which are thus capable of forming DNA adducts (Harvey, 1995). Substances such as polycyclic aromatic hydrocarbons, aromatic amines, azo compounds, nitroaryl compounds and nitrosamines are non-polar lipophilic components, which would build up in the organism if they were not actively transformed into water-soluble derivatives and excreted (Sipes & Gandolfi, 1991). This cellular detoxification mechanism produces intermediates, which are more reactive than the parent compound or their metabolites, and may therefore act as genotoxins forming DNA adducts (Harvey, 1995). A direct relationship between exposure to polycyclic aromatic compounds and the level of DNA adducts has been shown in several fish species, including English sole (*Pleuronectes vetulus*), winter flounder (*Pseudopleuronectes americanus*), and oyster toadfish (*Opsanis tau*), (Varanasi *et al.*, 1986; Collier *et al.*, 1993).

DNA adducts may also occur naturally and have been found in apparently unexposed populations (Randerath *et al.*, 1992; Nath *et al.*, 1996). The occurrence of these adducts may vary depending on environment, sexual maturity, history of stress and gene regulation and expression (Nestmann *et al.*, 1996). Such endogenous adducts are as yet chemically uncharacterised, though difunctional carbonyl compounds produced by lipid peroxidation may be responsible for their formation (Marnett, 1994; Burcham, 1998). The often high abundance of endogenous adducts may sometimes be a problem for non-specific methods of detecting genotoxin exposure such as ^{32}P labelling. However, initial studies suggest that endogenous adduct formation is rarer in fish than in mammals (Stein *et al.*, 1994). Furthermore, it is sometimes difficult to assess the biological significance of DNA adduct data considering levels of endogenous adducts that can be as high as one adduct in 10^5 normal nucleotides (Nestmann *et al.*, 1996).

2.2.1.2 Protection mechanisms

The induction of biotransformation enzyme systems such as P450 has been used widely as a biomarker of xenobiotic exposure (e.g. Courtenay *et al.*, 1994). Nevertheless, such systems may actually increase the genotoxic effects of xenobiotics by transforming them into electrophilic compounds capable of forming DNA adducts (Harvey, 1995). As such, the excretion of such hydrophilic metabolites and xenobiotics may be of primary importance to prevent DNA damage. Indeed, many marine organisms appear to have mechanisms capable of directly excreting xenobiotics, either by membrane glycoproteins such as P170 which bind to xenobiotics and thus facilitate their excretion, or by lysosomal accumulation (Moore *et al.*, 1986; Kurelec, 1992). Furthermore, xenobiotics may be excreted by the rectal glands of elasmobranchs, which are normally responsible for osmoregulation and NaCl excretion (Miller *et al.*, 1998).

Genetic variability within and between populations may also play an important role in the protection from adduct formation. Studies in humans have found evidence for genetic variation in the cytochrome P450 genes influence inducibility of enzyme production and thus susceptibility to some cancers (Courtenay *et al.*, 1994). Similarly, high interindividual variability has been shown in CYP1A mRNA inducibility in Atlantic tomcod (*Microgadus tomcod*), which are likely to be due to genetic variation (Courtenay *et al.*, 1994). The investigation of such interindividual and interpopulation genetic differences in the response to xenobiotics will not only be vital for the prediction of population level effects of pollution, but also for an assessment of the scope for adaptation increasing tolerance. These issues are discussed in greater detail in Chapter 7.

There also appears to be a considerable influence of abiotic and biotic factors on enzyme inducibility and thus genotoxic damage. In killifish (*Fundulus heteroclitus*), for example, CYP1A production was strongly affected by temperature, even though no temperature effect on CYP1A mRNA expression was detected (Kloepper-Sams & Stegeman, 1992). Similarly, enzyme inducibility may also depend on sex and reproductive status (Courtenay *et al.*, 1994; Troxel *et al.*, 1997). In addition, natural and anthropogenic compounds, such as organosulphur, may provide protection against genotoxic compounds by their antagonistic effects during metabolism (De Flora *et al.*, 1991; Harvey, 1995).

Other mechanisms protecting against detrimental effects of DNA adducts are tolerance and retrieval systems (Lewin, 1995). Tolerance mechanisms provide a means for damaged template sequences to be copied, probably with a high frequency of errors. Retrieval mechanisms use recombination to obtain undamaged copies from another source if replication has been forced to bypass a damaged site (Friedberg *et al.*, 1995).

Another mechanism of protection against genetic damage in an organism is cell cycle arrests (Friedberg *et al.*, 1995). Normally, the cell cycle is regulated by checkpoint controls

(Hartwell & Weinert, 1989), which ensure that a stage in the cell cycle does not begin before the previous stage has ended. For example, it would be detrimental for a cell to initiate mitosis in the presence of unreplicated DNA or an incompletely assembled mitotic spindle (Friedberg *et al.*, 1995). Therefore, eukaryotic organisms can arrest the cell cycle transiently at discrete stages to allow completion of biosynthetic processes associated with each phase. Although the link between DNA damage and checkpoint controls is poorly understood, there are examples where DNA damage in mammalian cells can lead to cell cycle arrest and subsequent programmed cell death.

2.2.1.3 Determination of adduct formation

The vast majority of DNA adduct studies in fish to date have used the ^{32}P post-labelling assay (Stein *et al.*, 1994). The primary reason for use of ^{32}P post-labelling is its high sensitivity, the possibility of assaying very small samples, and its non-specificity allowing the analysis of adducts of unknown structure. While this non-specificity allows the assessment of genotoxin exposure without knowledge on the exact composition of xenobiotics in the environment, it also makes the characterisation of specific adducts difficult (Stein *et al.*, 1994). Furthermore, quantitative determination of adducts is a problem as the efficiency of labelling steps is difficult to determine and may differ between specific adducts. The method involves the digestion of DNA into 3′-monophosphates, labelling with radioactive ^{32}P, and separating normal nucleotides from adducts by thin layer chromatography (Santella & Perera, 1994).

More specific methods include immunoassays, which have been widely used in human cancer research, because they do not require radiolabelling and can be easily applied to a large number of samples (reviewed in Poirier & Weston, 1996). The requirement of adduct specific antibodies, however, limits the general application of immunoassays to ecotoxicological studies, though they may become useful for more specific assessments of specific pollutants.

Fluorescence spectroscopy exploits the fluorescing properties of many xenobiotic compounds, such as PAHs and aflatoxins (Phillips & Farmer, 1995). In combination with high performance liquid chromatography (HPLC), fluorescence spectroscopy can be used to confirm the presence of specific adducts with known structure (Santella & Perera, 1996). The technique has a similar sensitivity to ^{32}P labelling, but requires much larger quantities of DNA (Phillips & Farmer, 1995). It has been used to detect B[a]P adducts in Beluga whales (Martineau *et al.*, 1988) though its use in fish has been limited (Stein *et al.*, 1994).

Finally, physicochemical methods such as gas chromatography or mass spectroscopy can be used to both quantify and characterise DNA adducts (Poirier & Weston, 1996). However, their high cost, low sample throughput and requirement for relatively large amounts of DNA may limit their application to biomonitoring studies. The potential of such methods has been demonstrated in a study correlating the incidence of 8-hydroxyguanine and 8-hydroxyadenine in English sole with pathologic lesions (Malins *et al.*, 1996).

2.2.1.4 Consequences of damage

The occurrence of endogenous DNA adducts often renders it difficult to define biologically effective doses of DNA adducts. These are doses of DNA adducts that have consequences to the cell or the organism (Santella & Perera, 1994; Nestmann *et al.*, 1996). The toxicological significance of DNA adduct data is further complicated by the high sensitivity of assays (1 per 10^{10}–10^{8} normal nucleotides) compared to background levels of endogenous adducts (1 in 10^{5} normal nucleotides) (Nestmann *et al.*, 1996). However, the vast majority of DNA adduct data stems from mammalian cells (Friedberg *et al.*, 1995), and there is preliminary evidence that levels of endogenous adducts may be lower in fish than in mammals (Stein *et al.*, 1994). Nevertheless, background levels of DNA adducts should be determined by using fish from uncontaminated control sites, that are matched for species, gender, age and reproductive stage (Pfau, 1997).

Despite such discussions about the toxicological significance of low levels of adducts, the strong correlation between adducts and higher level damage such as cell death, lesions and cancers is well documented (Johnson *et al.*, 1992; Stein *et al.*, 1994; but see Wirgin & Waldman, 1998). Unrepaired DNA adducts may lead to misincorporation, inhibition of DNA transcription (Choi *et al.*, 1996) or blockage in DNA replication, and may thus generate sites for frameshift and base substitution mutation (Nestmann *et al.*, 1996). The final consequence of adducts, however, will depend on various parameters, such as the time of exposure, state of cell cycle, the kinetics of repair mechanisms and the specific features of the adducts, such as quantity, mutageneity, repairability and stability. Thus it has been proposed that assays providing quantitative estimates of particular adducts evaluate end-points quite different from mutation and may not serve to directly relate an adduct to its mutagenic properties (Nestmann *et al.*, 1996). Indeed, studies in North American fish populations suggest that levels of adducts are not always predictive of the vulnerability to neoplasia of populations and species from polluted sites (Wirgin & Waldmann, 1998). Nevertheless, other studies have demonstrated a correlation between hepatic DNA adducts and prevalence of hepatic lesions in black croaker (*Cheilotrema saturnum*), and winter flounder (*Pseudopleuronectes americanus*) (Johnson *et al.*, 1992; Stein *et al.*, 1994; Reichert *et al.*, 1998). Moreover, elevated levels of hepatic DNA adducts have been shown to be a significant risk factor for certain degenerative and preneoplastic lesions occurring early in the histogenesis of hepatic neoplasms in feral English sole (*Pleuronectes vetulus*) exposed to polycyclic aromatic compounds (Myers *et al.*, 1998; Reichert *et al.*, 1998). Thus, DNA adducts are useful biomarkers for exposure, though mechanistic links between their occurrence and higher level effects, such as lesions, cell death and health need further research (Nestmann *et al.*, 1996).

2.2.2 Mutations

Heritable changes in genomic DNA, or mutations, comprise the ultimate source of genetic variability in natural populations. Mutations may occur spontaneously, with different genomic sequences exhibiting characteristic rates, or they may be promoted by various environmental agents. Genotoxic agents include natural and anthropogenically-released chemicals, radiation and ultraviolet light. Spontaneous mutations arise from alterations in the chemistry of genetic material as a consequence of natural processes such as replication, recombination and DNA repair (Friedberg *et al.*, 1995). Instability of chemical bonds resulting in such phenomena as tautomeric shifts (formation of structural isomers), deaminations of bases (loss of exocyclic amino group), depurination or depyrimidination, can lead to mismatches of bases during DNA replication. Genotoxic-induced damage, on the other hand, arises from direct interactions between environmental agents and DNA, either in their original form (direct acting genotoxins), or after biotransformation to a reactive intermediate (indirect acting genotoxins). Additionally, there are epigenetic processes, where DNA damage arises through such mechanisms as disruption of cellular macromolecules essential for the production and replication of new DNA.

Although many mutations are rare and detrimental, under certain conditions mutants may increase in frequency, and through selective mortality may lead to the evolution of adaptation. Theoretical models suggest that natural selection will adjust mutation rates to intermediate levels, resulting in a balance between levels that minimise mortality and reduced fitness (mutational load), while maximising genotypic variability to promote population persistence in changing environments (Gillespie, 1981). In natural environments, however, numerous agents interact to disrupt the mutation-selection balance through the process of mutagenesis, leading to impaired performance and reduced fitness (Turelli, 1984, 1986). In natural populations, however, it is often difficult to determine the intensity of selection and mutation rates for quantitative characters. Mutation-selection balance can only maintain substantial genetic variation if a trait is affected by a large number of loci or if the mutation rate for loci influencing a trait is high. It is unclear how many loci affect stress resistance traits, though it has been shown in some cases to be determined by polymorphisms at a single gene, gene complex, or by multiple copies of a single gene (Depledge, 1994). It is essential that further studies are conducted to locate genes related to resistance traits, as well as documenting mutation rates in identifiable genes under conditions of stress.

Genotoxin-induced mutations in gametes may impact on subsequent generations, greatly accelerating the evolutionary consequences of genotoxic damage (Shimas & Shimada, 1994). Such general associations are well established (Wirgin & Waldman, 1998). However, it has proven difficult to develop reliable methods for mutation quantification, to relate mutational damage to changes in allele frequencies and the structure of gene pools, and to assess the fitness consequences of genotoxin-induced mutations in terms of individual and population-level effects.

Although there is an increased awareness of the importance of exploring linkages between DNA damage at the nucleotide level through the emergence of so-called 'evolutionary toxicology' (Bickham & Smolen, 1994), the majority of studies continue to focus on the direct effects of contaminants on DNA, rather than exploring the population consequences of contamination by molecular genetic monitoring of allele frequencies.

2.2.2.1 Contaminants

Contaminants may impact genetic material either indirectly, through impacting natural cellular function, such as impeding DNA replication, or directly, through interaction with nucleotides. Contaminants include chemical agents (natural and anthropogenic emissions), ultraviolet light, radiation and viruses. Natural emissions in the marine environment arise from such phenomena as oil seeps, the erosion of sedimentary rocks, and atmospheric deposition of incomplete combustion products from volcanic activity and forest fires. Marine organisms may themselves produce toxic compounds (Payne & Rahimtula, 1989). Direct acting genotoxins include chemical compounds that are electrophilic, and therefore can potentially react directly with nucleophilic sites within DNA molecules, and include such compounds as carbonium ions, episulfonium ions, free radicals, diazonium ions, epoxides, azaridinium ions, strained lactones, halo ethers and enals (Williams & Weisburger, 1991).

2.2.2.2 Production mechanisms

Major forms of damage to DNA include damage to the phosphodiester backbone, the ribose sugars and the purine and pyrimidine bases. Damage to DNA consists of changes at two levels: single-base changes and structural distortions. Single-base changes affect the sequence, but not the overall structure, of DNA. They do not impact on transcription or replication, and thus such damage exerts its effects on future generations through the consequences of the change in DNA sequence. Structural distortions provide a physical impediment to replication or transcription.

Mutational damage may lead to carcinogenesis, whereby a chemically-induced change in genetic material results in the production of neoplasms. Neoplasms arise as a result of mutations in critical genes that control normal cell division, differentiation and cell death such as oncogenes and tumour suppresser genes (Wirgin *et al.*, 1990; Cosma *et al.*, 1992). Mutations comprise the first stage of neoplasia, the so-called 'initiation', whereby an irreversible change, or mutation, arises in the nucleotides. It may be followed by the stages of 'promotion', where initiated cells may be enhanced by proliferating agents, which increase the probability of further spontaneous or chemically-induced mutations (Vogelstein & Kinzler, 1993). This proceeds to 'progression', the final stages of carcinogenesis, where preneoplastic cells can develop and constitute a neoplasm. The latter stage may be further enhanced by mutations in critical target genes, resulting in the change from a benign, non-invasive neoplasm to a malignant form which may invade surrounding tissue (neoplasm). Neoplasms often lead to a pathological disturbance of cellular function and growth, characterised by excessive cell proliferation. Such proliferating cells are thought to contain heritable changes that enable the cell to ignore normal cellular signals that regulate growth (Payne & Rahimtula, 1989; Myers *et al.*, 1990).

The mutational process is therefore central to the initiation and progression of genotoxic damage, and as such has resulted in the development of a vast array of methodologies to detect their incidence under conditions of contamination. Effective detection of mutations may not only provide the basis for biomonitoring, but also serve to identify vulnerable stages in the life history of a species, the nature and dynamics of causal agents and associated phenotypic and population-level effects (Depledge, 1994; Hose, 1994; Shugart & Theodorakis, 1994). There are four principal mutational processes:

- *Point mutations*: These involve a change in the nucleotide sequence that can occur by the replacement of one nucleotide with another, or 'substitutions'. Substitutions can be either transitions or transversions. The consequences of such mutations will depend on the position in a nucleotide sequence, and may affect gene expression. For example, 'Missense' mutations arise when they cause codon changes at a critical site in the structure of a polypeptide, resulting in defective proteins and altered gene expression. 'Nonsense' mutations result in the termination of polypeptide synthesis (stop codons).

Alternatively, due to the degenerative nature of the genetic code ('sense' mutations), or through changes in non-coding regions ('silent' mutations), mutations may have no phenotypic effect.

It is generally recognised that the majority of adaptive evolutionary change results from point mutations, representing relatively small genomic changes, which may lead to the production of novel proteins. Such events may be detected using molecular genetic techniques and are likely, in part, to underpin adaptive responses to environmental stress (Hoffmann & Parsons, 1991; Hillis *et al.*, 1996). Such mutations may generate variants that lead to observable intraspecific genetic variability in susceptibility to pollutants in natural populations (Depledge, 1994). An understanding of the dynamics and significance of point mutations is therefore central to assessing the impacts of genotoxins and associated genotypic response.

- *Frameshift mutations*: A frameshift mutation alters the reading frame of the genetic code through the addition or deletion of one or more bases. Such changes may modify an entire sequence and hence alter the transcription of a gene, frequently leading to the production of non-functional gene products.
- *Chromosomal mutations*: Chromosomal mutations (aberrations) involve changes to the gross structure of chromosomes, resulting from the loss, breakage and reunion of genetic material during cell replication. Such events can give rise to deletions, inversions and translocations, and occasionally gene amplification. The latter has been shown to underlay genetic adaptation to chemical stress (Field *et al.*, 1989). Strand breakages occur under normal conditions but exposure to genotoxins can increase the amount (Shugart & Theodorakis, 1994). For example, ionising radiation can cause strand breakage directly, whereas other physical agents such as UV light or genotoxic chemicals can modify DNA molecules that are involved in DNA repair (e.g. photoproducts, adducts, modified bases), and thus promote strand breaks.
- *Genomic mutations*: Genomic mutations produce changes in the number of chromosomes (aneuploidy), and usually result from exposure to a substance that interferes with the mitotic apparatus during cell division. The majority of aneuploidies are lethal, but a small proportion do survive with reduced viability and may play an important role in the generation of genetic diseases such as neoplasia (Dixon & Clarke, 1982).

2.2.2.3 Detection of mutations

There are several short-term bacterial mutagenicity tests that are available for the detection of mutationally active compounds in the tissues of marine organisms. Despite their general ease of use, the assays provide no indication of the potential for mutation induction in the species under study. Several methods have been developed to detect point mutations *in vivo* in the exposed species, though there have been limitations in the study of aquatic organisms because of the relative scarcity of sequence information (Cotten, 1993).

Among the various impacts of DNA adducts is the induction of mutational change, which typically has been considered under three main categories: genomic, chromosomal and gene sequence mutations. A variety of cytogenetic methods are available for the detection of chromosomal aberrations (Stein *et al.*, 1994). Here the focus will fall on the detection of DNA sequence mutations. Direct assays involve the analysis of sequence variation using a variety of mutation assays, whereas indirect detection involves the employment of molecular genetic techniques.

Gene mutation analysis systems initially involved the use of molecular techniques such as southern hybridisation and the direct sequencing of cloned cDNA. These methods are progressively being replaced by procedures which incorporate the polymerase chain reaction (PCR). Several methods have been developed which incorporate the PCR (Cotten, 1993; Hayashi, 1994; Rossiter & Casket, 1994), and usually are accompanied by the direct sequencing of nucleotides.

The detection of unknown mutations involves the identification of heteroduplexes or mismatches between mutated and wild type sequences, based either upon the electrophoretic properties of the sequences or upon the selective chemical modification of such sequences. The two main electrophoretic methods are the denaturing gradient gel electrophoresis (DGGE) assay, and the single stranded conformational polymorphism (SSCP) assay. The DGGE can separate wild type and mutant DNA heteroduplexes, whereas the SSCP separates single stranded wild type and mutant DNA sequences due to differences in secondary structure. Although such procedures detect a variety of base substitutions, frame shifts and deletions, the methods fail to detect all mutations present. Approaches which exploit chemical differences between mutant and wild type sequences include carbodiimide modification, assay and the chemical cleavage mismatch assay. The former involves the addition of the reagent, thus changing the electrophoretic and PCR amplification properties of the heteroduplex, whereas the latter involves the cleavage of the heteroduplex by chemical reagents, followed by direct sequencing of the cleaved strands. These systems are capable of detecting 100% of the mutations in the targeted sequence.

The detection of known mutations involves mismatched primer techniques such as the allele-specific oligonucleotide technique, or the allele-specific amplification method. Both of these involve the amplification of mutant and wild type sequences. These approaches are based on the successful amplification of mutant sequences with primers specific to the suspected mutation, and therefore require sequence information of targeted areas. Despite the efficacy of the established techniques that frequently require the selection of mutant genotypes by artificial cell culture, they do not enable the direct analysis of cellular DNA of the tissues exposed, or of the study of DNA in non-dividing cells. Advances in transgenic approaches (Gossen & Vijg, 1993; Bailey *et al.*, 1994; Gossen *et al.*, 1994) have proven powerful assays for mutational change, whereby transgenes introduced at the zygote stage of development act as target genes capable of a phenotypic response to mutational events. These are subsequently screened using a bacterial system.

A valuable addition to the battery of detection methods is the restriction site mutation assay (RSM) (Parry *et al.*, 1990; Felley-Bosco *et al.*, 1991), which possibly provides the greatest potential for the detection of genotoxin-induced mutations in bioindicator species. Unlike other methods, the RSM does not depend on the selection of a mutated phenotype, thus allowing identification of dominant, recessive or silent mutations. The RSM is based on the detection of DNA sequence variation using a combined restriction enzyme and PCR approach. The wild type enzyme recognises sequences in the target sequence, and if any bases have mutated, the sequence will not undergo restriction cleavage. The second stage involves the preferential amplification of mutant molecules resistant to digestion since the cleaved wild type sequences will not serve as templates for amplification. The mutant region is then sequenced, yielding the nature of mutant genotype.

Because of its central role in cell cycle arrests and thus the repression of tumour formation, the p53 gene has been of central interest in human cancer research (Friedberg *et al.*, 1995). Its inactivation by mutations appears to be a prerequisite for neoplastic transformation in many human cancers (Hollstein *et al.*, 1991). There is limited information in fish, though initial trials identified the conserved and thus probably functionally important regions for use as genotoxin biomarkers (Defromentel *et al.*, 1992; Cheng *et al.*, 1997; Krause *et al.*, 1997; Bhaskaran *et al.*, 1999).

An overall estimate of mutations in the genome without the identification of individual changes can be obtained by gas chromatography-mass spectrometry with selected ion monitoring (GC-MS/SIM) and Fourier-transform infrared (FT-IR) spectroscopy, and has indeed revealed surprisingly high levels of structural DNA damage in exposed fish populations (Malins & Gunselman, 1994; Malins *et al.*, 1997a, b).

Molecular genetic analyses using markers such as allozymes and DNA polymorphism provide an indirect approach for the detection of mutants. Comparison of allelic diversity in samples taken from contaminated and control sites can provide estimates of genotoxic-induced mutants, though the efficacy of detection will depend markedly on the genomic areas assayed, as well as localised differences in microevolutionary forces and population history (Hoffmann & Parsons, 1991; Guttman, 1994). Allozymes have for example been used to examine the frequency of mutations and mutation-like events in populations of Scots pine (*Pinus sylvestris*) in areas of air pollution (Bakhtiyarova *et al.*, 1995; Bakhtiyarova, 1997). The frequency of rare electrophoretic variants of allozymes was significantly higher in two populations growing under industrial air pollution conditions.

The Chernobyl accident has served as a model system for exploring mutagenic events. Germline mutation at human minisatellite loci has been studied among children born in heavily polluted areas after the accident and in a control population (Dubrova *et al.*, 1996). The frequency of mutation was found to be twice as high in the exposed families as in the control group, and mutation rates in the contaminated population were correlated with the level of caesium-137 surface contamination, consistent with radiation induction of germline mutation. An increased frequency of partial albinism, a morphological aberration associated with the loss of fitness, was reported among barn swallows breeding close to Chernobyl (Ellegren *et al.*, 1997). Heritability studies indicated that mutations causing albinism were at least partially of germline origin. Furthermore, evidence of an increased germline mutation rate was obtained from segregation analysis at two hypervariable microsatellite loci, indicating that mutation events in these birds were two to tenfold higher than populations from control areas. Allozyme analysis of the fingernail clam (*Musculium transversum*) showed high frequencies of a pollution tolerant allele at the glucose-6-phosphate isomerase-2 locus (Sloss *et al.*, 1998). Polluted sites exhibited elevated frequencies of Gpi-2(100) whereas non-polluted sites exhibited elevated frequencies of Gpi-2(74), suggesting that natural selection was occurring in populations under severe toxic pressures, leading to an increase in its frequency. Thus, Gpi-2(100) is a possible pollution-tolerant mutation.

A technique with great potential for mutation assays is arbitrarily-primed polymerase chain reaction (AP-PCR). Despite problems concerning reproducibility and complexity of patterns (Atienzar *et al.*, 1998; Singh & Roy, 1999), the technique has several advantages for the detection of genomic mutations, such as ease, speed and low costs of experiments and the ability to clone aberrant fragments (Navarro & Jorcano, 1999). While the technique has so far been mainly used for investigation of human cancer tissues, its potential has been shown in a study on Japanese medaka (*Oryzias latipes*) where a correlation between g-ray-induced genomic damage and embryo malformations could be demonstrated (Kubota *et al.*, 1992).

There are relatively few published studies utilising the molecular genetic approach, though such analyses have the advantage of providing a rapid and relatively simple assay of genetic variation in natural populations. Additionally, observations on the incidence of mutations can be related directly to data on genetic structure that provides direct information on genotypic responses to environmental stress. Future studies could perhaps employ a combined mutation assay and molecular genetic marker approach to facilitate opportunities for relating genotoxic damage to population genetic structure and phenotypic estimates of fitness.

2.2.2.4 Consequences of damage

Consequences of DNA damage are wide-ranging and include alterations of enzyme function and protein turnover rates resulting in impaired metabolism, the production of cytotoxic injuries, inhibition of cellular growth, increased rates of tissue ageing, suppression of immune response, reduced reproductive fitness, and increased incidence of disease and neoplasia. Although the consequences at the cellular levels have been well documented there is considerably less data on the impact at higher biological levels such as fecundity and viability (Shugart *et al.*, 1992; Shugart & Theodorakis, 1994; Wirgin & Waldman, 1998). Neoplasms, for example, have been observed in numerous marine organisms including molluscs, amphibians and fishes (Payne & Rahimtula, 1989). However, their effects on physiology, growth and reproduction have been poorly defined. Nevertheless, mutations have been shown to be associated with gamete loss, abnormal development, embryonic mortality and heritable mutations (Shugart & Theodorakis, 1994). For example, embryonic mortality in Beluga whales has been attributed, in part, to lethal mutations (Martineau *et al.*, 1988), which may provide more sensitive indicators of reproductive impairment than changes in fecundity.

2.2.3 Repair mechanisms

Once DNA adducts or mutations are formed, there is a whole array of DNA repair mechanisms to amend the damage. DNA repair mechanisms comprise multiple reactions that recognise and remove DNA lesions induced by genotoxins. These processes maintain the genetic integrity of a species following genotoxin-induced damage, thus providing a balance between the generation of genetic diversity and the process of adaptation. Thus, the measured rate of mutation reflects a balance between the number of damaging events occurring in DNA and the number that have been corrected (or miscorrected). Having emphasised the potential long-term evolutionary importance of mutational input to evolutionary change and adaptation, it is evident that the effectiveness of repair mechanisms depends on the frequency and nature of mutations, especially the extent to which they may be heritable (occurring in gametes) and disruptive.

The most direct mode is the direct reversal of the damage, for example, when alkylated DNA bases are repaired by alkyltransferases (Friedberg *et al.*, 1995). This form of repair is highly effective because it occurs more rapidly than multistep biochemical pathways such as excision repairs, and because it produces relatively few errors. Nevertheless, this mode of repair may be energetically quite expensive, as an entire protein molecule is expended in each reaction.

The types of DNA damage that can be repaired by direct reversal are limited, and the most common mode of repair involves excision of the damaged bases and resynthesis of DNA (reviewed by Friedberg *et al.*, 1995). Base excision repair (BER) operates mainly on small DNA adduct complexes such as irreversibly alkylated DNA bases, and is carried out by DNA glycosylases which catalyse the hydrolysis of N-glycosylic bonds linking damaged bases to the deoxyribose-phosphate backbone of DNA. Subsequently, sites lacking their bases are removed by specific endonucleases, and the resynthesis and ligation of the excised region. Nucleotide excision repair (NER) involves the removal of whole nucleotides including bases and deoxyribose-phosphate backbone of DNA, and thus excised fragments are usually oligonucleotide fragments rather than free bases. As in base excision repair, the resulting gap is filled by repair synthesis using the alternate strand as template and ligation. However, nucleotide excision repair is more complex than base excision repair and involves many more gene products. Nucleotide excision repair is considered to repair DNA at different rates, thus allowing fast preferential repair of active cells (Harvey, 1995).

In addition to BER and NER, several other pathways for DNA repair exist, including direct reversal by photoreactivation of pyrimidine dimers, alkyltransferases, purine insertion and the ligation of strand breaks.

The efficiency of DNA repair processes will depend to a large degree on the development of DNA adducts and associated damage (Espina & Weis, 1995). Chemical agents with proliferation or inhibitive capabilities have been shown to influence the repair capacity of cells, by modulating the balance between repair and replication (Barrett, 1995). Efficiency of repair will not only depend on various physical factors such as position in the DNA sequence, chemical stability of adduct complex, and accessibility to repair complexes, but there is also apparent intrinsic variability depending on species and exposure (Anderson & Harrison, 1990). Data for example has indicated that DNA damage may be cumulative in oocytes, leading to the hypothesis that organisms with long synchronous periods of gametogenesis may be more vulnerable to chronic exposure, based on reduced repair capacities. In the mouse, data suggests that DNA repair is not active in postmeiotic cells (Russell *et al.*, 1990), and thus a precedent exists for low DNA repair capacities in gametogenic stages of some organisms. Studies that explore the effect of long-term, low-level exposures to mutagens on gametes would require direct assessments of the kinetics of absorbed dose and of DNA repair in gametes. The lability of repair systems, and the extent to which they may be associated with long-term exposure, have been relatively neglected in marine organisms (Wirgin & Waldman, 1998), and yet will have major impacts on the evolution of tolerance and long-term evolutionary impacts of genotoxicity.

Most studies concerning repair systems have been carried out in the bacteria *Escherichia coli* and the yeast *Saccharomyces cerevisiae*. Studies on aquatic organisms are rare, though repair systems have been reported in fish and invertebrates, including the teleost, *Fundulus heteroclitus* and mussels (Sikka *et al.*, 1990; Harvey & Parry, 1997). In some of the investigated fish species, repair mechanisms appear to be less efficient than in mammalian systems (e.g. rainbow trout (Bailey *et al.*, 1996), *Fundulus heteroclitus* (Espina & Weis, 1995)). A DNA polymerase likely to be a repair enzyme has been isolated from leach (*Misgurnus fossilis* (Sharova *et al.*, 1994)).

2.3 Direct chemical effects on chromosomes

Genotoxicity is a general term referring to alterations to the gross structure or content of chromosomes (clastogenicity) or base-pair sequences of DNA (mutagenicity) by exposure to toxic agents. Clastogenic activity may lead to genetic disease, teratogenesis, or carcinogenesis in fish populations (Al-Sabti, 1995a). The genotoxic effects of some pollutants may occur at cellular concentrations well below those causing gross cytotoxicity (Al-Sabti, 1994). Thus, consumption of contaminated fish can induce genotoxic damages such as chromosomal damage in lymphocytes of consumers (Al-Sabti & Metcalfe, 1995). Marine fish and shellfish often contaminated with high concentrations of pollutants can be major vectors for contaminant transfer to humans, especially in countries in which marine fish and shellfish are a major source of protein (Al-Sabti, 1994).

2.3.1 Contaminants and production mechanisms

Genotoxicity can result in three types of genetic lesions (Casciano, 1991; Zakrzewski, 1991). First, single-gene mutations, also called point mutations, which include alterations in the nucleotide sequence of DNA, and may involve either base substitution or frame-shift mutation. These have already been described in section 2.2.2 and will not be considered further here. Second are structural chromosomal mutations or genomic mutations which include changes in chromosomal structure, such as breaking of chromosome, or translocation of an arm (sister chromatid exchange), known as clastogenesis. Third are numerical changes in the genome (aneuploidy and hyperploidy), formed by a mis-separation of chromosomes during cell division. Many hereditary disorders are caused by this phenomenon.

Chromosome alterations can either be originated by direct DNA damage induced by chemicals or can be a consequence of the misrepair of chemically-induced DNA damage (Preston, 1990; Geard, 1992). It has been proposed that aberrations induced by radiation or chemical agents result from errors either during S-phase synthesis, during the resynthesis step of excision repair, or during the synthesis required for recombination repair of double strand breaks. 'Spontaneous' aberrations would be formed by the same mechanism, namely errors of replication, and thus radiation or chemically-induced DNA damages simply cause an enhanced probability of such errors occurring (Preston, 1990).

Replicating cells are more vulnerable to the action of DNA damaging agents than non-replicating cells, because error-free repair of DNA lesions must occur before cell division, and proliferating cells may not have enough time for this repair (De Flora & Ramel, 1988). Some genotoxic agents induce DNA and chromosomal damage in all phases of the cell cycle while others tend to be S-phase specific (Geard, 1992).

There are two main types of genotoxic agents which may induce changes in chromosomal structure and number: physical mutagens (UV light, ionising radiation), and chemical mutagens (organic and inorganic). In most of the cases, mutation of DNA or changes in the genetic information of the cell induced by electrophilic reactants is the primary event in the initiation of carcinogenesis by chemicals; however, it is possible that in some cases, non-genetic changes may be primary events. The most important organic carcinogens include the polycyclic aromatic hydrocarbons (PAHs), aromatic amines and aminoazo dyes, dialkyl nitrosamines, alkyl nitrosamines, polychlorinated aliphatic and alicyclic hydrocarbons, aflatoxins, pyrrolizidine alkaloids, ethionine, urethane, cycasins and a large array of other alkylating agents. Inorganic carcinogens include certain metals (such as chromium, cadmium, lead and mercury) and complex silicates (Casciano, 1991).

The induction of chromosome damages is one of the primary events in the initiation of carcinogenesis by chemicals. Several chemical pollutants can produce carcinogenic effects in fish species through the induction of genetic lesions. Indeed, most of these chemicals cause tumours at specific or multiple sites in fish (Harshbarger & Clark, 1990).

Carcinogens are divided into two categories: genotoxic and epigenetic. Compounds that react directly or indirectly with DNA are, in most cases, mutagens (polycyclic aromatic hydrocarbons, alkylating agents, specific metals), and they are designated as genotoxic because they have the potential to alter the genetic material. Epigenetic carcinogens, such as organochlorides, estrogens, clofibrate, phthalate esters, nitriloacetic acid, etc. are those carcinogens that are not classified as genotoxic, and a multitude of mechanisms may be involved in the induction of chromosomal damage by these carcinogens (Weisburger & Williams, 1991; Zakrzewski, 1991). The epigenetic carcinogens comprise a wide variety of compounds, such as metal ions (nickel, chromium, lead, cobalt, manganese and titanium); solid-state carcinogens (asbestos and silica); immunosuppressors (azathioprine and 6-mercaptopurine); and promoters (tetradecanoylphorbol acetate, phenobarbital, PCBs, tetrachlorodibenzodioxin, and chlorinated hydrocarbon pesticides) (Zakrzewski, 1991).

Several genotoxic effects like DNA adducts, DNA breakage, chromosome aberrations and sister chromatid exchange can be observed in aquatic organisms exposed *in situ* to xenobiotics (Al-Sabti, 1985, 1986a; Batel *et al.*, 1985; Al-Sabti *et al.*, 1994; Pacheco & Santos, 1996; Das & John, 1997; Venier *et al.*, 1997; Marlasca *et al.*, 1998). DNA strand breaks, chromosomal aberrations and sister chromatid exchanges have been detected in embryonic, larval or adult stages of *Mytilus* sp. after exposure to environmental or known genotoxic agents such as benzo[a]pyrene, bleomycin-Fe(II), bromodeoxyuridine, cyclophosphamide, mitomycin C, methylmethanesulphonate, or 4-nitroquinoline-*N*-oxide (Al-Sabti & Kurelec, 1985; Bihari *et al.*, 1990; Vukmirovic *et al.*, 1994).

Many xenobiotics enter the body as innocuous compounds and become carcinogens after metabolic activation. Such xenobiotics are referred to as precarcinogens (Zakrzewski, 1991). The majority of chemical carcinogens require metabolic biotransformation to produce their ultimate genotoxic metabolite(s), reactive electrophiles that combine with nucleophilic groups in nucleic acids and proteins (Batel *et al.*, 1985; Casciano, 1991). The high nucleophilic reactivity of many carcinogens results in genotoxic properties but also in other toxic reactions in the cells (Nielsen, 1993). Some of these reactions with nucleic acids and/or proteins are crucial to the initiation of the carcinogenic process.

Metabolic activation and detoxification is carried out by a variety of inducible detoxifying enzymes, such as the phase I and phase II enzyme systems (Doherty *et al.*, 1996 and see also Chapter 3). Fish and other marine organisms present enzymes involved in the activation and detoxification of xenobiotics (De Flora *et al.*, 1989; Rodriguez-Ariza *et al.*, 1994). Fish cytochrome P450 is only one of the multiple enzymes involved in xenobiotic transformation and metabolises many carcinogens in a manner analogous to mammalian organisms (Rodriguez-Ariza *et al.*, 1994; Stegeman & Lech, 1991; sections 2.1.2.2 and Chapter 3). It has been proved that hepatic S9 fraction from mullet increases the metabolic activation of several pollutants, such as benzo[a]pyrene, 2-acetylaminofluorene, 2-aminoanthracene, and aflatoxin B1 (Rodriguez-Ariza *et al.*, 1991, 1994). There is also an additional important mechanism leading to long-term consequences in marine organisms, such as the endogenous formation of genotoxic products resulting from the chemical reaction between inactive precursors, as reported *in vivo* for endogenous nitrosation of nitrosatable precursors to form mutagenic and/or carcinogenic diazo and N-nitroso compounds in fish (De Flora *et al.*, 1989).

Some organic mutagens, such as PAHs, aromatic and heterocyclic amines, aflatoxins, benzidine and azo compounds, usually present in complex mixtures in seawater and other contaminated environments, are frameshift mutagens of medium/high potency, which require metabolic activation. Other mutagens like inorganics, aliphatic compounds, epoxides, and hydrazines are direct-acting, base-substituting agents of medium/low potency, and their mutagenicity is often decreased by metabolic systems (De Flora *et al.*, 1989).

Environmental pollutants are normally present as complex mixtures rather than pure chemicals, like oil dispersants used in oil spills or various hazardous industrial wastes and pesticides (De Marini, 1991). These mixtures may give rise to synergistic, additive or antagonistic effects. The clastogenic effects of mercury and methylmercury are significantly decreased in the presence of selenium IV (Al-Sabti, 1994). Besides the interactions between different chemical compounds and mixture components, interactions can occur between physical and chemical agents. An example relevant to the marine environment is the interaction between sunlight or UV light and chemical compounds, which may have various effects: irradiation can decompose and deactivate noxious substances, or the opposite, conversion of inactive compounds into genotoxic products or the activation of promutagens/procarcinogens such as PAHs (De Flora *et al.*, 1989). Activated ROS species interact with DNA to cause strand breaks, or damage the purine or pyrimidine bases (Zakrzewski, 1991), or react with DNA polymerases, which results in a decrease of the fidelity of replication repair (De Flora & Ramel, 1988).

2.3.2 Protection mechanisms

Since DNA alteration is the primary event for the induction of chromosomal damage, protection mechanisms against DNA damage will also prevent chromosomal damage (sections 2.2.1.2 and 2.2.3).

Several chemical compounds have protective properties against genotoxic and/or carcinogenic hazards. Thus, hydroquinone derivatives isolated from a marine urochordate have antioxidant activity and can reduce *in vitro* the mutagenicity of benzo[a]pyrene, aflatoxin B_1 and UV radiation (De Flora *et al.*, 1989). Anticarcinogenic effects have been demonstrated in fish species by testing a variety of compounds. For example, indole-3-carbinol inhibits aflatoxin B_1-induced genotoxicity in trout; cytochrome P450 modulators such as alpha-naphthoflavone inhibit benzo[a]pyrene monooxygenase activity of microsomes from toadfish and eels; and polychlorinated biphenyls protect trout from aflatoxin B_1 carcinogenicity (De Flora *et al.*, 1989).

Inactivation of DNA damaging agents can be carried out by inhibiting the xenobiotic activation to electrophilic metabolites or by stimulating enzymatic systems involved in xenobiotic detoxification (De Flora & Ramel, 1988). Thus, suppression of PAH mutagenicity by complex mixtures due to inhibition of metabolic activation by the microsomal monooxygenase system has been reported (De Flora *et al.*, 1989).

Therefore, even those seawater pollutants usually cited for their harmful toxicological effects can behave as antimutagens and anticarcinogens. Indeed, this feature is rather common for several inhibitors of mutagenesis and carcinogenesis that often share noxious and protective properties, depending on many factors (De Flora *et al.*, 1989).

Different metabolites (glutathione, NADH, NADPH, vitamin A, vitamin C, thiols and compounds containing sulphured functional groups), and enzyme systems (DT diaphorase, cytochrome P450 reductase, glutathione S-transferase, superoxide dismutases) are involved in the inactivation of chemical inducers of DNA damage (De Flora *et al.*, 1989). Antioxidant agents, both natural (e.g. reduced glutathione) or synthetic (e.g. N-acetyl-cysteine) inactivate free radicals and also stimulate various cytosolic detoxifying enzyme activities, as well as enzymes involved in DNA repair (De Flora & Ramel, 1988).

2.3.3 Consequences of damage

Mutations, chromosome structural aberrations (such as deletions and translocations) and aneuploidy in somatic cells are all related to the induction of carcinogenesis, cell death and decreased individual survival (Tucker & Preston, 1996; Geard, 1992). Damage to DNA and chromosomes from germ cells can lead to reduced fertility, abortion, malformations (altered gene product), and heritable genetic diseases (Nielsen, 1993).

Damage to DNA by genotoxic compounds can result mainly in three different events that are indicative of chromosomal damage: sister chromatid exchange, chromosomal aberrations (numerical and structural), and micronuclei production.

2.3.3.1 Sister chromatid exchange

Sister chromatid exchange (SCE) involves the breakage and rejoining of chromosomal DNA and yields to the reciprocal interchange between chromatids. SCE indicates either 'spontaneous' or induced errors in DNA replication, and results from misreplication of a damaged DNA template through recombination at a stalled replication fork. Thus recombination between two stalled replication forks on separate chromosomes can result in chromatid interchange (Preston, 1991). Therefore, SCE induced by radiation or chemicals can be considered as a recombination process (frequently within homologous DNA regions) that occurs during the repair of damaged DNA or the replication of a damaged template. Since recombination is more likely to occur within a single replication fork, the frequency of SCE is presumably higher than that of chromatid interchanges between different chromosomes.

The incomplete SCE formation, i.e. when only one of the two DNA helices involved in the misreplication rejoins, can lead to chromatid deletions. Thus recombination at a stalled replication fork can also result in chromosome aberrations (Preston, 1991). The points of separation between early and late replicating DNA, visible by the limit between light and dark bands in chromosome preparations stained with Giemsa, represent the sites where 'stalled' replication forks are located, and can therefore be considered as 'hot-spots' for SCE and aberration induction (Preston, 1991). SCE analysis is a rapid and sensitive tool for the assessment of genetic damage induced by subtoxic doses of carcinogens and mutagens (Carrano *et al.*, 1978). Several studies have examined chromosomal aberrations and sister chromatid exchanges on lymphocytes from rats, humans and invertebrates exposed to different xenobiotics. These indicate that there is a correlation between the frequency of SCE and exposure to mutagenic agents (Zhang *et al.*, 1998).

It has been observed that embryo-larval polychaetes exposed to mitomycin C, methanesulphonate, cyclophosphamide and benzo[a]pyrene showed a dose-related increase in SCE frequency (Jha *et al.*, 1996). Similarly, Das & John (1997) recorded significant increases in SCE in gill tissues from bloch (*Etroplus suratensis*) exposed by intramuscular injection to three different dose levels of methylmethane sulphonate and cyclophosphamide, and observed that long chromosomes had more exchanges than short chromosomes. Although SCEs are generally more sensitive indicators of genotoxic effects than structural aberrations, they lack specificity, i.e. no direct association can be established between SCE induction and adverse cellular or health outcome, and SCEs do not indicate a mutagenic effect. Thus, the analysis of SCEs is a useful biomarker of exposure in short-term assays, but has a limited value in risk assessment (Tucker & Preston, 1996).

2.3.3.2 Chromosomal aberrations

Aneuploidy constitutes a numerical chromosome aberration. It can arise when chromosomes do not segregate correctly at mitotic or meiotic anaphase, giving place to hyperploid and hypoploid daughter cells. There are different mechanisms that can give rise to aneuploidy, such as alterations in cellular physiology, damage to the mitotic spindle and associated elements (which results in the failure of particular chromosomes to associate with the mitotic spindle), damage to chromosomal substructures (such as absence of a kinetochore or presence of a non-functional one), chromosome rearrangements, formation of a mutant topoisomerase II, or failure of centromere separation that can result in non-disjunction (Degrassi & Tanzarella, 1988; Preston, 1991; Tucker & Preston, 1996).

Clastogenic agents can induce functional aneuploidy as a result of chromosomal rearrangements and subsequent chromosome segregation. However, few chemicals (colchicine, vinblastine, nocodazole) have been identified to induce aneuploidy, and in general they affect the microtubule cytoskeleton interfering with the normal formation of the mitotic (meiotic) spindle (Preston, 1991).

Several specific aneuploidies have been associated with tumour development in humans (Tucker & Preston, 1996). Genetic defects induced in animal systems can be transmitted via sperm to offspring. The types of genetic damage transmitted by sperm include numerical aneuploidy, structural abnormalities, and gene mutations (MacGregor *et al.*, 1995). To the best of our knowledge, there is no report in the literature on chemically-induced aneuploidy in fish.

Chromosomal damage can also arise by the misrepair or misreplication of damaged DNA. This constitutes a structural chromosome aberration which is the primary target for the induction of chromosomal damage.

All types of damage in chromosome structure such as asymmetrical exchanges (dicentrics and rings) and deletions (terminal and interstitial) lead to the loss of chromosomal material at mitosis, or the inhibition of accurate chromosome segregation at anaphase, which can result finally in cell death (Tucker & Preston, 1996). Some of these events such as deletions and ring and dicentric chromosomes may yield chromatin pieces, also called acentric fragments that lack a centromere and are incorporated into micronuclei (section 2.3.3.3).

Jha *et al.* (1996) observed that exposure of embryo-larval polychaetes to mitomycin C, methanesulphonate, cyclophosphamide and benzo[a]pyrene leads to a dose-related induction of chromosomal aberrations. Al-Sabti and Kurelec (1985) detected chromosomal aberrations in gill cells from mussels that were transferred from a clean site to a site polluted with untreated domestic and harbour wastes. In the same study they observed a dose-response of chromosome aberrations induction in mussels exposed to benzo[a]pyrene in the laboratory. Different chromosomal aberrations, such as breaks, ring chromosomes and dicentric chromosomes, have been detected in kidney cells after the injection of three fish species (common carp, *Cyprinus carpio*; tench, *Tinca tinca*; grass carp, *Ctenopharyngodon idella*) with aflatoxin B1, aroclor 1254, benzidine, benzo[a]pyrene and 20-methylcholanthrene. Besides, these chromosomal aberrations are induced in a dose-dependent manner in the three fish species tested, although the level of chromosomal aberrations induced by each chemical differed in each fish species (Al-Sabti, 1985).

2.3.3.3 Micronucleae production

Micronuclei (MN) are the final expression of the molecular damage induced by genotoxic agents. In addition, micronuclei have been more frequently employed as a genotoxicity index than chromosomal aberrations and sister chromatid exchanges in cytogenetic studies performed in fish. Micronuclei are formed during mitotic anaphase, when acentric chromatid(s) and chromosomal fragments lag behind the centric elements move towards the spindle poles (Doherty *et al.*, 1996). A portion of the lagging elements form one or several secondary nuclei in the daughter cells much smaller than the principal nucleus (1/5 to 1/20), containing chromosomal fragments or acentric chromosomes that are not incorporated into daughter nuclei, and are therefore called micronuclei (Al-Sabti & Metcalfe, 1995). The effect of chiasmata at meiosis can also be important; for example, a chiasma within a paracentric inversion will generate an acentric fragment which can form a micronucleus (Heddle *et al.*, 1991).

There are four recognised mechanisms by which micronuclei can arise (Heddle *et al.*, 1991):

(1) Mitotic loss of acentric fragment
(2) A variety of mechanical consequences of chromosomal breakage and exchange

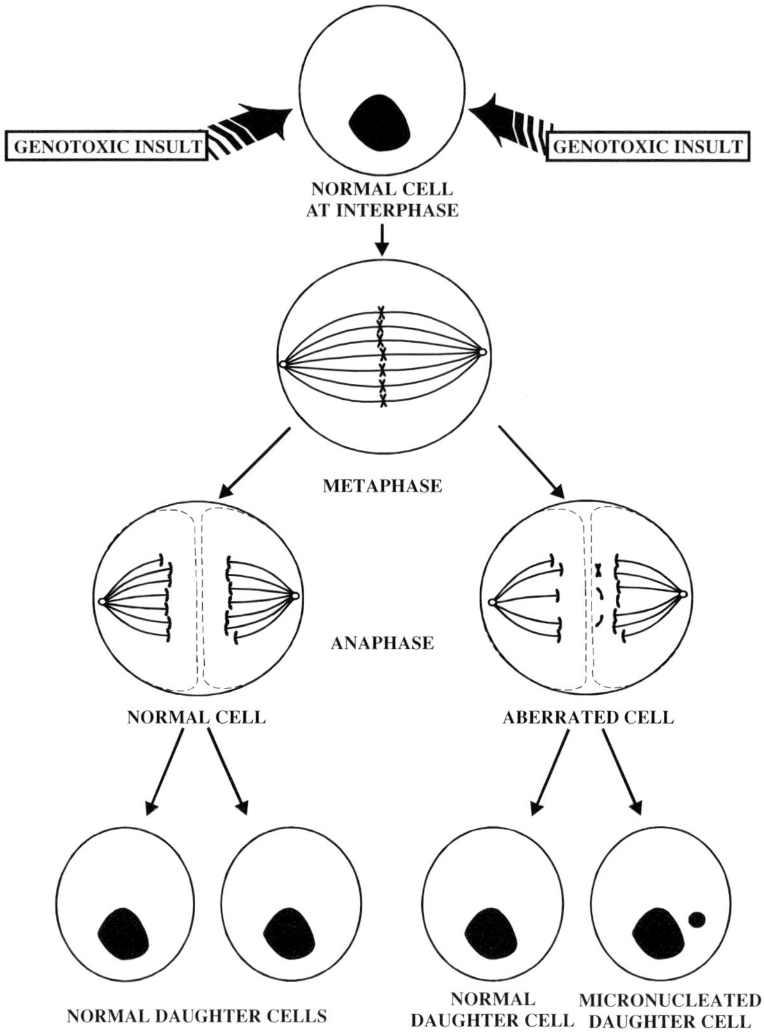

Fig. 2.7 Schematic illustration of the mechanism of micronuclei formation in cells after one cell replication following the DNA damaging event. Modified from Al-Sabti & Metcalf (1995).

(3) Mitotic loss of whole chromosomes
(4) Apoptosis (Fig. 2.7).

The latter is a form of nuclear destruction in which the nucleus disintegrates and nuclear fragments are formed. Apoptosis occurs both naturally and in response to chemically-induced cellular damage (which need not be genetic in nature, i.e. the inhibition of protein synthesis). While acentric fragments may originate from misrepaired DNA lesions (Fenech *et al.*, 1994) as well as from direct induction of double-strand breaks, disturbances of the mitotic cycle may cause chromosome misdistribution during the cell division, and appearance of micronuclei, finally giving rise to aneuploidy (DeGrassi & Tanzarella, 1988).

Micronuclei induced by clastogen chemicals (those inducing chromosome structural changes) can be distinguished morphologically as a class from those induced by aneugens (those inducing chromosome numerical changes) because they are smaller and by the frequency with which centromeres are present. Micronuclei induced by apoptosis may not be distinguishable morphologically from other micronuclei, since all micronuclei are pycnotic in these cells. It is noteworthy that some but not all micronuclei arising from apoptosis would be expected to contain centromeres (Heddle *et al.*, 1991). Micronuclei in fish are smaller when compared with micronuclei from mammalian cells, because most fish chromosomes are much smaller than mammalian chromosomes (Al-Sabti & Metcalfe, 1995).

Excluding apoptosis, at least one cell replication is necessary for micronuclei appearance after the DNA damaging event (Heddle *et al.*, 1991; Al-Sabti & Metcalfe, 1995). However, not all acentric fragments become micronuclei at the first cell division; some can survive, replicate, and become micronuclei at the second or subsequent division. It has additionally been suggested that micronuclei frequency decrease with cell division because chromosomes in micronuclei may continue to replicate and reattach to the spindle at a subsequent mitosis, producing a normal daughter cell, while the micronucleus remains associated with the originally produced hypoploid nucleus. This mechanism requires that there are minimal adverse effects when the chromosomes are in the micronucleus and that the cell can survive to mitosis (Tucker & Preston, 1996). Since micronuclei cannot be observed until after the first cell cycle, the frequencies of these within a cell population is highly dependent on the kinetics of cell proliferation. Rates of cell proliferation probably vary widely, depending on fish species, target tissue and environmental conditions (e.g. temperature) (Al-Sabti & Metcalfe, 1995).

There has been an increasing interest towards the use of micronuclei as an index of cytogenetic damage in fish and other marine organisms exposed to a variety of toxic and genotoxic pollutants under laboratory (Al-Sabti, 1986a,b; Al-Sabti, 1994; Al-Sabti *et al.*, 1994; Al-Sabti, 1995b; Burgeot *et al.*, 1995; Venier *et al.*, 1997; Marlasca *et al.*, 1998) and field conditions (Al-Sabti & Hardig, 1990; Al-Sabti, 1992a,b; Burgeot *et al.*, 1996a; Rao *et al.*, 1997). Micronuclei detection assay has been employed in genotoxicity studies carried out in invertebrates (Brunetti *et al.*, 1992; Burgeot *et al.*, 1995, 1996b; Venier *et al.*, 1997), fish (Al-Sabti *et al.*, 1994; Al-Sabti, 1995b; Rao *et al.*, 1997; Marlasca *et al.*, 1998) and humans (Fenech *et al.*, 1994; Vral *et al.*, 1994). Currently, MN detection represents a widely used parameter, easily performed, which also allows molecular approaches in studying the effects of many clastogenic or aneugenic agents (Venier *et al.*, 1997).

Various studies have shown that the peripheral erythrocytes of fish have a high incidence of micronuclei after exposure to different pollutants under field and laboratory conditions. Al-Sabti (1994) observed that selenium, mercury, methylmercury and their mixtures induce micronuclei under laboratory conditions in the binucleated erythrocytes of Prussian carp (*Carassius auratus gibelio*) in a dose-dependent manner. The exposure by injection of five carcinogenic-mutagenic chemicals (aflatoxin B1, aroclor 1254, benzidine, benzo[a]pyrene and 20-methylcholanthrene) of three species of cyprinids (common carp, *C. carpio*; tench, *T. tinca*; and grass carp, *C. idella*) enhanced the frequency of micronuclei in their erythrocytes (Al-Sabti, 1986a). Pacheco and Santos (1996) observed a significant increase in micronuclei in erythrocytes from eels (*Anguilla anguilla*) exposed under laboratory conditions to cyclophosphamide (a standard mutagenic compound) and bleached kraft pulp mill effluent (which induced higher MN frequencies than cyclophosphamide). Marlasca *et al.* (1998) found a significant increase in the frequencies of micronucleated erythrocytes from rainbow trout (*Oncorhynchus mykiss*) exposed in the laboratory to a textile industry effluent. It has also been observed that exposure of Prussian carp to various concentrations of chromium under laboratory and field conditions causes an increase in the frequency of micronuclei compared with the control groups (Al-Sabti *et al.*, 1994). Al-Sabti (1992a,b) reported an induction of the frequency of micronuclei in erythrocytes of four fish species (pike, *Esox lucius*; perch, *Perca fluviatilis*; roach *Rutilus rutilus*; and bream, *Abramis brama*) from Swedish lakes environmentally exposed to radiocaesium. Erythrocytes from perch (*Perca fluviatilis*) sampled from areas contaminated by pulp mill wastewater products showed a higher frequency of micronuclei compared to those sampled far from waste discharge points (Al-Sabti & Hardig, 1990).

The micronucleus assay has also been applied to hepatic cells from fish. Hepatocytes are generally exposed to high concentrations of xenobiotics since liver is the major site of xenobiotic metabolism and transformation in the body (Al-Sabti, 1995a; Rao *et al.*, 1997). Rao *et al.* (1997) observed an elevated incidence of hepatic micronuclei in brown bullheads (*Ameiurus nebulosus*) collected from Hamilton harbour (Ontario), a site contaminated with elevated concentrations of PAHs and showing also visible lesions in fish after environmental exposure to genotoxic substances, relative to the micronucleus incidence in bullheads from reference sites with no external pathologies. In the same study, rainbow trout (*O. mykiss*) injected with an extract from a pulp mill effluent exhibited an elevated incidence of hepatic micronuclei compared to controls. Hepatocytes from rainbow trout exposed *in vitro* to selenium, mercury, methylmercury and their mixtures, showed a dose-dependent increase in MN frequencies when compared to the relevant controls (Al-Sabti, 1995a).

2.3.4 Detection of chromosome damage

Several molecular and cytogenetic techniques originally developed for the assessment of genotoxicity in mammals have been applied to fish. However, many of those procedures using metaphase techniques, such as sister chromatid exchange and chromosomal aberration assays, are not practical for many fish species (e.g. salmonids, cyprinids, ictalurids) because the fish karyotype consists of large numbers of small irregular chromosomes (Al-Sabti 1995a; Zhang *et al.*, 1998). Although species of mudminnow (*Umbra* sp.) have a suitable karyotype for metaphase analysis of genotoxicity, these species are of little use for *in situ* monitoring studies because they are relatively rare and of no commercial value (Al-Sabti, 1995a).

2.3.4.1 Sister chromatid exchange

Interchanges between the chromatids of individual chromosomes and sister chromatid exchanges are detectable after two or more rounds of replication post the initiation of damage in DNA (Geard, 1992; Zhang *et al.*, 1998). Incubation with BrdU of mitotically active cells (by addition to the cell culture or by *in vivo* exposure) for two consecutive replication rounds and arresting cells in metaphase, yields sister chromatids that can be stained differentially and allows the identification of the exchanged segments (Preston, 1991; Tucker & Preston, 1996).

2.3.4.2 Chromosomal aberrations

Both structural and numerical chromosomal aberrations can be detected by cytogenetic techniques involving the visual analysis of slides of cells in metaphase and counting the number of metaphase chromosomes. Unbanded chromosomes have been used in the detection of all types of chromatid aberrations, such as asymmetrical exchanges (dicentrics and rings) and deletions (terminal and interstitial) (Tucker & Preston, 1996). Chromosome banding allows the analysis of all types of structural aberrations, specifically including symmetrical exchanges (reciprocal translocations, inversions and insertions). Low visibility of replication banding patterns has been obtained in most fish species. Another drawback of this technique is that it requires the construction and analysis of the karyotype for each cell scored. Moreover, it is slower and more expensive than the use of unbanded chromosomes and requires experimented observation.

Alternative methods to structural banding have been developed and applied to fish chromosomes. Replication banding of chromosomes is based on the use of a DNA base analog such as BrdU, which is incorporated to DNA of mitotically active cells (Preston, 1996; Zhang *et al.*, 1998), followed by the use of fluorescein tagged antibodies against BrdU and allows a better visualisation of chromosome structure. Potential solutions for this drawback also include densitometric analysis of chromosomes and immunochemical detection methods (Zhang *et al.*, 1997; Zhang & Tiersch, 1998a). Objective and quantitative analysis of weak bands found in fish chromosomes is possible by computer assisted analysis (Zhang *et al.*, 1998; Zhang & Tiersch, 1998b). Different techniques have been used in the analysis of chromosome numerical aberrations such as measurement of DNA content by flow cytometry or using microfluorimetry and microdensitometry (Al-Sabti, 1995b).

The preparation of metaphase chromosomes is time-consuming and can be difficult due to technical problems like chromosome loss during the procedure (which limits the analysis of aneuploidy) (Al-Sabti, 1986c), or reduced cell proliferation due to chemical exposure (MacGregor *et al.*, 1995; Tucker & Preston, 1996). The analysis of cytogenetic abnormalities in cells and tissues has been facilitated by the use of DNA probes. The development of fluorescent-based staining methods (FISH, fluorescent *in situ* hybridisation) has lead to a significant improvement in the metaphase-based cytogenetics. FISH provides fast, precise and sensitive localisation of DNA sequences since it involves a hybridisation reaction between a labelled nucleotide probe and a complementary strand of target DNA or RNA (Zhang *et al.*, 1999). Therefore, this technique permits the labelling of chromosomes along their entire length in a procedure commonly known as ‘chromosome painting’, and allows the identification of the location and number of copies of a particular chromosome in either metaphase or interphase cells (MacGregor *et al.*, 1995). FISH is commonly used for diagnosis of chromosomal abnormalities since structural and numerical alterations of chromosomes can be detected using this technique.

The chromosome painting technique offers several advantages compared to conventional cytogenetic analyses, such as increased speed of analysis, increased ease and efficiency of analysis and improved specificity and sensitivity for detecting both subtle and complicated alterations, particularly reciprocal translocations (MacGregor *et al.*, 1995). Probes for different chromosomes can be labelled in different colours and can be combined to analyse cells for alterations in chromosome structure and number. One deficiency of chromosome painting is that the method only detects exchanges that occur between chromosomes painted with different colours or between chromosomes painted in the same colour but with sufficiently different staining intensity (Preston, 1996). Another limitation of FISH is that multiple copies of a target sequence are needed for detection. The use of the *in situ* polymerase chain reaction that yields to the multiplication of target DNA sequences in combination with FISH (ISPCR) has enabled the detection of single copies of DNA (Zhang *et al.*, 1997; Engelen *et al.*, 1998). These procedures, although well developed for mammals, are not widely applied in fish. Zhang *et al.* (1999) developed an ISPCR technique able to detect a single-locus gene on catfish chromosomes.

Painting probes for human chromosomes have been widely used and are available from several commercial sources. Many approaches have been developed for testing aneuploidy in human sperm using DNA probes (for chromosome-specific repetitive sequences or high complexity probes) repetitive, multiple dyes and FISH (MacGregor *et al.*, 1995). However, development of painting probes for chromosomes of other species is more recent, because of the greater difficulty in obtaining pure individual chromosomes as a source for probe development. No information is available about painting probes for fish chromosomes.

Chromosome painting can be incorporated into existing toxicology studies without altering the exposure protocols normally employed and can provide important information about tissue-specific genotoxic effects (MacGregor *et al.*, 1995).

2.3.4.3 Micronuclei production

Micronuclei assays, originally developed with mammalian species, have been used extensively to test for the genotoxic activity of chemicals in fish (Al-Sabti, 1986a, 1992a,b, 1994, 1995b; Al-Sabti & Hardig, 1990; Rao *et al.*, 1997; Marlasca *et al.*, 1998). Scoring of micronuclei in the interphase is technically much easier and more rapid than the scoring of chromosomal aberrations during metaphase (Al-Sabti & Metcalfe, 1995).

The micronucleus assay consists basically of microscopical examination of fixed cells or tissue stained with Giemsa. It has been proved that the micronucleus assay works well in tests with fish, but it is necessary to score at least 1000 cells from each fish to evaluate clastogenicity (Al-Sabti, 1995a). Venier *et al.* (1997) applied the micronucleus assay to gill cells from mussels and concluded that at least 2000 cells per animal must be scored. The micronucleus assay using any type of cell requires that target cells treated with a genotoxic agent must undergo mitosis so that the micronuclei are visible in the cytoplasm after the first cell cycle or subsequent cell cycles (Doherty *et al.*, 1996). Thus, the frequencies of micronuclei observable within a cell population depend on the kinetics of cell proliferation. In general, the length of the cycle in organisms depends on the time needed to replicate DNA and perform nuclear division, and probably varies widely, depending on fish species, the target tissue and environmental conditions (e.g. temperature). There is little data on the duration of the cell cycle in the tissues of teleost species, partly because the cell cycle varies with temperature in these poikilotherms (Al-Sabti, 1994). Therefore, considerable work is needed to establish a time for optimum yield of MN after exposure to genotoxic agents and to standardise assay procedures (Al-Sabti & Metcalfe, 1995).

Cytokinesis can be blocked in cell cultures (without inhibiting nuclear division) by adding cytochalasin-B, so that micronuclei can be easily scored one cell division after genotoxic insult (Al-Sabti, 1994). The *in vitro* cytokinesis-block micronucleus assay has been applied to human lymphocytes (Fenech *et al.*, 1994; Vral *et al.*, 1994) and fish hepatic cells (Al-Sabti, 1995a,b) and erythrocytes (Al-Sabti, 1994).

Two main cell types from fish have been used in micronucleus assays: hepatocytes and erythrocytes. Since teleost erythrocytes are nucleated, *in vitro* methods using fish erythrocytes (Al-Sabti, 1994) have been developed, and micronuclei have been scored in fish erythrocytes as a measure of clastogenic activity (Al-Sabti, 1994; Al-Sabti & Metcalfe, 1995). Rao *et al.* (1996) described a very detailed procedure for the quantification of the number of micronuclei in hepatocytes of the teleost liver applicable either to field and laboratory studies. However, one of the drawbacks of using liver as a target tissue is that hepatocytes are not continually dividing and liver injury must be induced to stimulate proliferation of the hepatocytes (for example exposing fishes to allyl formate, a chemical hepatic necrogen) so that clastogenic end-points can be visualised (Al-Sabti & Metcalfe, 1995). Al-Sabti (1995a) described an *in vitro* micronucleus assay using hepatocytes as cell targets to evaluate the genotoxicity of single chemicals or complex environmental mixtures, without the need to injure the liver with allyl formate to induce cell proliferation.

The *in vitro* micronucleus assay may be used to assess the induction of both structural and numerical aberrations. Because micronuclei can arise from both structural and numerical chromosome aberrations through different mechanisms (chromosome breakage, spindle disruption, apoptosis), two molecular approaches have been developed in order to discern the process that induced micronuclei formation. First is the use of antikinetochore antibodies that label centromeric regions through binding to proteins present at the site where chromosomes attach to the spindle (Fenech *et al.*, 1994). Therefore, micronuclei can be distinguished that contain one or more whole chromosomes (the number of which can be often determined) arisen by disruption of mitotic spindle or other components of mitosis, from micronuclei formed by clastogenic processes which contain fragments of chromosomes. The approach is fast, simple and relatively inexpensive; it could be applied for routine screening of the induction of aneuploidy in genetic toxicology testing, both *in vitro* and *in vivo* (Heddle *et al.*, 1991).

Secondly, similar determination of the contents of micronuclei can be made by the use of DNA hybridisation probes. Most probes hybridise to the repetitive DNA adjacent to the centromere of a single pair of chromosomes. Others consist of pools of unique sequence DNA which label whole chromosomes. The centromeric probes yield significantly brighter signal compared with that achievable with the antikinetochore antibody. However, these probes also suffer from several disadvantages, including:

(1) A greater amount of work required to accomplish the staining
(2) Chromosome-to-chromosome variability with respect to the amount of centromeric heterochromatin, at least for humans

(3) Chromosome breaks which may occur within the heterochromatin such that fragments may contain enough labelled DNA to give the impression that a micronucleus contains a whole chromosome (Heddle *et al.*, 1991).

The possible subjectivity of the microscopic analysis for micronuclei can be avoided by using automatic systems approached either by flow cytometry or by computerised image analysis. A more sensitive and selective micronucleus assay system has been developed in order to improve some drawbacks of the assay, such as lack of sensitivity and the possibility to confound nuclear damage from viral erythrocytic necrosis as a clastogenic response (Al-Sabti, 1994; Al-Sabti & Metcalfe, 1995).

2.4 Higher level consequences of genetic damage

2.4.1 Germ line effects

While much of the research on genetic damage is focused on somatic effects, such as tumour formation and embryo malformation, germ line effects may be more significant under low exposure in the long term. This is significant because it may result both from the likelihood of occurrence of genetic damage in gametes, as well as its potential effects on population viability. There is evidence that DNA repair mechanisms may not be active in gametes (Anderson & Wild, 1994). This may be one of the causes for increased embryo mortality and malformation after exposure to genotoxins. Other consequences of such restricted DNA repair, however, include heritable effects of genotoxins. These may affect the next generation directly, by causing inherited diseases such as certain forms of cancer. Alternatively, if the mutation is recessive and thus has no effect on the phenotype of the embryo, they will be passed on to subsequent generations. The effects of the accumulation of such recessive mutations may then result in increased 'mutational load' and thus reduce population viability in the long term, issues which are discussed in greater detail in Chapter 7.

2.4.2 Somatic effects

Tumour formation is one of the possible somatic effects of xenobiotic-induced genetic damage. Whilst there is an extensive literature on neoplastic disorders in finfish and shellfish, none of it considers the disease as a possible consequence of genetic damage, with the few exceptions noted in previous sections 2.1. to 2.3. Indeed, whilst many of the studies originating from the USA correlate neoplastic disorders with tissue or substrate contaminant burdens, the literature from Europe, with the notable exception of Lowe and Moore (1978), tends to view many of the pathologies as being responses to pathogens. General aspects of tumour formation and prevalence in European fish are considered further in Chapter 4.

There is a class of mammalian pathologies referred to as storage diseases that are considered to result from a genetic mutation or genetic damage. The condition manifests itself as large deposits of, among others, glycogen and lipids in tissues of the body, and is the result of a breakdown in the lysosomally mediated degradative mechanisms for those substances.

Whilst abnormal accumulations of lipids have been observed in finfish (McCain *et al.*, 1978; Solangi & Overstreet, 1982; Köhler *et al.*, 1992; Lowe *et al.*, 1992) and shellfish (Lowe *et al.*, 1981; Wolfe *et al.*, 1981; Pipe & Moore, 1986; Cajaraville *et al.*, 1990), these are not attributed to genetic mutation or damage but rather to the cytotoxicity of contaminant chemicals.

2.4.3 Developmental effects

In a study on English sole it was shown that carcinogenic polycyclic aromatic hydrocarbons and their metabolites could accumulate in reproductive tissues and chemically modify gonadal macromolecules (DNA), and it has been shown in mammalian studies that this can lead to such effects as mutagenesis and teratogenesis (Varanasi *et al.*, 1982). This, together with data reported in sections 2.1 to 2.3, demonstrates that there is evidence to indicate that the DNA of marine species can be affected by contaminants which in all probability would translate into some type of abnormality in the offspring, or death.

Several studies have demonstrated developmental abnormalities in finfish species which could result from direct toxicity or as a consequence of damage to the DNA. Cameron and Berg (1992) examined embryos of dab collected from a series of sample stations along a transect extending from the inner German Bight out onto the Dogger Bank (the Bremerhaven Workshop transect). The results showed that whereas some 32% of embryos from the inner, more polluted, site had malformations, the figure dropped to 9% offshore and then increased again in samples taken from the Dogger Bank, which is known to have high levels of contaminants. Similarly, von Westernhagen *et al.* (1988) observed malformations in fish embryos, including cod, flounder and plaice, in the western Baltic and concluded that anthropogenic inputs may have been the cause. von Westernhagen *et al.* (1988) were unable to say, however, whether the embryonic malformations were as a direct consequence of contaminant exposure on the eggs or through the accumulation of toxicants in the parental gonad. However, in laboratory studies where winter flounder were exposed to DDT and dieldrin, prior to spawning, decreasing fertilisation success was observed (Smith & Cole, 1973) suggesting that parental exposure to contaminants can have serious consequences for the offspring (Weis & Weis, 1989). By contrast, investigations with brown trout demonstrated that whilst oogenesis was delayed following exposure to cadmium, the eggs and fry that were produced developed normally after fertilisation (Brown *et al.*, 1994).

Information about developmental effects of peroxisome proliferators on fish species is scarce. In mammals peroxisome proliferators are known to adversely affect reproduction and development, in addition to their ability to cause hepatocellular carcinogenesis (section 2.1.5.1). It has been shown that some of the contaminants causing peroxisome proliferation such as phthalate esters are estrogenic (Jobling *et al.*, 1995; see also Chapter 5) and produce adverse reproductive effects disrupting normal male development (IPCS, 1992; Wine *et al.*, 1997). For instance, the phthalate ester plasticiser diethylhexyl phthalate (DEHP) causes testicular atrophy and shows teratogenic properties in rodents and other laboratory mammals (IPCS, 1992). In fish, DEHP administration has been related to a reduced survival of rainbow trout and zebrafish fry, and to decreased production of fry in guppies (IPCS, 1992).

In conclusion, as reviewed extensively by Weis and Weis (1989), there is ample evidence in the literature to show that exposure to some contaminants, whether parental or early life-stage, results in malformations in a diverse range of finfish species, many of which are known to be of commercial significance to European fisheries. Those contaminants include chemicals known to damage DNA either directly or through ROS production (i.e. some pesticides, PCBs, mutagenic PAHs such as benzo[a]pyrene, oil derivatives, transition metals), although in most studies the links between contaminant-induced DNA damage and further reproductive and developmental effects are not demonstrated. In addition, environmental contaminants can vary greatly in their effects on different fish species (Weis & Weis, 1989). The effects of xenobiotics on larval development are covered in more detail in Chapter 3.

2.5 Conclusions

Mechanisms of genetic damage and their link to molecular responses in wild fish populations have been reviewed, with special consideration of higher level effects and gaps in current knowledge. Damage to DNA may occur by oxygen radicals, by adduct formation or directly by mutagenic chemicals and radiation. The production of 'reactive oxygen species' by cytochrome P450-driven reactions and by peroxisome proliferation, and the relevance of protection mechanisms such as oxyradical scavengers, lysosomal sequestration and the induction of antioxidant enzymes, stress proteins and metallothioneins have been discussed.

DNA adducts may be formed by many hydrophilic compounds, or by metabolites of detoxification systems such as cytochrome P450. Although such adducts potentially lead to mutations and tumour formation, the empirical demonstration of a link between elevated levels of DNA adducts and higher level effects has been difficult. Direct genetic damage may occur by mutagenic chemicals or radiation, and may affect a wide range of cellular functions. DNA repair mechanisms revert some DNA damage, though their efficiency may be affected by physiological factors and life-story stage. Direct chemical effects on chromosomes including sister chromatid exchange, micronucleae production and other nuclear abnormalities are also considered. There is still limited knowledge on quantitative links between damage at the genetic and molecular level, and individual health, fecundity and population productivity and viability.

2.6 Acknowledgements

Work in the laboratory of MP Cajaraville has been funded by the Spanish Ministry of Science and Technology through project AMB99-0324 (CICYT) and by the European Commission (Research Directorate General, Environment Programme-Marine Ecosystems) through the BEEP project 'Biological Effects of Environmental Pollution in Marine Coastal Ecosystems' (contract EVK3-CT2000-00025). BEEP project is part of the EC IMPACTS cluster.

2.7 References

Abe, T., T. Konishi, T. Hirano, H. Kasai, K. Shimizu, M. Kashimura & K. Higashi (1995) Possible correlation between DNA damage induced by hydrogen peroxide and translocation of heat shock

70 protein into the nucleus. *Biochemical and Biophysical Research Communications*, **206**, 548–555.

Abel, J. & N. de Ruiter (1989) Inhibition of hydroxyl-radical-generated DNA degradation by metallothionein. *Toxicology Letters*, **47**, 191–196.

Adema, C.M., W.P.W. van der Knnap & T. Sminia (1991) Molluscan hemocyte-mediated cytotoxicity: the role of reactive oxygen intermediates. *Reviews of Aquatic Sciences*, **4**, 201–223.

Aldridge, T.C., J.D. Tugwood & S. Green (1995) Identification and characterization of DNA elements implicated in the regulation of CYP4A1 transcription. *Biochemical Journal*, **306**, 473–479.

Almar, M., L. Otero, C. Santos & J. Gonzalez-Gallego (1998) Liver glutathione content and glutathione-dependent enzymes of two species of freshwater fish as bioindicators of chemical pollution. *Journal of Environmental Science and Health. Part B*, **33** (6), 769–783.

Aloj Totaro, E., F.A. Pisanti, P. Glees & A. Continillo (1986) The effect of copper pollution on mitochondrial degeneration. *Marine Environmental Research*, **18**, 245–253.

Al-Sabti, K. (1985) Carcinogenic-mutagenic chemicals induced chromosomal aberrations in the kidney cells of three cyprinids. *Comparative Biochemistry and Physiology. Part C. Pharmacology, Toxicology and Endocrinology*, **82** (2), 489–493.

Al-Sabti, K. (1986a) Comparative micronucleated erythrocyte cell induction in three cyprinids by five carcinogenic-mutagenic chemicals. *Cytobios*, **47** (190–191), 147–154.

Al-Sabti, K. (1986b) Clastogenic effects of five carcinogenic-mutagenic chemicals on the cells of the common carp, *Cyprinus carpio L. Comparative Biochemistry and Physiology. Part C. Pharmacology, Toxicology and Endocrinology*, **85** (1), 5–9.

Al-Sabti, K. (1986c) Karyotypes of Cyprinus carpio and Leuciscus cephalus. *Cytobios*, **47** (188), 19–25.

Al-Sabti, K. (1992a) Monitoring the genotoxicity of radiocontaminants in Swedish lakes by fish micronuclei. *Cytobios*, **70** (281), 101–106.

Al-Sabti, K. (1992b) Micronuclei induction in pike (*Esox lucius*) in Swedish lakes contaminated with radiocaesium. *Cytobios*, **70** (280), 27–32.

Al-Sabti, K. (1994) Micronuclei induced by selenium, mercury, methylmercury and their mixtures in binucleated blocked fish erythrocyte cells. *Mutation Research*, **320** (1–2), 157–163.

Al-Sabti, K. (1995a) An *in vitro* binucleated blocked hepatic cell technique for genotoxicity testing in fish. *Mutation Research*, **335** (2), 109–120.

Al-Sabti, K. (1995b) Detection of triploidy in fish using the cytokinesis-blocked method for erythrocyte and hepatic cells. *Cytobios*, **82** (330), 181–187.

Al-Sabti, K. & J. Hardig (1990) Micronucleus test in fish for monitoring the genotoxic effects of industrial waste products in the Baltic Sea, Sweden. *Comparative Biochemistry and Physiology. Part C. Pharmacology, Toxicology and Endocrinology*, **97** (1), 179–182.

Al-Sabti, K. & B. Kurelec (1985) Induction of chromosomal aberrations in the mussel *Mytilus galloprovincialis* watch. *Bull. Environmental Contamination and Toxicology*, **35**, 660–665.

Al-Sabti, K. & C.D. Metcalfe (1995) Fish micronuclei for assessing genotoxicity in water. *Mutation Research*, **343** (2–3), 121–135.

Al-Sabti, K., M. Franko, B. Andrijanic, S. Knez & P. Stegnar (1994) Chromium-induced micronuclei in fish. *Journal of Applied Toxicology*, **14** (5), 333–336.

Anderson, R. (1989) Early warnings of stress. *New Scientist*, **7**, 50–52.

Anderson, S.L. & F.L. Harrison (1990) Predicting the ecological significance of exposure to genotoxic substances. In: (ed. Sandhu, S.S.) *First Symposium on in situ evaluation of biological hazards of environmental pollutants*. Plenum Press, UK, pp. 81–93.

Anderson, S.L. & G.C. Wild (1994) Linking genotoxic responses and reproductive success in ecotoxicology. *Environmental Health Perspectives*, **102** (Suppl 12), 9–12.

Andersson, T., L. Förlin, J. Härdig & Å. Larsson (1988) Physiological disturbances in fish living in coastal water polluted with bleached kraft mill effluents. *Canadian Journal of Fisheries and Aquatic Sciences*, **45**, 1525–1536.

Arnold, H., H.-J. Pluta & T. Braunbeck (1995) Simultaneous exposure of fish to endosulfan an disulfoton in vivo: ultrastructural, stereological and biochemical reactions in hepatocytes of male rainbow trout (*Oncorhynchus mykiss*). *Aquatic Toxicology*, **33**, 17–43.

Atienzar, F., P. Child, A. Evenden, A. Jha, D. Savva, C. Walker & M. Depledge (1998) Application of the arbitrarily primed polymerase chain reaction for the detection of DNA damage. *Marine Environmental Research*, **46**, 331–335.

Au, D.W.T., R.S.S. Wu, B.S. Zhou & P.K.S. Lam (1999) Relationship between ultrastructural changes and EROD activities in liver of fish exposed to benzo[a]pyrene. *Environmental Pollution*, **104**, 235–247.

Awashi, Y.C., S.V. Singh, S.K. Goel & J.K. Reddy (1984) Irreversible inhibition of hepatic glutathione-S-transferase by ciprofibrate, a peroxisome proliferator. *Biochemical and Biophysical Research Communications*, **123**, 1012–1018.

Bailey, G.S., J.D. Hendricks & D.E. Williams (1994) *Improved food safety through discovery and control of natural and induced toxicants and antioxicants*. United States Department of agriculture/cooperative State, Fedrip Database, National Technical Information Service (NTIS).

Bailey, G.S., D.E. Williams & J.D. Hendricks (1996) Fish models for environmental carcinogenesis: The rainbow trout. *Environmental Health Perspectives*, **104**, 5–21.

Bakhtiyarova, R.M. (1997) Genetic variation in Scots pine under industrial pollution in the southern Urals: mutagenesis. *Genetika*, **33**, 644–649.

Bakhtiyarova, R.M., R.M. Starova & Y.A. Yanbaev (1995) Genetic changes in populations of Scots pine growing under industrial air pollution conditions. *Silvae Genetica*, **44**, 157–160.

Barrett, C.J. (1995) Role of mutagenesis and mitogenesis in carcinogenesis. In: (eds Phillips, D.H. & S. Venitt) *Environmental Mutagenesis*. Bios Scientific Publishers, Oxford, pp. 24–30.

Bartsch, R., D. Klein & K.H. Summer (1990) The Cd-Chelex assay: a new sensitive method to determine metallothionein containing zinc and cadmium. *Archives of Toxicology*, **64**, 177–180.

Batel, R., N. Bihari, B. Kurelec & R.K. Zahn (1985) DNA damage by benzo[a]pyrene in the liver of mosquito fish Gambusia affinis. *Science of the Total Environment*, **41**, 275–283.

Baumgart, E., A. Völkl, J. Pill & H.D. Fahimi (1990) Proliferation of peroxisomes without simultaneous induction of the peroxisomal fatty acid b-oxidation. *FEBS Letters*, **264**, 5–9.

Beier, K. & H.D. Fahimi (1991) Environmental pollution by common chemicals and peroxisome proliferation: efficient detection by cytochemistry and automatic image analysis. *Progress in Histochemistry and Cytochemistry*, **23**, 150–163.

Bell, D.R., R.G. Bars & C.R. Elcombe (1992) Differential tissue-specific expression and induction of cytochrome P450IVA1 and acyl-CoA oxidase. *European Journal of Biochemisty*, **206**, 979–986.

Bentley, P., I. Calder, C. Elcombe, P. Grasso, D. Stringer & H.-J. Wiegand (1993) Hepatic peroxisome proliferation in rodents and its significance for humans. *Food and Chemical Toxicology*, **31**, 857–907.

Bhaskaran, A., D. May, M. RandWeaver & C.R. Tyler (1999) Fish p53 as a possible biomarker for genotoxins in the aquatic environment. *Environmental and Molecular Mutagenesis*, **33**, 177–184.

Bickham, J. & M. Smolen (1994) Somatic and heritable effects of environmental genotoxins and the emergence of evolutionary toxicology. *Environmental Health Perspectives*, **102**, 25–28.

Bihari, N., R. Batel & R.K. Zahn (1990) DNA damage determination by tha alkaline elution technique in the haemolymph of mussel Mytilus galloprovincialis trated with benzo[a]pyrene and 4-niroquinoline-N-oxide. *Aquatic Toxicology*, **18**, 13–22.

Böck, P., R. Kramar & M. Pavelka (1980) *Peroxisomes and related particles in animal tissues*. Cell Biology Monographs ed. Vol. 7. Springer-Verlag, Wien, New York, 239p.

Braunbeck, T. & A. Völkl (1991) Induction of biotransmformation in the liver of eel (*Anguilla anguilla* L.) by sublethal exposure to Dinitro-o-cresol: an ultrastructural and biochemical study. *Ecotoxicology and Environmental Safety*, **21**, 109–127.

Braunbeck, T., K. Gorgas, V. Storch & A. Völkl (1987) Ultrastructure of hepatocytes in golden ide (*Leuciscus idus melanotus* L.; Cyprinidae: Teleostei) during thermal adaptation. *Anatomy and Embryology*, **175**, 303–313.

Bremner, I. (1991a) Nutritional and physiological significance of metallothionein. In: (eds Riordan, J.F. & B.L. Vallee) *Metallobiochemistry, Part B: metallothionein and related molecules*. Academic Press, London, pp. 25–35.

Bremner, I. (1991b) Metallothionein and copper metabolism in liver. In: (eds Riordan, J.F. & B.L. Vallee) *Metallobiochemistry, Part B: metallothionein and related molecules*. Academic Press, London, pp. 584–591.

Bremner, I. (1993) Involvement of metallothionein in the regulation of mineral metabolism. In: (eds Suzuki, K.T., N. Imura & M. Kimura) *Metallothionein III*. Basel, Birkhäuser Verlag, pp. 112–124.

Brown, V., D. Shurben, W. Miller & M. Crane (1994) Cadmium toxicity to rainbow-trout *Oncorhynhus mykiss walbaum* and brown-trout *Salmo trutta* L. over extended exposure periods. *Ecotoxicology and Environmental Safety*, **29** (1), 38–46.

Brunetti, R., O. Fumagalli, P. Valerio & M. Gabriele (1992) Genotoxic effects of anoxia on *Mytilus galloprovincialis*. *Marine Ecology Progress Series*, **83**, 71–74.

Brunk, U. & E. Cadenas (1988) The potential intermediate role of lysosomes in oxygen free radical pathology. *APMIS*, **96**, 3–13.

Brunk, U.T., H. Zhang, K. Roberg & K. Öllinger (1995) Lethal hydrogen peroxide toxicity involves lysosomal iron-catalyzed reactions with membrane damage. *Redox Report*, **1**, 267–277.

Bucheli, T. & K. Fent (1995) Induction of cytochrome-P450 as a biomarker for environmental contamination in aquatic ecosystems. *Critical Reviews in Environmental Science and Technology*, **25**, 201–268.

Bucher, F., R. Hofer & W. Salvenmoser (1992) Effects of treated paper mill effluents on hepatic morphology in male bullhead (*Cottus gobio* L.). *Archives of Environmental Contamination and Toxicology*, **23**, 410–419.

Bunton, T.E. & J.M. Frazier (1989) Hepatocellular ultrastructure in white perch (*Morone americana*) with abnormal hepatic copper storage. *Marine Environmental Research*, **28**, 375–382.

Bunton, T.E., S.M. Baksi, S.G. George & J.M. Frazier (1987) Abnormal hepatic copper storage in a teleost fish (*Morone americana*). *Veterinary Pathology*, **24**, 515–524.

Burcham, P.C. (1998) Genotoxic lipid peroxidation products: their DNA damaging properties and role in formation of endogenous DNA adducts. *Mutagenesis*, **13** (3), 287–305.

Burgeot, T., E. His & F. Galgani (1995) The micronucleus assay in *Crassostrea gigas* for the detection of seawater genotoxicity. *Mutation Research*, **342**, 125–140.

Burgeot, T., G. Bocquené, C. Porte, J. Dimeet, R.M. Santella, L.M. Garcia de la Parra, A. Pfhol-Leszkowicz, C. Raoux & F. Galgani (1996a) Bioindicators of pollutant exposure in the northwestern Mediterranean Sea. *Marine Ecology Progress Series*, **131** (1–3), 125–141.

Burgeot, T., S. Woll & F. Galgani (1996b) Evaluation of the micronucleus test on *Mytilus galloprovincialis* for monitoring applications along French coasts. *Marine Pollution Bulletin*, **32** (1), 39–46.

Cai, L. & M.G. Cherian (1996) Adaptive response to ionizing radiation-induced chromosome aberrations in rabbit lymphocytes: effect of pre-exposure to zinc, and copper salts. *Mutation Research*, **369**, 233–241.

Cajaraville, M.P. (1991) *Efectos histopatológicos y citotóxicos de los hidrocarburos derivados del petróleo, y su cuantificación en el mejillón Mytilus galloprovincialis (Lmk.)*. Ph.D. Dissertation, University of the Basque Country, Bilbao.

Cajaraville, M.P., J.A. Marigómez & E. Angulo (1990) Ultrastructural study of the short-term toxic effects of naphthalene on the kidney of the marine prosobranch *Littoria littorea*. *Journal of Invertebrate Pathology*, **55**, 215–224.

Cajaraville, M.P., Y. Robledo, M. Etxeberria & I. Marigómez (1995) Cellular biomarkers as useful tools in the biological monitoring of environmental pollution: molluscan digestive lysosomes. In: (ed. Cajaraville, M.P.) *Cell Biology in Environmental Toxicology*. University of the Basque Country Press Service, Bilbao, pp. 29–55.

Cajaraville, M.P., I. Olabarrieta & I. Marigómez (1996) *In vitro* activities in mussel hemocytes as biomarkers of environmental quality: a case study in the Abra estuary (Biscay Bay). *Ecotoxicology and Environmental Safety*, **35**, 253–260.

Cajaraville, M.P., A. Orbea, I. Marigómez & I. Cancio (1997) Peroxisome proliferation in the digestive epithelium of mussels exposed to the water accommodated fraction of three oils. *Comparative Biochemistry and Physiology*, **117C**, 233–242.

Cajaraville, M.P., I. Cancio, A. Ibabe & A. Orbea (2003) Peroxisome proliferation as biomarker in environmental pollution assessment. In: (ed. Cajaraville, M.P.) *Microscopy Research and Techniques*. Special issue on 'Cell Biology of Peroxisomes'. John Wiley & Sons Inc., New York in press.

Cameron, P. & J. Berg (1992) Morphological and chromosomal-aberrations during embryonic-development in dab *Limanda limanda*. *Marine Ecology-Progress Series*, **91** (1–3), 163–169.

Cancio, I. & M.P. Cajaraville (2000) Cell biology of peroxisomes and their characteristics in aquatic organisms. *International Review of Cytology*, **199**, 201–293.

Cancio, I., I. ap Gwynn, M.P. Ireland & M.P. Cajaraville (1995) The effect of sublethal lead exposure on the ultrastructure and on the distribution of acid phosphatase activity in chloragocytes of earthworms (Annelida, Oligochaeta). *Histochemical Journal (London)*, **27**, 965–973.

Cancio, I., A. Orbea, A. Völkl, H.D. Fahimi & M.P. Cajaraville (1998) Induction of peroxisomal oxidases in mussels: comparison of effects of lubricant oil and benzo[a]pyrene with two typical peroxisome proliferators on peroxisome structure and function in *Mytilus galloprovincialis*. *Toxicology and Applied Pharmacology*, **149**, 64–72.

Cancio, I., A. Ibabe & M.P. Cajaraville (1999) Seasonal variation of peroxisomal enzyme activities and peroxisomal structure in mussels *Mytilus galloprovincialis* and its relationship with the lipid content. *Comparative Biochemistry and Physiology*, **123**, 135–144.

Carginale, V., R. Scudiero, C. Capasso, A. Capasso, P. Kille, G. di Prisco & E. Parisi (1998) Cadmium-induced differential accumulation of metallothionein isoforms in the Antarctic icefish, which exhibits no basal metallothionein protein but high endogenous mRNA levels. *Biochemical Journal*, **332**, 475–481.

Carpenè, E. & M. Vasak (1989) Hepatic metallothioneins from goldfish (*Crassius auratus* L.). *Comparative Biochemistry and Physiology*, **92B**, 463–468.

Carrano, A., V.L.H. Thompson, P.A. Lindl & J.L. Minkler (1978) Sister chromatid exchange as an indicator of mutagenesis. *Nature*, **271**, 551–553.

Casciano, D.A. (1991) Introduction: historical perspectives of genetic toxicology. In: (eds Li, A.P. & R.H. Heflich) *Genetic Toxicology*. CRC Press, Boca Raton, Florida, pp. 1–12.

Cattley, R.C., J.G. Conway & J.A. Popp (1987) Association of persistent peroxisome proliferation and oxidative injury with hepatocarcinogenicity in female F-34 rats fed Di(2-ethylhexyl)phthalate for 2 years. *Cancer letters*, **38**, 15–22.

Chang, L.-Y., J.W. Slot & H.J. Geuze (1988) Molecular immunocytochemistry of the CuZn superoxide dismutase in rat hepatocytes. *Journal of Cell Biology*, **107**, 2169–2179.

Cheng, R., B.L. Ford, P.E. Oneal, C.Z. Mathews, C.S. Bradford, T. Thongtan, D.W. Barnes, J.D. Hendricks & G.S. Bailey (1997) Zebrafish (*Danio rerio*) p53 tumor suppressor gene: cDNA sequence and expression during embryogenesis. *Molecular Marine Biology and Biotechnology*, **6**, 88–97.

Choi, D.J., R.B. Roth, T.M. Liu, N.E. Geacintov & D.A. Scicchitano (1996) Incorrect base insertion and prematurely terminated transcripts duping T7 RNA polymerase transcription elongation past benzo[a]pyrenediol epoxide-modified DNA. *Journal of Molecular Biology*, **264** (2), 213–219.

Chubatsu, L.S. & R. Meneghini (1993) Metallothionein protects DNA from oxidative damage. *Biochemical Journal*, **291**, 193–198.

Ciriolo, M.R., I. Mavelli, G. Rotilio, V. Borzatta, M. Cristofari & L. Stanzani (1982) Decrease of superoxide dismutase and glutathione peroxidase in liver of rats treated with hypolipidemic drugs. *FEBS Letters*, **144**, 264–268.

Collier, T.K., J.E. Stein & A. Goksøyr (1993) Biomarkers of PAH exposure in oyster toadfish (*Opsanus tau*) from the Elizabeth river, Virginia. *Environmental Science*, **2**, 161–177.

Collier, T.K., B.F. Anulacion, J.E. Stein, A. Goksøyr & U. Varanasi (1995) A field evaluation of cytochrome P4501A as a biomarker of contaminant exposure in three species of flatfish. *Environmental Toxicology and Chemistry*, **14** (1), 143–152.

Comporti, M. (1985) Biology of disease: Lipid peroxidation and cellular damage in toxic liver injury. *Laboratory Investigation*, **53**, 599–623.

Cosma, G., F. Crofts, D. Currie, I. Wirgin, P. Toniolo & S.J. Garte (1992) Racial differences in restriction fragment length polymorphisms and messenger RNA inducibility of the human CYP1A1 gene. *Cancer and Epidemiology and Biomarkers Prevention*, **2**, 53.

Cosson, R.P., C. Amiard-Triquet & J.C. Amiard (1991) Metallothioneins and detoxification. Is the use of detoxification protein for MTs a language abuse? *Water Air and Soil Pollution*, **57–58**, 555–567.

Cotten, R.G.H. (1993) Current methods of mutation detection. *Mutation Research*, **285**, 125–144.

Courtenay, S., J. Williams, C. Grunwald, B. Konkle, T.-L. Ong & I. Wirgin (1994) Assessment of within group variation in CYP1A mRNA inducibility in environmentally exposed and chemically treated Atlantic tomcod. *Environmental Health Perspectives*, **102** (12), 85–90.

Crapo, J.D., T. Oury, C. Rabouille, J.N. Slot & Y.-N. Chang (1992) Copper, zinc-superoxide dismutase is primarily a cytosolic protein in human cells. *Proceedings of the National Academy of Sciences of the USA*, **89**, 10405–10409.

Das, P. & G. John (1997) *In vivo* induction of sister chromatid exchanges (SCE) in a tropical fish *Etroplus suratensis* (Bloch). *Acta Biologica Hungarica*, **48** (2), 167–172.

Davies, K.J.A. (1986) Intracellular proteolytic systems may function as secondary antioxidant defenses: a hypothesis. *Free Radical Biology and Medicine*, **2**, 155–173.

de Duve, C. (1965) Functions of microbodies (peroxisomes). *Journal of Cell Biology*, **27**, 25A–26A.

de Duve, C. & P. Baudhuin (1966) Peroxisomes (microbodies and related particles). *Physiological Reviews*, **46**, 323–357.

De Flora, S. & C. Ramel (1988) Mechanisms of mutagenesis and carcinogenesis. Classification and overview. *Mutation Research*, **202**, 285–306.

De Flora, S., P. Zanachi, C. Bennicelli, A. Camoirano, C. Basso, M. Bagnasco, A. Izzotti & G.S. Badolati (1989) Genotoxicity, biotransformations, and interactions of marine pollutants as related to genetic and carcinogenic hazards. In: (ed. Grandjean, E.) *Carcinogenic, mutagenic, and teratogenic marine pollutants: impact on human health and the environment*. Vol. 5. (Advances in Applied Biotechnology.) Gulf Publishing Company, Houston, Texas, pp. 3–31.

De Flora, S., M. Bagnasco & P. Zanacchi (1991) Genotoxic, carcinogenic and teratogenic hazards in the marine environment, with special reference to the Mediterranean Sea. *Mutation Research*, **258**, 285–320.

De Marini, D.M. (1991) Environmental mutagens/complex mixtures. In: (eds Li, A.P. & R.H. Heflich) *Genetic toxicology*. CRC Press, Boca Raton, Florida, pp. 285–302.

de Pomerai, D.I. (1996) Heat shock proteins as biomarkers of pollution. *Human and Experimental Toxicology*, **15**, 279–285.

Defromentel, C.C., F. Pakdel, A. Chapus, C. Baney, P. May & T. Soussi (1992) Rainbow trout p53 -Cdna cloning and biochemical characterization. *Gene*, **112**, 241–245.

Degrassi, F. & C. Tanzarella (1988) Immunofluorescent staining of kinetochores in micronuclei: a new assay for the detection of aneuploidy. *Mutation Research*, **203** (5), 339–345.

Demoz, A., A. Svardal & R.K. Berge (1993) Relationship between peroxisome-proliferating sulfur-substituted fatty acid analogs, hepatic lipid peroxidation and hydrogen peroxide metabolism. *Biochemistry and Pharmacology*, **45**, 257–259.

Depledge, M. (1994) Genotypic toxicity: implications for individuals and populations. *Environmental Health Perspectives*, **102** (12), 101–104.

Dhaunsi, G.S., S. Gulati, A.K. Singh, J.K. Orak, K. Asayama & I. Singh (1992) Demonstration of CuZn superoxide dismutase in rat liver peroxisomes: Biochemical and immunochemical evidence. *Journal of Biological Chemistry*, **267**, 6870–6873.

Dhaunsi, G.S., I. Singh & C.D. Hanevold (1993) Peroxisomal participation in the cellular response to the oxidative stress of endotoxin. *Molecular and Cellular Biochemistry*, **126**, 25–35.

Di Giulio, R.T. (1991) Indices of oxidative stress as biomarkers for environmental contamination. In: (eds Mayes, M.A. & M.G. Barron) *Aquatic Toxicology and Risk Assessment*. Vol. 14. *ASTM STP*, 1124, American Society for Testing and Materials, Philadelphia, pp. 15–31.

Di Giulio, R.T., P.C. Washburn, R.J. Wenning, G.W. Winston & C.S. Jewell (1989) Biochemical responses in aquatic animals: a review of determinants of oxidative stress. *Environmental Toxicology and Chemistry*, **8**, 1103–1123.

Di Giulio, R.T., C. Habig & E.P. Gallagher (1993) Effects of black rock harbor sediments on indices of biotransformation, oxidative stress, and DNA integrity in channel catfish. *Aquatic Toxicology*, **26**, 1–22.

Dixon, D.R. & K.R. Clarke (1982) Sister chromatid exchange. A sensitive method for detecting damage caused by exposure to environmental mutagens in the chromosomes of adult *Mytilus edulis*. *Marine Biology Letters*, **3**, 163–172.

Doherty, A.T., S. Ellard, E.M. Parry & J.M. Parry (1996) An investigation into the activation and deactivation of chlorinated hydrocarbons to genotoxins in metabolically competent human cells. *Mutagenesis*, **11** (3), 247–274.

Donohue, M., L.A. Baldwin, D.A. Leonard, P.T. Kostecki & E.J. Calabrese (1993) Effect of hypolipidemic drugs gemfibrozil, ciprofibrate, and clofibric acid on peroxisomal beta-oxidation in primary cultures of rainbow trout hepatocytes. *Ecotoxicology and Environmental Safety*, **26** (2), 127–132.

Dubrova, Y.E., V.N. Nesterov, N.G. Krouchinsky, V.A. Ostapenko, R. Neumann, D.L. Neil & A.J. Jeffreys (1996) Human minisatellite mutation rate after Chernobyl accident. *Nature*, **380**, 683–686.

Dunlap, D.Y. & F. Matsumura (1997) Development of broad spectrum antibodies to heat shock protein 70s as biomarkers for detection of multiple stress by pollutants and environmental factors. *Ecotoxicology and Environmental Safety*, **37**, 238–244.

Ellegren, H., G. Lindgren, C.R. Primmer & A.P. Moller (1997) Fitness loss and mutations in barn swallows breeding in Chernobyl. *Nature*, **389**, 593–596.

Engelen, J.J.M., J.C.M. Albrechts, J.H. Hamers & J.P.M. Geraedts (1998) A simple and efficient method for microdissection and microFISH. *Journal of Medical Genetics*, **35** (4), 265–268.

Espina, N.G. & P. Weis (1995) DNA repair in fish from polluted estuaries. *Marine Environmental Research*, **39** (1–4), 309–312.

Fahimi, H.D. & M.P. Cajaraville (1995) Induction of peroxisome proliferation by some environmental pollutants and chemicals. In: (ed. Cajaraville, M.P.) *Cell Biology in Environmental Pollution*. University of the Basque country Press Service, Bilbao, pp. 221–255.

Farber, J.L., M.E. Kyle & J.B. Coleman (1990) Biology of disease: mechanisms of cell injury by activated oxygen species. *Laboratory Investigation*, **62**, 670–679.

Felley-Bosco, E., J. Zijlstra, P. Amstad & P. Cerutti (1991) A genotypic mutation system measuring mutations in restriction recognition sequences. *Nucleic Acids Research*, **19**, 2913–2919.

Fenech, M., J. Rinaldi & J. Surrales (1994) The origin of micronuclei induced by cytosine arabinoside and its synergistic interaction with hydroxyurea in human lymphocytes. *Mutagenesis*, **9** (3), 273–277.

Field, L.M., M.S. Williamson, G.D. Moores & A.L. Devonshire (1989) Cloning and analysis of the esterase genes conferring insecticide resistance in the peach-potato aphid, *Myzus persicae* (Sulzer). *Biochemical Journal*, **194**, 569–574.

Friedberg, E.C., G.C. Walker & W. Siede (1995) *DNA Repair and Mutagenesis*. American Society for Microbiology, Washington, DC. 698 pp.

Furukawa, K., S. Numoto, K. Furuya, N.T. Furukawa & G.M. Williams (1985) Effects of the hepatocarcinogen nafenopin, a peroxisome proliferator, on the activities of rat liver glutathione requiring enzymes and catalase in comparison to the action of phenobarbital. *Cancer Research*, **45**, 5011–5019.

Geard, C.R. (1992) Cytogenetic assays for genotoxic agents. *Lens and Eye Toxicity Research*, **9** (3–4), 413–428.

Gedamu, L., R. Foster, N. Jahroudi, S. Samson, N. Shworak & M. Zafarullah (1993) Regulation of human and rainbow trout metallothionein genes. In: (eds Suzuki, K.T., N. Imura & M. Kimura) *Metallothionein III*. Basel, Birkhäuser Verlag, pp. 361–380.

George, S.G. (1994) Enzymology and molecular biology of phase II xenobiotic-conjugating enzymes in fish. In: (eds Malins, D.C. & G.K. Ostrander) *Aquatic Toxicology: Molecular, biochemical and cellular perspectives*. Lewis Publishers, Boca Raton, FL, pp. 37–85.

George, S.G. & P.-E. Olsson (1994) Metallothioneins as indicators of trace metal pollution. In: (ed. Kramer, K.J.M.) *Biomonitoring of coastal waters and estuaries*. Boca Raton, CRC Press, pp. 151–178.

Gillespie, J.H. (1981) Mutation modification in a random environment. *Evolution*, **35**, 468–471.

Goksøyr, A. (1995) Cytochrome P450 in marine mammals: isozyme forms, catalytic functions, and physiological regulations. In: *Whales, Seals, Fish and Man*. Elsevier Science BV, London, pp. 629–639.

Goksøyr, A. & L. Förlin (1992) The cytochrome P450 system in fish, aquatic toxicology and environmental monitoring. *Aquatic Toxicology*, **22**, 287–311.

Goksøyr, A. & A.-M. Husøy (1998) Immunochemical approaches to studies of CYP1A localization and induction by xenobiotics in fish. In: (eds Braunbeck, T., D.E. Hinton & B. Streit) *Fish Ecotoxicology*. Birkhäuser Verlag, Basel, pp. 165–202.

Gossen, J. & J. Vijg (1993) Transgenic mice as model systems for studying gene mutations *in vivo*. *Trends in Genetics*, **9**, 27–31.

Gossen, J.A., W.J.F. Leeuw & J. Vijg (1994) Lac Z transgenic mouse models; their application in genetic toxicology. *Mutation Research*, **307**, 451–459.

Grasso, P. (1993) Hepatic changes associated with peroxisome proliferation. In: (eds Gibson, G. & B. Lake) *Peroxisomes: Biology and Importance in Toxicology and Medicine*. Taylor & Francis, London, pp. 639–652.

Groopman, J. & P. Skipper (1991) *Molecular Dosimetry and Human Cancer: Analytical, Epidemological and Social Considerations*. CRC Press, Boca Raton, FL.

Grøsvik, B.E. & A. Goksøyr (1996) Biomarker protein expression in primary cultures of salmon (*Salmo salar* L.) hepatocytes exposed to environmental pollutants. *Biomarkers*, **1**, 45–53.

Gulati, S., L. Ainol, J. Orak, A.K. Singh & I. Singh (1993) Alterations of peroxisomal function in ischemia-reperfusion injury of rat kidney. *Biochimica et Biophysica Acta*, **1182**, 291–298.

Gutteridge, J.M.C. & B. Halliwell (1990) The measurement and mechanism of lipid peroxidation in biological systems. *TIBS*, **15**, 129–135.

Guttman, S.I. (1994) Population genetic structure and ecotoxicology. *Environmental Health Perspectives*, **102** (12), 97–100.

Haasch, M.L. (1996) Induction of anti-trout lauric acid hydroxylase immunoreactive proteins by peroxisome proliferators in bluegill and catfish. *Marine Environmental Research*, **42**, 287–291.

Haasch, M.L., M.C. Henderson & D.R. Buhler (1998) Induction of CYP2M1 and CYP2K1 lauric acid hydroxylase activities by peroxisome proliferating agents in certain fish species: possible implications. *Marine Environmental Research*, **46** (1–5), 37–40.

Halliwell, B. & J.M.C. Gutteridge (1986) Oxygen free radicals and iron in relation to biology and medicine: some problems and concepts. *Archive for Biochemistry and Biophysics*, **246**, 501–514.

Hamilton, S.J. & P.M. Mehrle (1986) Metallothionein in fish: review of its importance in assessing stress from metal contaminants. *Transactions of the American Fisheries Society*, **11**, 596–609.

Harshbarger, J.C. & J.B. Clark (1990) Epizootiology of neoplasms in bony fish of North America. *Science of the Total Environment*, **94**, 1–32.

Hartwell, L.H. & T.A. Weinert (1989) Checkpoints: controls that ensure the order of cell cycle events. *Science*, **246** (4930), 629–634.

Harvey, J.S. (1995) *Genotoxins in the Marine Environment*. Ph.D. Dissertation, University of Wales, Swansea.

Harvey, J.S. & J.M. Parry (1997) 32P-postlabelling analysis of DNA adduct formation and persistence in the common mussel *Mytilus edulis* exposed to 4-nitroquinoline 1-oxide. *Mutation Research*, **12**, 153–158.

Hayashi, K. (1994) Manipulation of DNA by PCR. In: (eds Mullis, K.B., F. Ferre & R.A. Gibbs) *The Polymerase Chain Reaction*. Birkhauser Press, Boston.

Heddle, J.A., M.C. Cimino, M. Hayashi, F. Romagna, M.D. Shelby, J.D. Tucker, P. Vanparys & J.T. MacGregor (1991) Micronuclei as an index of cytogenetic damage: past, present, and future. *Environmental and Molecular Mutagenesis*, **18**, 277–291.

Hemminki, K. (1990) Environmental Carcinogens. In: (eds Cooper, C.S. & P.L. Grover) *Chemical Carcinogenesis and Mutagenesis I*. Springer Verlag, Berlin, pp. 33–61.

Hendricks, J.D., T.R. Meyers, D.W. Shelton, J.L. Casteel & G.S. Bailey (1985) Hepatocarcinogenicity of benzo[a]pyrene to rainbow trout by dietary exposure and intraperitoneal injection. *Journal of the National Cancer Institute*, **74**, 839–851.

Hidalgo, J., L. Campmany, M. Borrás, J.S. Garvey & A. Armario (1988) Metallothionein response to stress in rats: role in free radical scavenging. *American Journal of Physiology*, **255**, E518–E524.

Hillis, D., D. Moritz & B.K. Mable (1996) *Molecular Systematics*. Second ed. Sinauer Associates, Massachusetts.

Hoffmann, A.A. & P.A. Parsons (1991) *Evolutionary Genetics and Environmental Stress*. Oxford Scientific Publications, UK.

Hogstrand, C. & C. Haux (1990) A radioimmunoassay for perch (*Perca fluviatilis*) metallothionein. *Toxicology and Applied Pharmacology*, **103**, 56–65.

Hogstrand, C. & C. Haux (1996) Naturally high levels of zinc and metallothionein in liver of several species of the squirrelfish family from Queensland, Australia. *Marine Biology*, **125**, 23–31.

Hogstrand, C., N.J. Gassman, B. Popova, C.M. Wood & P.J. Walsh (1996) The physiology of massive zinc accumulation in the liver of female aquirrelfish and its relationship to reproduction. *jeb*, **199**, 2543–2554.

Hollstein, M., P. Sidransky, P. Vogelstein & C.C. Harris (1991) p53 mutations in human cancer. *Science*, **253**, 49–53.

Hose, J.E. (1994) Large-scale genotoxicity assessments in the marine environment. *Environmental Health Perspectives*, **102**, 29–32.

Hruban, Z. & M. Rechcigl (1969) *Microbodies and related particles: morphology, biochemistry and physiology*. Aca Press, New York, London, 296 pp.

Huang, P.C. (1993) Metallothionein structure/function interface. In: (eds Suzuki, K.T., N. Imura & M. Kimura) *Metallothionein III*. Basel, Birkhäuser Verlag, pp. 405–426.

Hylland, K. (1999) Biological effects of contaminants: quantification of metallothionein in fish liver tissue. *ICES Techniques in Marine Environmental Sciences*, No. 26, 18 pp.

Ibabe, A. (1998) *Ziklo marealaren eragina muskuiluen peroxisomen egituran eta aktibitate entzimatikoetan*. Licenciature Thesis, University of the Basque Country, Bilbo. 106 pp.

IPCS (1992) *International Programme on Chemical Safety*. World Health Organization, Geneva, 141 pp. (Environmental Health Criteria, 131).

Issemann, I. & S. Green (1990) Activation of a member of the steroid hormone receptor superfamily by peroxisome proliferators. *Nature*, **347**, 645–650.

Iwama, G.K., P.T. Thomas, R.H. Forsyth & M.M. Vijayan (1998) Heat shock protein expression in fish. *Reviews in Fish Biology*, **8**, 35–56.

James, M.O. (1987) Conjugation of organic pollutants in aquatic species. *Environmental Health Perspectives*, **71**, 97–103.

James, S.I. & J.T. Ahokas (1992) Effect of peroxisome proliferators on glutathione-dependent sulphobromophthalein excretion. *Xenobiotica.*, **22**, 1425–1432.

James, N.H. & R.A. Roberts (1995) Species differences in the clonal expansion of hepatocytes in response to the coaction of epidermal growth factor and nafenopin, a rodent hepatocarcinogenic peroxisome proliferator. *Fundamental and Applied Toxicology*, **26**, 143–149.

Jha, A.N., T.H. Hutchinson, J.M. Mackay, B.M. Elliot & D.R. Dixon (1996) Development of an in vivo genotoxicity assay using the marine worm *Platynereis dumerilii* (Polychaeta: Nereidae). *Mutation Research*, **359**, 141–150.

Jobling, S., T. Reynolds, R. White, M.G. Parker & J.P. Sumpter (1995) A variety of environmentally persistent chemicals, including some phthalate plasticisers, are weakly estrogenic. *Environmental Health Perspectives*, **103**, 582–587.

Johnson, L.L., J.E. Stein, T.K. Collier, E. Casillas, B. McCain & U. Varanasi (1992) *Bioindicators of contaminant exposure, liver pathology, and reproductive development in prespawning female winter flounder (*Pseudopleuronectes americanus*) from urban and nonurban estuaries on the Northeast Atlantic coast*. National Marine Fisheries Service, NOAA, USA (August 1992, NOAA Technical Memorandum).

Kägi, J.H.R. (1993) Evolution, structure and chemical activity of class I metallothioneins: an overview. In: (eds Suzuki, K.T., N. Imura & M. Kimura) *Metallothionein III*. Basel, Birkhäuser Verlag, pp. 30–55.

Kasai, H. & S. Nishimura (1986) Hydroxylation of guanine in nucleosides and DNA at the C-8 position by heated glucose and oxygen radical-forming agents. *Environmental Health Perspectives*, **67**, 111–116.

Kehrer, J.P. (1993) Free radicals as mediators of tissue injury and disease. *Critical Reviews in Toxicology*, **23**, 21–48.

Keller, G., T.G. Warner, K.S. Steimer & R.A. Halewell (1991) Cu-Zn superoxide dismutase is a peroxisomal enzyme in human fibroblasts and hepatoma cells. *Proceedings of the National Academy of Sciences of the USA*, **88**, 7381–7385.

Kille, P., J. Kay & G.E. Sweeney (1993) Analysis of regulatory elements flanking metallothionein genes in Cd-tolerant fish (pike and stone loach). *Biochimica et Biophysica Acta*, **1216**, 55–64.

Kling, P. & P.-E. Olsson (1995) Regulation of the rainbow trout metallothionein-A gene. *Marine Environmental Research*, **39**, 117–120.

Kling, P., L.J. Erkell, P. Kille & P.-E. Olsson (1996) Metallothionein induction in rainbow trout gonadal (RTG-2) cells during free radical exposure. *Marine Environmental Research*, **42**, 33–36.

Kloepper-Sams, P.J. & J.J. Stegeman (1992) Effects of temperature acclimation on the expression of hepatic cytochrome P4501A mRNA and protein in the fish *Fundulus heteroclitus*. *Archive for Biochemistry and Biophysics*, **299**, 38–46.

Köhler, A. (1991) Lysosomal perturbations in fish liver as indicators for toxic effects of environmental pollution. *Comparative Biochemistry and Physiology*, **100C** (1/2), 123–127.

Köhler, A. & H.J. Pluta (1995) Lysosomal injury and MFO activity in the liver of flounder (*Platichthys flesus* L.) in relation to histopathology of hepatic degeneration and carcinogenesis. *Marine Environmental Research*, **39**, 255–260.

Köhler, A., H. Deisemann & B. Lauritzen (1992) Histological and cytochemical indices of toxic injury in the liver of dab *Limanda limanda*. *Marine Ecology Progress Series*, **91** (1/3), 141–153.

Krause, M.K., L.D. Rhodes & R.J. Van Beneden (1997) Cloning of the p53 tumor suppressor gene from the Japanese medaka (*Oryzias latipes*) and evaluation of mutational hotspots in MNNG-exposed fish. *Gene*, **189**, 101–106.

Krishnakumar, P.K., E. Casillas & U. Varanasi (1995) Effects of chemical contaminants on the health of *Mytilus edulis* from Puget Sound, Washington. II. Cytochemical detection of subcellular changes in digestive cells. *Marine Biology*, **124**, 251–259.

Kubota, Y., A. Shimada & A. Shima (1992) Detection of gamma-ray induced DNA damages in malformed dominant lethal embryos of the Japanese Medaka (*Oryzias latipes*) using Ap-Pcr fingerprinting. *Mutation Research*, **283**, 263–270.

Kurelec, B. (1992) The Multixenobiotic Resistance Mechanism in Aquatic Organisms. *Critical Reviews in Toxicology*, **22**, 23–43.

Kwak, H.J., C.D. Jun, H.O. Pae, J.C. Yoo, Y.C. Park, B.M. Choi, Y.G. Na, R.K. Park, H.T. Chung, H.Y. Chung, W.Y. Park & J.C. Seo (1998) The role of inducible 70-kDa heat shock protein in cell cycle control, differentiation, and apoptotic cell death of the human myeloid leukemic HL-60 cells. *Cellular Immunology*, **187**, 1–12.

Lackner, R. (1998) 'Oxidative stress' in fish by environmental pollutants. In: (eds Braunbeck, T., D.E. Hinton & B. Streit) *Fish Ecotoxicology*. Birkhäuser Verlag, Basel, pp. 203–224.

Lake, B.G. (1995) Mechanisms of hepatocarcinogenicity of peroxisome-proliferating drugs and chemicals. *Annual Review of Pharmacology and Toxicology*, **35**, 483–503.

Lawrence, A.J. & B. Nicholson (1998) The use of stress proteins in *Mytilus edulis* as indicators of chlorinated effluent pollution. *Water Science and Technology*, **38**, 253–261.

Leaver, M.J. & S.G. George (1998) A piscine glutathione S-transferase which efficiently conjugates the end-products of lipid peroxidation. *Marine Environmental Research*, **46** (1–5), 71–74.

Leaver, M.J., J. Wright & S.G. George (1997) Structure and expression of a cluster of glutathione S-transferase genes from a marine fish, the plaice (*Pleuronectes platessa*). *Biochemical Journal*, **321**, 405–412.

Leaver, M.J., J. Wright & S.G. George (1998) A peroxisomal proliferator-activated receptor gene from the marine flatfish, the plaice (*Pleuronectes platessa*). *Marine Environmental Research*, **46** (1–5), 75–79.

Lekube, X., M.P. Cajaraville & I. Marigómez (1998) Application of the B5 system: Use of specific antibodies for the detection of changes induced by environmental contaminants in lysosomes. *Cuad. Inv. Biol.*, **20**, 237–239.

Lekube, X., M.P. Cajaraville, & I. Marigómez (2000) Use of polyclonal antibodies for the detection of changes induced by cadmium in lysosomes of aquatic organisms. *Science of the Total Environment*, **247**, 201–212.

Lemaire, P., A. Mathieu, S. Carriere, J.F. Narbonne, M. Lafaurie & J. Giudicelli (1992) Hepatic biotransformation enzymes in aquaculture European sea bass (*Dicentrarchus labrax*): kinetic parameters and induction with benzo[a]pyrene. *Comparative Biochemistry and Physiology*, **103B**, 847–853.

Lemaire, P., L. Förlin & D.R. Livingstone (1996) Responses of hepatic biotransformation and anti-oxidant enzymes to CYP1A-inducers (3-methylcholanthrene, b-naphthoflavone) in sea bass (*Dicentrarchus labrax*), dab (*Limanda limanda*) and rainbow trout (*Oncorhynchus mykiss*). *Aquatic Toxicology*, **36**, 141–160.

Lenártová,, V, K. Holovská, J.R. Pedrajas, E. Martínez Lara, J. Peinado, J. López Barea, I. Rosival & P. Kosúth (1997) Antioxidant and detoxifying fish enzymes as biomarkers of river pollution. *Biomarkers*, **2** (4), 247–252.

Lewin, B. (1995) *Genes V*. Oxford University Press, Oxford, 1272 pp.

Lindquist, S. & E.A. Craig (1988) The heat-shock proteins. *Annual Reviews in Genetics*, **22**, 631–677.

Livingstone, D.R., P. García Martínez & G.W. Winston (1989) Menadione-stimulated oxyradical formation in digestive gland microsomes of the common mussel, *Mytilulus edulis* L. *Aquatic Toxicology*, **15**, 213–236.

Livingstone, D.R., S. Archibald, J.K. Chipman & J.W. Marsh (1992) Antioxidant enzymes in liver of dab *Limanda limanda* from the North sea. *Marine Ecology Progress Series*, **91** (1–3), 97–104.

Livingstone, D.R., P. Lemaire, A. Matthews, L. Peters, D. Bucke & R.J. Law (1993) Pro-oxidant, antioxidant and 7-ethoxyresorufin o-deethylase (EROD) activity responses in liver of dab (*Limanda limanda*) exposed to sediment contaminated with hydrocarbons and other chemicals. *Marine Pollution Bulletin*, **26** (11), 602–606.

Lorez Arnaiz, S., M. Travacio, S. Llesuy & A. Boveris (1995) Hydrogen peroxide metabolism during peroxisome proliferation by fenofibrate. *Biochimica et Biophysica Acta*, **1272**, 175–180.

Loschen, G. & L. Flohé (1971) Respiratory chain linked H202 production in pigeon heart mitochondria. *FEBS Letters*, **18**, 261–263.

Lowe, D.M. & M.N. Moore (1978) Cytology and quantitative cytochemistry of a proliferative atypical hemocytic condition in *Mytilus edulis* (Bivalvia, Mollusca). *Journal of the National Cancer Institute*, **60** (6), 1455–1459.

Lowe, D.M., M.N. Moore & K.R. Clarke (1981) Effects of oil on digestive cells in mussels: quantitative alterations in cellular and lysosomal structure. *Aquatic Toxicology*, **1**, 213–226.

Lowe, D.M., M.N. Moore & B.M. Evans (1992) Contaminant impact on interactions of molecular probes with lysosomes in living hepatocytes from dab *Limanda limanda*. *Marine Ecology Progress Series*, **91** (1/3), 135–140.

Lowe, D.M., V.U. Fossato & M.H. Depledge (1995) Contaminant induced lysosomal damage in blood cells of mussel *Mytilus galloprovincialis* from the Venice Lagoon: an *in vitro* study. *Marine Ecology Progress Series*, **129**, 189–196.

MacGregor, J.T., J.D. Tucker, D.A. Eastmond & A.J. Wyrobek (1995) Integration of cytogenetic assays with toxicology studies. *Environmental and Molecular Mutagenesis*, **25**, 328–337.

Malins, D.C. & S.J. Gunselman (1994) Infrared spectroscopy and gas chromatography mass-spectrometry reveal a remarkable degree of structural damage in the DNA of wild fish exposed to toxic chemicals. *Proceedings of the National Academy of Sciences of the USA*, **91**, 13038–13041.

Malins, D.C. & R. Haimanot (1991) The etiology of cancer: hydroxyl radical-induced DNA lesions in histologically normal livers of fish from a population with liver tumors. *Aquatic Toxicology*, **20**, 123–130.

Malins, D.C., B.B. McCain, J.T. Landahl, M.S. Myers, M.M. Krahn, D.W. Brown, S.L. Chan & W.T. Roubal (1988) Neoplastic and other diseases in fish in relation to toxic chemicals: An overview. *Aquatic Toxicology*, **11** (1/2), 43–67.

Malins, D.C., N.L. Polissar, M.M. Garner & S.J. Gunselman (1996) Mutagenic DNA base modifications are correlated with lesions in non-neoplastic hepatic tissue of the English sole carcinogenesis model. *Cancer Research*, **56** (24), 5563–5565.

Malins, D.C., N.L. Polissar, Y.Z. Sy, H.S. Gardner & S.J. Gunselman (1997a) A new structural analysis of DNA using statistical models of infrared spectra. *Nature Medicine*, **3**, 927–930.

Malins, D.C., N.L. Polissar & S.J. Gunselman (1997b) Infrared spectral models demonstrate that exposure to environmental chemicals leads to new forms of DNA. *Proceedings of the National Academy of Sciences of the USA*, **94**, 3611–3615.

Maracine, M. & H. Segner (1998) Cytotoxicity of metals in isolated fish cells: importance of the cellular glutathione status. *Comparative Biochemistry and Physiology*, **120**, 83–88.

Marafante, E. (1976) Binding of mercury and zinc to cadmium-binding protein in liver and kidney of goldfish (*Carassius auratus* L.). *Experientia.*, **32**, 149–152.

Marafante, E., G. Pozzi & P. Scoppa (1972) Detossicazione dei metall pesanti pesci: Isolamento delle metallothioneina dal fegato di *Carassius auratus*. *Boll. Soc. Ital. Biol. Speriment.*, **48**, 109–111.

Marlasca, M.J., C. Sanpera, M.C. Riva, R. Sala & S. Crespo (1998) Hepatic alterations and induction of micronuclei in rainbow trout (*Oncorhynchus mykiss*) exposed to a textile industry effluent. *Histology and Histopathology*, **13** (3), 703–712.

Marnett, L.J. (1994) DNA adducts of I,J-unsaturated aldehydes and dicarbonyl compounds. In: (eds Hemminki, K., A. Dipple, D.E.G. Shuker, F.F. Kadlubar, D. Segerback & D. Bartsch) *DNA Adducts: Identification and Biological Significance*. Vol. 125. IARC Scientific Publications, Lyon, pp. 151–163.

Martineau, D.A., R. Lagace, P. Beland, R. Higgings, D. Armstrong & L.R. Shugart (1988) Pathology of stranded beluga whales (*Delphinapterus leucas*) from the St. Lawrence estuary, Quebec, Canada. *Journal of Comparative Pathology*, **9** (3), 287–311.

Mather-Mihaich, E. & T. Di Giulio (1991) Oxidant, mixed-function oxidase and peroxisomal responses in channel catfish exposed to a bleached kraft mill effluent. *Archives of Environmental Contamination and Toxicology*, **20**, 391–397.

McCain, B.B., H.O. Hodgins, W.D. Gronlund, J.W. Hawkes, D.W. Brown, M.S. Myers & J.H. Vandermeulen (1978) Bioavailability of crude oil from experimentally oiled sediments to English sole (*Parophrys vetulus*), and pathological consequences. *Journal of the Fisheries Research Board of Canada*, **35**, 657–664.

Miller, D. (1989) Heat-shock proteins to the rescue. *New Scientist*, **7**, 47–50.

Miller, D.S., R. Masereeuw, J. Henson & K.J. Karnaky (1998) Excretory transport of xenobiotics by dogfish shark rectal gland tubules. *American Journal of Physiology-regulatory Integrative and Comparative Physiology*, **44** (3), R697–R705.

Min, K.-S., N. Itoh, H. Okamoto & K. Tanaka (1993) Indirect induction of metallothionein by organic compounds. In: (eds Suzuki, K.T., N. Imura & M. Kimura) *Metallothionein III*. Basel, Birkhäuser Verlag, pp. 157–174.

Minisini, M.P., S. Kantengwa & B.S. Polla (1994) DNA damage and stress protein synthesis induced by oxidative stress proceed independently in the human premonocytic line U937. *Mutation Research*, **315**, 169–179.

Moore, M.N. (1980) Cytochemical determination of cellular responses to environmental stressors in marine organisms. *Rapports et Procès-verbaux de Réunions du Conseil International pour l'Exploration de la Mer*, **179**, 7–15.

Moore, M.N. (1990) Lysosomal cytochemistry in marine environmental monitoring. *Histochemical Journal (London)*, **22**, 187–191.

Moore, M.N., D.M. Lowe, D.R. Livingstone & D.R. Dixon (1986) Molecular and cellular indices of pollutant effects and their use in environmental impact assessment. *Water Science and Technology*, **18**, 223–232.

Muerhoff, A.S., K.J. Griffin & E.F. Johnson (1992) The peroxisome proliferator-activated receptor mediates the induction of CYP4A6, a cytochrome P450 fatty acid w-hydroxylase, by clofibric acid. *Journal of Biological Chemistry*, **267**, 19051–19053.

Myers, M.S., J.T. Landahl, M.N. Krahn, L.L. Johnson & B.B. McCain (1990) Overview of studies on liver carcinogenesis in English sole from Puget Sound; evidence for a xenobiotic chemical etiology I: Pathology and Epizootiology. *Science of the Total Environment*, **94**, 33–50.

Myers, M.S., L.L. Johnson, T. Hom, T.K. Collier, J.E. Stein & U. Varanasi (1998) Toxicopathic lesions in subadult English sole (*Pleuronectes vetulus*) from Puget Sound, Washington, USA: relationships with other biomarkers of contamiant exposure. *Marine Environmental Research*, **45** (1), 47–67.

Nath, R.G., K. Randerath, D. Li & F.-L. Chung (1996) Detection of endogenous DNA adducts. *Regulatory Pharamcology and Toxicology*, **23**(1), 22–28.

Navarro, J.M. & J.L. Jorcano (1999) The uses of arbitrarily primed polymerase chain reaction in cancer research. *Electrophoresis*, **20**, 283–290.

Nemali, M.R., K. Reddy, N. Usuda, P.G. Reddy, L.D. Comeau, M.S. Rao & J.K. Reddy (1989) Differential induction and regulation of peroxisomal enzymes: predictive value of peroxisome proliferation in identifying certain nonmutagenic carcinogens. *Toxicology and Applied Pharmacology*, **97**, 72–87.

Nestmann, E.R., D.W. Bryant & C.J. Carr (1996) Toxicological significance of DNA adducts: summary of discussions with an expert panel. *Regulatory Toxicology and Pharmacology*, **24**, 9–18.

Nielson, K.B., C.L. Atkin & D.R. Winge (1985) Distinct metal-binding configurations in metallothionein. *Journal of Biological Chemistry*, **260**, 5342–5350.

Nielsen, P.A. (1993) Chemical carcinogens in the environment: risk assessment for the environment. *Pharmacology and Toxicology*, **72**, 46–50.

Nishimoto, M., W.T. Roubal, J.E. Stein & U. Varanasi (1991) Oxidative damage in tissues of English sole (*Parophrys vetulus*) exposed to nitrofurantoin. *Chemico-Biological Interactions*, **80**, 317–326.

Nohl, H. & D. Hegner (1978) Do mitochondira produce oxygen radicals *in vivo*? *European Journal of Biochemisty*, **82**, 563–567.

Novi, S., C. Pretti, A.M. Cognetti, V. Longo, S. Marchetti & P.G. Gervasi (1998) Biotransformation enzymes and their induction by beta-naphthoflavone in adult sea bass (*Dicentrarchus labrax*). *Aquatic Toxicology*, **41** (1–2), 63–81.

Olafson, R.W. & P.-E. Olsson (1991) Electrochemical detection of metallothionein. In: (eds Riordan, J.F. & B.L. Vallee) *Metallobiochemistry. Part B: metallothionein and related molecules*. Academic Press, London, pp. 205–213.

Olsson, P.E. (1993) Metallothionein gene expression and regulation in fish. In: (eds Hochachka, P.W. & T.P. Mommsen) *Molecular biology frontiers*. Elsevier, Amsterdam, pp. 259–278.

Olsson, P.E. & P. Kille (1997) Functional comparison of the metal-regulated transcriptional control regions of metallothionein genes from cadmium-sensitive and tolerant fish species. *Biochimica et Biophysica Acta*, **1350**, 325–334.

Olsson, P.E. & P. Kling (1995) Regulation of hepatic metallothionein in estradiol-treated rainbow trout. *Marine Environmental Research*, **39**, 127–129.

Olsson, P.E., P. Kling, L.J. Erkell & P. Kille (1995) Structural and functional analysis of the rainbow trout (*Oncorhynchus mykiss*) metallothionein-A gene. *European Journal of Biochemisty*, **230**, 344–349.

Orbea, A., K. Beier, A. Völkl, H.D. Fahimi & M.P. Cajaraville (1998a) Baseline study on the structure and function of fish peroxisomes, a possible target of certain environmentally relevant organic pollutants. *Cuad. Invest. Biol.*, **20**, 405–408.

Orbea, A., K. Beier, A. Völkl, H.D. Fahimi & M.P. Cajaraville (1998b) Fish liver peroxisomes: a cell compartment involved in lipid metabolism and oxyradical homeostasis. 5th Joint Meeting Japan Soc Histochem Cytochem & Histochem Soc, San Diego, CA.

Orbea, A., K. Beier, A. Völkl, H.D. Fahimi & M.P. Cajaraville (1999) Ultrastructural, immunocytochemical and morphometric characterization of liver peroxisomes in gray mullet *Mugil cephalus* L. *Cell Tissue Research*, **297**, 493–502.

Orbea, A., Fahimi, H.D. & Cajaraville, M.P. (2000) Immunolocalisation of four antioxidant enzymes in digestive glands of molluscs and crustaceans and fish liver. *Histochemistry and Cell Biology*, **114**, 393–404.

Orrenius, S. & P. Nicotera (1987) On the role of calcium in chemical toxicity. *Archives of Toxicology*, **11**, 11–19.

Ortiz de Montellano, P.R. (1995) *Cytochrome P450: structure, mechanism, and biochemistry.* Plenum Press, New York.

Oulmi, Y., R.D. Negele & T. Braunbeck (1995a) Cytopathology of liver and kidney in rainbow trout (*Oncorhynchus mykiss*) after a long term exposure to sublethal concentrations of linuron. *Diseases of Aquatic Organisms*, **21**, 35–52.

Oulmi, Y., R.-D. Negele & T. Braunbeck (1995b) Segment specificity of the cytological response in rainbow trout (*Oncorhynchus mykiss*) renal tubules following prolonged exposure to sublethal concentrations of atrazine. *Ecotoxicology and Environmental Safety*, **32**, 39–50.

Pacheco, M. & M.A. Santos (1996) Induction of micronuclei and nuclear abnormalities in the erythrocytes of *Anguilla anguilla* L. exposed either to cyclophosphamide or to Bleached Kraft Pulp Mill Effluent. *Fresenius Environmental Bulletin*, **5** (11–12), 746–751.

Palace, V.P. & J.F. Klaverkamp (1993) Variation of hepatic enzymes in three species of freshwater fish from precambrian shield lakes and the effect of cadmium exposure. *Comparative Biochemistry and Physiology*, **104C** (1), 147–154.

Palace, V.P., C.L. Baron & J.F. Klaverkamp (1998) An assessment of Ah-inducible phase I and phase II enzymatic activities and oxidative stress indices in adult lake trout (*Salvelinus namaycush*) from Lake Ontario and Lake Superior. *Aquatic Toxicology*, **42** (2), 149–168.

Parsell, D.A, & S. Lindquist (1993) The function of heat shock proteins in stress tolerance: degradation and reactivation of damaged proteins. *Annual Review of Genetics*, **23**, 437–469.

Parry, J., M. Shamsheer & D. Skibinski (1990) Restriction site mutation analysis, a proposed methodology for the detection and study of DNA base changes following mutagen exposure. *Mutagenesis*, **5**, 209–212.

Payne, J.F. & A. Rahimtula (1989) Monitoring for mutagens and carcinogens in the aquatic environment. An overview. In: *World Health Organisation Regional Office for Europe and United Nations Environment Program, Carcinogenic, Mutagenic and Teratogenic Marine Pollutants. Impact on Human Health and the Environment.* Advances in Applied Biotechnology ed. Vol. 5. Portfolio, pp. 227–248.

Pedrajas, J.R., J. Peinado & J. López-Barea (1995) Oxidative stress in fish exposed to model xenobiotics. Oxidatively modified forms of Cu, Zn-superoxide dismutase as potential biomarkers. *Chemico-Biological Interactions*, **98**, 267–282.

Pedrajas, J.R., J. López-Barea & J. Peinado (1996) Dieldrin induces peroxisomal enzymes in fish (*Sparus aurata*) liver. *Comparative Biochemistry and Physiology*, **115C** (2), 125–131.

Pedrajas, J.R., F. Gavilanes, J. López Barea & J. Peinado (1998) Incubation of superoxide dismutase with malondialdehyde and 4-hydroxy-2-nonenal forms new active isoforms and adducts. An evaluation of xenobiotics in fish. *Chemico-Biological Interactions*, **116**, 1–17.

Peters, L.D., C. Porte, J. Albaigés & D.R. Livingstone (1994) 7-Ethoxyresorufin O-deethylase (EROD) and antioxidant enzyme activities in larvae of sardine (*Sardina pilchardus*) from the North Coast of Spain. *Marine Pollution Bulletin*, **28**, 299–304.

Peterson, L.A. & F.P. Guengerich (1988) Comparison of and relationships between glutathione S-transferase and cytochrome P-450 systems. In: (eds Sies, H. & B. Ketterer) *Glutathione conjugation: mechanisms and biological significance*. Academic Press, London, pp. 193–233.

Pfau, W. (1997) DNA adducts in marine and freshwater fish as biomarkers of environmental contamination. *Biomarkers*, **2** (3), 145–151.

Phillips, D.H. & P.B. Farmer (1995) Protein and DNA adducts as biomarkers of exposure to environmental mutagens. In: (eds Phillips, D.H. & S. Venitt) *Environmental Mutagenesis*. Bioscientific Publishers, pp. 367–395.

Piotrowski, J.R., W. Bolanowska & A. Sapota (1973) Evaluation of metallothionein content in animal tissues. *Acta Biochem. Pol.*, **20**, 207–215.

Pipe, R.K. & M.N. Moore (1986) An ultrastructural study on the effects of phenanthrene on lysosomal membranes and distribution of the lysosomal enzyme b-glucuronidase in digestive cells of the periwinkle *Littorina littorea*. *Aquatic Toxicology*, **8**, 65–76.

Poirier, M.C. & A.W. Weston (1996) Human DNA adduct measurements. State of the art. *Environmental Health Perspectives*, **104** (5), 883–893.

Preston, R.J. (1990) Mechanisms of induction of specific chromosomal alterations. In: (eds Sutherland, B.M. & A.D. Woodhead) *DNA Damage and Repair in Human Tissues*. Plenum Press, New York, pp. 329–336.

Preston, R.J. (1991) Mechanisms of chromosomal alterations and sister chromatid exchanges: presentation of a generalized hypothesis. In: (eds Li, A.P. & R.H. Heflich) *Genetic Toxicology*. CRC Press, Boca Raton, Florida, pp. 41–65.

Preston, R.J. (1996) Aneuploidy in germ cells: disruption of chromosome mover components. *Environmental and Molecular Mutagenesis*, **28** (3), 176–181.

Pretti, C., S. Novi, V. Longo & P.G. Gervasi (1999) Effect of clofibrate, a peroxisome proliferator, in sea bass (*Dicentrarchus labrax*), a marine fish. *Environmental Research*, **80** (3), 294–296.

Radi, A.A.R. & B. Matkovics (1988) Effects of metal ions on the antioxidant enzyme activities, protein content and lipid peroxidation of carp tissues. *Comparative Biochemistry and Physiology*, **90C**, 69–72.

Radi, A.A., B. Matkovics & I. Csengeri (1987) Comparative studies of the phospholipid fatty acids and the antioxidant enzyme activities in fish with different feeding habits. *Comparative Biochemistry and Physiology*, **87B**, 49–54.

Randerath, K., D. Li, R. Nath & E. Randerath (1992) Exogenous and endogenous DNA modifications as monitored by 32P-postlabelling: relationships to cancer and ageing. *Experimental Gerontology*, **27**, 533–549.

Rao, S.S., T. Neheli, J.H. Carey & V.W. Cairns (1996) Fish hepatic micronuclei as an indication of exposure to genotoxic environmental contaminants. *Environmental Toxicology and Water Quality Journal*, **12**, 217–222.

Rao, S.S., T. Neheli & C.D. Metcalfe (1997) Hepatic micronucleus assay for the assessment of genotoxic responses in fish. *Environmental Toxicology and Water Quality Journal*, **11**, 167–170.

Reddy, J.K. & N.D. Lalwani (1983) Carcinogenesis by hepatic peroxisome proliferators: evaluation of the risk of hypolipidemic drugs and industrial plasticizers to humans. *Critical Reviews in Toxicology*, **12**, 1–53.

Reddy, J.K. & G.P. Mannaerts (1994) Peroxisomal lipid metabolism. *Annual Review of Nutrition*, **14**, 343–370.

Reddy, J.K. & M.S. Rao (1989) Oxidative DNA damage caused by persistent peroxisome proliferation: its role in hepatocarcinogenesis. *Mutation Research*, **214**, 63–68.

Reddy, J.K., D.L. Azarnoff & C.E. Hignite (1980) Hypolipidemic hepatic peroxisome proliferators from a new class of carcinogens. *Nature*, **283**, 397–398.

Reichert, W.L., M.S. Myers, K. PeckMiller, B. French, B.F. Anulacion, T.K. Collier, J.E. Stein & U. Varanasi (1998) Molecular epizootiology of genotoxic events in marine fish: linking contaminant exposure, DNA damage, and tissue-level alterations. *Mutation Research Reviews in Mutation Research*, **411** (3), 215–225.

Richards, E.H., E. Hickey, L. Weber & J.R. Master (1996) Effect of overexpression of the small heat shock protein HSP27 on the heat and drug sensitivities of human testis tumor cells. *Cancer Research*, **56**, 2446–2451.

Roberg, K. & K. Öllinger (1998) A pre-embedding technique for immunocytochemical visualization of cathepsin D in cultured cells subjected to oxidative stress. *Journal of Histochemistry and Cytochemistry*, **46**, 411–418.

Robledo, Y. & M.P. Cajaraville (1996) Isolation and morphofunctional characterization of mussel digestive gland cells *in vitro*. *European Journal of Cell Biology*, **72**, 362–369.

Rocha, E., A. Lobo-da-Cunha, R.A.F. Monteiro, M.W. Silva & M.H. Oliveira (1999) A stereological study along the year on the hepatocytic peroxisomes of brown trout (*Salmo trutta*). *Journal of Submicroscopic Cytology and Pathology*, **31** (1), 91–105.

Rodríguez-Ariza, A., G. Dorado, J. Peinado & C. Pueyo (1991) Biochemical effects of environmental pollution in fishes from the Spanish South-Atlantic littoral. *Biochemical Society Transactions*, **19**, 301S.

Rodríguez-Ariza, A., J. Peinado, C. Pueyo & J. López-Barea (1993) Biochemical indicators of oxidative stress in fish from polluted littoral areas. *Canadian Journal of Fisheries and Aquatic Sciences.*, **50**, 2568–2573.

Rodríguez-Ariza, A., G. Dorado, J.I. Navas, C. Pueyo & J. Lopez-Barea (1994) Promutagen activation by fish liver as a biomarker of littoral pollution. *Environmental and Molecular Mutagenesis*, **24** (2), 116–123.

Rossiter, J.F.B. & C.T. Casket (1994) Clinical Applications of the Polymerase Chain Reaction. In: (eds Mullis, B.K., F. Ferre & A.R. Gibbs) *PCR: The Polymerase Chain Reaction*. Birkhauser, Boston, pp. 395–405.

Russell, L.B., W.L. Russell, E.M. Rinchik & P.R. Hunsicker (1990) Factors affecting the nature of induced mutations. In: *Biology of Mammalian Germ Cell Mutagenesis*. Bambury Report 34, Cold Spring Harbor, pp. 271–289.

Ruyter, B., O. Andersen, A. Dehli, A.-K. Östlund, T. GjÝen & M.S. Thomassen (1997) Peroxisome proliferator activated receptors in Atlantic salmon (*Salmo salar*): effects on PPAR transcription and acyl-Co-A oxidase activity in hepatocytes by peroxisome proliferators and fatty acids. *Biochimica et Biophysica Acta*, **1348**, 331–338.

Samali, A. & T.G. Cotter (1996) Heat shock proteins increase resistance to apoptosis. *Experimental Cell Research*, **223** (1), 163–170.

Sanders, B.M. & L.S. Martin (1993) Stress proteins as biomarkers of contaminant exposure in archived environmental samples. *Science of the Total Environment*, **139/140**, 459–470.

Santella, R.M. & F.P. Perera (1994) Molecular epidemologic approaches in environmental carcinogenesis. In: (Ed. Garte, S.J.) *Molecular Environmental Biology*. Lewis Publishers, Boca Raton, pp. 153–176.

Sato, M. (1991) Metallothionein synthesis in rats exposed to radical generating agents. In: (Eds. Klaassen, C.D. & K.T. Suzuki) *Metallothionein in biology and medicine*. CRC Press, Boca Raton, Florida, pp. 221–235.

Scarano, S.C., E.J. Calabrese, P.T. Kostecki, L.A. Baldwin & D.A. Leonard (1994) Evaluation of a rodent peroxisome proliferator in two species of freshwater fish: a rainbow trout (*Oncorhynchus mykiss*) and Japanes medaka (*Oryzias latipes*). *Ecotoxicology and Environmental Safety*, **29**, 13–19.

Scheuhammer, A.M. & M.G. Cherian (1986) Quantification of metallothioneins by a silver-saturation method. *Toxicology and Applied Pharmacology*, **82**, 417–425.

Schiewe, M.H., D.D. Weber, M.S. Myers, F.J. Jacques, W.L. Reichert, C.A. Krone, D.C. Malins, B.B. McCain, S.-L. Chan & U. Varanasi (1991) Induction of foci cellular alteration and other hepatic lesions in English sole (*Parophrys vetulus*) exposed to an extract of an urban marine sediment. *Canadian Journal of Fisheries and Aquatic Sciences*, **48**, 1750–1760.

Schlenk, D. & C.D. Rice (1998) Effect of zinc and cadmium treatment on hydrogen peroxide-induced mortality and expression of glutathione and metallothionein in a teleost hepatoma cell line. *Aquatic Toxicology*, **43**, 121–129.

Schroeder, J.J. & R.J. Cousins (1990) Interleukin 6 regulates metallothionein gene expression and zinc metabolism in hepatocyte monolayer cultures. *Proceedings of the National Academy of Sciences of the USA*, **87**, 3137–3141.

Scudiero, R., P.P. De Prisco, L. Camardella, R. D'Avino, G. Di Prisco & E. Parisi (1992) Apparent deficiency of metallothionein in the liver of the Antarctic icefish *Chionodraco hamatus*. Identification and isolation of a zinc-containing protein unlike metallothionein. *Comparative Biochemistry and Physiology*, **103B**, 201–207.

Scudiero, R., V. Carginale, M. Riggio, C. Capasso, A. Capasso, P. Kille, G. diPrisco & E. Parisi (1997) Difference in hepatic metallothionein content in Antarctic red-blooded and haemoglobinless fish: Undetectable metallothionein levels in haemoglobinless fish is accompanied by accumulation of untranslated metallothionein mRNA. *Biochemical Journal*, **322**, 207–211.

Sharova, N.P., E.D. Eliseeva & V.S. Mikhailov (1994) Presence of zeta-type DNA-polymerase in eggs of the teleost fish *Misgurnus fossilis* L. *Biochemistry Moscow*, **59**, 799–805.

Shimas, A. & A. Shimada (1994) The Japanese Medaka, *Oryzias latipes*, as a new model organism for studying environmental germ-line mutagenesis. *Environmental Health Perspectives*, **102**, 33–36.

Shugart, L. & C. Theodorakis (1994) Environmental genotoxicity: probing the underlying mechanisms. *Environmental Health Perspectives*, **102**, 13–18.

Shugart, L., J.F. McCarthy & R.S. Halbrook (1992) Biological markers of environmental and ecological contamination: an overview. *Risk Analysis*, **12**, 353–360.

Sies, H. & H. de Groot (1992) Role of reactive oxygen species in cell toxicity. *Toxicology Letters*, **64–65** (1), 547–551.

Sikka, H.C., J.P. Rutkowski, C. Kandaswami, S. Kumar, K. Earley & R.C. Gupta (1990) Formation and persistence of DNA adducts in the liver of brown bullheads exposed to benzo[a]pyrene. *Cancer letters*, **49**, 81–87.

Simpson, A.E.C.M. (1997) The cytochrome P450 4 (CYP4) family. *General Pharmacology*, **28**, 351–359.

Singh, I. (1997) Biochemistry of peroxisomes in health and disease. *Molecular and Cellular Biochemistry*, **167**, 1–29.

Singh, K.P. & D. Roy (1999) Detection of mutation(s) or polymorphic loci in the genome of experimental animal and human cancer tissues by RAPD/AP-PCR depend on DNA polymerase. *International Journal of Oncology*, **14**, 753–758.

Singh, A.K., G.S. Dhaunsi, K. Asayama, J.K. Orak & I. Singh (1994) Demonstration of glutathione peroxidase in rat liver peroxisomes and its intraorganellar distribution. *Archives of Biochemistry and Biophysics*, **315**, 331–338.

Sipes, I.G. & A.J. Gandolfi (1991) Biotransformation of toxicants. In: (eds Amdur, M.O., J. Doull & C.D. Klaassen) *Cassaret and Doull's Toxicology. The Basic Science of Poisons*. Pergamon Press, New York, pp. 88–126.

Sloss, B.L., M.A. Romano & R.V. Anderson (1998) Pollution-tolerant allele in fingernail clams. *Archives of Environmental Contamination and Toxicology*, **35**, 302–308.

Smith, R.M. & C.F. Cole (1973) Effects of egg concentrations of DDT and Dieldrin on development in winter flounder (*Pseudopleuronectes americanus*). *Journal of the Fisheries Research Board of Canada*, **30**, 1894–1898.

Sohal, R.S. & U.T. Brunk (1990) Lipofuscin as an indicator of oxidative stress and aging. In: (Ed.: Porta, E.A.) *Lipofuscin and ceroid pigments*. Plenum Press, New York, pp. 17–29.

Solangi, M.A. & R.M. Overstreet (1982) Histopathological changes in two estuarine fishes, *Menidia beryllina* (Cope) and *Trinectes maculatus* (Bloch and Schneider), exposed to crude oil and its water-soluble fractions. *Journal of Fish Diseases*, **5**, 13–35.

Soto, M. & I. Marigómez (1997) Metal bioavailability assessment in 'mussel-watch' programmes by automated image analysis of autometallographical black silver deposits (BSD) in digestive cell lysosomes. *Marine Ecology Progress Series*, **156**, 141–150.

Soto, M., M.P. Cajaraville & I. Marigómez (1996) Tissue and cell distribution of copper, zinc and cadmium in the mussel *Mytilus galloprovincialis* determined by autometallography. *Tissue & Cell*, **28**, 557–568.

Srivastava, S.K., N.H. Ansari, S. Liu, A. Izban, B. Das, G. Szabo & A. Bhatnagar (1989) The effect of oxidants on biomembranes and cellular metabolism. *Molecular and Cellular Biochemistry*, **91**, 149–157.

Stegeman, J.J. & M.H. Hahn (1994) Biochemistry and molecular biology of monooxygenases: current perspectives on forms, functions, and regulation of cytochrome P450 in aquatic species. In: (eds Malins, D.C. & G.K. Ostrander) *Aquatic Toxicology: Molecular, Biochemical and Cellular Perspectives*. Lewis Publishers, Boca Raton, Florida.

Stegeman, J.J. & J.J. Lech (1991) Cytochrome P-450 monooxygenase systems in aquatic species: carcinogen metabolism and biomarkers for carcinogen and pollutant exposure. *Environmental Health Perspectives*, **90**, 101–109.

Stein, J.E., W.L. Reichert & U. Varanasi (1994) Molecular epizootiology: assessment of exposure to genotoxic compounds in teleosts. *Environmental Health Perspectives*, **102** (Suppl. 12), 19–23.

Stott, W.T. (1988) Chemically induced proliferation of peroxisomes: implications for risk assessment. *Regulatory Toxicology and Pharmacology*, **8**, 125–159.

Sunderman, F.W.J. (1986) Metals and lipid peroxidation. *Acta Pharmacology and Toxicology*, **59**, 248–255.

Takagi, A., K. Sai, T. Umemura, R. Hasegawa & Y. Kurokawa (1990) Significant increase of 8-hydroxydeoxyguanosine in liver DNA of rats following short-term exposure to the peroxisome proliferators di(2-thylhexyl)phthalate and di(2-ethylhexyl)adipate. *Japanese Journal of Cancer Research*, **81**, 213–215.

Thornally, P.J. & M. Vasak (1985) Possible role for metallothionein in protection against radiation-induced oxidative stress. Kinetics and mechanism of its reaction with superoxide and hydroxyl radicals. *Biochimica et Biophysica Acta*, **827**, 36–44.

Troxel, C.M., A.P. Reddy, P.E. O'Neal, J.D. Hendricks & G.S. Bailey (1997) *In vivo* aflatoxins B1 metabolism and hepatic DNA adductions in zebrafish (*Dania rerio*). *Toxicology and Applied Pharmacology*, **143**, 213–220.

Tucker, J.D. & Preston R.J. (1996) Chromosome aberrations, micronuclei, aneuploidy, sister chromatid exchanges, and cancer risk assessment. *Mutation Research*, **365**, 147–159.

Turelli, M. (1984) Heritable genetic variation via mutation-selection balance: Lerch's zeta meets the abdominal bristle. *Theoretical Population Biology*, **25**, 138–193.

Turelli, M. (1986) Gaussian versus non-Guassian genetic analysis of polygenic mutation-selection balance. In: (eds Karlin, S. & E. Nevo) *Evolutionary processes and theory*. Academic Press, New York, pp. 607–628.

Varanasi, U., M. Nishimoto, W.L. Reichert & J.E. Stein (1982) Metabolism and subsequent covalent binding of benzo[a]pyrene to macromolecules in gonads and liver of ripe English sole (*Parophrys vetulus*). *Xenobiotica*., **12** (7), 417–425.

Varanasi, U., M. Nishimoto, W.L. Reichert & B.T. Leeberhart (1986) Comparative Metabolism of Benzo[a]pyrene and covalent binding to hepatic DNA in English sole, starry flounder and rat. *Cancer Research*, **46** (8), 3817–3824.

Venier, P., S. Maron & S. Canova (1997) Detection of micronuclei in gill cells and haemocytes of mussels exposed to benzo[a]pyrene. *Mutation Research*, **390**, 33–44.

Viarengo, A. (1989) Heavy metals in marine invertebrates: mechanisms of regulation and toxicity at the cellular level. *CRC Criical Reviews in Aquatic Sciences*, **1**, 295–317.

Viarengo, A., E. Ponzano, F. Dondero & R. Fabbri (1997) A simple spectrophotometric method for metallothionein evaluation in marine organisms: an application to Mediterranean and Antarctic molluscs. *Marine Environmental Research*, **44**, 69–84.

Vogelstein, B. & K.W. Kinzler (1993) The multistep nature of cancer. *Trends in Genetics*, **9**, 138–141.

von Westernhagen, H., V. Dethlefsen, P. Cameron, J. Berg & G. Fürstenberg (1988) Developmental defects in pelagic fish embryos from the western Baltic. *Helgoländer. Meeresun.*, **42**, 13–36.

Vral, A., F. Verhaegen, H. Thierens & L. de Ridder (1994) The *in vitro* cytokinesis-block micronucleus assay: a detailed description of an improved silde preparation technique for the automated detection of micronuclei in human lymphocytes. *Mutagenesis*, **9** (5), 439–443.

Vukmirovic, M., N. Bihari, R.K. Zhan & W.E.G. Müller (1994) DNA damage in marine mussel Mytilus galloprovincialis as a biomarker of environmental contamination. *Marine Ecology Progress Series*, **109**, 165–171.

Weinstein, J.B. (1988) The origins of human cancer: molecular mechanisms of carcinogensis and their implications for cancer prevention. *Cancer Research*, **48**, 4135–4143.

Weis, J.S. & P. Weis (1989) Effects of environmental pollutants on early fish development. *CRC Critical Reviews in Aquatic Sciences*, **1**, 45–73.

Weisburger, J.H. & G.M. Williams (1991) Critical effective methods to detect genotoxic carcinogens and neoplasm-promoting agents. *Environmental Health Perspectives*, **90**, 121–126.

Wientjes, F.B. & A.W. Segal (1995) NADPH oxidase and the respiratory burst. *Sem. Cell Biol.*, **6**, 357–365.

Williams, G.M. & J.H. Weisburger (1991) Chemical carcinogenesis. In: (eds Amdur, M.O., J. Doull & C.D. Klaassen) *Casarett and Doull's Toxicology. The Basic Science of Poisons.* 4th ed. Pergamon Press, New York, pp. 127–200.

Wine, R.N., L.-H. Li, L.H. Barnes, D.K. Gulati & R.E. Chapin (1997) Reproductive toxicity of di-n-butylphthalate in a continuous breeding protocol in Sprague-Dawley rats. *Environmental Health Perspectives*, **105**, 102–107.

Winston, G.W., M.N. Moore, I. Straatsburg & M.A. Kirchin (1991) Decreased stability of digestive gland lysosomes from the common mussel *Mytilus edulis* L. by *in vitro* generation of oxygen-free radicals. *Archives of Environmental Contamination and Toxicology*, **21**, 401–408.

Wirgin, I. & J.R. Waldman (1998) Altered gene expression and genetic damage in North American fish populations. *Mutation Research-Fundamental and Molecular Mechanisms of Mutagenesis*, **399** (2), 193–219.

Wirgin, I.I., M. D'Amore, C. Grunwald, A. Goldman & S.J. Garte (1990) Genetic diversity at an oncogene locus and in mitochondrial DNA between populations of cancer-prone Atlantic tomcod. *Biochemical Genetics*, **28**, 459.

Wolfe, D.A., R.C. Clark, C.A. Foster, J.W. Hawkes & W.D. MacLeod (1981) Hydrocarbon accumulation and histopathology in bivalve molluscs transplanted to the Baie de Morlaix and the Rade de Brest. In: *Amoco Cadiz, fates and effects of the oil spill. Proc. Int. Symp.*, Centre Oceanologique de Bretagne, Brest, pp. 599–616.

Wolff, S.B., A. Garner & R.T. Dean (1986) Free radicals, lipids and protein degradation. *TIBS*, **11**, 27–31.

Yang, J.-H., P.T. Kostecki, E.J. Calabrese & L.A. Baldwin (1990) Induction of peroxisome proliferation in rainbow trout exposed to ciprofibrate. *Toxicology and Applied Pharmacology*, **104**, 476–482.

Zafarullah, M., P.-E. Olsson & L. Gedamu (1989) Rainbow trout metallothionein gene structure and regulation. In: (ed. Anonymous) *Oxford Survey of Eukaryotic Genes*, Oxford, pp. 111–143.

Zakrzewski, S.F. (ed.) (1991) *Principles of Environmental Toxicology*. America Chemical Society Books, Washington.

Zhang, Q. & T.R. Tiersch (1998a) Identification and analysis of weak linear banding patterns of fish chromosomes with a computer-based densitometric method. *Biotechniques*, **24**, 996–997.

Zhang, Q. & T.R. Tiersch (1998b) Standardization of the channel catfish karyotype with localization of constitutive heterochromatin and restriction enzyme banding. *Transactions of the American Fisheries Society*, **127**, 551–559.

Zhang, Q., R.K. Cooper & T.R. Tiersch (1997) Detection of a single-locus gene on channel catfish chromosomes by In-Situ Polymerase Chain Reaction. *Comparative Biochemistry and Physiology B.*, **118** (4), 793–796.

Zhang, Q., W.R. Wolters & T.R. Tiersch (1998) Replication banding and sister-chromatid exchange of chromosomes of channel catfish (Ictalurus punctatus). *Journal of Heredity*, **89** (4), 348–353.

Zhang, Q., R.K. Cooper & T.R. Tiersch (1999) Detection by In Situ Polymerase Chain Reaction of a channel catfish gene within cells and nuclei. *App. Immunohistochem. Mol. Morphol.*, **7** (1), 66–72.

Zwacka, R.M., A. Reuier, E. Plaff, J. Moll, K. Gorgas, M. Karasawa & H. Weiher (1994) The glomerulosclerosis gene MPV17 encodes a peroxisomal protein producing reactive oxygen species. *EMBO J.*, **13**, 5129–5134.

Chapter 3

Molecular/Cellular Processes and the Physiological Response to Pollution

A.J. Lawrence, A. Arukwe, M. Moore, M. Sayer and J. Thain

3.1 Induction of specific proteins

As seen in Chapter 2, protein mediated responses play an important role in the protection of organisms exposed to a wide variety of chemical or physical stressors. In addition, it is possible to demonstrate a link between the induction of these proteins and increased protein degradation and turnover. Protein turnover may be linked with lysosome function and have important physiological consequences on the energy balance and physiology of an organism. Evidence for these links is presented here.

3.1.1 Phase I and II detoxification enzymes

Biotransformation or metabolism of lipophilic chemicals to more water soluble compounds is a prerequisite for detoxification and excretion (Goksøyr & Förlin, 1992). In addition, certain steps in the biotransformation pathway are responsible for the activation of foreign compounds to the reactive intermediates that ultimately result in toxicity, carcinogenicity and other adverse effects (Guengerich, 1987; Nebert & Gonzalez, 1987; Varanasi, 1989; and Chapter 2). Biotransformation is divided into phase I and phase II according to the terminology of Williams (1974).

The cytochrome P450 (CYP) monooxygenase system, participates in the phase I (usually oxidative and functionalisation step) biotransformation process. It is also engaged in critical physiological functions such as steroid hormone synthesis and inactivation, metabolism of fatty acids (Fitzpatrick & Murphy, 1989) and of prostaglandins (Zimniak & Waxman, 1993) among other functions, making interactions between foreign chemicals and physiological processes possible.

In phase II (conjugation and detoxification), larger endogenous groups are conjugated to the activated (oxygenated) xenobiotic with the aid of different families of transferase enzymes such as UDP glucuronosyltransferase (UDPGT) and glutathione S-transferase (GST) (George, 1994), thereby transforming a lipophilic xenobiotic into a polar and water-soluble end-product which can be excreted from the organism through bile or urine or over the gill.

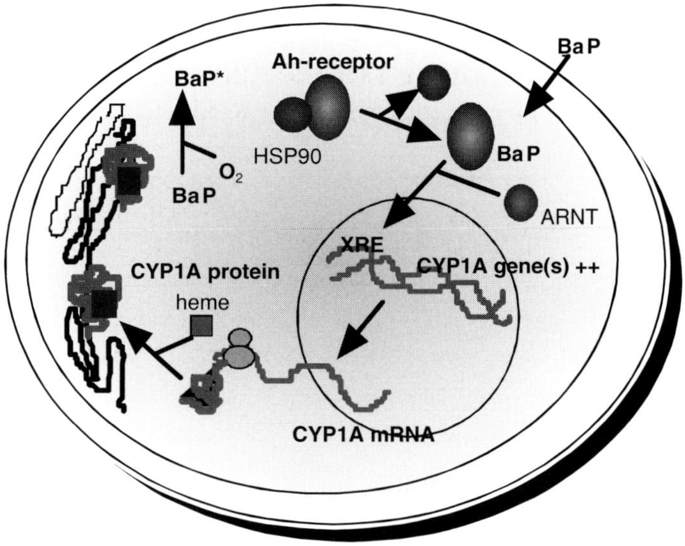

Fig. 3.1 The induction of CYP1A. This involves the binding of a planar aromatic ligand (e.g. TCDD) to the cytosolic AhR, translocation of the complex (including the AhR nuclear translocator, ARNT protein) into the nucleus, and activation of the CYP1A gene(s) by binding upstream to the xenobiotic responsive elements (XREs). The CYP1A mRNA is translated to protein in the ribosomal machinery, binds heme, and is inserted into the membrane of the endoplasmic reticulum, where it performs a monooxygenase activity on a xenobiotic substrate BaP (Benzo[a]pyrene. Modified from Goksøyr, 1995).

Some of the isozymes in the P450 gene superfamily are constitutively expressed in the cell. Other, inducible isozymes are expressed only after stimulation by specific hormonal or chemical compounds. Many chemically different compounds are known to induce *de novo* synthesis of cytochrome P450 (Nebert & Gonzalez, 1987; Nebert *et al.*, 1989). Inducers of the P450 system are classically divided into polyaromatic hydrocarbon (PAH)-type and phenobarbitol (PB)-type inducers. However, it has become evident that many other compounds induce specific patterns of cytochrome P450 isozymes. Today, inducers are classified according to the family or subfamily of P450 genes that they activate.

Generally, the induction response is a process by which a chemical stimulates the rate of gene transcription, resulting in increased levels of messenger RNA and new synthesis of cytochrome P450 protein. Subsequent processing involves heme insertion and folding (post translational modification) yielding the catalytically active enzyme (e.g. CYP1A; Fig. 3.1).

In fish, 3-methylcholantrene (3-MC), polychlorinated biphenyl (PCB) mixtures (Aroclor 1254 and Clophen A50) and β-naphthoflavone (BNF) are known to induce hepatic and extra-hepatic UDPGT synthesis (Kleinow *et al.*, 1987; Pesonen *et al.*, 1987; Clarke *et al.*, 1992; Gadagbui *et al.*, 1996). The alkylphenolic xenoestrogen, 4-nonylphenol (NP), has been shown to increase hepatic UDPGT activity by 20% in juvenile Atlantic salmon (*Salmo salar*) at 1 mg NP/kg fish. At higher doses, apparent gradual decreases (albeit non-significant) in mean UDPGT activity were observed (Arukwe *et al.*, 1997a).

In conjunction with CYP1A induction, the effects of inducing agents on total GST activity towards 1-chloro-2,-4-dinitrobenzene (CDNB) in fish liver have been reported in several studies (Sinclair & Eales, 1972; Chatterjee & Bhattacharya, 1984; George & Young, 1986; Goksøyr *et al.*, 1987; James, 1988; Van Veld *et al.*, 1990; Zhang *et al.*, 1990; Leaver *et al.*, 1992; Gadagbui & Goksøyr, 1996). However, UDPGT and GST are also enzymes of multigene families, and comparatively less is known about their function and regulation in fish species (George, 1994).

In addition to exposure to certain types of environmental pollutants, several other biotic and abiotic factors are known to influence the cytochrome P450 monooxygenase system in fish. These include sex, reproductive status and steroid levels (Förlin & Hansson, 1982; Stegeman *et al.*, 1982; Andersson, 1990; Förlin & Haux, 1990; Larsen *et al.*, 1992; Arukwe & Goksøyr, 1997), changes in season and temperature (Lindström-Seppä *et al.*, 1985; Snegaroff & Bach, 1990; Lange *et al.*, 1994; Sleiderink *et al.*, 1995).

3.1.2 Multidrug resistance protein

Multiple resistance is a phenomenon representing a complex group of cellular processes that are of importance in toxicology and oncology. Several mechanisms can account for reduced xenobiotic toxicity observed in various organisms. These include impaired uptake, sequestration into a non-target compartment, target alteration, biotransformation and enhanced excretion. Among these mechanisms, the importance of enhanced xenobiotic exportation has been recognised as contributing significantly to antibiotic and drug resistance in organisms from microbes to man. Xenobiotic expulsion, mediated by membrane-associated drug efflux pumps, can protect cells from a range of toxic compounds and, therefore, may confer single-step multixenobiotic resistance (MXR) (Higgins, 1992).

One of the most studied multiple resistance mechanisms is known as multidrug resistance (MDR). First described in mammalian cancer cell lines, this mechanism is related to the expression of a membrane permeability glycoprotein (Pgp) that confers the ability to lower the intracellular concentration of many different structurally and functionally unrelated toxic compounds below their toxic level (Gottesman & Pastan, 1993). This phenomenon is a major problem in cancer chemotherapy and tumours found to express the P-glycoprotein have been shown to have a poor prognosis (Chan *et al.*, 1990). Nevertheless, multiple resistance is not restricted to cancer cell lines. A related phenomenon occurs in tissues of a wide range of natural species in order to prevent xenobiotic accumulation by transporting toxic xenobiotics or endogenous metabolites out of the cell (Thiebaut *et al.*, 1987; Ouellette & Borst, 1991; Wu *et al.*, 1991).

Marine organisms possess one or several proteins related to this transport system. Studies using radiolabelled, photolabelled and fluorescent compounds have shown that cells expressing this protein share some similar pharmacological behaviour with MDR-positive cancer cells (Kurelec, 1992; Holland-Toomey & Epel, 1993; Cornwall *et al.*, 1995). The proposed role for this MXR mechanism in these marine animals is to serve as a defence system against environmental xenobiotics (Kurelec, 1992; Kurelec *et al.*, 1996). In accordance with this hypothesis, some environmental xenobiotics, mainly hydrophobic pesticides, have been reported to interact with the mussel MXR-protein (Cornwall *et al.*, 1995; Galgani *et al.*, 1996) and differential expression levels of the MXR-protein have been found in mussels living in polluted and unpolluted waters (Minier *et al.*, 1993).

The MDR phenomenon occurs in mussels and other invertebrates. The process is associated with a specific glycoprotein which is induced by exposure to the *Vinca* alkaloid vincristine, as well as by exposure to complex mixtures of environmental contaminants (Minier & Moore, 1996; Smital & Kurelec, 1998). Consequently, MXR should be considered in the larger biological context of detoxication and protection processes in mussels and other marine invertebrates. These include the cytochrome P450 (CYP) detoxication enzymes, glutathione and glutathione-conjugating enzymes, other plasma-membrane drug efflux pumps, as well as the lysosomal accumulation of a structurally diverse range of drugs and xenobiotics (Moore & Willows, 1998). Any link between MDR/MXR induction and pollutant-induced pathologies remains as yet unknown, although MDR-glycoprotein is increasingly important as a biomarker in clinical oncology. MDR/MXR does appear to have potential merit as a diagnostic biomarker of organic micropollutant exposure in other marine species.

3.1.3 Stress proteins/chaperonins, metallothioneins

When cells experience unfavourable conditions, proteins become denatured, and more stress proteins (chaperonins) are synthesised to help with cellular repair and protection (Chapter 2). The proteins synthesised under stress conditions are highly conserved, and may be present at detectable levels in unstressed cells. Indeed, 'stress proteins' are essential for normal cellular homeostasis, and are known to promote thermotolerance.

The way in which stress proteins confer tolerance to extreme environments is directly relevant to understanding the physiology and ecology of marine organisms. For example, Smerdon *et al.* (1995) investigated the relationships between stress protein accumulation, natural seasonal changes in environmental temperature, and thermotolerance in the blue mussel (*Mytilus edulis*). Using Western analysis and a monoclonal antibody, they developed a protocol that enabled the simultaneous detection of four isoforms within the 70-kDa family of stress proteins. This family is the most abundant and conserved subset of eukaryotic stress proteins, acting as molecular chaparones that direct the folding, assembly and degradation of cellular proteins. They showed significant seasonal variation in endogenous levels of the 70, 72 and 78 kDa isoforms within gill tissue of mussels, which each correlated positively with local seasonal changes in both air and sea temperatures. In addition, seasonal changes in sea temperature and the abundances of 70, 72 and 78 kDa isoforms each correlated with thermotolerance measured experimentally as the time to 50% mortality at 28.5°C.

The high levels of stress-70 proteins detected during the summer months suggest that thermal stress in the natural environment was sufficient to cause protein damage in *M. edulis*. These findings also suggest that seasonally increased levels of stress-70 protein confer enhanced thermotolerance in mussels, indicating that stress proteins may reflect or influence survival and distribution limits of eurythermal ectotherms.

The effects of contaminants on stress proteins has mostly focused on metallothioneins (Pedersen & Lundebye, 1996). The impact of organic micropollutants on other stress proteins has not as yet shown any readily usable biomarkers although Lawrence & Nicholson (1998) have used Rat HSP70 monoclonal antibody to demonstrate sublethal induction of stress-70 protein in response to exposure to chlorine by-products.

3.1.4 Antioxidant enzymes

Partial reduction of molecular oxygen results in the formation of potentially toxic reactive species (ROS), such as the super oxide anion radical, hydrogen peroxide and hydroxyl radical. Such radicals are produced continuously in biological systems as by-products of normal metabolism: they are generally detoxified by antioxidant defence processes (Livingstone *et al.*, 1992; see Chapter 2). Lemaire and Livingstone (1997) have demonstrated that a widespread potential for oxyradical production exists in fish liver via redox cycling of AH-quinones. The significance of enzymes such as DT-diaphorase, which detoxifies AH-quinones in mammals, is not clear since there is evidence of this enzyme leading to enhanced radical production in fish (Lemaire *et al.*, 1996).

3.2 Protein degradation

3.2.1 Direct effects on protein catabolism

When marine molluscs such as mussels are exposed to contaminant chemicals, the lysosomes in the digestive gland epithelial cells show fairly rapid and characteristic pathological alterations (Lowe, 1988; Moore, 1988). These include swelling of the digestive cell lysosomes, accumulation of unsaturated neutral lipid in the lysosomes, increased fragility of the lysosomal membrane, and excessive build-up of lipofuscin in the lysosomal compartment. These changes are accompanied by atrophy of the digestive epithelium, apparently involving augmented autophagic processes, although there is also evidence of increased cell deletion (probably analogous to apoptosis in mammals) and the relationship between the two processes, if any, is unclear (Lowe, 1988; Pipe & Moore, 1986). For example, it is not known whether the autophagic-type changes predispose the cells to deletion. Linked biochemical and cytochemical investigations have demonstrated that increased fragility of the lysosomes, induced by phenanthrene, corresponds directly with increased catabolism of cytosolic proteins (Moore & Viarengo, 1987).

Experimental studies have clearly demonstrated that the lysosomal alterations described above can be induced by single toxicants such as copper and polycyclic aromatic hydrocarbons (Moore *et al.*, 1984). At first sight this finding is perhaps surprising given that many thousands of individual chemicals are often present in a contaminated situation. However, it would appear that the pattern of lysosomal response observed is essentially very generalised and can be induced by non-chemical stressors such as hypoxia, hypothermia, osmotic shock and dietary depletion (Moore, 1985). Thus it would appear that many adverse conditions are capable of inducing autophagic-type changes. This non-specificity of the lysosomal reactions is therefore of value as a general indicator of deterioration in the health of the animal. It does not, however, identify the nature of the particular contaminants that are causing cell injury. More specific information about the causative agents can be obtained through the use of tests for lysosomal accumulation of sulphydryl-rich metal-binding proteins (e.g. metallothioneins) which are induced by exposure to particular metals, and cytochrome P450 reductase which is induced by many lipophilic xenobiotics (Viarengo *et al.*, 1985; Moore,

1988). Considered as a package, the use of cytochemical tests as subcellular pathological probes can provide relatively specific information.

Autophagic (self-eating) processes play an important role in the degradation of intracellular proteins, particularly under conditions of stress or induced cell injury (Moore, 1990). There is abundant evidence of stress-induced autophagy in animal cells, including those of invertebrates, fish and mammals. In particular, the exposure of animals to pollutant chemicals (both metals and organic xenobiotics) is known to induce cell injury and ensuing pathological change, which frequently involves autophagy and lysosomal alterations (Moore, 1990; Lowe *et al.*, 1995a,b). The latter include increased fragility of the lysosomal membrane, enlargement and in some cases lipidosis. Enhanced catabolism of cytosolic proteins has also been indicated in some experimental studies. However, there is still considerable debate concerning the relative importance of lysosomal as opposed to non-lysosomal pathways of intracellular protein catabolism. A great variety of xenobiotics are taken up by lysosomes (Rashid *et al.*, 1991), and more recent work has indicated that lysosomotropic chemicals can enhance traffic of intracellular proteins to the degradative lysosomal compartment (Moore *et al.*, 1996a).

3.2.2 Radical damage to proteins and production of protein adducts

Aquatic organisms are sensitive to oxidative stress associated with exposure to environmental contaminants (Kirchin *et al.*, 1992). Pre-exposure to copper causes elevated levels of protein carbonyl groups in the digestive glands of mussels. There are indications that this may occur particularly in the lysosomes where there is active production of oxyradicals (Winston *et al.*, 1991). This area is certainly one area that deserves further study in order to determine the consequences of oxidative damage to proteins for cells.

3.2.3 Lysosomal damage in relation to protein turnover

Studies of lysosomal membrane fragility have been carried out in fish and invertebrate species. Exposure to a variety of contaminant effluents such as sewage sludge, pulp-mill waste, oil spillages and mixed wastes from industry have all been found to increase the fragility of fish hepatocyte lysosomes and molluscan digestive cell lysosomes (Moore, 1985, 1988, 1990; Köhler, 1989; Köhler *et al.*, 1992; Lowe *et al.*, 1992, 1995a,b; Moore *et al.*, 1996a). In general, the reduction in lysosomal stability is accompanied by enlargement or swelling. Fatty change is also a frequent reaction to xenobiotics in the digestive cells, leading to apparent autophagic uptake of the unsaturated neutral lipid into the often already enlarged lysosomes (Moore, 1988).

In order to better understand both the metabolic basis and functional consequences of differences in whole-body protein turnover, procedures have been developed to study the component activities of different proteolytic pathways (Bayne & Hawkins, 1997). Traditionally, it has been thought that requirements for biosynthesis dominate energy expenditure. Nevertheless, among animals generally, a large component of about 30% of the empirical costs of protein deposition cannot be attributed to known synthetic processes, and it has been suggested that costs of protein turnover may contribute to the discrepancy. Bayne and Hawkins (1997) have shown that separate whole-body activities of the four main lysosomal proteases were collectively associated with as much as 73% of the variation in maintenance energy expenditure between individual *M. edulis*. These associations were positive for cathepsin B, cathepsin D and the aminoacyl peptidase *Lap-2*. Conversely, higher whole-body activity of cathepsin L was associated with lower maintenance energy expenditure, apparently because cathepsin L was most active in the main tissue of nutrient storage, thereby mobilising energy reserves and reducing the need for protein turnover in remaining tissues. These findings indicate profound physiological consequences of lysosomal proteolysis, and that consequences vary according to functional differences between separate proteolytic pathways. They also suggest that the relative balance between proteolytic pathways will prove a major determinant of growth efficiency and other performance traits (Bayne & Hawkins, 1997).

3.2.4 Stress pigment formation

The uptake and toxicity of organic micropollutants in aquatic organisms are governed by the physical chemical speciation of these contaminants. Since lipophilic pollutants are largely bound to particulate and colloidal organic carbon, it is probable that contaminant entry into cells is directly related to the extracellular and intracellular behaviour of particulates/colloids with adsorbed chemicals. The aim here is to consider the cellular mechanisms of accumulation of organic chemical micropollutants, with emphasis on bulk transport into cells, via endocytic uptake into membrane enclosed vesicles, of particulate organic carbon with sorbed contaminant ligands. In this context, lysosomal accumulation of toxic metals and organic xenobiotics is a well-documented cellular phenomenon, and it has been repeatedly demonstrated that induced lysosomal damage is also a significant factor in cell injury. Sequestration in lysosomes has also been postulated to have a protective role through the physical detoxication of pollutants. Physical chemical binding of ligands to lysosomal lipofuscin (generated by the interaction of oxyradicals and protein breakdown) is also considered in relation to pollutant storage capacity and thresholds for cell injury. It has been suggested that animals with highly developed cellular lysosomal systems are more tolerant of pollutants (Moore, 1990; Moore & Willows, 1998).

Stress pigment or lipofuscin is a characteristic complex macromolecular lipopigment found in lysosomes (Moore, 1990). Lipofuscin is produced by the action of lipid peroxides on intralysosomal peptides and proteins, which in turn are produced by the action of reactive oxygen species (ROS) on the intralysosomal unsaturated lipids (Moore & Willows, 1998).

Lipofuscin is characterised by repetitive conjugated Schiff bases on the molecule (Moore & Willows, 1998). These conjugated sites on the molecules, together with substituent groups on the peptide chain, will provide binding sites for free contaminant ligands within the lysosomal microenvironment. Such binding by lipofuscin will essentially provide a trapping mechanism which represents a detoxication and protection process. In lower organisms, such as invertebrates, cells can eject lipofuscin by exocytosis of residual bodies (tertiary lysosomes) (Moore, 1990). This released lipofuscin will become incorporated into faecal material or else will be lost into the urine if it is produced in kidney or pericardial gland epithelia (e.g. in molluscs) (Moore, 1990). For organisms which have only a very limited capability for metabolising the contaminant ligands, such as molluscs, this mechanism may be the primary pathway for detoxication and excretion (Moore & Willows, 1998).

3.2.5 Cellular pathology and repair processes

3.2.5.1 Cell injury and carcinogenesis

Tissue and organ structure is an integration of the many biochemical, cellular and physiological processes occurring within it, as well as any pathological disturbances to these processes (Hinton & Lauren, 1990; Moore *et al.*, 1994). Hence, histopathology provides a potentially powerful tool for the assessment of cell injury by environmental pollutants and in the prediction of higher level consequences of such injury. Extensive histopathological studies have been conducted on the impact of pollutants on fish and shellfish (see Moore *et al.*, 1994 and Chapter 4). A number of characteristic changes have been noted in the digestive glands of molluscs and the livers of fish (Moore *et al.*, 1994).

In fish livers the cellular changes preceding the formation of phenotypically altered foci (preneoplastic lesions) have been described (Köhler *et al.*, 1992). These involve changes in the lysosomes and endoplasmic reticulum of the hepatocytes, as well as phospholipidosis and other evidence of autophagy. The relationship of foci to neoplastic change in fish liver cannot at present be determined and requires further experimental investigation. It should be noted that neoplastic change in itself is of little significance in ecological terms unless the incidence in the population is extremely high. The value of pursuing the 'neoplastic pathway' is that this type of disease is probably indicative of exposure to carcinogenic organic micropollutants, although whether this occurs in embryos and larvae associated with the xenobiotic-rich surface microlayer or in juveniles and adults in contact with bottom sediment and contaminated prey organisms is still an open question (see Moore *et al.*, 1994). However, the findings that the prevalence of ras-oncoprotein positivity and foci of altered cells in the livers of adult dab are similarly distributed would support the hypothesis that the initial steps in the process are occurring in adult fish, as the involvement of ras-oncoprotein in the process of carcinogenesis is believed to occur at a very early stage (Moore *et al.*, 1994).

An integrated pathological approach is required in order to identify the processes involved in cellular changes leading to liver damage and tumour growth in fish. Such an approach is likely to generate effective indicators of the harmful changes that can be used as biomarker tests for impact assessment.

3.3 Physiological effects: whole body responses/regulation

3.3.1 Energetics and energy budgets

Cellular and organism energetics and energy budgets are clearly one of the mechanisms through which cellular responses to pollution can be linked to higher order impacts at physiological/reproductive and population levels. The energy that an organism gains from its food is appointed between various biological functions. When resources are abundant, the energy remaining to the organism, after excretion and metabolism, is available for growth and reproduction. At other times energy may be used to accommodate environmental stress thereby reducing that available to production. Examination of this allocation of energy to various internal compartments can give a detailed indication of the organism's energetic status.

3.3.1.1 Scope for growth

Simple energy budgets have been developed to characterise the allocation of energy between various compartments. These budgets are an account of all the energy gained, stored and lost by an individual animal. The overall equation for balance of energy is:

$$C = P + E + F + M + W$$

Energy consumed (C) equals energy stored in tissue growth or production (P) plus energy lost in excretion (E) and faeces (F) plus energy used in metabolism (M) and external work (W). Energy used in production is either as growth and repair (Pg) or gametes (Pr).

From this equation the energy assimilated (A) can be calculated as: $A = C - F$ and the energy used in metabolism $(M) = R$ (respiration) $+ E$.

Each of the parameters C, E, F, M and W can be determined experimentally, and consequently the scope for growth (SfG) can be calculated. Scope for growth is defined as the energy available for production (somatic or reproductive) and is given by:

$$\text{SfG or } P = A - M$$

Scope for growth has been used to assess the energetic cost of environmental stress, including pollution burden, directly. Stress engages homeostatic mechanisms in the organism which attempt to restore the equilibrium. In the case of pollution this homeostatic mechanism includes the induction of detoxication mechanisms involving the proteins described in section 3.1. The response has a metabolic cost to the organism and without an equivalent rise in energy assimilated, the SfG is reduced. Bayne (1989) has described how a number of measurable impacts combine to reduce SfG in *Mytilus edulis* and a modification of this is shown in Fig. 3.2.

The SfG of an organism under stress is determined based on the energy budget of an individual and has been used extensively with the molluscs. A reduction in scope for growth has been demonstrated in *Mytilus edulis* at tributyltin (TBT) concentrations above 4 ug l^{-1} (Widdows & Page, 1993), and in natural populations around the Sullom Voe oil terminal a consistent relationship was found between SfG and level of aromatic hydrocarbons in the tissue (Donkin & Widdows, 1986).

There is some evidence from a number of studies in the literature to demonstrate the link between a reduction in SfG and induction of stress homeostasis. For example, a correlation between reduced SfG and induction of HSP60 was determined in *Mytilus edulis* exposed to copper (Sanders *et al.*, 1991) and reduced SfG has been recorded in mussels which produce HSPs to protect them from chlorine residual oxidants (Lawrence & Nicholson, 1998).

SfG has been used to monitor changes in environmental quality along the North Sea coastline of the UK (Widdows *et al.*, 1995). SfG was found to decline in mussels from north to south. A large contribution towards the observed decline was caused by toxic polyaromatic hydrocarbons, polar organic compounds and TBT.

Less work and literature appears to be available on the use of SfG in fish. One study examined the toxic effect of waterborne nitrate on the energy budget of grass carp (*Ctenopharyngodon idella*) in which it was found that nitrate caused a reduction in

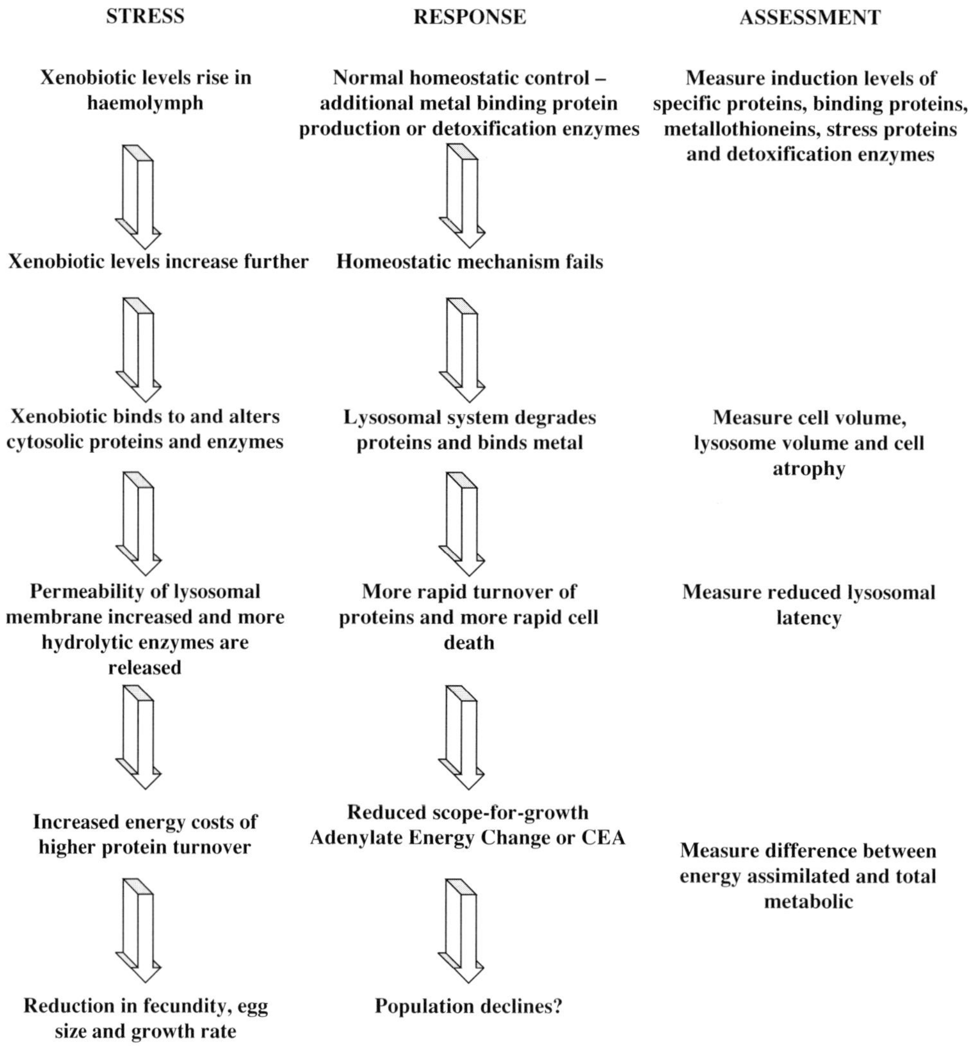

Fig. 3.2 The response of an organism to rise in levels of xenobiotics and the assessment of stress response. Modified from Bayne, 1989.

assimilation efficiency, respiration rate and scope for growth (Alcaraz & Espina, 1997). The lack of research relating xenobiotic effects on SfG in fish may be indicative of the intensive nature of the work or problems associated with experimental design.

Whilst SfG presents a clear mechanism that links subcellular responses to pollution to whole organism physiological parameters such as growth and reproduction, this should be accepted with some caution.

3.3.1.2 Adenylate energy charge

An alternative to SfG, which has been used to assess the effects of pollution and environmental stress on energy status, is the adenylate energy charge (AEC). This is defined as the amount of energy available to an organism from the adenylate pool (Atkinson, 1977). It is calculated from the measured concentrations of the three adenine nucleotides ATP, ADP and AMP which are integral to the energy metabolism of all organisms (Ivanovici, 1980) and is calculated from the formula:

$$\frac{\text{ATP} + 1/2\ \text{ADP}}{\text{ATP} + \text{ADP} + \text{AMP}}$$

In a review of the use of AEC, Ivanovici (1980) highlighted a number of benefits of this against other measures. These included a consistent reduction in AEC related to stressful conditions; a relationship between high AEC and high growth rates and the ability to reproduce; an inability to recover from stressful conditions at AEC below 0.5 and a lower variability in AEC compared to measures of each of the individual nucleotides.

Liver nucleotides and AEC have been used as measures of stress in rainbow trout (*Oncorhynchus mykiss*) subjected to a range of dissolved oxygen concentrations. These varied significantly and there was a reduction in AEC in hypoxic and hyperoxic fish (Caldwell & Hinshaw, 1994). A similar reduction in AEC in response to hypoxic conditions has been reported in the mussel *Mytilus galloprovincialis* (Isani *et al.*, 1997). Significant differences in AEC between more and less polluted sites have also been reported in the polychaete *Lanice conchilega* (Pires *et al.*, 1995) and scallop and sea urchin (Lukyanova, 1994). There is some evidence, using cultured human respiratory epithelium, that induction of HSPs, particularly HSP70, can confer protection against cytotoxicity by preserving the cellular energetics systems (Wong *et al.*, 1997). However, studies using cultured mammal cell lines have shown that the relationship between AEC and stress is not as consistent as first thought. For example, Chinese hamster ovary incubated with cytotoxic doses of copper-putrescine-pyridine showed reduced survival caused by oxidation and depletion of glutathione but AEC remained constant (Nagele, 1995). Similarly, reactive oxygen metabolites had no effect on AEC on carcinoma cell line Caco-2 (Baker *et al.*, 1995).

This inconsistency extends to pollution studies on marine invertebrates and fish. Sublethal cadmium caused no variation in ATP, ADP or the AEC in the shrimp *Palaemon serratus*. Only the LC50 concentration impaired energetic metabolism (Thebault *et al.*, 1996). Similarly, the red abalone (*Haliotis rufescens*) showed no change in AEC in response to PCP and sodium azide exposure (Shofer & Tjeerdema, 1998), although Asian sea bass (*Lates calcarifer*) exposed to nitrite maintained its AEC by producing ammonia via the degradation of AMP to IMP (Woo & Chiu, 1997).

3.3.1.3 Cellular energy allocation

Cellular energy allocation (CEA) has been developed as a biomarker technique to assess the effect of toxic stress on the energy budget of test organisms (De Coen & Janssen, 1997). This is based on short-term changes in energy reserves measured as total carbohydrate, protein and lipid content and energy consumption by electron transport activity. Using *Daphnia magna*, the ecological relevance of CEA was assessed by comparing the response to population parameters – intrinsic rate of natural increase (rm) and the mean total offspring per female. Results showed that the CEA-based LOAEC was mostly more sensitive than either population parameter although the response was toxicant specific. There was a significant linear relationship between CEA and the population level effects, demonstrating a linking of energy-based suborganism effect criteria with effects emerging at higher levels of organisation.

At the time of writing no study comparable to this has been performed with fish. However, temporary and short-term reductions in energy reserves (metabolic disorders) have been recorded in fish during and following episodic exposures to sublethal contaminants (Sancho *et al.*, 1998). In particular, heavy metals have been shown to exert a wide range of effects on fish metabolism (Soengas *et al.*, 1996). Some marine fish species subjected to rapid decreases in water temperature enter a hypometabolic state to resist the challenge (Sayer & Davenport, 1996). Whether this physiologically-driven behaviour would also be protective against contaminant exposure has not been tested. Muscle energy metabolism can sustain prolonged effects following shock contaminant exposure, with, in some cases, full recovery to pre-exposure levels never being attained (De Boeck *et al.*, 1997).

Clearly more studies are needed to examine the effects of xenobiotics on SfG, AEC and CEA. These measures can be useful although their use must be related to the fish's natural history. Work also needs to be undertaken to clarify some of the inconsistencies highlighted in the literature to date. The benefit of each of the methods is that a reduction in SfG or AEC can be correlated to the body burden of xenobiotic or induction of detoxification system, and in the case of CEA has been linked to population level responses. Cellular energetics and energy budgets, therefore, offer an important link between these and higher order effects.

3.3.2 Osmoregulation and ionoregulation

3.3.2.1 Ionoregulation

Marine teleost fishes maintain their internal body fluids at optimal concentrations through a process of, usually, hypo-osmoregulation, by continually drinking and continually excreting excess salts. By the nature of hypo-osmoregulation, the fish must actively pump salts from the body by energetically-expensive methods against concentration gradients. When a fish is then stressed through exposure to pollutants, two things can happen. If the pollutant exhibits a non-specific whole animal effect then the whole process of osmoregulation can be affected with the consequences described below in section 3.3.2.2. However, some pollutants can act in a more targeted inhibitory manner, which can disrupt specific physiological processes (e.g. Thaker *et al.*, 1996; Webb & Wood, 1998). In marine teleosts, the two most obvious ionoregulatory processes are the excretion of sodium and chloride ions. However, there are instances where osmolarity and levels of sodium and chloride ions remain constant in stressed fish, but disruption of the balance of potassium ions can disrupt the fish haematology (Alkindi *et al.*, 1996).

Whole-body ionic concentrations can be employed as indicators of sublethal pollution stress. For example, whole body sodium concentration is a frequently used indicator of whole-animal stress in freshwater fish (Sayer *et al.*, 1991a,b; Dennis & Bulger, 1995). Ionic disturbance can also be a reliable indicator of sublethal stress in marine species, although it is more apparent in dying fish (Sayer & Reader, 1996; LeRuyet *et al.*, 1998).

There is inferred evidence to suggest that slight changes in the ionic balance of marine fishes can be accompanied by metabolic and behavioural shifts (Sayer & Davenport, 1996; Sayer & Reader, 1996). If those changes in behaviour affect locomotory or reproductive performance, then there could be marked ecological consequences. However, the extent to which it is the ionic imbalance *per se* which causes the subsequent changes in behaviour is not tested. An influencing factor in any form of experiment designed to examine the links between ionic disturbance and ecological consequences must take into account interspecific differences in seasonality in the ability to resist contaminant challenge. In some cases fish become less resistant during the winter (Lemly, 1996), but they are more resistant in others (Sayer & Reader, 1996).

In severe cases, ionic imbalances have caused osmotic stresses resulting in secondary perturbations (such as blood fluid decreases and increases in blood cell volume), leading to enhanced mortality (Sayer *et al.*, 1991a,b; Webb & Wood, 1998). Sodium regulation can be used to predict the relative sensitivity of various life-stages and different species of aquatic fauna in acid sensitive situations (Havas & Advokaat, 1995). However, a similar predictive tool based on sublethal ionic disturbances does not, as yet, exist for the more complex marine ecosystems.

3.3.2.2 Osmoregulation

Nearly all marine teleosts hypo-osmoregulate in order to maintain internal fluid osmolarities at levels optimal for sustained life. This necessity can yield quantitative indications of physiological stress through relatively simple osmolarity analysis against time, and examining for any departure from the hypo-osmotic, usually towards the iso-osmotic. Significant and marked loss of hypo-osmoregulatory ability of some north temperate fish has been recorded in species subjected to decreases in seawater temperature and/or salinity (Provencher *et al.*, 1993; Sayer & Reader, 1996). However, this methodology does not appear to have been adopted as a way of detecting sublethal physiological stress during exposure to pollutants in marine circumstances in a similar manner to that adopted for freshwater fishes.

Stressors increase the permeability of the surface epithelia, including the gills, to water and ions, and thus induce systemic hydromineral disturbances (Wendelaar Bonga, 1997). However, seasonal variation in physiological responses to environmental challenges has been recorded for north temperate marine fishes (Dutil *et al.*, 1992; Sayer & Reader, 1996). The maintenance of hypo-osmotically regulating marine fish in iso-osmotic media during a period of contaminant exposure can have a protective effect (Wilson & Taylor, 1993).

3.3.2.3 Excretion/respiration

Elevated or depressed total ammonia nitrogen plasma concentrations can be indicative of sublethal stress, though only where ambient ammonia concentrations are the applied stressors (Knoph, 1995; Le Ruyet *et al.*, 1998). Loss of nitrogenous excretion regulation can be induced by sublethal concentrations of copper (De Boeck *et al.*, 1995). Regulation of nitrogenous excretion taken in association with measured oxygen consumption rates can be an effective indicator of sublethal stress in fish (De Boeck *et al.*, 1995).

Blood oxygen content has been demonstrated to be severely affected by acute sublethal exposures to contaminants (Alkindi *et al.*, 1996). Sublethal toxicity of some contaminants can result in significant decreases in oxygen consumption and transport caused predominantly by reduced haematological oxygen-carrying capacity (Reddy & Bashamohideen, 1995; Powell & Perry, 1997). Contaminant exposure can elicit changes in the respiratory behaviour of fish, quantified through measurements of ventilation and cough (gill purge) frequency (Atchison *et al.*, 1987).

However, various studies have provided evidence for either increased or decreased respiratory activity in a variety of organisms exposed to xenobiotics. For example, an increase in respiratory activity was noted in flounder (*Pseudopleuronectes americanus*) exposed to mercury chloride (Vernberg *et al.*, 1975). Striped bass (*Morone saxatilus*), on the other hand, showed slightly reduced oxygen consumption when exposed to the same pollutant for the same time period. Exposure of *Tilapia mossambica* to lindane showed a biphasic response with increased oxygen consumption initially but decreased consumption at the end (Basha *et al.*, 1984). A similar biphasic response to copper, benzo[a]pyrene and pentachlorophenol has been demonstrated in the invertebrate *Gammarus duebeni* (Lawrence & Poulter, 1996, 1998). In this instance, the response was found to be both concentration and time dependent.

Elevated plasma NH^+ and HCO_3^- concentrations during sublethal toxic exposure are indicative that some aspects of gill ion transport involved with nitrogenous waste excretion are vulnerable to disruption (Wilson & Taylor, 1993). Disruptions of this type have been facilitated by the inclusion of copper into sea water, although the response is partly mediated at higher-strength sea waters (Wilson & Taylor, 1993). In addition, fish gills have been used to quantify the ultrastructural effects of environmental stressors (Mallat *et al.*, 1995; see Chapter 4).

In testing for any contaminant effects the fact that respiration rates can vary with the magnitude of salinity variation, and that sensitivity can be size-dependent (Moser & Miller, 1994), should be taken into account.

3.3.3 *Effects on growth*

3.3.3.1 Genotypic dependant effects

One of the most pressing requirements both for effective conservation and for fisheries management is improved understanding of the functional value of genetic polymorphism, the evolutionary processes that determine genetic diversity, and the ecological processes that determine species diversity. Yet little has been understood of the processes by which genotype may confer consequences for fitness.

Bayne and Hawkins (1997) have shown how studies of protein metabolism may help to understand those processes among animals generally. To identify differences in the intensity of protein metabolism, the separate processes of protein synthesis and protein breakdown must be measured. Only then may the imbalance effecting either net protein gain or net protein loss be demonstrated, and protein turnover, that is defined as the continuous degradation and renewal or replacement of cellular proteins, be quantified. Protein turnover is essential for life, providing the metabolic flux that enables repair and cellular sanitation, regulation, development and adaptation (Hawkins, 1991; Hawkins & Hilbish, 1992). However, it is also energetically expensive, representing the major component of all energy required for maintenance processes (Hawkins *et al.*, 1989a). Whether comparing individuals, in response to selection, or in heterozygosity-associations, reduced whole-body protein turnover consistently underlies lower energy expenditure, with beneficial consequences that include higher growth efficiency and longer survival following the general inhibition of energy intake in response to environmental stressors (Hawkins *et al.*, 1987, 1989b; Hawkins, 1988, 1991; Hawkins & Bayne, 1991; Carter *et al.*, 1993a,b; McCarthy *et al.*, 1994; Hawkins & Day, 1996; Bayne & Hawkins, 1997).

Past work has established that advantages of multilocus heterozygosity and heterosis are associated with slower intensities with which proteins are renewed and replaced (= protein turnover) (Bayne & Hawkins, 1997). Slower turnover results in lower energy requirements and reduced metabolic sensitivity to environmental change, together representing the mechanistic basis for evolutionary consequences of genetic polymorphism. In order to determine the genetic and functional basis of differences in whole-body protein turnover, different proteolytic pathways have started to be resolved, searching for genetic polymorphism with a direct effect upon proteolysis, and assessing the metabolic and physiological consequences of those genetic influences in bivalve shellfish. Findings have confirmed the physiological importance of proteolytic enzymes under normal conditions of basal proteolysis, showing significant associated effects on energy flux that vary according to functional differences between separate pathways (Bayne & Hawkins, 1997). In particular, they have established that separate polymorphisms at loci coding for the two lysosomal aminopeptidases *Lap-1* and *Lap-2* have direct influences on protein metabolism, including associated influences on energy flux and animal condition. These findings strongly suggest that the energy requirements for protein turnover represent the functional basis for a growing body of evidence that the phenotypic effects of genetic polymorphism are greatest at loci coding for enzymes acting in protein catabolism and energy provision (Bayne & Hawkins, 1997).

3.3.3.2 Optimal strategies (age/size trade-offs)

In organisms, production is defined as the result of anabolic activities during a certain period of time, thus ensuring a constant renewal of molecular structures within cells and cells within tissues (Pascaud, 1989). Generally, overall production is relatively complex and depends on the organism studied. Nevertheless, ingested energy that is neither lost as faecal or excretory products, nor used for metabolism, is available for growth. Growth can take two forms: somatic growth or reproductive growth (Jobling, 1994). In fish, growth has usually been recorded in terms of weight gains, and it has often been assumed that an increase in body weight is synonymous with increase in energy gain; thus assuming that the composition of fish tissue is constant and that a change in weight will accurately reflect a change in the energy content of the body (Jobling, 1994). Apparently, a reduction in growth rate or decreased energetic commitment to reproduction may suggest that there is a decreased conversion of energy into somatic and reproductive tissue.

The optimal size-to-age at maturity depends on growth and mortality rates, which vary with environment. Therefore, organisms in spatial or temporal environments should develop adaptive phenotypic plasticity for these traits (Perrin & Rubin, 1990). The life history of many animals is divided into well-defined stages. Many fish species go through most of the following stages; egg, one or more larval stages, juvenile (sexually immature) and adult (sexually mature) stages. Understanding the factors that determine the timing of transition between life-history stages in fishes is crucial to an understanding of their demography, since behaviour of populations will be different for different timing mechanisms (Policansky, 1983). For example, if a transition is triggered by the attainment of a certain age, conditions unfavourable to growth will result in a population of smaller individuals. On the contrary, if the transition is triggered by size, conditions unfavourable to growth will delay the transition. However, if one of the stages involves dispersal, conditions unfavourable to growth will increase dispersal, decrease local recruitment, and perhaps gene flow. If the transition is triggered by both age and size, the effects of combinations unfavourable to growth will be different from those associated with purely size or age-triggered transition (Policansky, 1983). Theoretically, the onset of these stages is considered to be age determined (Leslie, 1945). This is because it is easier to treat ages than sizes, since age increases linearly with time while size need not.

Constraints relate a decision variable to a currency, and optimality models of life histories generally incorporate two kinds of constraints. The first is the direct relationship between fitness and the value of a trait. The second is the relationships between different traits of the same individual which result from varying the allocation of resources between the traits, in other words a trade-off. However, it is important to note at this point that life-history traits are frequently phenotypically plastic, i.e. a single genotype produces a range of phenotypes depending on the environment. Phenotypic variation may be continuous, in which case the relationship between phenotype and the environment for each genotype is called a reaction norm (Perrin & Rubin, 1990), or gradual environmental change may be accompanied by sudden switches between discrete phenotypes (polyphenism). Phenotypic plasticity may be irreversible, for instance when it involves developmental changes, or reversible, as in the case of clutch size variation in iteroparous species. Several biotic and abiotic environmental features may invoke phenotypic plasticity. For example, growth rates are often phenotypically plastic (Stearns & Koella, 1986).

In fish, the onset of sexual maturation is apparently influenced by both fish size and by its age depending on the condition. There is also great genetic variability in the age and size at maturation, probably due to allelic substitutions at a single locus (Leary *et al.*, 1984; Varnavskaya & Varnavsky, 1988). In nature, fish live in extremely variable environments (e.g. Gordon & Gordon, 1954; Kallman *et al.*, 1973). Therefore, it is not surprising that there is a great deal of genetic variability and phenotypic plasticity for age and size at maturation. Under stable conditions with abundant food, the fish should grow rapidly and mature as soon as they are developmentally able to do so. If conditions are still stable, but with less abundant food, then it should be advantageous for a fish to grow slowly and delay maturation. Under very poor conditions, or strongly fluctuating, unpredictable conditions, such as might be found in water bodies that are not permanent, a fish that matures at a small size is better off genetically than one that waits for the attainment of a larger size that may never be reached (Policansky, 1983). Following this line of reasoning, the distribution of maturation genotypes should be predictable on an environmental basis. Thus, it is expected that maturation should be age determined in rich or unpredictable conditions, and size determined otherwise.

3.3.3.3 Growth impacts

Compared with the amount of work which has been carried out examining the effects of pollution on the immediate post-hatch development of fish larvae/postlarvae in freshwater species (see Sayer *et al.*, 1993, for review), there have been relatively few marine case studies. Waring *et al.* (1996) recorded that although hatching success was not significantly affected, larval weight and yolk volume were. Where larval weight is lower in contaminant-exposed fish compared with control fish, and yolk-sac volumes are higher, this can suggest that development has been retarded which, by prolonging the period of reliance on egg reserves, can have a protective effect (Sayer *et al.*, 1993). Similar effects have been described in the estuarine amphipod *Chaetogammarus marinus*. In this study it was found that exposure of embryos to copper and pentachlorophenol (PCP) significantly extended the period of larval development whilst exposure to benzo[a]pyrene resulted in significantly smaller juveniles being hatched at the normal hatching time (Lawrence & Poulter, 2001).

The implications for pollutant exposure effects on growth can be complicated by the relative ambient temperature regime, where more optimal temperatures can negate any deleterious pollution-caused growth effects (Linton *et al.*, 1997). Sublethal stress can induce reduced growth rates in fish, possibly as a result of energy reallocation (Wendelaar Bonga, 1997). In some cases, poor growth rates are caused by a reduction in predation effectiveness (Bryan *et al.*, 1995). In addition, it is possible that pollution-related factors and contaminant bioavailability are important factors influencing skeletal deformities quantified in some fish species (Lindesjöö & Thulin, 1992; Vethaak, 1992; Lindesjöö *et al.*, 1994).

3.3.3.4 Condition indices

Condition indices can vary intraspecifically with season and geographical location (e.g. Sayer *et al.*, 1995, 1996), and so identifying quantitative trends in effected fish condition within this variation can be problematical. However, it is possible that ionoregulatory over-compensation, caused as a result of pollutant stress, can necessitate the diversion of energy from somatic growth, explaining the poorer condition of fish from polluted waters (Dennis & Bulger, 1995). Condition factor does not always correlate with concentrations of contaminant in sediments or tissues (Vethaak & Jol, 1996), and some studies have noted sexual differences in the effects of contaminants on the somatic condition of fish (Perkins *et al.*, 1997). In general it appears as if the somatic condition factor (K_S) is not as reliable an indicator of contaminant-derived stress as is the hepato-somatic condition index (e.g. Hoque *et al.*, 1998), although again seasonality, geographical location and sex may be additional parameters of variation (Sayer *et al.*, 1995).

3.3.4 Impact on developmental processes

3.3.4.1 Skeletal calcification

During the developmental transition from the larval to the postlarval stages of fish, the skeletal material becomes partially calcified (Sayer *et al.*, 1993). Disruption of this process by pollutants can either be through an upset to the calcium uptake/mobilisation mechanism(s) or through retarded development caused solely by a decrease in the developmental rate (Sayer *et al.*, 1989, 1991b). Retarded or affected skeletal calcification has been utilised widely as a quantitative assessor of pollution effects in freshwater fish because of the relative ease of bleaching and calcium-specific staining in the early postlarval stage (Sayer *et al.*, 1991b, 1993). However, similar quantitative studies do not appear to have been undertaken in marine species. Compensatory energy expenditure promoted by contaminant exposure during development can cause incomplete or disrupted skeletal development resulting in asymmetric developmental appearances (Campbell *et al.*, 1998).

The consequences for retarded skeletal development are not always deleterious for the immediate postlarval fish. However, if it is an extended effect then retarded calcification can reduce the locomotory ability of postlarval fish with unfavourable consequences for subsequent survival (Sayer *et al.*, 1993). Where the retarded development is as a result of restricted growth rates, then this can convey protection against pollution incidents (Sayer, 1991).

3.3.4.2 Muscle development

Much of the work on the environmental effects on muscle development have concentrated on the effects of temperature (Johnston, 1993). However, changes in muscle quality have been recorded during the developmental stage caused by contaminant exposure (Handeland *et al.*, 1996).

3.3.5 Nutrition

There do not appear to have been any studies undertaken which look to examine the direct effects of contaminant exposure on the ability of marine fish to assimilate nutritional intake.

3.3.6 Neuroendocrine and immune responses

Stress response in teleosts show many similarities to those of higher vertebrates. These relate to the principal messengers of the brain-sympathetic-chromaffin cell axis and the brain-pituitary-interrenal axis.

Activation of the hypothalmic-pituitary-interrenal axis results in secretion of the steroid hormone, cortisol (Pickering, 1993) and the catecholamines noradrenaline and adrenaline (Alkindi *et al.*, 1996). Cortisol is synthesised in the interrenal cells of the teleost head kidney and has a major role in the physiological response to physical and chemical stressors. Plasma levels of cortisol increase in physiologically competent fish exposed to pollutants such as cadmium and mercury and PAHs (Alkindi *et al.*, 1996; Hontela, 1998). Cortisol is involved in the stimulation of oxygen uptake and transfer, mobilisation of energy substrates (Wright *et al.*, 1989), reallocation of energy away from growth and reproduction, ionic regulation and suppressive effects on immune function. However, if levels of plasma cortisol are chronically elevated, this can result in damage to the fish, particularly with regard to the defence system and reproduction. In salmonid populations, this in turn can lead to increased mortality and reduced recruitment (Pickering, 1993).

Reduced levels of blood cortisol and thyroxin have been reported in several species of fish chronically exposed to mixtures of pollutants including heavy metals, PAHs, PCBs and bleached craft mill effluent. Hontela *et al.* (1995) reported lower levels of blood cortisol and thyroxin in sexually immature and mature male and female yellow perch (*Perca flavescens*), from a site contaminated by organic and heavy metal contaminants compared to fish from a reference site. In addition, the contaminated fish exhibited greater liver glycogen stores and had smaller gonads and lower condition factor than fish from the clean site. This endocrine impairment, characterised by a reduced ability to elevate plasma cortisol in response to stress, has also been described in northern pike (*Esox lucius*) from contaminated sites. In each case the fish showed reduced ability to respond to adrenocorticotropic hormone (ACTH), indicating disruption to the normal neuroendocrine response (Hontela, 1998). It is suggested that lifelong exposure to chemical pollutants may lead to an exhaustion of the cortisol producing endocrine system, possibly as a result of prolonged hyperactivity of the system (Hontela *et al.*, 1992).

Catecholamines have been reported to be released in response to conditions which give rise to hypoxaemia (Thomas & Perry, 1992). This may have a number of benefits including stimulation of splenic release of erythrocytes to aid oxygen carrying capacity (Pearson *et al.*, 1992; Alkindi *et al.*, 1996). However, high catecholamine levels as well as structural damage to the gill are also prime causal factors for induced systemic hydromineral disturbance (Bonga, 1997). This is associated with increased cellular turnover in these organs. In fish, cortisol combines glucocorticoid and mineralocorticoid actions, with the latter being essential for the restoration of hydromineral homeostasis. An inability to raise blood cortisol levels may therefore indicate a breakdown in this homeostatic mechanism.

Two opposite behavioural coping strategies to stress appear to be associated with this neuroendocrine mechanism. Van Raaij *et al.* (1996) subjected rainbow trout to severe hypoxia and measured blood levels of catecholamines, cortisol, glucose, FFA, lactate and electrolytes. Approximately 60% of the fish survived the experiment. Behavioural strategy appeared to be highly related to survival. Non-surviving fish displayed strenuous avoidance behaviour whereas surviving fish did not panic and remained quiet. Behavioural differences were associated with marked differences in plasma catecholamine levels which were four to five times higher in non-surviving fish. The cortisol response tended to be lower in the non-surviving fish.

There is also growing supporting evidence for interaction between the neuroendocrine system and the immune system in fish. For example, rainbow trout (*Oncorhynchus mykiss*) subjected to acute hyperosmotic stress showed high blood cortisol and prolactin levels which were correlated with a weak antiYersinia ruckeri antibody response compared to normal fish (Betoulle *et al.*, 1995). However, fish subjected to chronic stress showed no difference in blood cortisol or prolactin levels despite low antibody titres. Betoulle *et al.* (1995) suggest that in acute stress, cortisol and prolactin levels might exert immunosuppressive effects on antibody production, whereas in chronic stress other neuroendocrine hormones might result in reduced humoral immunity.

Evidence for the role of serotonin as a regulator of hypothalamic-pituitary-interrenal activity in teleost fish has been presented. The presence of a serotonin (1A), (5-HT1A) receptor subtype has been reported in the salmonid fish brain (Winberg *et al.*, 1997). In addition, it was shown that administration of a 5-HT1A receptor agonist raised plasma cortisol levels by a factor of 10 in rainbow trout (*Oncorhynchus mykiss*). This supports the theory that the brain serotonergic system plays a key role in integrating autonomic, behavioural and neuroendocrine stress response in fish as well as mammals (Winberg *et al.*, 1997).

Furthermore, it has been shown that a PCB mixture (Aroclor 1254) fed to male Atlantic croaker (*Micropogonias undulatus*) significantly reduced 5-HT (serotonin) and dopamine concentrations and increased their metabolites in both the preoptic-anterior hypothalamus and the medial and posterior hypothalamus (Khan & Thomas, 1996). In addition, Arochlor 1254 exposure resulted in the loss of the gonadotropin response to stimulation by luteinising-hormone releasing hormone analogue (LHRHa). This would indicate that Arochlor 1254 induced alteration in pituitary gonadotropin release may be mediated partially by altered hypothalamic serotonergic activity (Khan & Thomas, 1996, 1997).

Levels of other neuroendocrine factors norepinephrine and vanillylmandelic acid levels have been shown to be altered by Pb exposure. Whereas removing Pb did not facilitate a return to control values, adding DMSA did (Weber *et al.*, 1997). Both norepinephrine and serotonin have inhibitory actions on growth hormone release, whilst secretion is stimulated by a number of neuroendocrine factors including growth hormone releasing factor, dopamine, gonadotropin-releasing hormone, neuropeptide Y, thyrotropin-releasing hormone (Peng & Peter, 1997). Any stress which reduces serotonin release may, therefore, increase the release of growth hormone. Growth hormone is known to inhibit the expression of some P450 enzymes in mammals and has been shown to significantly decrease the level of hepatic cytochrome P450 in rainbow trout (Cravedi *et al.*, 1995a,b).

Control of plasma cortisol levels is not only controlled by serotonin. Melanin-concentrating hormone (MCH) is a neurohypophysial peptide that induces pigmentary pallor in teleosts. In addition, the peptide depresses ACTH and hence cortisol secretion during moderate stress. Plasma MCH concentrations can be raised by repeated stress in the rainbow trout (Green & Baker, 1991). This supports the suggestion that the modulatory effect of MCH on the hypothalamo-pituitary-interrenal axis of fish might be enhanced under conditions of stress.

It is known that the spawning cycle of biweekly spawning killifish (*Fundulus heteroclitus*) is synchronised with tides and coincident with the new and full moon. Changes in ovarian development are correlated with changes in dopamine and serotonin in the telencephalon, hypothalamus and pituitary (Subhedar *et al.*, 1997). It would seem likely, therefore, that sublethal pollution may not simply affect reproductive cycles but may take them out of phase with spawning time set by environmental factors.

3.3.7 Impact on neurosensory physiology

Chemical communication in fish plays an important role in synchronising reproductive physiology and behaviour. It has been hypothesised that contaminants could affect the neurosensory system of fish, impairing the lateral line and olfactory sensory capabilities and resulting in alterations to the effectiveness of feeding behaviour (Sindermann, 1996). While there is little evidence for pollutant-controlled neurosensory disruption affecting feeding, significant effects on the olfactory system of mature male Atlantic salmon parr have been recorded during the exposure of the organophosphate diazinon, suppressing the pheromone-driven induction of spawning (Moore & Waring, 1996). Electrophysiological recordings of olfactory epithelium in fish exposed to carbofuran indicate reduced response levels or detection abilities to priming pheromones (Waring & Moore, 1997). In these cases it is the ability to detect chemical cues emitted by ovulated and nesting salmonids which is being suppressed in males, causing reductions in the spawning readiness and success of the males. Where contaminants are affecting these gradual changes in fish neurosensory physiology, there are potential deleterious long-term implications for individuals and populations.

3.3.8 Rhythmicity

Many invertebrates and vertebrates exhibit circadian and circannual rhythmicity to certain aspects of their behaviour, physiology and reproductive biology. For example, rainbow trout show circadian feeding and locomotory rhythms (Sanchez-Vazquez & Tabata, 1998) and many fish have lunar or semi-lunar reproductive cycles (Duston & Bromage, 1988, 1991; Omori, 1995; Fujita *et al.*, 1997). Photoperiodic control of reproduction is believed to increase the rate of mating and fertilisation (Olive *et al.*, 1990; Omori, 1995; Lawrence, 1996).

Setting reproductive cycles and spawning times to a photoperiodic cycle rather than temperature ensures that juveniles are released at a precise time of the year. It is assumed that timing of juvenile release coincides with periods of high food availability, again increasing the likelihood of survival (Olive *et al.*, 1990; Lawrence, 1996).

There is recent evidence in invertebrates that gonadotropic hormones may act as the transducer system between the environment and the developing oocyte (Olive & Lawrence, 1990; Lawrence & Olive, 1995; Lawrence, 1996). Furthermore, in fish it has been shown that certain stages of vitellogenesis are photosensitive and that reduced egg size brought about by premature photoinduction of oogenesis could not be accounted for by low levels or circulating vitellogenin (Bon *et al.*, 1997). Additionally, changes in ovarian development of killifish (*Fundulus heteroclitus*), which synchronises spawning to lunar cycles, are correlated with changes in dopamine and serotonin in the telencephalon, hypothalamus and pituitary (Subhedar *et al.*, 1997).

Concern about the potential impact of global climate change on species that use photoperiod to synchronise their reproductive cycle has been highlighted (Olive *et al.*, 1990; Norse, 1994; Lawrence, 1996). The problem for these species is that the time of year as set by photoperiod will come out of phase with the time set by temperature. This may have severe consequences for larval survival if, for example, the food supply is no longer available when they are released. Given that there must be high selective pressure on individuals to set their reproductive cycle to the time of year that others in the population spawn, and as set by photoperiod, future survival of the population/species may depend on how quickly this trait can be modified in relation to how quickly the climate changes (Lawrence, 1996).

Pollution may also have a more direct impact on photoperiodic control of reproduction and other physiological processes. There is now limited evidence for the involvement of the neuroendocrine system acting as a transducer between the environment and gamete (Lawrence, 1996). Together with evidence for the direct impact of pollution on the neuroendocrine system (see section 3.3.6) it is likely that pollution will affect any photoperiodic cycle control mechanism, again taking the reproductive cycle out of phase with spawning time set by environmental factors.

This is an area which requires much more detailed research. Work to date has clearly shown that the natural reproductive and spawning cycle of any test species must be thoroughly understood (Norberg *et al.*, 1991). The impact of global climate change on commercial fish stocks has been highlighted as an area of critical importance by the EU at the 3rd MAST Conference, Lisbon, 1998, and given the preliminary understanding of the mechanisms linking photoperiod to oogenesis this impact can be tested.

One final development also highlights the need to consider natural rhythmicity in any study. It has been shown that the toxicity of drugs and other substances can show a circadian or circannual variation. It is assumed that this is attributable to quantitative changes in metabolism, receptor sensitivity and kinetics (Heinze *et al.*, 1993).

3.3.9 Lysosome damage and reduced immune competence

Immunological defences have proven to be sensitive markers of exposures to environmental contaminants (Bayne & Moore, 1997). Such internal defences are not restricted to vertebrates. Natural history traits of several invertebrate species predispose them to serving as excellent sentinels for pollutants. For bivalve molluscs in particular, combined ecological and physiological traits predispose them to serve as ideal models for studies in immunotoxicology (Bayne & Moore, 1997). Mussels, clams and oysters have proven suitable for assays at the levels of Tier I (molecules and cells), Tier II (cells and tissues) and Tier III (host resistance challenge) (see Chapter 4).

Here, Tier I assays in molluscs are considered as they yield the most clear-cut and interpretable data on effects of xenobiotics. Measurements of lysosomal accumulation and retention of foreign chemicals have proven to be easy means to obtain data on the health status of cells, and on the level of expression of multixenobiotic resistance transporter proteins. Both of these are prognostic for more debilitating effects of prolonged and heavier exposures to toxic chemicals (Bayne & Moore, 1997). Additional assays which have been used productively in environmental toxicology with these animals include measurements of induced metal-binding proteins (metallothioneins), induced enzymes with oxygen-scavenging activities (superoxide dismutase, catalase) and metabolising activities for polychlorinated hydrocarbons (cytochromes P450), phagocytosis, respiratory burst and the plasma concentrations of various humoral factors (see Bayne & Moore, 1997).

In the specific context of molluscan blood cells, these latter are generally rich in lysosomes, phagocytically active and highly responsive to pollutant chemicals (Grundy *et al.*, 1996a,b; Lowe *et al.*, 1995a,b; Moore *et al.*, 1996a,b; Viarengo *et al.*, 1994; Winston *et al.*, 1996). Lysosomes form an important part of the haemocyte's physiological apparatus, for it is in the lysosomal compartment that foreign cells are killed and degraded to monomeric chemical constituents (i.e., cell feeding). Consequently, any pollutant-induced damage to lysosomal function will impact directly on the cellular immune process which is dependent on effective phagocytic engulfment of invading organisms or abnormal 'self' followed by intracellular digestion/degradation. In fact, such effects have been clearly demonstrated both *in vitro* and *in vivo* in the haemocytes of the marine mussel (*Mytilus edulis*) by Grundy *et al.* (1996a,b).

Lysosomal injury is a sensitive and reliable biomarker of pollutant-induced damage and has been shown to be effective in this capacity in the haemocytes of marine mussels and the coelomocytes of earthworms. Given the important role of the endocytotic-lysosomal system in cellular immunity, it seems reasonable to propose that evidence of lysosomal damage in cells of the invertebrate immune system should provide a biomarker of immunodeficiency in invertebrates (Grundy *et al.*, 1996b).

3.3.10 Effects on reproduction

3.3.10.1 Reduced energy for reproduction

An organism can only acquire a limited amount of energy for which several processes compete directly. The trade-off concept assumes that an increase in the energetic allocation to one process must result in a decrease in energy allocation to others (Ware, 1980, 1982; Sibly & Calow, 1983), as illustrated in Fig. 3.3. The concepts of optimal foraging and life history provide the physiological basis for the fate of food energy ingested by animals. The sexual maturation process is energetically expensive and is reflected in the general finding that female fish mature later than males (Thorpe, 1994). In general, it requires a greater energy accumulation to develop ovaries and eggs than to develop testes and sperm. In order to meet their standard metabolism (maintenance) and activity costs, fish must transform ingested food into net (useable) energy (Ware, 1980).

Figure 3.3 illustrates the fate of surplus energy (i.e. absorbed energy minus energy used for respiration and standard metabolism) in organisms. If a proportion, q, of surplus energy in food is allocated to growth, then $1 - q$ can be allocated to reproduction (Sibly & Calow, 1983). There is the existence of power allocation trade-offs between reproduction and growth, condition and survival, current and future reproduction, quantity and quality of progeny. Both growth and reproduction has the optimal aim of maintaining parental and offspring fitness (see Chapter 5). Comparatively, growth will either stop or gradually diminish with age as increasingly more energy is invested in reproduction (Schaffer, 1974; Ware, 1982; Stearns, 1983; Thorpe, 1994). Stored energy reserve is wastefully used during xenoestrogen-induced Vtg synthesis outside normal reproductive period; however, this energy will not be readily available for normal reproduction when environmental variables are optimal for embryo survival. Additionally, xenobiotic detoxification/biotransformation

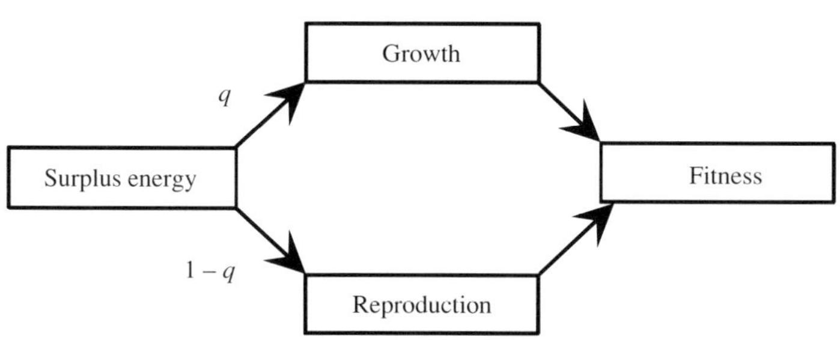

Fig. 3.3 Fate of surplus energy (i.e. absorbed energy minus energy used for respiration and standard metabolism). If a proportion q of surplus energy in food is allocated to growth, then $1 - q$ can be allocated to reproduction. Both growth and reproduction increase fitness.

are energetically very expensive and the energy invested for these processes will obviously reduce surplus energy.

3.3.10.2 Induced or reduced vitellogenesis and zonagenesis

Vitellogenesis and zonagenesis are defined as oestradiol-17β (E_2)-regulated hepatic synthesis of the egg yolk protein precursor, vitellogenin (Vtg) and eggshell *zona radiata* proteins (*Zr*p), respectively, their secretion and transport in blood to the ovary and its uptake into maturing oocytes in oviparous vertebrates. The liver of oviparous vertebrates has proved to be an excellent model for the studies of molecular mechanisms of steroid hormone action (Tata & Smith, 1979; Wahli, 1988). Vtg is a bulky (MW; 250–600 kDa) and complex calcium-binding phospholipoglycoprotein (Tyler *et al.*, 1991a,b; Schneider, 1996). The classification of Vtg as a phospholipoglycoprotein indicates the crucial functional groups that are carried on the protein backbone of the molecule: lipids, some carbohydrates, and phosphate groups (Mommsen & Walsh, 1988). In addition, the ion-binding properties of Vtg serve as a major supply of minerals to the oocytes. Oocyte growth in fish is due to the uptake of circulating Vtg, which is then modified by and deposited as yolk in the oocyte (Wallace, 1985).

The molecular mechanisms that lead to the production of Vtg and *Zr*-proteins in the hepatocyte will not be presented in detail here. Briefly, E_2 produced by the ovarian follicular cells in response to gonadotropin (GtH I) enters the cell by diffusion. In the cell, the E_2 is retained in target cells by high affinity binding to a specific steroid-receptor protein (such as the E_2-receptor, ER). The hormone-receptor complex binds tightly in the nucleus at oestrogen responsive elements (ERE) located upstream of, or within, the oestrogen-responsive genes in DNA. This results in the activation or enhanced transcription of Vtg (and possibly *Zr*-proteins) genes and subsequent increase and stabilisation of Vtg and *Zr*-proteins messenger RNA (mRNA). Vtg and *Zr*-proteins precursors are modified extensively in the rough endoplasmic reticulum (RER) and secreted into the serum for transport to the ovary. In the ovary Vtg is incorporated by receptor-mediated endocytosis, and processed by enzymatic cleavage into lipovitellin I and II and phosvitin (Lazier & MacKay, 1993) that serve as nutrient reserves for the embryo.

Several metabolic changes occur during vitellogenesis in the maturing female fish. This is reflected in the pronounced increases in liver weight, RNA contents, lipid deposition, glycogen depletion, increases in plasma protein, calcium and magnesium and phosphoprotein contents (Weigand, 1982; Björnsson *et al.*, 1986). These parameters can be used as markers of plasma Vtg. In addition, Vtg and gonadal maturation are energetically very expensive processes, since the full grown gonads account for about 25% of the total weight of a mature female fish.

Laboratory and field studies have been conducted to evaluate the impact of fish exposure to toxicants on vitellogenesis and zonagenesis (for review see Lam, 1983; Susani, 1986; Kime, 1995; Arukwe & Goksøyr, 1998). In some reports, it has been shown that fish exposed to xenobiotic oestrogens (xenoestrogen) or sewage treatment work (STW) effluent show high serum or plasma Vtg levels (Arukwe *et al.* 1997a,b; Wester & Canton, 1986; Jobling & Sumpter, 1993; Pelissero *et al.*, 1993; Purdom *et al.*, 1994; White *et al.*, 1994; Sumpter & Jobling, 1995; Donohoe & Curtis, 1996; Harries *et al.*, 1996, 1997; Jobling *et al.*, 1996; Gray & Metcalfe, 1997; Lye *et al.*, 1997). In addition, reduced ovarian development has been reported by Jobling *et al.* (1996). In other studies, Anderson *et al.* (1996a,b) have reported reduced liver synthesis of Vtg (i.e. antioestrogenic effects) in juvenile fish treated with cytochrome P4501A inducing compounds.

Given the energetic cost of reproduction and the long decision time, it seems most likely that xenobiotically-induced hepatic Vtg synthesis may cause an imbalance in the reproductive strategy of a given fish population (see Chapter 5). Thorpe (1994) suggested that during maturation, the internal responses that are synchronised by external signals depend on some genetically determined performance threshold, and that maturation processes will continue if this performance exceeds a set point at this critical time. For example, in salmonids survival after spawning implies a chance dependent balance between stored energy and that spent on reproduction, because maturation has developmental priority over somatic growth (Policansky, 1983). Therefore, xenoestrogen-induced Vtg synthesis outside normal maturation period may result in wasteful use of stored energy resources. The ecological implication of this might be failure in the reproduction of affected individual fish, and in the long term affecting recruitment in the entire population (see review by Arukwe & Goksøyr, 1998). Another possible deleterious effect is that high Vtg concentrations might cause kidney failure and increased mortality rates as a result of metabolic stress (Herman & Kincaid, 1988).

Furthermore, although not yet demonstrated, there is a possibility that reduced testicular growth could reduce fertility (Jobling *et al.*, 1996). Continued synthesis of Vtg diverts available energy resources (lipids, proteins), thereby reducing chances of juvenile survival before they start feeding. Loss of calcium from bones and also from the scales during active Vtg synthesis (Carragher & Sumpter, 1991) may increase the susceptibility of fish to disease.

In a report by Arukwe *et al.* (1997b), it was shown that the alkylphenol, 4-nonylphenol (NP), induced the production of eggshell *zona radiata* protein (*Zr*p) in juvenile fish. Also in this report, *Zr*p was shown to be comparatively more sensitive to the xenoestrogen than Vtg. Xenoestrogen-induced changes in *Zr*p synthesis appear to have a higher potential for ecologically adverse effects than Vtg induction, because critical population parameters such as offspring survival and recruitment may be more directly affected. The argument for this is that, whereas subtle changes in Vtg content would not be of great significance to the survival of the offspring, small changes in *Zr*p synthesis might alter the thickness and mechanical strength of the eggshell, thus causing a loss in its ability to prevent polyspermy during fertilisation and to protect the embryo during development (Arukwe *et al.*, 1997a,b; Arukwe & Goksøyr, 1998).

3.3.10.3 Impacts on fecundity

In fisheries biology there are two principle definitions of fecundity: absolute fecundity, which is defined as the total number of eggs ovulated per fish; and relative fecundity, which is the number of eggs ovulated per unit (kg) body weight. However, other terms occur in the literature and these are not always well defined by the authors. Batch fecundity, for example, is the number of eggs produced per spawning. From this, the number of eggs that a female spawns will depend on the number of eggs per spawning, which in itself will depend on the size of the eggs and the number of spawnings per season. These distinctions are important because the impact of pollution on fecundity, and therefore future population size, will depend on the life history and reproductive cycle of the particular species of fish.

In addition, most teleosts are iteroparous, that is they spawn over several years. The total number of eggs produced over a fish's life will, therefore, depend on the total number of eggs spawned in a breeding season and the reproductive longevity of the fish. However, not all iteroparous fish spawn every year. Some fish show non-annual and or irregular periodicity (Burton & Idler, 1987; Burton, 1991). Consequently, when looking at the long-term effects of low-level pollution on fecundity, it is important to distinguish between batch fecundity, breeding season fecundity and lifetime fecundity.

Within a species and an individual, fecundity can vary within and between spawning and seasons. Fecundity can be affected by growth rate and nutritional status. For example, winter flounder has been shown to have a nutritionally sensitive period for gametogenesis, and insufficient energy reserves at this time cause it to switch off gonad development (Burton, 1994). Similar effects have been observed in the sea bass (*Dicentrarchus labrax*), in which the effect of food ration on oestrogen and VtG plasma levels, fecundity and larval survival were compared (Cerda *et al.*, 1994). It may be possible, therefore, for low-level pollution to have no adverse effect on a fish but to still affect its survival at a population level, by knocking out its food supply, particularly at important times.

Fish, like many vertebrates and invertebrates, can also exhibit atresia (oocyte degeneration) in response to a number of factors including poor nutritional state. This can occur at any stage of development and will theoretically affect the number of oocytes that form mature eggs and, therefore, fecundity. Atresia, however, appears to be relatively uncommon in fish held under optimal conditions (Tyler *et al.*, 1990). The problem again may be in separating pollution effects on atresia from nutritional effects in natural populations.

A fecundity gene (FecB gene) has been identified as playing a role in determining fecundity in mammals (Braw-tal *et al.*, 1993; Montgomery *et al.*, 1994). It is likely that a similar genetic basis to fecundity exists in fish. This gene appears to affect the plasma level of the gonadotropin FSH and elevated plasma levels of this are associated with higher fecundity (McNatty *et al.*, 1994). In rainbow trout recruitment of oocytes into the maturing vitellogenic pool is accompanied by elevated levels of plasma GtH 1 which plays a similar role in fish to FSH in mammals (Tyler & Sumpter, 1996).

Pollution can increase the level of atresia seen in marine invertebrates and fish and has been shown experimentally (Widdows *et al.*, 1982; Lawrence *et al.*, in prep; Hanson *et al.*, 1985; see Chapter 2). Incidences of atresia have been related to degenerative follicles from previous sexual cycles that had failed to be ovulated (Wallace & Selman, 1979). However, elevated ovarian follicular apoptosis and HSP70 expression has also been observed in white sucker (*Catostomus commersoni*) exposed to bleached kraft pulp mill effluent. This was associated with reduced ovary size, decreased plasma testosterone, increased plasma beta-oestradiol but not induction of EROD activity (Janz *et al.*, 1997). Apoptosis is regulated by several hormonal factors and conserved gene products. Therefore, this study indicates that BKME can increase ovarian cell apoptosis by stimulating cell death signalling, but the mechanism is unclear.

The impact of sublethal pollution on fecundity has been demonstrated in a number of laboratory-based experimental studies. In many of these studies, exposure to xenobiotics has been related to a reduction in fecundity. For example, zebrafish exposed to 2,3,7,8 tetrachlorodibenzo-p-dioxin (TCDD) showed a dose-related reduction in egg numbers (Wannemacher, 1992). Specifically, TCDD impaired development of previtellogenic and vitellogenic oocytes. Zebrafish have also been used in life cycle studies. Those exposed to a mixture of dichloroaniline and lindane stopped spawning irreversibly, whilst fish exposed to the same xenobiotics after reaching maturity showed reduced fecundity (Ensenbach & Nagel, 1997). Fathead minnows (*Pimephales promelas*) exposed to lead showed a reduction in number of eggs oviposited despite no differences in GSI between treated and control fish (Weber, 1993).

A negative correlation between concentration of the pesticide esfenvalerate and the fecundity of Australian crimson spotted rainbow fish (*Melanotaenia fluviatilis*) has also been reported despite no effect on hepatic EROD, ECOD or EFCOD activities (Barry *et al.*, 1995). Furthermore, radionuclides and chemical genotoxicants can affect fecundity. Mosquitofish (*Gambusia affinis*) exposed to radionuclides showed a negative correlation between fecundity and the level of double strand breaks in the DNA of fish from contaminated sites (Theodorakis *et al.*, 1997).

This relationship between xenobiotic exposure and reduced fecundity has been supported in some field studies. In redbreast sunfish (*Lepomis auritus*) elevated levels of detoxification enzymes were associated with decreased fecundity which it was suggested was due to the reduced capacity of the liver to manufacture yolk proteins (Adams *et al.*, 1992a,b).

The inducibility of spawning in English sole (*Parophrys vetulus*) from four areas in Puget Sound varying in chemical contamination showed that those fish from the site with the highest levels of hydrocarbons and PCBs showed highest reproductive impairment. This was linked to low initial plasma oestradiol and ALP, high measures of contaminant exposure and a prevalence of pollution-associated liver lesions (Casillas *et al.*, 1991). In winter flounder (*Pleuronectes americanus*) sampled from sites on the north-east coast of the USA, decreased egg weight and increased atresia was found in fish with high tissue concentrations of PCB (Johnson *et al.*, 1994).

However, this negative relationship is not always clear, particularly in field studies in which the fish are subjected to both natural and anthropogenic impacts which can affect fecundity. For example, in an examination of the effects of five di-ortho PCB congeners on fathead minnows, it was found that no significant impact was observed on reproductive success in terms of total number of eggs, clutch size, number of clutches or percent hatchability despite a significant reduction in growth of females and a significant body burden of the congeners (Suedel *et al.*, 1997). In the winter flounder study performed by Johnson *et al.* (1994), despite increased atresia, contaminant exposure had no clear negative impact on GSI, plasma oestradiol concentration or fecundity.

An examination of the effects of bleached kraft mill effluent on reproductive parameters of white sucker (*Catostomus commersoni*) found that exposed fish showed strong induction of EROD activity. In females testosterone and 17beta-oestradiol levels were significantly reduced but GSI was not affected. The effect on fecundity was more variable. Consequently, the authors could not clearly relate perturbation in plasma steroid levels to impaired reproduction as measured by gonad weight and fecundity (Gagnon *et al.*, 1994). Similarly, rocky mountain whitefish (*Prosopium williamsoni*) and longnose sucker (*Catostomus catostomus*), when exposed to BMKE showed no reduction in relative gonad size or fecundity despite the induction of cytochrome P4501A in both species (Kleoppersams *et al.*, 1994). It would seem, therefore, that production of this biochemical biomarker of exposure does not have to be associated with any discernible adverse effects on individual fish health or reproductive capacity.

Furthermore, in some field studies there is evidence that xenobiotic exposure can have a positive effect on fecundity. In a study in Canada the fecundity, egg diameter, fish length and weight of Brown bullhead (*Ameiurus nebulosus*) from three river systems of varying pollution load were compared. Fish from the contaminated sites were larger and fecundity was significantly different between the river systems. Those from the most polluted river had the greatest number of eggs per female. It is suggested that this increased fecundity may have been the result of reduced competition for an invertebrate food source (Lesko *et al.*, 1996). Similarly, in a study examining the effect of acidification on populations of perch (*Perca fluviatilis* L.) it was found that the fish in most acidified systems showed higher length specific fecundity and higher reproductive potential relative to stock density (Linlokken *et al.*, 1991).

In comparing the effects of xenobiotics on winter flounder and English sole in Puget sound, Johnson *et al.* (1994) suggest that the difference in susceptibility of the fish to contaminant-induced reproductive dysfunction could be related to a number of factors including the migratory behaviour of the two species. English sole reside in contaminated estuaries throughout vitellogenesis and move offshore to spawn, whilst winter flounder often remain offshore during vitellogenesis and move into contaminated estuaries before spawning.

3.3.10.4 Fertilisation impairment

Many studies have been performed to examine the effect of pollution on the development of fertilised eggs (Dethlefsen *et al.*, 1996). These studies involve subjecting previously fertilised eggs to xenobiotic and examining effects on development, hatch success and embryo abnormalities; they are reviewed in detail in Chapter 4. Much less, however, has been published on the effects of xenobiotics on the fertilisation process itself. Furthermore, little has been done to examine the effect of pollution on sperm maturation or function, although the literature in this field is beginning to develop with the concern over oestrogen mimics.

One possible reason for the lack of work relating pollution effects on fertilisation, is the process by which the majority of fish, including all of the European commercial fish, reproduce. Rather than forming pairs, many species congregate in dense shoals at a particular time and spawn millions of eggs and sperm into the sea. Fertilisation, therefore, takes place in the sea rather than in a body cavity. The likelihood of fertilisation taking place between eggs and sperm is increased by the gametes being generally positively buoyant and, therefore, floating near the surface of the sea. There is inevitably a great waste of sex cells and this is associated with the production of a huge number of gametes. Despite this, it has been estimated that in cod for example, only one egg in every million released becomes an adult fish (Norman & Greenwood, 1963).

Separating the effects of pollution on the fertilisation process, from all of the other parameters that affect egg survival, must therefore be very difficult. It should, however, be possible to perform experiments to determine any mechanistic problems associated with the fertilisation process. However, very little work has been published on this. This may be why, for example, hatch success is used as a measure of reproductive success much more frequently than fertilisation rate. However, it should also be noted that in several studies it has been found that fertilisation is not affected by xenobiotic but that pollution-induced effects only become evident between fertilisation and hatching (Crane *et al.*, 1992).

Field studies have shown that, along with a number of other reproductive parameters, fertilisation can be affected by exposure to xenobiotics in fish. In English sole (*Parophrys vetulus*) from Puget Sound fertilisation success was positively correlated with ALP (vitellogenin) concentrations which were low in fish from sites with high sediment loads of PAHs and PCBs (Casillas *et al.*, 1991). Low egg fertility has been reported in salmonids from the great lakes (Leatherland, 1993) and reduced sperm counts have been reported in stressed rainbow trout (Campbell *et al.*, 1992).

There has been much publicity and concern about the effect of oestrogen mimics on male reproduction. Nonylphenol has been found to be oestrogenic and in male eelpout (*Zoarces viviparus*) has been shown to affect GSI, and significantly reduce milt in treated fish. Microscopically, seminiferous lobules were degenerated and Sertoli cells contained phagocytosed spermatozoa (Christiansen *et al.*, 1998). The effect of pollution on sperm development has been examined using computer assisted sperm analysis (Kime *et al.*, 1996). This has shown that the progressive motility of catfish sperm decreased after exposure to cadmium and zinc at concentrations found in the gonad as a result of bioaccumulation. It seems reasonable to infer from this type of study that any impact on the motility of sperm may reduce the likelihood of successful fertilisation by reducing the chance of a sperm swimming to an egg. This relationship has not been proven but preliminary studies indicate that they are closely linked (Kime, 1998). Alternatively, given the general pattern of external fertilisation, this problem may be offset by the huge number of gametes spawned at a particular time.

The effect of organic compounds on reproductive performance of male American plaice (*Hippoglossoides platessoides*) has also been examined in laboratory experiments. Maturing fish were exposed to sediment of varying level of contamination. Semen was collected and used to fertilise eggs from a non-exposed female. Eggs fertilised with sperm from males maintained on the most contaminated sediment produced 48% less larvae than controls. There was no difference between groups with respect to the number of sperm produced or GSI but there was a negative correlation between male CYP1A1 levels and hatch success (Nagler & Cyr, 1997).

3.3.10.5 Embryonic and larval abnormalities and genotoxic damage during gametogenesis

There is a great range and diversity of papers on the effect of contaminants on embryos and larvae. These have been extensively reviewed by Rosenthal and Alderdice (1976); Laale and Lerner (1981); von Westernhagen (1988); Weis and Weis (1989); and Bodammer (1993).

In most cases, embryos and larvae are used in screening tests for aquatic toxicity testing to derive maximum acceptable toxicant concentrations. These studies generally use endpoints such as hatching success, early larval survival and growth. Very few studies include observations on the occurrence of developmental abnormalities in embryos and larvae. This is possibly due to the fact that most investigators regard mortality as an easily measured end-point which is key to the survival of a species. It is generally accepted that embryos and larvae are very sensitive to contaminants and as a result whole life cycle toxicity tests have often been replaced by early life-stage tests. Weis and Weis (1989) reviewed 194 experimental data sets on the effects of environmental pollutants on early fish development. Of these, 164 were investigations on the effects of exposure of fertilised eggs to aquatic toxicants and 30 were investigations on the effects on reproduction and subsequent survival of offspring following exposure of adult fish or their gametes to aquatic pollutants; four only were based on field collected investigations.

Fish embryos in the natural environment can be exposed to contaminants in three ways:

(1) Via the yolk which is synthesised during oogenesis
(2) During the brief period between shedding of the gametes, fertilisation and formation of the chorion proper
(3) As embryos and larvae.

Fish eggs have a large amount of yolk and a protective membrane, the chorion, which is composed of a polysaccharide and proteinaceous material. The chorion becomes completely toughened after fertilisation and acts as a physical and possibly a chemical barrier to the influx of chemicals (Tesoriero, 1977). After fertilisation, the cytoplasm of the egg cell becomes segregated from the yolk and forms a blastodisc. The blastodisc further subdivides during cleavage to form the blastoderm, which later forms the body of the fish embryo. Towards the end of cleavage the blastomeres spread, which is followed by the process of epiboly, during which the primary germ layers of the embryo are established and the embryonic axis is defined. As a result of cell movements the embryonic shield develops, within which the primary organs of the embryo are formed including the neural tube, the notochord and the somites.

Contaminants can affect any of the developmental processes described above in a number of non-specific ways. These may be characterised as follows, after Weis and Weis (1989):

(1) Morphogenetic; failure of cells to orientate and migrate during gastrulation leading to severe neurological defects and incomplete axial development
(2) Tissue interactions; two different tissues become associated with each other, resulting in altered development of one or both of the tissues, e.g. partial fusion of eyes or cyclopia or no lens development
(3) Growth; effects on hormones and growth factors leading to growth inhibition, overgrowth, misplaced growth and uncontrolled growth and formation of tumours; such effects can be systemic or localised to specific organs
(4) Degeneration; cell death is a normal part of embryonic development; if inhibited or accelerated by a chemical contaminant the embryo will be defective
(5) General development; fish embryos in general tend to show 'natural' abnormalities. The skeletal, circulatory and optical systems and rates of development to specific stages appear to be very sensitive. Chemical contaminants may increase the incidence of these abnormalities and increase or retard the rate of development
(6) Mutagenic effects; mutagenic materials can damage chromosomes, causing cytogenetic defects, which could ultimately result in morphological abnormalities.

Rosenthal and Alderdice (1976) state that gonadal tissue, the early embryo, and the stage of larval transition between endogenous and exogenous food sources are the most sensitive stages to pollution. In general, most fish are highly fecund and embryo and larval survival is naturally low (<5%). In most species the embryo larval stage lasts for several weeks. Mortality and abnormal development can occur from poor quality of eggs, low food reserves in the embryo, predation, hostile environmental conditions (physical disturbance, low dissolved oxygen, variable salinity) and contaminant effects (Rosenthal & Alderdice, 1976; Wiegand *et al.*, 1989; Purceli *et al.*, 1990). In some instances high 'natural' abnormalities have been observed. For example, Loning (1977) indicated that up to 30% of Atlantic cod (*Gadus morhua*) embryos cultured under control conditions may undergo aberrant development, leading to death during cleavage and gastrulation.

As previously noted, nearly all of the data that has been reported in the literature is derived from experiments carried out on embryos exposed after fertilisation for short time periods to environmentally unrealistic concentrations of contaminants. However, embryos in the natural environment are exposed to contaminants in two additional ways: via the yolk which is synthesised during oogenesis by exposed females, and during the brief period between the shedding of the gametes and formation of the chorion. For organic contaminants there is a high correlation between maternal transfer during oogenesis and the lipid content of the egg (Nimi, 1983). There are few studies of the impact of contaminants on gametes prior to fertilisation but examples do exist; reduced sperm motility in trout after exposure to Hg (McIntyre, 1973) reduced hatch in trout eggs when sperm from male trout exposed to Hg were used to fertilise the eggs (Birge *et al.*, 1979).

It is clear from the literature that the precise details of the mechanisms by which contaminants influence the developing embryo and larvae are not known. This is compounded by the fact that there appears to be a similarity in morphological defects observed for embryos exposed to the major classes of contaminants (i.e. heavy metals, chlorinated hydrocarbons, petroleum hydrocarbons). The reviews by Rosenthal and Alderdice (1976), von Westernhagen (1988) and Weis and Weis (1989) clearly show that notochord abnormalities, crano-facial defects, brain and eye defects, cardiovascular defects and spinal abnormalities may be induced in developing embryos exposed to a variety of contaminants. This led Rosenthal and Alderdice (1976) to suggest that an embryo responds to toxic insults with a generalised 'stress' response. This may be clearly seen in an example given by Bodammer (1993) where the initial treatment of the egg and continued exposure of the embryo to cadmium can:

(1) Modify the permeability of the egg membrane prior to and after fertilisation
(2) Disrupt gastrulation and axiation during the mid to later stages of embryogenesis
(3) Retard the growth, development and organogenesis
(4) Reduce embryonic heart rate
(5) Reduce or modify embryo motility
(6) Decrease the activity of several biosynthetic enzymes in late-stage embryos
(7) Disrupt normal osteogenesis, resulting in skeletal abnormalities
(8) Reduce yolk-sac size via osmotic effects on perivitilline fluid, resulting from cadmium affected membranes
(9) Result in premature or delayed hatching.

A further important and complicating factor which must also be considered is the possibility of stage-dependent sensitivity to contaminants (Marty *et al.*, 1990).

There are few examples in the literature on field-related effects of contaminants causing larval abnormalities and poor recruitment. However, the examples that do exist provide convincing evidence that this is an important area for further investigation. Longwell and Hughes (1981), studying the mackerel (*Scomber scombrus*) in the New York Bight, showed significantly lower egg viability in highly impacted Bight areas (disposal sites) than in areas offshore, and that the health of the embryos correlated with contaminant concentrations. Several investigators have worked on the sea surface microlayer (Hardy *et al.*, 1987; Cross *et al.*, 1987; Kocan *et al.*, 1987). These studies have demonstrated that sea surface waters from contaminated areas produce significantly higher morphological or chromosome abnormalities than waters from control areas.

Perry *et al.* (1991) showed that winter flounder caught at polluted sites in Long Island Sound and spawned in the laboratory, produced embryos having varying degrees of chromosomal damage and other forms of abnormal cell damage. Cameron *et al.* (1992, 1996; Cameron & von Westernhagen, 1997) have investigated malformation rates in embryos of North Sea fish between 1984 and 1987 and in 1991 and 1992. In the 1991 and 1992 study up to 19 fish species were investigated over the whole of the southern North Sea (Lat 52° to Lat 57°N). Malformation rates varied (from below 10% to 80%) depending on the species, the time of year and the development stages considered. For example, during one winter cruise a 70% malformation rate was recorded for whiting, while in summer pilchard and sprat displayed the greatest effects (40–50%). Malformation rates were highest in embryos taken in the German Bight (close to the Elbe River) and on the UK coast by the Humber estuary. Common defects were blister proliferation in early and late embryos, failure to close the blastopore and deformation of the notochord. Supporting studies showed that abnormal embryos taken back to the laboratory and held under optimum conditions did not survive. Although in specific areas the fish embryo malformation rates were high, no attempt was made to relate abnormality rates with contaminant concentrations in environmental water or sediment samples, nor in the embryos or parent fish.

3.3.11 Behavioural responses

Behavioural responses of fish and invertebrates to pollution have been suggested as a possible early warning method of monitoring xenobiotic impact. This is because a behavioural response requires the integration of a suite of physiological processes including neurosensory, neurosecretory, endocrine and physiological energetics (Lawrence & Poulter, 1996, 1998; Miller, 1980). Consequently, pollution impact on any one of these processes should be manifest at a behavioural level, if normal compensatory mechanisms are overridden.

3.3.11.1 Locomotion

Contaminants can affect the locomotory activities of fish by eliciting movement toward (attracting) or away from (avoidance) a contaminated area; alter sensory perception; cause alterations in free locomotory activity; alter locomotory components such as turning frequency or angular orientation; and reduce swimming performance and/or endurance (see

Atchison *et al.*, 1987, for review). The use of any of these responses is limited by interspecific and intraspecific variability, as well as the environmental relevance of much of the laboratory assays, but any detrimental effects on locomotory ability could affect escape capabilities, predator/prey interactions, migratory patterns, and reproduction (Atchison *et al.*, 1987). More recently a metabolic trade-off has been proposed between locomotion and detoxification in rainbow trout exposed to copper (Handy *et al.*, 1999).

3.3.11.2 Escape

Groups of osmotically-challenged Atlantic salmon demonstrated reduced escape distance and schooling behaviour, and suffered higher predatory mortality (Handeland *et al.*, 1996). However, where the stress was mild, basic predator avoidance behaviour which was modified during the challenge returned to pre-exposure levels quickly (Olla *et al.*, 1992).

3.3.11.3 Foraging models

Pollution exposure has been demonstrated to alter the efficiency of foraging behaviour in fish (e.g. Bryan *et al.*, 1995). Fish subjected to pollutants have been shown both to attack fewer prey items and to be less effective in capturing the smallest prey (Bryan *et al.*, 1995). Exposure to metals can impair feeding performance in fish by reducing the appetite and changing the reaction distance and prey handling time (Atchison *et al.*, 1987). No attempts appear to have been made to apply optimal foraging theory to toxicity testing or to verify in the field the effects of metals on the foraging behaviour of fish (Atchison *et al.*, 1987).

3.3.11.4 Reproductive behaviour

A number of studies have noted changes in the duration, intensity and format of courtship displays during the preliminary stages of reproduction (see Jones & Reynolds, 1997, for review). It is suggested that reduced and/or affected courtship displays can disrupt the process of sexual selection resulting in delayed and/or reduced reproduction with subsequent reductions in population fitness and success (Colgan *et al.*, 1982; Stafford & Ward, 1983; Matthiessen & Logan, 1984; Schröder & Peters, 1988; Kime, 1995).

Exposure to masculinising contaminants can cause male-like courtship patterns in female fish, though with little residual effects on aggression or sexual success (Krotzer, 1990). However, parental care behaviour can be substantially disrupted with pollutants causing nest abandonment, reduced maintenance levels, irrational swimming actions and/or increased aggression (Breitburg, 1992; Lorenz & Taylor, 1992; Weber, 1993; Tanner & Knuth, 1995, 1996). Enhanced nest maintenance behaviour has been recorded with increased pollutant exposure where the contaminant is smothering in nature (Potts *et al.*, 1988).

3.3.11.5 Consequences of behavioural change

There are a number of behavioural changes elicited by contaminant exposure which taken either separately or as a whole could have marked effects on the ecological balance. Any detrimental effects on locomotory ability could effect escape capabilities, predator/prey interactions, migratory patterns, and reproduction (Atchison *et al.*, 1987) and effects on antipredatory behaviour have been shown to result in enhanced predatory mortality (Handeland *et al.*, 1996). There have also been recorded incidences of pollution affecting reproductive behaviour both at the level of mate recognition and/or attraction, and in the process of parental care. All of these behavioural changes could have significant effects on the subsequent ecological and population balances, although there appear to have been no attempts to quantify these consequences.

3.3.12 Conclusions

From this review it can be seen that there are clearly potential links at the physiological level between subcellular responses to xenobiotic exposure and higher level consequences on population and yield.

Links can be demonstrated between the induction of proteins employed in detoxification and protection, and increased protein degradation and protein turnover. This has consequences on the energy balance of an organism and the allocation of this energy between detoxification, repair and growth and reproduction. However, it has also been demonstrated that xenobiotics can have a direct impact on various physiological processes including osmoregulation and ionoregulation, excretion, respiration, neuroendocrinology and immune response and developmental processes. Each of these responses may further impact on individual fitness and reproductive capacity.

Of particular interest is the link between induction of detoxification systems, lysosomal compartmentalisation of xenobiotics, protein turnover and cellular/organism energetics. The recognition that reduced whole-body protein turnover consistently underlies lower energy expenditure, with beneficial consequences that include higher growth efficiency and longer survival following pollution impact, is central to understanding how species survive stressed environments. In addition, the observation that multilocus heterozygosity within populations is associated with slower intensities with which proteins are renewed and replaced, again provides a mechanistic link between physiology and population fitness and survival.

Evidence for several of the mechanistic links between cellular and physiological effects has been developed in particularly marine invertebrates and there is a real need to develop investigations that repeat these studies in marine fish. Particularly, and in relation to the previous comments, there is a need to demonstrate pollution impacts on fish energetics and it may be that the more recently developed approach investigating cellular energy allocation makes this feasible. Where studies have employed fish, these tend to be freshwater and whilst demonstrating physiological impact, again this requires extrapolation to marine species.

Also crucial to the link between cellular responses to pollution and population level consequences is the impact of xenobiotics on reproduction and fecundity. At the moment, there is a great deal of effort focused on the impact of endocrine disrupters on the reproduction of fish. Whilst this process is important (see Chapter 5), the present chapter has additionally highlighted the importance of other processes by which reproduction can be impacted. Of particular importance here are direct effects on fecundity and the fertilisation process. More work needs to focus on these aspects of the reproductive process to fully understand population level consequences of sublethal xenobiotics.

3.4 References

Adams, S.M., W.D. Crumby, M.S. Greeley, M.G. Ryon & E.M. Schilling (1992a) Relationships between physiological and fish population responses in a contaminated stream. *Environmental Toxicology and Chemistry*, **11** (11), 1549–1557.

Adams, S.M., W.D. Crumby, M.S. Greeley Jr, L.R. Shugart & C.F. Saylor (1992b) Responses of fish populations and communities to pulp mill effluents: a holistic assessment. *Ecotoxicology and Environmental Safety*, **24**, 347–360.

Alcaraz, G. & S. Espina (1997) Scope for growth of juvenile grass carp *Ctenopharyngodon idella* exposed to nitrite. *Comparative Biochemistry and Physiology C-Pharmacology Toxicology & Endocrinology*, **116** (1), 85–88.

Alkindi, A.Y.A., J.A. Brown, C.P. Waring & J.E. Collins (1996) Endocrine, osmoregularoty, respiratory and haematological parameters in flounder exposed to the water soluble fraction of crude oil. *Journal of Fish Biology*, **49**, 1291–1305.

Anderson, M.J., M.R. Miller & D.E. Hinton (1996a) *In vitro* modulation of 17-β-estradiol-induced vitellogenin synthesis: Effects of cytochrome P4501A1 inducing compounds on rainbow trout (*Oncorhynchus mykiss*) liver cells. *Aquatic Toxicology*, **34** (4), 327–350.

Anderson, M.K., H. Olsen, F. Matsumura & D.E. Hinton (1996b) *In vivo* modulation of 17β-estradiol-induced vitellogenin synthesis and estrogen receptor in rainbow trout (*Oncorhynchus mykiss*) liver cells by b-naphthoflavone. *Toxicology and Applied Pharmacology*, **137**, 210–218.

Andersson, T. (1990) Sex differences in cytochrome P-450 dependent xenobiotic and steroid metabolism in the mature rainbow trout kidney. *Journal of Endocrinology*, **126**, 9–16.

Arukwe, A. & A. Goksøyr (1997) Changes in three hepatic cytochrome P450 subfamilies during a reproductive cycle in Turbot (*Scophthalmus maximus* L). *Journal of Experimental Zoology*, **277** (4), 313–325.

Arukwe, A. & A. Goksøyr (1998) Xenobiotics, xenoestrogens and reproduction disturbances in fish. *Sarsia*, **83** (3), 225–241.

Arukwe, A., L. Forlin & A. Goksøyr (1997a) Xenobiotic and steroid biotransformation enzymes in Atlantic salmon (*Salmo salar*) liver treated with an estrogenic compound, 4-nonylphenol. *Environmental Toxicology and Chemistry*, **16** (12), 2576–2583.

Arukwe, A., F.R. Knudsen & A. Goksøyr (1997b) Fish *zona radiata* (eggshell) protein: a sensitive biomarker for environmental estrogens. *Environmental Health Perspectives*, **105**, 418–422.

Atchison, G.J., M.G. Henry & M.B. Sandheinrich (1987) Effects of metals on fish behaviour: a review. *Environmental Biology of Fishes*, **18**, 11–25.

Atkinson, D.E. (1977) *Cellular Energy Metabolism and its Regulation.* Academic Press, London, 293 pp.

Baker, R.D., S.S. Baker & K. Larosa (1995) Polarised Caco-2 cells – effects of reactive oxygen metabolites on enterocyte barrier function. *Digestive Diseases & Sciences*, **40**, 510–518.

Barry, M.J., K. Ohalloran, D.C. Logan, J.T. Ahokas & D. Holdway (1995) Sublethal effects of esfenvalerate pulse-exposure on spawning and non-spawning Australian Cimson-spotted Rainbowfish (*Melanotaenia fluviatilis*). *Archives of Environmental Contamination and Toxicology*, **28** (4), 459–463.

Basha, S.M., K.S.P. Rao, K.R.S.S. Rao & K.V.R. Rao (1984) Respiratory potentials of the fish (*Tilapia mossambica*) under malathion, carbaryl and lindane intoxication. *Bulletin of Environmental Contamination and Toxicology*, **32** (5), 570–574.

Bayne, B.L. (1989) Measuring the biological effects of pollution: the mussel watch approach. *Water Science and Technology*, **21**, 1089–1100.

Bayne, B.L. & A.J.S. Hawkins (1997) Protein metabolism, the costs of growth and genomic heterozygosity: experiments with the mussel *Mytilus galloprovincialis* L. *Physiological Zoology*, **70**, 391–402.

Bayne, B.L. & M.N. Moore (1997) Non-lymphoid immunologic defenses in aquatic invertebrates and their value as indicators of aquatic pollution. In: (ed. Zelikoff, J.T.) *Ecotoxicology: Responses, Biomarkers and Risk Assessment*. SOS Publications, Fair Haven, New Jersey, pp. 243–261.

Betoulle, S., D. Troutaud, N. Khan & P. Deschaux (1995) Antibody response, cortisol and prolactin levels in rainbow trout. *Comptes Rendus de l'Academie des Sciences Serie III – Sciences de la Vie-Life Sciences*, **318**, 677–681.

Birge, W.J., J.A. Black, A.G. Westerman & J.E. Hudson (1979) The effects of mercury on reproduction of fish and amphibians. In: (ed. Nriagu, J.) *The Biogeochemistry of Mercury in the Environment*. Elsevier, Amsterdam, Holland, 629 pp.

Björnsson, B.T.H., C. Haux, L. Förlin & L.J. Deftos (1986) The involvement of calcitonin in the reproductive physiology of the rainbow trout. *Journal of Endocrinology*, **108**, 17–23.

Bodammer, J.E. (1993) The teratological and pathological effects on embryonic and larval fishes exposed as embryos: a brief review. *American Fisheries Society Symposium*, **14**, 77–84.

Bon, E., G. Corraze, S. Kaushik & F. Le Menn (1997) Effects of accelerated photoperiod regimes on the reproductive cycle of the female rainbow trout. 1. Seasonal variations of plasma lipids correlated with vitellogenesis. *Comparative Biochemistry & Physiology A – Physiology*, **118** (1), 183–190.

Bonga, S.E.W. (1997) The stress response in fish. *Physiological Reviews*, **77** (3), 591–625.

Braw-tal, R., K.P. McNatty, P. Smith, D.A. Heath, N.L. Hudson, D.J. Philips, B.J. McCleod & G.H. Davis (1993) Ovaries of ewes homozygous for the X-linked Inverdale gene (Fex XI) are devoid of secondary and tertiary follicles but contain many abnormal structures. *Biology of Reproduction*, **49**, 895–907.

Breitburg, D.L. (1992) Episodic hypoxia in Chesapeake Bay: interacting effects of recruitment, behavior, and physical disturbance. *Ecological Monographs*, **62**, 525–546.

Bryan, M.D., G.J. Atchison & M.B. Sandheinrich (1995) Effects of cadmium on the foraging behavior and growth of juvenile bluegill, *Lepomis macrochirus*. *Canadian Journal of Fisheries and Aquatic Sciences*, **52**, 1630–1638.

Burton, M.P.M. (1991) Induction and reversal of the non-reproductive state in winter flounder *Pseudopleuronectes americanus* Walbaum, by manipulating food availability. *Journal of Fish Biology*, **39**, 909–910.

Burton, M.P.M. (1994) A critical period for nutritional control of early gametogenesis in female winter flounder, *Pleuronectes americanus* (Pisces, Teleostei). *Journal of Zoology*, **233** (3), 405–415.

Burton, M.P. & D.R. Idler (1987) An experimental investigation of the non-reproductive, post-mature state of winter flounder. *Journal of Fish Biology*, **30**, 643–650.

Caldwell, C.A. & J.M. Hinshaw (1994) Nucleotides and the adenylate energy charge as indicators of stress in rainbow trout (*Oncorhynchus mykiss*) subjected to a range of dissolved oxygen concentrations. *Comparative Biochemistry and Physiology B – Biochemistry & Molecular Biology*, **109**, 313–323.

Cameron, P. & H. von Westernhagen (1997) Malformation rates in embryos of North sea fishes in 1991 and 1992. *Marine Pollution Bulletin*, **34**, 129–134.

Cameron, P., J. Berg, V. Dethlefsen & H. Vonwesternhagen (1992) Developmental defects in pelagic embryos of several flatfish species in the southern North-Sea. *Netherlands Journal of Sea Research*, **29** (1–3), 239–256.

Cameron, P., J. Berg & H. Vonwesternhagen (1996) Biological effects monitoring of the North Sea employing fish embryological data. *Environmental Monitoring and Assessment*, **40** (2), 107–124.

Campbell, P.M., T.G. Pottinger & J.P. Sumpter (1992) Stress reduces the quality of gametes produced by rainbow trout. *Biology of Reproduction*, **47** (6), 1140–1150.

Campbell, W.B., J.M. Emlen & W. Hershberger (1998) Thermally induced chronic development stress in coho salmon: integrating measures of mortality, early growth and developmental instability. *Oikos*, **81**, 398–410.

Carragher, J.F. & J.P. Sumpter (1991) The mobilization of calcium from calcified tissues of rainbow trout (*Oncorhynchus mykiss*) induced to synthesize vitellogenin. *Comparative Biochemistry and Physiology*, **A99** (1–2), 169–172.

Carter, C.G., D.F. Houlihan, J. Brechin & I.D. McCarthy (1993a) The relationships between protein intake and protein accretion, synthesis, and retention efficiency for individual grass carp, *Ctenopharyngodon idella* (Valenciennes). *Canadian Journal of Zoology*, **71**, 392–400.

Carter, C.G., D.F. Houlihan, B. Buchanan & A.I. Mitchell (1993b) Protein-nitrogen flux and protein growth efficiency of individual Atlantic salmon (*Salmo salar* L.). *Fish Physiology and Biochemistry*, **12**, 305–315.

Casillas, E., D. Misitano, L.L. Johnson, L.D. Rhodes, T.K. Collier, J.E. Stein, B.B. McCain & U. Varanasi (1991) Inducibility of spawning and reproductive success of female English sole (*Parophrys vetulus*) from urban and non-urban areas of Puget Sound, Washington. *Marine Environmental Research*, **31** (2), 99–122.

Cerda, J., M. Carrillo, S. Zanuy & J. Ramos (1994) Effect of food ration on estrogen and vitellogenin plasma levels, fecundity and larval survival in captive sea bass, *Dicentrarchus labrax* – preliminary observations. *Aquatic Living Resources*, **7**, 255–266.

Chan, H.S.L., P.S. Thorner, G. Haddad & V. Ling (1990) Immunohistochemical detection of P-glycoprotein: prognostic correlaiton in soft tissue sarcoma of childhood. *Journal of Clinical Oncology*, **8**, 689–704.

Chatterjee, S. & S. Bhattacharya (1984) Detoxication of industrial pollutants by the glutathione glutathione-s-transferase system in the liver of *Anabas testudineus* (Bloch). *Toxicology Letters*, **22** (2), 187–198.

Christiansen, T., B. Korsgaard & A. Jespersen (1998) Effects of nonylphenol and 17 beta-oestradiol on vitellogenin synthesis, testicular structure and cytology in male eelpout *Zoarces viviparus*. *Journal of Experimental Biology*, **201** (2), 179–192.

Clarke, D.J., B. Burchell & S.G. George (1992) Differential expression and induction of UDP-glucuronosyltransferase isoforms in hepatic and extrahepatic tissues of fish, *Pleuronectes platessa*; immunochemical and functional characterization. *Toxicology and Applied Pharmacology*, **115**, 130–136.

Colgan, P.W., J.A. Cross & P.H. Johansen (1982) Guppy behaviour during exposure to a sub-lethal concentration of phenol. *Bulletin of Environmental Contamination and Toxicology*, **28**, 20–27.

Cornwall, R., B.H. Toomey, S. Bard, C. Bacon, W.M. Jarman & D. Epel (1995) Characterization of multixenobiotic multidrug transport in the gills of the mussel *Mytilus californianus* and identification of environmental substrates. *Aquatic Toxicology*, **31** (4), 277–296.

Crane, M., T. Flower, D. Holmes & S. Watson (1992) The toxicity of selenium in experimental freshwater ponds. *Archives of Environmental Contamination and Toxicology*, **23** (4), 440–452.

Cravedi, J.P., A. Paris, E. Perdudurand & P. Prunet (1995a) Influence of growth-hormone on the hepatic mixed-function oxidase and transferase systems of rainbow-trout. *Fish Physiology and Biochemistry*, **14** (4), 259–266.

Cravedi, J.P., E. Perdudurand, A. Paris & P. Prunet (1995b) Growth-hormone effect on xenobiotic-metabolizing activities of rainbow-trout. *Marine Environmental Research*, **39** (1–4), 89–92.

Cross, J.N., J.T. Hardy, J.E. Hose, G.P. Hershelman, L.D. Antrim, R.W. Gosselt & E.A. Crecelius (1987) Contaminant concentrations and toxicity of sea-surface microlayer near Los Angeles, California. *Marine Environmental Research*, **23**, 307–323.

De Boeck, G., H. Desmet & R. Blust (1995) The effect of sublethal levels of copper on oxygen consumption and ammonia excretion in the common carp, *Cyprinus carpio*. *Aquatic Toxicology*, **32**, 127–141.

De Boeck, G., R. Borger, A. VanderLinden & R. Blust (1997) Effects of sublethal copper exposure on muscle energy metabolism of common carp, measured by P^{31} nuclear magnetic resonance spectroscopy. *Environmental Toxicology and Chemistry*, **16**, 676–684.

De Coen, W.M. & C.R. Janssen (1997) The use of biomarkers in *Daphnia magna* toxicity testing. IV. Cellular Energy Allocation: a new methodology to assess the energy budget of toxicant-stressed *Daphnia* populations. *Journal of Aquatic Ecosystem Stress and Recovery*, **6**, 43–55.

Dennis, T.E. & A.J. Bulger (1995) Condition factor and whole-body sodium concentrations in a freshwater fish: evidence for acidification stress and possible ionoregulatory over-compensation. *Water Air and Soil Pollution*, **85**, 377–382.

Dethlefsen, V., H. von Westernhagen & P. Cameron (1996) Malformations in North Sea pelagic fish embryos during the period 1984–1995. *ICES Journal of Marine Science.*, **53** (6), 1024–1035.

Donkin, P. & J. Widdows (1986) Scope for growth as a measure of environmental pollution and its interpretation using structure-activity relationships. *Chemistry and Industry*, **21**, 732–735.

Donohoe, R.M. & L.R. Curtis (1996) Estrogenic activity of chlordecone, o,p'-DDT and o,p'-DDE in juvenile rainbow trout: induction of vitellogenesis and interaction with hepatic estrogen binding sites. *Aquatic Toxicology*, **36**, 31–52.

Duston, J. & N. Bromage (1988) The entrainment and gating of the endogenous circannual rhythm of reproduction in the female Rainbow trout (*Salmo gairdneri*). *Journal of Comparative Physiology A-Sensory Neural and Behavioural Physiology*, **164** (2), 259–268.

Duston, J. & N. Bromage (1991) Circannual rhythms of gonadal maturation in female Rainbow trout (*Oncorhynchus mykiss*). *Journal of Biological Rhythms*, **6** (1), 49–53.

Dutil, J.D., J. Munro, C. Audet & M. Besner (1992) Seasonal variation in the physiological response of Atlantic cod (*Gadus morhua*) to low salinity. *Canadian Journal of Fisheries and Aquatic Sciences*, **49**, 1149–1156.

Ensenbach, U. & R. Nagel (1997) Toxicity of binary chemical mixtures: Effects on reproduction of zebrafish (*Brachydanio rerio*). *Archives of Environmental Contamination and Toxicology*, **32** (2), 204–210.

Fitzpatrick, F.A. & R.C. Murphy (1989) Cytochrome P-450 metabolism of arachidonic acid: Formation and biological actions of 'Epoxygenase' derived eicosanoids. *Pharmacological Reviews*, **40**, 229–241.

Förlin, L. & T. Hansson (1982) Effects of oestradiol-17-beta and hypophysectomy on hepatic mixed function oxidases in rainbow trout. *Journal of Endocrinology*, **95** (2), 245–252.

Förlin, L. & C. Haux (1990) Sex differences in hepatic cytochrome-P-450 monooxygenase activities in Rainbow trout during an annual reproductive cycle. *Journal of Endocrinology*, **124** (2), 207–213.

Fujita, T., W. Hamaura, A. Takemura & K. Takano (1997) Histological observations of annual reproductive cycle and tidal spawning rhythm in the female porcupine fish *Diodon holocanthus*. *Fisheries Science*, **63** (5), 715–720.

Gadagbui, B.K.M. & A. Goksøyr (1996) CYP1A and other biomarker responses to effluents from a textile mill in the Volta River (Ghana) using caged tilapia (*Oreochromis niloticus*) and sediment-exposed mudfish (*Clarias anguillaris*). *Biomarkers*, **1** (4), 252–261.

Gadagbui, B.K.M., M. Addy & A. Goksøyr (1996) Species characteristics of hepatic biotransformation enzymes in two tropical freshwater teleosts, Tilapia (*Oreochromis niloticus*) and mudfish (*Clarias anguillaris*). *Comparative Biochemistry & Physiology C – Pharmacology, Toxicology & Endocrinology*, **114** (3), 201–211.

Gagnon, M.M., J.J. Dodson, P.V. Hodson, G. van der Kraak & J.H. Carey (1994) Seasonal effects of bleached kraft mill effluent on reproductive parameters of white sucker (*Catostomus commersoni*) populations of the St. Maurice river, Quebec, Canada. *Canadian Journal of Fisheries and Aquatic Sciences*, **51** (2), 337–347.

Galgani, F., R. Cornwall, B.H. Toomey & D. Epel (1996) Interaction of environmental xenobiotics with a multixenobiotic defense mechanism in the bay mussel *Mytilus galloprovincialis* from the coast of California. *Environmental Toxicology and Chemistry*, **15**, 325–331.

George, S.G. (1994) Enzymology and molecular biology of phase II xenobiotic-conjugating enzymes in fish. In: (eds Malins, D.C. & G.K. Ostrander) *Aquatic Toxicology: Molecular, biochemical and cellular Perspecitves*. Lewis Publishers, Boca Raton, FL, pp. 37–85.

Georgy, S.C. & P. Young (1986) The time course of effects of cadmium and 3-methylcholanthrene on activities of enzymes of xenobiotic metabolism and metallothionein levels in the plaice *Pleuronectes platessa*. *Comparative Biochemistry and Physiology*, **83C**, 37–44.

Goksøyr, A. (1995) Cytochrome P450 in marine mammals: isozyme forms, catalytic functions, and physiological regulations. In: *Whales, Seals, Fish and Man*. Elsevier Science BV, London, pp. 629–639.

Goksøyr, A. & L. Förlin (1992) The cytochrome P450 system in fish, aquatic toxicology and environmental monitoring. *Aquatic Toxicology*, **22**, 287–311.

Goksøyr, A., T. Andersson, T. Hansson, J. Klungoyr, Y. Zhang & L. Forlin (1987) Species characteristics of the hepatic xenobiotic and steroid biotransformation systems of two teleost fish, Atlantic cod (*Gadus morhua*) and Rainbow trout (*Salmo gairdneri*). *Toxicology and Applied Pharmacology*, **89**, 347–360.

Gordon, H. & M. Gordon (1954) Bionetry of seven natural populations of the platyfish, *Xiphophorus maculatus*, from Central America. *Zoologica*, **39**, 37–59.

Gottesman, M.M. & I. Pastan (1993) Biochemistry of multidrug resistance mediated by the multidrug transporter. *Annual Review of Biochemistry*, **62**, 385–427.

Gray, M.A. & C.D. Metcalfe (1997) Induction of testis-ova in Japanese medaka (*Oryzias latipes*) exposed to *p*-nonylphenol. *Environmental Toxicology and Chemistry*, **16**, 1082–1086.

Green, J.A. & B.I. Baker (1991) The influence of repeated stress on the release of melanin-concentrating hormone in the Rainbow trout. *Journal of Endocrinology*, **128** (2), 261–266.

Grundy, M.M., M.N. Moore, S.M. Howell & N.A. Ratcliffe (1996a) Phagocytic reduction and effects on lysosomal membranes of polycyclic aromatic hydrocarbons, in haemocytes of *Mytilus edulis*. *Aquatic Toxicology*, **34** (4), 273–290.

Grundy, M.M., N.A. Ratcliffe & M.N. Moore (1996b) Immune inhibition in marine mussels by polycyclic aromatic hydrocarbons. *Marine Environmental Research*, **42** (1–4), 187–190.

Guengerich, F.P. (ed.) (1987) *Mammalian cytochromes P-450*. CRC Press Inc., Boca Raton, Florida.

Handeland, S.O., T. Jarvi, A. Ferno & S.O. Stefansson (1996) Osmotic stress, antipredator behaviour and mortality of Atlantic salmon (*Salmo salar*) smolts. *Canadian Journal of Fisheries and Aquatic Sciences*, **53**, 2673–2680.

Handy, R.D., D.W. Sims, A. Giles, H.A. Campbell & M.M. Musonda (1999) Metabolic trade-off between locomotion and detoxification for maintenance of blood chemistry and growth parameters by rainbow trout (*Oncorhynchus mykiss*) during chronic dietary exposure to copper. *Aquatic Toxicology*, **47**, 23–41.

Hanson, P.D., H. von-Westernhagen & H. Rosenthal (1985) Chlorinated hydrocarbons and hatching success in Baltic herring spring spawners. *Marine Environmental Research*, **15**, 59–76.

Hardy, J., S. Kiesser, L. Antrim, A. Stubin, R. Kocan & J. Strand (1987) The sea-surface microlayer of Puget Sound, Washington, USA. I. Toxic effects on fish eggs and larvae. *Marine Environmental Research*, **23**, 227–250.

Harries, J.E., D.A. Sheahan, S. Jobling, P. Matthiessen, P. Neall, E. Routledge, R. Rycroft, J.P. Sumpter & T. Tylor (1996) A survey of estrogenic activity in United Kingdom inland waters. *Environmental Toxicology and Chemistry*, **15**, 1993–2002.

Harries, J.E., D.A. Sheahan, S. Jobling, P. Matthiessen, P. Neall, J.P. Sumpter, T. Tyler & N. Zaman (1997) Estrogenic activity in five United Kingdom rivers detected by measurement of vitellogenesis in caged male trout. *Environmental Toxicology and Chemistry*, **16**, 534–542.

Havas, M. & E. Advokaat (1995) Can sodium regulation be used to predict the relative acid-sensitivity of various life stages and different species of aquatic fauna? *Water Air and Soil Pollution*, **85**, 865–870.

Hawkins, A.J.S. (1988) Genetic variations in the physiology of marine shellfish. In: (ed. Iturrondobeitia, J.C.) *Actas del Congreso de Biologia Ambiental; II Congreso Mundial Vasco*. Vol. 1. Servicio Editorial de la Universidad del Pais Vasco, Bilbao, pp. 121–131.

Hawkins, A.J.S. (1991) Protein turnover: a functional appraisal. *Functional Ecology*, **5**, 222–233.

Hawkins, A.J.S. & B.L. Bayne (1991) Nutrition of marine mussels: factors influencing the relative utilizations of protein and energy. *Aquaculture*, **94** (2–3), 177–196.

Hawkins, A.J.S. & A.J. Day (1996) The metabolic basis of genetic differences in growth efficiency among animals. *Journal of Experimental Marine Biology and Ecology*, **203**, 93–115.

Hawkins, A.J.S. & T.J. Hilbish (1992) The costs of cell volume regulation: protein metabolism during hyperosmotic adjustment. *Journal of the Marine Biological Association of the United Kingdom*, **72**, 569–578.

Hawkins, A.J.S., N.R. Menon, R. Damodaran & B.L. Bayne (1987) Metabolic responses of the mussels *Perna viridis* and *Perna indica* to declining oxygen tension at different salinities. *Comparative Biochemistry& Physiology A – Physiology*, **88** (4), 691–694.

Hawkins, A.J.S., J. Widdows & B.L. Bayne (1989a) The relevance of whole-body protein metabolism to measured costs of maintenance and growth in *Mytilus edulis*. *Physiological Zoology*, **62**, 745–763.

Hawkins, A.J.S., J. Rusin, B.L. Bayne & A.J. Day (1989b) The metabolic/physiological basis of genotype-dependent mortality during copper exposure in *Mytilus edulis*. *Marine Environmental Research*, **28** (1–4), 253–257.

Heinze, W., S. Kruger & H.W. Kamp (1993) Circadian and circannual rhymicity – effects on toxicity of substances. *Monatshefte Fur Veterinarmedizin*, **48** (1), 37–43.

Herman, R.L. & H.L. Kincaid (1988) Pathological effects of orally administered estradiol to rainbow trout. *Aquaculture*, **72**, 165–172.

Higgins, C.F. (1992) ABC transporters: from micro-organisms to man. *Annual Review of Cell Biology*, **8**, 67–113.

Hinton, D.E. & D.J. Lauren (1990) Liver structural alterations accompanying chronic toxicity in fishes: potential biomarkers of exposure. In: (eds McCarthy, J.F. & L.K. Shugart) *Biomarkers of Environmental Contamination*. Lewis Publishers, Boca Rota, Ann Arbor, London, pp. 17–37.

Holland-Toomey, B.H. & D. Epel (1993) Multixenobiotic resistance in *Urechis caup* embryos: Protection from environmental toxins. *Biological Bulletin*, **185**, 355–364.

Hontela, A. (1998) Interrenal dysfunction in fish from contaminated sites: *in vivo* and *in vitro* assessment. *Environmental Toxicology and Chemistry*, **17** (1), 44–48.

Hontela, A., J.B. Rasmussen, C. Audet & G. Chevalier (1992) Impaired cortisol stress response in fish from environments polluted by PAHs, and mercury. *Archives of Environmental Contamination and Toxicology*, **22** (3), 278–283.

Hontela, A., P. Dumont, D. Duclos & R. Fortin (1995) Endocrine and metabolic dysfunction in Yellow perch, *Perca flavescens*, exposed to organic contaminants and heavy-metals in the St. Lawrence river. *Environmental Toxicology and Chemistry*, **14** (4), 725–731.

Hoque, M.T., F.M. Yusoff, A.T. Law & M.A. Syed (1998) Effect of hydrogen sulphide on liver somatic index and Fulton's condition factor in *Mystus nemurus*. *Journal of Fish Biology*, **52**, 23–30.

Isani, G., R. Serra, O. Cattani, P. Cortesi & E. Carpene (1997) Adenylate energy charge and metallothionein as stress indices in *Mytilus galloprovincialis*. *Journal of the Marine Biological Association of the United Kingdom*, **77** (4), 1187–1197.

Ivanovici, A.M. (1980) Adenylate energy charge: an evaluation of applicability to assessment of pollution effects and directions for future research. *Rapports et Procès-verbaux de Réunions du Conseil International pour l'Exploration de la Mer.*, **179**, 23–28.

James, W.H. (1988) Testosterone levels, handedness and sex ratio at birth. *Journal of Theoretical Biology*, **133**, 261–266.

Janz, D.M., M.E. McMaster, K.R. Munkittrick & G. van der Kraak (1997) Elevated ovarian follicular apoptosis and heat shock protein 70 expression in white sucker exposed to bleached kraft pulp mill effluent. *Toxicology and Applied Pharmacology*, **147**, 391–398.

Jobling, M. (1994) *Fish Bioenergetics*. Chapman & Hall, London, 309 pp.

Jobling, S., D. Sheahan, J.A. Osborne, P. Matthiessen & J.P. Sumpter (1996) Inhibition of testicular growth in rainbow trout (*Oncorhynchus mykiss*) exposed to estrogenic alkylphenolic chemicals. *Environmental Toxicology and Chemistry*, **15** (2), 194–202.

Jobling, S.J. & J.P. Sumpter (1993) Detergent components in sewage effluent are weakly oestrogenic to fish: an *in vitro* study using rainbow trout hepatocytes. *Aquatic Toxicology*, **27**, 361–372.

Johnson, L.L., J.E. Stein, T.K. Collier, E. Casillas & U. Varanasi (1994) Indicators of reproductive development in prespawning female Winter flounder (*Pleuronectes americanus*) from urban and non-urban estuaries in the Northeast United States. *Science of the Total Environment*, **141**, 241–260.

Johnston, I.A. (1993) Phenotypic plasticity of fish muscle to temperature change. In: (eds Rankin, J.C. & F.B. Jensen) *Fish Ecophysiology*. Chapman and Hall, London, pp. 322–340.

Jones, J.C. & J.D. Reynolds (1997) Effects of pollution on reproductive behaviour of fishes. *Reviews in Fish Biology*, **7** (4), 463–491.

Kallman, K.D., M.P. Schreibman & V. Borkoski (1973) Genetic control of gonadotrop differentiation in the platyfish, *Xiphophorus masculatus* (poeciliidae). *Science*, **181**, 678–680.

Khan, I.A. & P. Thomas (1996) Disruption of neuroendocrine function in Atlantic croaker exposed to Aroclor-1254. *Marine Environmental Research*, **42** (1–4), 145–149.

Khan, I.A. & P. Thomas (1997) Aroclor 1254-induced alterations in hypothalamic momoamine metabolism in the Atlantic croaker (*Micropogonias undulatas*): Correlation with pituitary gonadotropin release. *Neurotoxicology*, **18** (2), 553–560.

Kime, D.E. (1995) The effects of pollution on reproduction in fish. *Reviews in Fish Biology*, **5** (1), 52–95.

Kime, D.E. (1998) A strategy for assessing the effects of xenobiotics on fish reproduction. *Science of the Total Environment*, **225** (1–2), 3–11.

Kime, D.E., M. Ebrahimi, K. Nysten, I. Roelants, E. Rurangwa, H.D.M. Moore & F. Ollevier (1996) Use of computer assisted sperm analysis (CASA) for monitoring the effects of pollution on sperm quality of fish; Application to the effects of heavy metals. *Aquatic Toxicology*, **36** (3–4), 223–237.

Kirchin, M.A., M.N. Moore, R.T. Dean & G.W. Winston (1992) The role of oxyradicals in intracellular proteolysis and toxicity in mussels. *Marine Environmental Research*, **34**, 315–320.

Kleinow, K.M., M.J. Melancon & J.J. Lech (1987) Biotransformation and induction: Implications for toxicity, bioaccumulation and monitoring of environmental xenobiotics in fish. *Environmental Health Perspectives*, **71**, 105–119.

Kloeppersams, P.J., S.M. Swanson, T. Marchant, R. Schryer & J.W. Owens (1994) Exposure of fish to biologically treated bleached-kraft effluent. I. Biochemical, physiological and pathological assessment of rocky-mountain whitefish (*Prosopium williamsoni*) and longnose sucker (*Catostomus catostomus*). *Environmental Toxicology and Chemistry*, **13** (9), 1469–1482.

Knoph, M.B. (1995) Effects of metomidate anaesthesia or transfer to pure sea water on plasma parameters in ammonia-exposed Atlantic salmon (*Salmo salar* L.) in sea water. *Fish Physiology and Biochemistry*, **14**, 103–109.

Kocan, R.M., H. von Westernhagen, M.L. Laandolt & G. Furstenburg (1987) Toxicity of sea surface microlayer: effects of hexane extracts on Baltic herring and Atlantic cod embryos. *Marine Environmental Research*, **23**, 291–305.

Köhler, A. (1989) Cellular effects of environmental contamination in fish from the River Elbe and the North Sea. *Marine Environmental Research*, **28**, 417–424.

Köhler, A., H. Deisemann & B. Lauritzen (1992) Histological and cytochemical indices of toxic injury in the liver of dab *Limanda limanda*. *Marine Ecology Progress Series*, **91** (1/3), 141–153.

Krotzer, M.J. (1990) The effects of induced masculinization on reproductive and aggressive behaviors of the female mosquitofish, *Gambusia affinis*. *Environmental Biology of Fishes*, **29**, 127–134.

Kurelec, B. (1992) The Multixenobiotic Resistance Mechanism in Aquatic Organisms. *Critical Reviews in Toxicology*, **22**, 23–43.

Kurelec, B., S. Krca & D. Lucic (1996) Expression of multixenobiotic resistance mechanism in a marine mussel *Mytilus galloprovincialis* as a biomarker of exposure to polluted environments. *Comparative Biochemistry & Physiology C – Pharmacology Toxicology & Endocrinology*, **113** (2), 283–289.

Laale, H.W. & W. Lerner (1981) Teratology and early fish development. *American Zoologist*, **21**, 517–533.

Lam, T.J. (1983) Environmental influences on gonadal activity in fish. In: (eds Hoar, W.S., D.J. Randall & E.M. Donaldson) *Fish Physiology*. Academic Press, New York, pp. 65–116.

Lange, U., A. Goksøyr, R. Stagg, D. Danischewski, D. Siebers, F. Buchholz & L. Karbe (1994) *Abiotic and biotic factors influencing the natural variation of the cytochrome P450 enzyme system in the liver of dab* (Limanda limanda). Berichte Zentr. Meeres-u. Klimaforsch, Interim Report.

Larsen, H.E., M. Celander & A. Goksøyr (1992) The cytochrome-P450 system of the Atlantic salmon (*Salmo salar*). 2. Variations in hepatic catalytic activities and isozyme patterns during an annual reproductive cycle. *Fish Physiology and Biochemistry*, **10** (4), 291–301.

Lawrence, A.J. (1996) Environmental and endocrine control of reproduction in two species of polychaete: Potential bio-indicators for global climate change. *Journal of the Marine Biological Association of the United Kingdom*, **76**, 247–250.

Lawrence, A.J. & P.J.W. Olive (1995) Gonadotrophic hormone in *Eulalia viridis* (Polychaeta, Annelida): Stimulation of vitellogenesis. *International Journal of Invertebrate Reproduction and Development*, **28**, 43–52.

Lawrence, A.J. & B. Nicholson (1998) The use of stress proteins in *Mytilus edulis* as indicators of chlorinated effluent pollution. *Water Science and Technology*, **38**, 253–261.

Lawrence, A.J. & C. Poulter (1996) The potential role of the estuarine amphipod *Gammarus duebeni* in sub-lethal ecotoxicology testing. *Water Science and Technology*, **34** (7–8), 93–100.

Lawrence, A.J. & C. Poulter (1998) Development of a sub-lethal pollution bioassay using the estuarine amphipod *Gammarus duebeni*. *Water Research*, **32**, 569–578.

Lawrence, A.J.C. & Poulter (2001) The impact of copper, PCP and benzo[a]pyrene on the reproduction of *Chaetogammarus marinus*. *Marine Ecology Progress Series*, **223**, 213–223.

Lazier, C.B. & M.E. MacKay (1993) Vitellogenin gene expression in teleost fish. In: (eds Hochachka, P.W. & T.P. Mommsen) *Biochemistry and Molecular Biology of Fishes*. Vol. 2. Elsevier Science, New York, pp. 391–405.

Le Ruyet, J.P., G. Boeuf, J.Z. Infante, S. Helgason & A. Le Roux (1998) Short-term physiological changes in turbot and seabream juveniles exposed to exogenous ammonis. *Comparative Biochemistry and Physiology*, **119A**, 511–518.

Leary, R.F., F.W. Allendorf & K.L. Knudsen (1984) Superior development stability for heterozygotes at enzyme loci in salmonid fishes. *The American Naturalist*, **124**, 540–551.

Leatherland, J.F. (1993) Field observations on reproductive and developmental dysfunction in introduced and native salmonids from the Great-Lakes. *Journal of Great Lakes Research*, **19** (4), 737–751.

Leaver, M.J., D.J. Clarke & S.G. George (1992) Molecular studies of the phase II of xenobiotic-conjugating enzymes of Pleuronectid flatfish. *Aquatic Toxicology*, **22**, 265–278.

Lemaire, P. & D.R. Livingstone (1997) Aromatic hydrocarbon quinone-mediated reactive oxygen species production in hepatic microsomes of the flounder (*Platichthys flesus* L.). *Comparative Biochemistry and Physiology*, **117C**, 131–139.

Lemaire, P., J. Sturve & L. Förlin (1996) Studies on aromatic hydrocarbon quinone metabolism and DT-diaphorase function in liver of fish species. *Marine Environmental Research*, **42** (1–4), 317–321.

Lemly, A.D. (1996) Wastewater discharges may be most hazardous to fish during winter. *Environmental Pollution*, **93**, 169–174.

Lesko, L.T., S.B. Smith & M.A. Blouin (1996) The effect of contaminated sediments on fecundity of the brown bullhead in three Lake Erie tributaries. *Journal of Great Lakes Research*, **22** (4), 830–837.

Leslie, P.H. (1945) On the use of matrices in certain population mathematics. *Biometrika*, **33**, 183–212.

Lindesjöö, E. & J. Thulin (1992) A skeletal deformity of northern pike (*Esox lucius*) related to pulp mill effluents. *Canadian Journal of Fisheries and Aquatic Sciences*, **49**, 166–172.

Lindesjöö, E., J. Thulin, B.E. Bengtsson & U. Tjarnlund (1994) Abnormalities of a gill cover bone, the operculum, in perch *Perca fluviatilis* from a pulp mill effluent area. *Aquatic Toxicology*, **28**, 189–207.

Lindström-Seppä, P., U. Koivusaari, O. Hänninen & H. Pyysalo (1985) Cytochrome P-450 and monooxygenase activities in the biomonitoring of aquatic environments. *Pharmazie*, **40**, 232–234.

Linlokken, A., E. Kleiven & D. Matzow (1991) Population structure, growth and fecundity of perch (*Perca fluviatilis* L.) in an acidified river system in southern Norway. *Hydrobiologia*, **220**, 179–188.

Linton, T.K., S.D. Reid & C.M. Wood (1997) The metabolic costs and physiological consequences to juvenile rainbow trout of a simulated summer warming scenario in the presence and absence of sublethal ammonia. *Transactions of the American Fisheries Society*, **126**, 259–279.

Livingstone, D.R., P. Lemaire, A. Matthews, L. Peters, J.K. Chipman, J.W. Marsh, S. Archibald, D. Bucke & P. Dixon (1992) Antioxidant enzymes in liver of dab (*Limanda limanda*) from the North Sea. *Marine Ecology Progress Series*, **91**, 97–104.

Longwell, A.C. & J.B. Hughes (1981) Cytologic, cytogenic, and embryological state of Atlantic mackerel eggs from surface waters of the New York Bight in relation to pollution. *Rapports et Procès-verbaux de Réunions du Conseil International pour l'Exploration de la Mer*, **178**, 76–78.

Loning, S. (1977) The effects of crude Ekofisk oil and oil products on marine fish larvae. *Astarte*, **10**, 37–47.

Lorenz, J.J. & D.H. Taylor (1992) The effects of low pH as a chemical stressor on the ability of convict cichlids to raise their young. *Copeia*, **1992**, 832–839.

Lowe, D.M. (1988) Alterations in cellular structure of *Mytilus edulis* resulting from exposure to environmental contaminants under field and experimental conditions. *Marine Ecology Progress Series*, **46**, 91–100.

Lowe, D.M., M.N. Moore & B.M. Evans (1992) Contaminant impact on interactions of molecular probes with lysosomes in living hepatocytes from dab *Limanda limanda*. *Marine Ecology Progress Series*, **91** (1/3), 135–140.

Lowe, D.M., V.U. Fossato & M.H. Depledge (1995a) Contaminant induced lysosomal damage in blood cells of mussel *Mytilus galloprovincialis* from the Venice Lagoon: an *in vitro* study. *Marine Ecology Progress Series*, **129**, 189–196.

Lowe, D.M., C. Soverchia & M.N. Moore (1995b) Lysosomal membrane damage in the blood and digestive cells of mussels exposed to fluoranthene. *Aquatic Toxicology*, **33**, 105–112.

Lukyanova, O.N. (1994) Some biochemical parameters of marine invertebrates from a region of anthropogenic pollution. *Russian Journal of Ecology*, **25**, 435–439.

Lye, C.M., C.L.J. Frid, M.E. Gill & D. McCormick (1997) Abnormalities in the reproductive health of flounder *Platichthys flesus* exposed to effluent from a sewage treatment works. *Marine Pollution Bulletin*, **34** (1), 34–41.

Mallat, J., J.F. Bailey, S.J. Lampa, M.A. Evans & S. Brumbaugh (1995) A fish gill system for quantifying the ultrastructural effects of environmental stressors: methylmercury, Kepone, and heat shock. *Canadian Journal of Fisheries and Aquatic Sciences*, **52**, 1165–1182.

Marty, G.D., J.M. Nunez, D.J. Lauren & D.E. Hinton (1990) Age-dependent changes in toxicity of N-nitroso compounds to Japanese medaka (*Oryziass latipes*) embryos. *Aquatic Toxicology*, **17**, 45–62.

Matthiessen, P. & J.W.M. Logan (1984) Low concentration effects of endosulfan insecticide on reproductive behavior in the tropical cichlid fish *Sarotherodon mossambicus*. *Bulletin of Environmental Contamination and Toxicology*, **33**, 575–583.

McCarthy, I.D., D.F. Houlihan & C.G. Carter (1994) Individual variation in protein turnover and growth efficiency in rainbow trout, *Oncorhynchus mykiss* (Walbaum). *Proceeedings of the Royal Society of London B*, **257**, 141–147.

McIntyre, A.D. (1973) Toxicity of methylmercury for steelhead trout sperm. *Bulletin of Environmental Contamination and Toxicology*, **9**, 98.

McNatty, K.P., N.L. Hudson, L. Shaw & L. Moore (1994) Plasma concentrations of FSH, LH, thyroid stimulating hormone and growth hormone after exogenous stimulation with GnRH, TRH and GHRH in Booroola ewes that are homozygous carriers for the FecB gene. *Journal of Reproduction and Fertility*, **102**, 177–183.

Miller, D.C. (1980) Some applications of locomotor response in pollution effects monitoring. *Rapports et Procès-verbaux de Réunions du Conseil International pour l'Exploration de la Mer*, **179**, 154–161.

Minier, C. & M.N. Moore (1996) Induction of multixenobiotic resistance in mussel blood cells. *Marine Environmental Research*, **42**, 389–392.

Minier, C., F. Akcha & F. Galgani (1993) P-glycoprotein expression in *Crassostrea gigas* and *Mytilus edulis* in polluted seawater. *Comparative Biochemistry and Physiology*, **106B**, 1029–1036.

Mommsen, P.T. & P.J. Walsh (1988) Vitellogenesis and oocyte assembly. In: (eds Hoar, W.S., D.J. Randall & E.M. Donaldson) *Fish Physiology*. Academic Press, New York, pp. 347–406.

Montgomery, G.W., E.A. Lord, J.M. Penty, K.G. Dodds, T.E. Broad, L. Cambridge, L.S.L.F. Sunden, R. Stone & A.M. Crawford (1994) The Booroola fecundity (FecB) gene maps to sheep chromosome 6. *Genomics*, **22**, 148–153.

Moore, A. & C.P. Waring (1996) Sublethal effects of the pesticide Diazinon on the olfactory function in mature male Atlantic salmon parr. *Journal of Fish Biology*, **48**, 758–775.

Moore, M.N. (1985) Cellular responses to pollutants. *Marine Pollution Bulletin*, **16**, 134–139.

Moore, M.N. (1988) Cytochemical responses of the lysosomal system and NADPH-ferrihemoprotein reductase in molluscs to environmental and experimental exposure to xenobiotics. *Marine Ecology Progress Series*, **46**, 81–89.

Moore, M.N. (1990) Lysosomal cytochemistry in marine environmental monitoring. *Histochemical Journal*, **22**, 187–191.

Moore, M.N. & A. Viarengo (1987) Lysosomal membrane fragility and catabolism of cytosolic proteins: evidence for a direct relationship. *Experientia.*, **43**, 320–323.

Moore, M.N. & R.I. Willows (1998) A model for cellular uptake and intracellular behaviour of particulate-bound micropollutants. *Marine Environmental Research*, **46** (1–5), 509–514.

Moore, M.N., J. Widdows, J.J. Cleary, R.K. Pipe, P.N. Salkeld, P. Donkin, S.V. Farrar, S.V. Evans & P.E. Thompson (1984) Responses of the mussel *Mytilus edulis* to copper and phenanthrene: interactive effects. *Marine Environmental Research*, **14**, 167–183.

Moore, M.N., A. Köhler, D.M. Lowe & M.G. Simpson (1994) An integrated approach to cellular biomarkers in fish. In: (eds Fossi, M.C. & C. Leonzio) *Non-destructive biomarkers in vertebrates*. Lewic/CRC, Boca Raton, pp. 171–197.

Moore, M.N., C. Soverchia & M. Thomas (1996a) Enhanced lysosomal autophagy of intracellular proteins by xenobiotics in living molluscan blood cells. Proceedings of the 10th International Congress. *Acta Histochemica et Cytochemica*, **29**, 947–948.

Moore, M.N., R.J. Wedderburn, D.M. Lowe & M.H. Depledge (1996b) Lysosomal reaction to xenobiotics in mussel haemocytes using BODIPY-FL-verapamil. *Marine Environmental Research*, **42**, 99–105.

Moser, M.L. & J.M. Miller (1994) Effects of salinity fluctuation on routine metabolism of juvenile spot, *Leiostomus xanthurus*. *Journal of Fish Biology*, **45**, 335–340.

Nagele, A. (1995) Unimpaired metabolism of pyridine dinucleotides and adenylates in Chinese hamster ovary cells during oxidative stress elicited by cytotoxic does of copper-putrescine-pyridine. *Biochemical Pharmacology*, **49**, 147–155.

Nagler, J.J. & D.G. Cyr (1997) Exposure of male American plaice (*Hippoglossoides platessoides*) to contaminated marine sediments decreases the hatching success of their progeny. *Environmental Toxicology and Chemistry*, **16** (8), 1733–1738.

Nebert, D.W. & F.J. Gonzalez (1987) P450 genes: Structure, evolution, and regulation. *Annual Review of Biochemistry*, **56**, 945–993.

Nebert, D.W., D.R. Nelson, M. Adesnik, M.J. Coon, R.W. Estabrook, F.J. Gonzalez, F.P. Guengerich, I.C. Gunsalus, E.F. Johnson, B. Kemper, W. Levin, I.R. Phillips, R. Sato & M.R. Waterman (1989) The P450 superfamily: updated listing of all genes and recommended nomenclature for the chromosomal loci. *DNA*, **8**, 1–13.

Nimi, A.J. (1983) Biological and toxicological effects of environmental contaminants in fish and their eggs. *Canadian Journal of Fisheries and Aquatic Sciences*, **40**, 306.

Norberg, B., V. Valkner, J. Huse, I. Karlsen & G.L. Grung (1991) Ovulatory rhythms and egg viability in the Atlantic halibut (*Hippoglossus hippoglossus*). *Aquaculture*, **97** (4), 365–371.

Norman, J.R. & P.H. Greenwood (1963) *A History of Fishes*. Hill & Wang, New York.

Norse, E.A. (ed.) (1994) *Global Marine Biological Diversity: A Strategy for Building Conservation into Decision Making*. Island Press, Washington DC.

Olive, P.J.W. & A.J. Lawrence (1990) Gonadotrophic hormone in Nephtyidae (Polychaeta, Annelida): Stimulation of ovarian protein synthesis. *International Journal of Invertebrate Reproduction and Development*, **18** (3), 189–195.

Olive, P.J.W., S. Clark & A.J. Lawrence (1990) Global warming and seasonal reproduction: Perception and transduction of environmental information. In: (eds Hoshi, M. & O. Yamashita) *Advances in Invertebrate Reproduction*. Vol. 5. Elsevier, Netherlands, pp. 265–270.

Olla, B.L., M.W. Davis & C.B. Schreck (1992) Comparison of predator avoidance capabilities with corticosteroid levels induced by stress in juvenile coho salmon. *Transactions of the American Fisheries Society*, **121**, 544–547.

Omori, K. (1995) The adaptive significance of a lunar or semilunar reproductive cycle in marine animals. *Ecological Modelling*, **82** (1), 41–49.

Ouellette, M. & P. Borst (1991) Drug resistance and P-glycoprotein gene amplification in the protozoan parasite *Leishmania*. *Res. Microbiol.*, **142**, 737–746.

Pascaud, M. (1989) Renouvellement biologique. Encyclopedia Universalis. *Corpus*, **19**, 813–816.

Pearson, M., G. van der Kraak & E.D. Stevens (1992) *In vivo* pharmacology of spleen contraction in rainbow trout. *Canadian Journal of Zoology*, **70** (3), 625–627.

Pedersen, S. & A.-K. Lundebye (1996) Metallothionein and stress protein levels in shore crabs (*Carcinus maenas*) along a trace metal gradient in the Fal Estuary (UK). *Marine Environmental Research*, **42**, 241–246.

Pelissero, C., G. Flouriot, J.L. Foucher, B. Bennetau, J. Dunogues, F.L. Gac & J.P. Sumpter (1993) Vitellogenin synthesis in cultured Hepatocytes; an *in vitro* test for the estrogenic potency of chemicals. *Journal of Steroid Biochemistry and Molecular Biology*, **44**, 263–272.

Peng, C. & R.E. Peter (1997) Neuroendocrine regulation of growth hormone secretion and growth in fish. *Zoological Studies*, **36** (2), 79–89.

Perkins, E.J., B. Griffin, M. Hobbs, J. Gollon, L. Wolford & D. Schlenk (1997) Sexual differences in mortality and sublethal stress in channel catfish following a 10 week exposure to copper sulfate. *Aquatic Toxicology*, **37**, 327–339.

Perrin, N. & J.F. Rubin (1990) On dome-shaped norms of reactions for size-to-age at maturity in fishes. *Functional Ecology*, **4**, 53–57.

Perry, D.M., J.B. Hughes & A.T. Hebert (1991) Sublethal abnormalities in embryos of winter flounder, pseudopleuronectes-americanus, from long-island sound. *Estuaries*, **14** (3), 306–317.

Pesonen, M., M. Celander, L. Förlin & T. Andersson (1987) Comparison of xenobiotic biotransformation enzymes in kidney and liver of rainbow trout (*Salmo gairdneri*). *Toxicology and Applied Pharmacology*, **91**, 75–84.

Pickering, A.D. (1993) Endocrine-induced pathology in stressed salmonid fish. *Fisheries Science*, **17** (1–2), 35–50.

Pipe, R.K. & M.N. Moore (1986) An ultrastructural study on the effects of phenanthrene on lysosomal membranes and distribution of the lysosomal enzyme β – glucuronidase in digestive cells of the periwinkle *Littorina littorea*. *Aquatic Toxicology*, **8**, 65–76.

Pires, A., J. Branco, A. Picado & E. Mendonca (1995) Metabolic responses in the assessment of pollution effects. *Environmetrics*, **6** (2), 155–163.

Policansky, D. (1983) Size, age and demography of metamorphosis and sexual maturation in fishes. *American Zoologist*, **23**, 57–63.

Potts, G.W., M.H.A. Keenleyside & J.M. Edwards (1988) The effect of silt on the parental behaviour of the sea stickleback, *Spinachia spinachia*. *Journal of the Marine Biological Association of the United Kingdom*, **68**, 277–286.

Powell, M.D. & S.F. Perry (1997) Respiratory and acid-base pathophysiology of hydrogen peroxide in rainbow trout (*Oncorhynchus mykiss* Walbaum). *Aquatic Toxicology*, **37**, 99–112.

Provencher, L., J. Munro & J.D. Dutil (1993) Osmotic performance and survival of Atlantic cod (*Gadus morhua*) at low salinities. *Aquaculture*, **116**, 219–231.

Purceli, J.E., D. Grosse & J.J. Grover (1990) Mass abundance of abnormal Pacific herring larvae at a spawning ground in British Columbia. *Transactions of the American Fisheries Society*, **119**, 463–469.

Purdom, C.E., P.A. Hardiman, V.J. Bye, N.C. Eno, C.R. Tyler & J.P. Sumpter (1994) Estrogenic effects of effluents from sewage treatment works. *Chemistry and Ecology*, **8**, 275–285.

Rashid, F., R.W. Horobin & M.A. Williams (1991) Predicting the behaviour and selectivity of fluorescent probes for lysosomes and related structures by means of structure – activity models. *Histochemical Journal*, **23**, 450–459.

Reddy, P.M. & M. Bashamohideen (1995) Modulations in the levels of respiration and ions in carp, *Cyprinus carpio* (L.), exposed to cypermethrin. *Environmental Monitoring and Assessment*, **35**, 221–226.

Rosenthal, H. & D.F. Alderdice (1976) Sublethal effects of environmental stressors, natural and pollutional, on marine fish eggs and larvae. *Journal of the Fisheries Research Board of Canada*, **33**, 2047–2065.

Sanchez-Vazquez, F.J. & M. Tabata (1998) Circadian rhythms of demand-feeding and locomotor activity in rainbow trout. *Journal of Fish Biology*, **52** (2), 255–267.

Sancho, E., M.D. Ferrando & E. Andreu (1998) Effects of sublethal exposure to a pesticide on levels of energetic compounds in *Anguilla anguilla*. *Journal of Environmental Science and Health*, **33B**, 411–424.

Sanders, B.M., L.S. Martin, W.G. Nelson, D.K. Phelps & W. Welch (1991) Relationships between accumulation of a 60 kDa stress protein and scope-for-growth in *Mytilus edulis* exposed to a range of copper concentrations. *Marine Environmental Research*, **31**, 81–97.

Sayer, M.D.J. (1991) Survival and subsequent development of brown trout, *Salmo trutta* L., subjected to episodic exposures of acid, aluminium and copper in soft water during embryonic and larval stages. *Journal of Fish Biology*, **38**, 969–972.

Sayer, M.D.J. & J. Davenport (1996) Hypometabolism in torpid goldsinny wrasse subjected to rapid reductions in seawater temperature. *Journal of Fish Biology*, **49**, 64–75.

Sayer, M.D.J. & J.P. Reader (1996) Exposure of goldsinny, rock cook and corkwing wrasse to low temperature and low salinity: survival, blood physiology and seasonal variation. *Journal of Fish Biology*, **49**, 41–63.

Sayer, M.D.J., J.P. Reader & R. Morris (1989) The effect of calcium concentration on the toxicity of copper, lead and zinc to yolk-sac fry of brown trout, *Salmo trutta* L., in soft, acid water. *Journal of Fish Biology*, **35**, 323–332.

Sayer, M.D.J., J.P. Reader, T.R.K. Dalziel & R. Morris (1991a) Mineral-content and blood parameters of dying brown trout (*Salmo trutta* L.) exposed to acid and aluminum in soft water. *Comparative Biochemistry and Physiology C-Pharmacology Toxicology & Endocrinology*, **99C**, 345–348.

Sayer, M.D.J., J.P. Reader & R. Morris (1991b) Embryonic and larval development of brown trout, *Salmo trutta* L. exposure to aluminum, copper, lead or zinc in soft, acid water. *Journal of Fish Biology*, **38** (3), 431–455.

Sayer, M.D., J.P. Reader & T.R.K. Dalziel (1993) Freshwater acidification. Effects on the early life stages of fish. *Reviews in Fish Biology*, **3**, 95–132.

Sayer, M.D.J., R.N. Gibson & R.J.A. Atkinson (1995) Growth, diet and condition of goldsinny on the west coast of Scotland. *Journal of Fish Biology*, **46**, 317–340.

Sayer, M.D.J., R.N. Gibson & R.J.A. Atkinson (1996) Growth, diet and condition of corkwing wrasse and rock cook on the west coast of Scotland. *Journal of Fish Biology*, **49**, 76–94.

Schaffer, W.M. (1974) Selection for optimal life histories: The effects of age structure. *Ecology*, **53**, 291–303.

Schneider, W.J. (1996) Vitellogenin receptors: Oocyte-specific members of the low-density lipoprotein receptor supergene family. *International Review of Cytology*, **166**, 103–135.

Schröder, J.H. & K. Peters (1988) Differential courtship activity and alterations of reproductive success of competing guppy males (*Poecilia reticulata* Peters: Pisces: Poeciliidae) as an indicator for low concentrations of aquatic pollutants. *Bulletin of Environmental Contamination and Toxicology*, **40**, 396–404.

Shofer, S.L. & R.S. Tjeerdema (1998) Effects of hypoxia and toxicant exposure on adenylate energy charge and cytosolic ADP concentrations in abalone. *Comparative Biochemistry and Physiology C-Pharmacology Toxicology & Endocrinology*, **119**, 51–57.

Sibly, R. & P. Calow (1983) An integrated approach to life-cycle evolution using selective landscapes. *Journal of Theoretical Biology*, **102**, 527–547.

Sinclair, D.A. & J.G. Eales (1972) Iodothyronine-glucuronide conjugates in the bile of brook trout, *Salvelinus fontinalis* (mitchill) and other freshwater teleosts. *General and Comparative Endocrinology*, **19**, 552–598.

Sindermann, C.J. (1996) *Ocean Pollution: effects on living resources and humans*. CRC Press, London.

Sleiderink, H.M., I. Oostingh, A. Goksøyr & J.P. Boon (1995) Sensitivity of cytochrome-p450 1a induction in dab (*Limanda limanda*) of different age and sex as a biomarker for environmental contaminants in the southern North sea. *Archives of Environmental Contamination and Toxicology*, **28** (4), 423–430.

Smerdon, G.R., J.P. Chapple & J.S. Hawkins (1995) The simultaneous immunological detection of four stress 70 protein isoforms in *Mytilus edulis*. *Marine Environmental Research*, **40** (4), 399–407.

Smital, T. & B. Kurelec (1998) The activity of multixenobiotic resistance mechanism determined by rhodamine B efflux method as a biomarker of exposure. *Marine Environmental Research*, **46** (1–5), 443–447.

Snegaroff, J. & J. Bach (1990) The effects of temperature on the basal activity of cytochrome P-450 in rainbow trout (*Salmo gairdneri*). *Comparative Biochemistry and Physiology*, **95B**, 515–519.

Soengas, J.L., M.J. Agralago, B. Carballo, M.D. Andres & J.A.R. Veira (1996) Effect of an acute exposure to sublethal concentrations of cadmium on liver carbohydrate metabolism of Atlantic salmon (*Salmo salar*). *Bulletin of Environmental Contamination and Toxicology*, **57**, 625–631.

Stafford, C.L. & J.A. Ward (1983) Effects of monochloramine on courtship and spawning in the cichlid fish *Etroplus maculatus*. In: (eds Noakes, D.L.G., D.G. Linquist, G.S. Helfman & J.A. Ward) *Predators and Prey in Fishes*. The Hague, Netherlands, pp. 213–220.

Stearns, S.C. (1983) The genetic basis of differences in life-history traits among six populations of mosquitofish (*Gambusia affinis*) that shared ancestors in 1905. *Evolution*, **37**, 618–627.

Stearns, S.C. & J. Koella (1986) The evolution of phenotypic plasticity in life history traits: predictions of reaction norms for age and size at maturity. *Evolution*, **40**, 893–913.

Stegeman, J.J., A.M. Pajor & P. Thomas (1982) Influence of estradiol and testosterone on cytochrome P-450 and monooxygenase activity in immature brook trout, *Salvelinus fontinalis*. *Biochemical Pharmacology*, **31**, 3979–3989.

Subhedar, N., J. Cerda, B.G. Calman & R.A. Wallace (1997) Changes in forebrain and pituitary dopamine and serotonin contents of female *Fundulus* during its biweekly reproductive cycle. *Comparative Biochemistry and Physiology A – Physiology*, **118** (3), 577–584.

Suedel, B.C., T.M. Dillon & W.H. Benson (1997) Subchronic effects of five di-ortho PCB congeners on survival, growth and reproduction in the fathead minnow *Pimephales promelas*. *Environmental Toxicology and Chemistry*, **16** (7), 1526–1532.

Sumpter, J.P. & S. Jobling (1995) Vitellogenesis as a biomarker or estrogenic contaminants of the aquatic environment. *Environmental Health Perspectives*, **103** (Suppl. 7), 173–178.

Susani, L. (1986) *Effects of contaminants on teleost reproduction: past and ongoing studies*. Vol. 29. NOAA Techn. Memo, NOS OMA, USA, 18 pp.

Tanner, D.K. & M.L. Knuth (1995) Effects of azinphos-methyl on the reproductive success of the bluegill sunfish, *Lepomis macrochirus*, in the littoral enclosures. *Ecotoxicology and Environmental Safety*, **32**, 184–193.

Tanner, D.K. & M.L. Knuth (1996) Effects of esfenvalerate on the reproductive success of the bluegill sunfish, *Lepomis macrochirus*, in the littoral enclosures. *Archives of Environmental Contamination and Toxicology*, **31**, 244–251.

Tata, J.R. & D.F. Smith (1979) Vitellogenesis: A versatile model hormonal regulation of gene expression. *Recent Progress in Hormone Research*, **35**, 47–90.

Tesoriero, J.V. (1977) Formation of the chorion (*zona pellucida*) in the teleost, *Oryzias latipes*. II Polysaccharide cytochemistry of early oogenesis. *Journal of Histochemistry and Cytochemistry*, **25**, 1376.

Thaker, J., J. Chaya, S. Nuzhat, R. Mittal, A.P. Mansuri & R. Kundu (1996) Effects of chromium (VI) on some ion-dependent ATPases in gills, kidney and intestine of a coastal teleosts. *Periophthalmus dipes*. *Toxicol.*, **112**, 237–244.

Thebault, M.T., A. Biegniewska, J.P. Raffin & E.F. Skorkowski (1996) Short term cadmium intoxication of the shrimp *Palaemon serratus*: effect on adenylate metabolism. *Comparative Biochemistry and Physiology C-Pharmacology Toxicology & Endocrinology*, **113**, 345–348.

Theodorakis, C.W., B.G. Blaylock & L.R. Shugart (1997) Genetic ecotoxicology 1. DNA integrity and reproduction in mosquitofish exposed *in situ* to radionuclides. *Ecotoxicology*, **6**, 205–218.

Thiebaut, F., T. Tsuruo, H. Hamada, M.M. Gottesman, I. Pastan & M.C. Willingham (1987) Cellular localization of the multidrug-resistance gene product P-glycoprotein in normal human tissues. *Proceedings of the National Academy of Sciences of the USA*, **84**, 7735.

Thomas, S. & S.F. Perry (1992) Control and consequences of adrenergic activation of red blood cell Na+/H+ exchange on blood oxygen and carbon dioxide transport in fish. *Journal of Experimental Zoology*, **263** (2), 160–175.

Thorpe, J.E. (1994) Reproductive strategies in Atlantic salmon, *Salmo salar* L. *Aquaculture and Fisheries Management*, **25**, 77–87.

Tyler, C.R. & J.P. Sumpter (1996) Oocyte growth and development in teleosts. *Reviews in Fish Biology*, **6**, 287–318.

Tyler, C.R., J.P. Sumpter & P.R. Witthames (1990) The dynamics of oocyte growth during vitellogenesis in the rainbow trout, *Salmo gairdneri*. *Biology of Reproduction*, **43**, 202–209.

Tyler, C.R., J.P. Sumpter & P.M. Campbell (1991a) Uptake of vitellogenin into oocytes during early vitellogenic development in the rainbow trout, *Oncorhynchus mykiss* (walbaum). *Journal of Fish Biology*, **38** (5), 681–689.

Tyler, C.R., J.P. Sumpter, H. Kawauchi & P. Swanson (1991b) Involvement of gonadotropin in the uptake of vitellogenin into vitellogenic oocytes of the rainbow trout, *Oncorhynchus mykiss*. *General and Comparative Endocrinology*, **84** (2), 291–299.

Van Raaij, M.T.M., D.S.S. Pit, P.H.M. Balm, A.B. Steffens & G.E.E.J.M. van der Thillart (1996) Behavioural strategy and the physiological stress response in rainbow trout exposed to severe hypoxia. *Hormones and Behaviour*, **30** (1), 85–92.

Van Veld, P.A., D.J. Westbrook, B.R. Woodin, R.C. Hale, C.L. Smith, R.J. Huggett & J.J. Stegeman (1990) Induced cytochrome P-450 in intestine and liver of spot (*Leiostomus xanthurus*) from a polycyclic aromatic hydrocarbon contaminated environment. *Aquatic Toxicology*, **17**, 119–132.

Varanasi, U. (ed.) (1989) *Metabolism of Polycyclic Aromatic Hydrocarbons in the Aquatic Environment*. CRC Press Inc., Boca Raton, Florida, 341 pp.

Varnavskaya, N.V. & V.S. Varnavsky (1988) The biology of the dwarf sockeye salmon of Dalnee Lake. *Biologiya Morya*, **10**, 16–23.

Vethaak, A.D. (1992) Diseases of flounder (*Platichtys flesus* L.) in the Dutch Wadden Sea, and their relation to stress factors. *Netherlands Journal of Sea Research*, **29** (1–3), 257–272.

Vethaak, A.D. & J.G. Jol (1996) Diseases of flounder (*Platichthys flesus* L.) in Dutch coastal and estuarine waters, with particular reference to environmental stress factors. I. Epizootiology of gross lesions. *Diseases of Aquatic Organisms*, **26**, 81–97.

Viarengo, A., M.N. Moore, M. Pertica, G. Mancinelli, G. Zanicchi & R.K. Pipe (1985) Detoxification of copper in the cells of the digestive gland of mussel: the role of lysosomes and thioneins. *Science of the Total Environment*, **44**, 135–145.

Viarengo, A., L. Canesi, M.N. Moore & M. Orunesu (1994) Effects of Hg^{2+} and CU^{2+} on the cytosolic Ca^{2+} level in molluscan blood cells evaluated by confocal microscopy and spectrofluorimetry. *Marine Biology*, **119**, 557–564.

Vernberg, F.J., A. Calabrese, F.P. Thurborg & W.B. Vernberg (eds) (1975) *Symposium on the Physiological Responses on Marine Biota to Pollution*. Academic Press, New York.

von Westernhagen, H. (1988) Sublethal effects on fish eggs and larvae. In: (eds Hoar, W.S. & D.J. Randall) *Fish Physiology. Vol. XI. The physiology of developing fish*. Academic Press, New York, pp. 253–346.

Wahli, W. (1988) Evolution and expression of vitellogenin genes. *TIG*, 4, 227–232.

Wallace, R.A. (1985) Vitellogenesis and oocyte growth in non-mammalian vertebrates. In: (ed. Browder, L.W.) *Developmental Biology*. Plenum Press, New York, pp. 127–177.

Wallace, R.A. & K. Selman (1979) Physiological aspects of oogenesis in two species of sticklebacks, *Gasterosteus aculeatus* (L.) and *Apeltes quadracus* (Mitchill). *Journal of Fish Biology*, **14**, 551–564.

Wannemacher, R., A. Rebstock, E. Kulzer, D. Schrenk & K.W. Bock (1992) Effects of 2,3,7,8-tetrachlorodibenzo-para-dioxin on reproduction and oogenesis in zebrafish (*Brachydanio rerio*). *Chemosphere*, **24** (9), 1361–1368.

Ware, D.M. (1980) Bioenergetics of stock and recruitment. *Canadian Journal of Fisheries and Aquatic Sciences*, **37**, 1012–1024.

Ware, D.M. (1982) Power and evolutionary fitness of teleosts. *Canadian Journal of Fisheries and Aquatic Sciences*, **39**, 3–13.

Waring, C.P. & A. Moore (1997) Sublethal effects of a carbomate pesticide on pheromonal mediated endocrine function in mature male Atlantic salmon (*Salmo salar* L.) parr. *Fish Physiology and Biochemistry*, **17**, 203–211.

Waring, C.P., R.M. Stagg, K. Fretwell, H.A. McLay & M.J. Costello (1996) The impact of sewage sludge exposure on the reproduction of the sand goby, *Pomatoschistus minutus*. *Environmental Pollution*, **93** (1), 17–25.

Webb, N.A. & C.M. Wood (1998) Physiological analysis of the stress response associated with acute silver nitrate exposure in freshwater rainbow trout (*Oncorhynchus mykiss*). *Environmental Toxicology and Chemistry*, **17**, 579–588.

Weber, D.N. (1993) Exposure to sublethal levels of waterborne lead alters reproductive-behaviour patterns in fathead minnows (*Pimephalespromelas*). *Neurotoxicology*, **14** (2–3), 347–358.

Weber, D.N., W.M. Dingel, J.J. Panos & R.E. Steinpreis (1997) Alterations in neurobehavioral responses in fishes exposed to lead and lead-chelating agents. *American Zoologist*, **37**, 354–362.

Weigand, M.D. (1982) Vitellogenesis in fishes. In: (eds Richter, C.J.J. & H.J.T. Goos) *Reproductive Physiology of Fish*. Pudoc, Wagenningen, pp. 136–146.

Weis, J.S. & P. Weis (1989) Effects of environmental pollutants on early fish development. *CRC Critical Reviews in Aquatic Sciences*, **1**, 45–73.

Wendelaar Bonga, S.E. (1997) The stress response in fish. *Physiological Reviews*, **77**, 591–625.

Wester, P.W. & J.H. Canton (1986) Histopathological study of *Oryzias latipes* (Medaka) after long-term β-hexachlorocyclohexane exposure. *Aquatic Toxicology*, **9**, 21–45.

White, R., S. Jobling, S.A. Hoare, J.P. Sumpter & M.G. Parker (1994) Environmentally persistent alkylphenolic compounds are estrogenic. *Endocrinology*, **135**, 175–182.

Widdows, J. & D.S. Page (1993) Effects of tributyltin and dibutyltin on the physiological energetics of the mussel, *Mytilus edulis*. *Marine Environmental Research*, **35** (3), 233–249.

Widdows, J., T. Bakke, B.L. Bayne, P. Donkin, D.R. Livingstone, D.M. Lowe, M.N. Moore, S.V. Evans & S.L. Moore (1982) Responses of *Mytilus edulis* on exposure to the water accommodated fraction of North Sea oil. *Marine Biology*, **67**, 15–31.

Widdows, J., P. Donkin, M.D. Brinsley, S.V. Evans, P.N. Salkeld, A. Franklin, R.J. Law & M.J. Waldock (1995) Scope for growth and contaminant levels in North Sea mussels *Mytilus edulis*. *Marine Ecology Progress Series*, **127** (1–3), 131–148.

Wiegand, M.D., J.M. Hately, C.L. Kitchen & L.G. Buchanan (1989) Induction of developmental abnormalities in larval goldfish *Carassius auratus* L., under cool incubation conditions. *Journal of Fish Biology*, **35**, 85–95.

Williams, R.T. (1974) Inter-species variations in the metabolism of xenobiotics. *Biochemical Society Transactions*, **2**, 359–377.

Wilson, R.W. & E.W. Taylor (1993) Differential responses to copper in rainbow trout (*Oncorhynchus mykiss*) acclimated to sea water and brackish water. *Journal of Comparative Physiology*, **163B**, 239–246.

Winberg, S., A. Nilsson, P. Hylland, V. Soderstom & G.E. Nilsson (1997) Serotonin as a regulator of hypothalamic-pituitary-interrenal activity in teleost fish. *Neuroscience Letters*, **230** (2), 113–116.

Winston, G.W., M.N. Moore, I. Straatsburg & M.A. Kirchin (1991) Decreased stability of digestive gland lysosomes from the common mussel *Mytilus edulis* L. by *in vitro* generation of oxygen-free radicals. *Archives of Environmental Contamination and Toxicology*, **21**, 401–408.

Winston, G.W., M.N. Moore, M.A. Kirchin & C. Soverchia (1996) Production of reactive oxygen species (ROS) by hemocytes from the marine mussel, *Mytilus edulis*. *Comparative Biochemistry and Physiology*, **113C**, 221–229.

Wong, H.R., M. Ryan, I.Y. Menendez, A. Deneberg & J.R. Wispe (1997) Heat shock protein induction protects human respiratory epithelium against nitric oxide mediated cytotoxicity. *Shock*, **8** (3), 213–218.

Woo, N.Y.S. & S.F. Chiu (1997) Metabolic and osmoregulatory responses of the sea bass *Lates calcarifer* to nitrite exposure. *Environmental Toxicology and Water Quality*, **12**, 257–264.

Wright, P.A., S.F. Perry & T.W. Moon (1989) Regulation of hepatic gluconeogenesis and glycogenolysis by catecholamines in rainbow trout during environmental hypoxia. *Journal of Experimental Biology*, **147**, 169–188.

Wu, C.T., M. Budding, M.S. Griffen & J.M. Croop (1991) Isolation and characterisation of *Drosophilia* multidrug resistance gene homologs. *Molecular and Cell Biology*, **11**, 3940–3948.

Zhang, Y.S., T. Andersson & L. Förlin (1990) Induction of hepatic xenobiotic biotransformation enzymes in rainbow trout by β-naphthoflavone. *Comparative Biochemistry and Physiology*, **95B**, 247–253.

Zimniak, P. & D.J. Waxman (1993) Liver cytochrome P450 metabolism of endogenous steroid hormones, bile acids and fatty acids. In: (eds Schenkman, J.B. & H. Greim) *Cytochrome P450*. Springer-Verlag, New York, pp. 123–144.

Chapter 4
Molecular/Cellular Processes and the Health of the Individual

K. Hylland, S. Feist, J. Thain and L. Förlin

4.1 Introduction

A fish can be considered healthy if it grows at a normal rate, if it fulfils its natural reproductive capacity and if it can be expected to live its full life span, i.e. as long as is normal in the relevant population. Disease in fish, viewed as the negation of health, is any deviation from normal life. Disease can be an abnormality in a physiological process, such as reproduction failure, growth retardation, organ dysfunction, perturbation of normal metabolism or pathological effects, such as skin ulcer development, parasite infection or neoplastic lesions. The health of fish can be determined using a range of biochemical, physiological and morphological methods, commonly referred to as biomarkers. Fish diseases and pathology, whether caused by infectious agents, environmental factors or xenobiotics are increasingly used as indicators of environmental stress at the population level (ICES, 1996, 1997). Physiological and toxicopathic lesions at the tissue and organ level are the consequence of changes at even more fundamental levels of biological organisation and these are considered in more detail in Chapters 2 and 3 of this book. Generally, a selection of both physiological and pathological methods would be used in combination with chemical analyses to diagnose or assess health in individual fish.

The underlying concept is based on the general understanding that interactions between biochemical and physiological functions in fish, detected as disruption of normal processes at subcellular, cellular or organ levels, may indicate or lead to adverse effect at the individual level. Such effects may then cause vital disturbances in reproduction, growth or survival of the fish (Haux & Förlin, 1988). Such individual health effects occur early in a damaging process. Adverse effects on fish populations or communities can then be avoided by taking appropriate measures.

Although it is generally understood that impacts on fish health may lead to effects on fish populations or communities, it is difficult to quantify such relationships. To overcome this difficulty, various test strategies have been employed. One such strategy is to couple biomarker responses to effects observed in parallel studies with ecological end-points such as growth and abundance of fry and adults, reproductive success and recruitment. Such an integrated strategy using ecological, biochemical/physiological and pathological end-points has been employed with fish in reference areas and polluted receiving waters, for example in monitoring adverse effects of pulp mill effluents (Sandström *et al.*, 1997). Studies of the same biomarkers in the laboratory and field are another strategy to overcome the difficulties encountered when interpreting the results of complex field exposures (Larsson *et al.*, 1985; Förlin *et al.*, 1986). Laboratory studies then provide the background upon which it is possible to understand and interpret the responses measured in fish in polluted waters. The link between individual health effects and populations is treated in Chapter 5.

Considerable effort has been devoted to develop experimental techniques for field studies including well-defined methods for fish capture and handling, sampling of tissues, storage of tissues, analyses and statistical treatment of data. Abiotic and biotic factors such as temperature, salinity, season, nutritional state, sex and sexual maturation influence biomarker responses. In addition, methodological factors including sampling and analytical techniques are sources of variability for the determination of toxic responses. Great variability also exists in biomarker responses between tissues and between species. Standardised protocols should include representative species and appropriate, selected biomarkers. Protocols need to include details on the selection of individuals (stratified sampling) and quality control measures for each step in the analytical process. By using such protocols the 'noise' contributed by confounding factors can be minimised.

In this chapter, a distinction has been made between physiological and pathological effects in fish. The reason for this is operational, as it is obvious that changes in a tissue will affect physiological parameters and vice versa. In addition to physiological and pathological conditions in adult fish, evidence of contaminant effects on larvae and embryonic development has been included in a separate section. Finally, to show some of the results in the 'real world', two specific case studies have been discussed.

4.2 Physiological aberrations

Biochemical and physiological functions contribute to maintain homeostasis in fish. As a consequence, many early responses to toxicants are often homeostatic responses that can be difficult to discriminate from responses caused by natural stressors. When toxicity occurs, such acclimation responses have evolved into adverse effects. Therefore, acclimation responses may serve as early warning signals to toxicant exposure and/or effects (Huggett *et al.*, 1992). This has become an increasingly accepted conceptual aid in assessing effects of pollutants in the aquatic environment even though it is generally difficult to link such acclimation responses to effects on populations or communities. Physiological aberrations differ from pathological changes in that they may occur within hours of exposure to a contaminant. Biochemical and physiological responses may therefore, to a larger extent than pathological changes, be used as early warning signals for toxicity, but the ecological significance of the response is on the other hand less clear. Physiological investigations give the opportunity to make a primary assessment of disturbed or threatened quality of fish including both the use of fish for human consumption and the ability of the fish to reproduce and survive in a polluted environment. These types of fish investigations can also form the basis for and be combined with other studies such as chemical and ecological studies.

In the following sections, biochemical and physiological aberrations are discussed in relation to specific tissues or organs in fish. The immune system is treated separately due to its relevance for health in general. Disruption of the metabolism of trace elements involves various tissues and is discussed separately. Another process involving most tissues in an organism, endocrine regulation, is treated in Chapter 6.

4.2.1 Effects on the immune system

The immune system of fish is similar to that found in mammals in many respects. As in mammals (and other vertebrates), there are signalling systems as well as non-specific and specific humoral and cellular components in the fish immune system. Although signalling systems for immune responses do exist in fish, there is still limited knowledge concerning the function and diversity of messengers, e.g. cytokines (Secombes *et al*., 1996). It appears probable that eicosanoids are involved in immune modulation in fish, as is the case for mammals (Rowley *et al*., 1995). One important aspect of fish immunology is the large diversity of both mechanisms and function: immune responses in different species vary substantially and it is therefore difficult to extrapolate from one species to another.

4.2.1.1 The non-specific components of the fish immune system

The non-specific immune response in fish has been studied in some detail for a small number of fish species. As for other vertebrates, it consists of humoral and cellular components. The humoral immune components include lysozyme, complement, C-reactive protein, transferrin, lectins, hemolysin and various other substances (see Yano, 1996, for an overview). These substances may be found in mucus and eggs as well as in plasma and their main role is to provide an immediate first, unspecific, line of defence against pathogenic microorganisms. For this reason, the substances are also commonly referred to as acute-phase reactants. The components of the acute-phase response are predominantly glycoproteins and other proteins that are excreted into the blood. Plasma levels of acute-phase reactants increase following infection (White & Fletcher, 1983; Szalai *et al*., 1994).

In addition to the humoral components of the non-specific immune system, there are cellular components, reviewed by Secombes *et al*. (1996). The non-specific cellular defence includes macrophages, granulocytes and non-specific cytotoxic (NCC) cells. Macrophages normally have high phagocytic activity, will migrate according to chemical stimuli, can secrete free radicals and can act as accessory cells for lymphocyte responses. Macrophage responses have been used to identify immunomodulation from various agents (see below). In fish, granulocytes are generally either neutrophils or eosinophils. As macrophages, granulocytes are mobile and phagocytotic, but their bactericidal activity is not as high as that of macrophages (Secombes *et al*., 1996). In some fish species there is a particularly well-developed cell-mediated cytotoxic capacity, mediated through non-specific cytotoxic cells (Pettey & McKinney, 1988; Haynes & McKinney, 1991; Inoue *et al*., 1998). In bony fish, the NCC cells are considered to be analogous to natural killer (NK) cells in mammals, whereas macrophages appear to have that function in sharks.

4.2.1.2 The specific components of the fish immune system

The cellular component of the specific immune system in fish really consists of three major parts: lymphocytes with the associated production of antibodies, helper cell functions and allograft rejection mechanisms. The antigen-specific response in fish does involve B-cells and T-cells, but there appear to be species-dependent differences in the nature and regulation of the specific immune response in fish (Manning & Nakanishi, 1996). In contrast to mammals, the immunoglobulins of fish are primarily tetrameric (referred to as IgM, as the mammalian pentameric Ig) (Glynn & Pulsford, 1990) and there does not appear to be an evolution of different globulin classes in the course of an antigen-response (Kaattari & Piganelli, 1996). Some studies do indicate a diversity of Igs in fish and the question of the roles of different Igs during phases of infection or involvement in other immune responses still remains unresolved (Bang *et al.*, 1996).

4.2.1.3 Methods to study fish immune responses to xenobiotics

The methods that have been used to identify effects of xenobiotics on fish immune response can be divided into three groups: non-specific assays, specific assays and either assay type used with antigen stimulation (Anderson, 1990). The non-specific assays include hematocrit, leukocrit, the content of various classes of leukocytes and thrombocytes, general assays of macrophage function. As the name implies, specific assays are functions related to or responses to specific antigens. These include agglutination assays, mitotic response (proliferation of B-cells), reduced ability to synthesise antigen-specific IgM and the ability to survive a challenge with a pathogen.

4.2.1.4 Natural modulation of the fish immune system

In addition to pathogens, immune responses in fish are modulated by various endogenous and environmental factors, such as season, temperature, food availability, food quality, age and gender. Lower temperature has been found to decrease immune responses in fish species, e.g. catfish (Dexiang & Ainsworth, 1991), carp (Schneider & Ambrosius, 1987) and rainbow trout (Hardie *et al.*, 1994). It is not clear whether temperature-dependent effects are direct or whether they are entirely or partly stress-mediated (Schreck, 1996). In addition to temperature, several studies indicate that immune responses may change through the year (White & Fletcher, 1983; Szalai *et al.*, 1994). Both innate immune function and the immune response following stress will differ through the life of a fish. Such age-dependent changes relate both to ontogenetic development and to ageing of tissues (Ourth & Ratts, 1991; Tatner, 1996). There are indications that there may be differences between juvenile fish and sexually mature individuals. In one study, Røed *et al.* (1992) found lower haemolytic activity in the blood of sexually mature fish compared to juveniles. In addition to being affected by endogenous factors, the diet of fish also affects their ability to combat disease. Components of the immune system appear to be particularly sensitive to deficiencies in trace metals (Scarpa *et al.*, 1992; Inoue *et al.*, 1998), lipid (Sheldon & Blazer, 1991) and vitamins (Hardie *et al.*, 1990, 1991) in the diet.

4.2.1.5 Effects of contaminants on non-specific immune responses

There is a large body of evidence that suggests that exposure to xenobiotics affects non-specific immune responses in marine and freshwater fish species (reviewed by Anderson *et al.*, 1996). The humoral parameters most commonly measured are lysozyme, C-reactive protein and complement. In an early study, several chlorinated compounds were found to increase the levels of C-reactive protein by one to two orders of magnitude (Winkelhake *et al.*, 1983). In one study, Karrow *et al.* (1999) found that plasma lysozyme in rainbow trout decreased following chronic exposure to creosote. A study by Albergoni & Viola (1995) in which catfish (*Ictalurus melas*) was exposed to Cd shows some of the difficulties found in relating xenobiotic exposure to immune responses: serum IgM decreased immediately following exposure, but then increased to control levels within two weeks.

Cellular non-specific responses have mainly been assessed through the determination of the phagocytotic or bactericidal activity of macrophages. In a study on English sole (*Pleuronectes vetulus*), Clemons *et al.* (1999) found increased superoxide anion production (e.g. oxidative burst) in fish exposed to polycyclic aromatic hydrocarbons. Exposure to contaminants does not always cause increased immune responses, however; pronephros leukocyte oxidative burst was found to decrease in creosote-exposed rainbow trout (Karrow *et al.*, 1999). Similarly, Rice *et al.* (1996) found decreased oxidative burst in PMA-stimulated leukocytes from fish held near a contaminant source compared to fish from a reference location. In the same study, no differences were found between sites regarding non-specific cytotoxic cell activity. Anderson and Brubacher (1993) also found decreased oxidative burst response in phagocytes from PCP-exposed medaka (*Oryzias latipes*) compared to control fish.

It is not easy to interpret such responses as many immune parameters may decrease or increase depending on the duration and nature of the exposure. Interpretation may easily be further confounded by the fact that non-contaminant stress will affect acute-phase responses, such as lysozyme activity (Demers & Bayne, 1997) and replace with the alternative pathway of complement activation system (Sunyer *et al.*, 1995).

4.2.1.6 Effects of contaminants on specific immune responses

There is increasing evidence that the specific components of the immune system are affected by xenobiotics. The bulk of such studies can be divided in two main groups: those studying general lymphocyte processes and those using challenge from a pathogen.

Various contaminants have been shown to affect lymphocyte proliferation and lymphocyte function in fish species. Thuvander (1989) found that Cd would increase the mitogenic response in lymphocytes from rainbow trout (*Oncorhynchus mykiss*). Similarly, in a study on catfish exposed to Cd, Albergoni and Viola (1995) observed increased haemolytic activity and antibody titer in exposed fish compared to control fish. The opposite was found in a study on the freshwater fish species spot (*Leiostomus xanthurus*). In this study, exposure to polycyclic aromatic hydrocarbons caused decreased lymphocyte mitogenic response (Faisal & Huggett, 1993). In another study on rainbow trout, Karrow *et al.* (1999) found only marginally decreased mitogenic response in head kidney leukocytes following treatment with LPS, whereas PHA or Con A had no effect. Both polycyclic aromatic hydrocarbons and chlorinated hydrocarbons suppressed B-cell function in chinook salmon (*Oncorhynchus tschawytscha*) (Arkoosh *et al.*, 1994), whereas only minor differences between populations were seen in a study on English sole (*Pleuronectes vetulus*) (Arkoosh *et al.*, 1996).

As for non-specific responses, there is clear evidence that stress will depress specific immune responses in for example, Atlantic salmon (*Salmo salar*) (Thompson *et al.*, 1993), rainbow trout (Betoulle *et al.*, 1995) and catfish (Ainsworth *et al.*, 1991).

4.2.1.7 The use of immune responses in fish for contaminant monitoring

There is a general lack of monitoring studies in which immune responses in fish have been determined alongside other markers for health or exposure. Only in a limited number of studies have immunotoxic effects in fish been assessed in relation to environmental contamination.

Red drum (*Sciaenops ocellatus*) from contaminated areas was found to have decreased serum antibody titres compared to red drum collected from clean reference areas (Evans *et al.*, 1997). In a series of studies, Secombes and co-workers assessed effects of sewage sludge on the immunocompetence of flatfish species under experimental (Houlihan *et al.*, 1994; Secombes *et al.*, 1991, 1992) and field (Secombes *et al.*, 1995) conditions. The immunocompetence of dab (*Limanda limanda*) and rainbow trout (*Oncorhynchus mykiss*) in response to oil drilling muds was assessed in a similar manner (Tahir *et al.*, 1993; Tahir & Secombes, 1995). The results from the series of studies indicated a complex response pattern in the parameters measured, which included condition indices, serum lysozyme, protein, immunoglobulin (Ig) and antiprotease, macrophage and leukocyte function, as well as lymphocyte mitogen response. In the field study with plaice (*Pleuronectes platessa*), some parameters increased close to the most contaminated site (sewage dump site), i.e. hepatosomatic index (LSI), serum lysozyme, total serum Ig and kidney leucocyte bactericidal activity (Secombes *et al.*, 1995).

The number of different types of white blood cells (WBC) including lymphocytes, granulocytes and thrombocytes has been used to indicate pollutant effects on the immune defence in fish (Larsson *et al.*, 1985; Förlin *et al.*, 1986; Andersson *et al.*, 1988). A reduced total WBC count or lymphocyte count seem to be a secondary stress response in fish subjected to acute stress (Ellis, 1981; Larsson *et al.*, 1985). In chronically exposed perch (*P. fluviatilis*) caught downstream of a kraft pulp mill the reduced lymphocyte count was considered to indicate impairment of the immune system (Andersson *et al.*, 1988). The authors hypothesised that the high frequency of fin alterations including eroded and shortened tail fin found in the most polluted sites might have been a direct consequence of the immune system impairment (Lindesjöö & Thulin, 1987).

4.2.2 Perturbed metabolism of vitamins, trace elements, etc.

In addition to direct costs through reduced food uptake or food utilisation, effects of xenobiotics may be mediated through their interactions with components required for normal metabolism.

4.2.2.1 Vitamin C (ascorbic acid)

Ascorbic acid plays a role in oxidant defence and therefore participates in the protection against certain contaminants. Important contaminant-induced abnormalities in fish such as skeletal deformities, fin erosions and skin lesions can be related to dietary ascorbic acid deficiency and/or overutilisation of ascorbic acid stores in the defence mechanisms against toxicants (Guha *et al.*, 1993; Thomas & Wofford, 1993; Palace *et al.*, 1996). For example, DDT cause haematological disturbances in fish which can be at least partly counteracted by dietary ascorbic acid (Guha *et al.*, 1993). PCB exposure to fish seems not to cause any effects on ascorbic acid content (Thomas & Wofford, 1993; Palace *et al.*, 1996), whereas oil and cadmium exposure may cause markedly decreased levels of ascorbic acid in fish (Thomas & Neff, 1984; Thomas, 1987; Thomas & Wofford, 1993). Other metals including lead, copper and zinc may cause a small or no effect on ascorbic acid. With copper, ascorbic acid seems to play different roles in prevention of dietary and water borne copper exposure. While ascorbic acid does not affect copper toxicity in dietary exposed fish (Lanno *et al.*, 1985), ascorbic acid prevents the toxicity against waterborne copper (Yamamoto *et al.*, 1977, 1981).

Relationships between ascorbic acid and reproductive success in fish are of particular ecotoxicological importance. In male fish (rainbow trout) ascorbic acid deficiency has been shown to reduce sperm concentration and motility (Cierszko & Dabrowksi, 1995). In female fish, deficiency in ascorbic acid has been shown to reduce hatchability of egg, increase the number of deformed fry and negatively affect growth, food utilisation and survival of fry (Sandnes *et al.*, 1984; Siolman *et al.*, 1986). In studies of feed supplemented with ascorbic acid to adult female fish it has been shown that maternal transfer of the vitamin can counteract ascorbic acid deficiency-related toxicity during early life-stages (Blom & Dabrowski, 1996).

4.2.2.2 Trace metal metabolism (Cu, Zn, Fe)

Exposure to xenobiotics may affect the metabolism of the trace elements Cu, Fe and Zn in fish tissues. Many cases of such perturbations involve the low-molecular weight, metal-binding protein metallothionein (MT). This protein is thought to be involved in the normal homeostatic control of intracellular Zn and Cu levels, but will also bind non-essential metals such as Cd, Hg and Ag (Bremner, 1991). Exposure to any of the above metals will cause increased synthesis of MT and increased binding sites for both essential and non-essential metals in the cells. There have been suggestions that MT may regulate both enzyme activities and gene transcription by increasing or decreasing Zn availability (Churchich *et al.*, 1988; Suzuki *et al.*, 1991; Thiele, 1992). Through such mechanisms, increased intracellular or nuclear levels of MT would have detrimental effects on the cell through perturbation of Zn metabolism. There is no clear evidence that MT could affect the activity of Cu-dependent enzymes, especially as the protein binds this metal with much higher affinity than it binds Zn.

Both Cu and Fe are redox-cycling metals that may give rise to free radicals in the cell (Halliwell & Gutteridge, 1989). As one component of many, MT is thought to be involved in the cellular antioxidant defence as a radical scavenger. This subject and involvement of MT with DNA damage is treated in more detail in Chapter 2. Iron is a trace element with a wide range of roles in fish physiology, including oxygen-transport, as cofactor in various catabolic enzymes and in the transfer of electrons in mitochondria. While not interacting directly with Fe, some xenobiotics will affect the metabolism of Fe-containing heme groups. The metal Pb will inhibit one step in the synthesis of heme, δ-aminolevulinic acid dehydratase (ALA-D) activity (Hodson, 1976; Haux *et al.*, 1986). In fish this step does not appear to be rate-limiting for heme synthesis, but in birds and mammals ALA-D inhibition will ultimately cause anaemia.

4.2.3 Organ dysfunction

Exposure to and accumulation of xenobiotics may lead to tissue damage and subsequent organ dysfunction. Organ dysfunction will affect the well-being of fish and thus the individual fish health statues. Histological methods are perhaps the most common tools in examination of tissue and organ dysfunction (see section 4.3). Physiological methods including measurements of subcellular and cellular processes can provide additional and important information of organ functions. Biochemical and physiological aberrations that indicate tissue and organs dysfunction in fish with emphasis on data from field studies are reviewed in this section.

4.2.3.1 Gills

In gills, histopathological methods are commonly used to study contaminant-induced abnormalities (see below). Fish gills have many important functions including exchange of gases, transport of many mono and divalent ions, excretion of waste nitrogen (ammonia, urea), and uptake and excretion of various xenobiotics. The biochemistry of many of these vital functions still needs elucidating. Therefore, biochemical markers reflecting these processes are relatively rare. One such biomarker that may serve as a marker of contaminant exposure is the inhibition of sodium/potassium ATPases. The activity of this enzyme is affected by exposure of metals and organochlorine (Cutcomp *et al.*, 1972; Desaiah & Koch, 1975; Watson & Beamish, 1980). Also other biomarker responses more frequently studied in fish liver have been tested in gills such as induction of EROD activity and metallothionein content and oxidative stress. Relatively few studies concern field experiments but rather effects and regulation (Rana *et al.*, 1995; Rodriguez-Ariza *et al.*, 1999), localisation (Husøy *et al.*, 1994) and organ-specific responses (George *et al.*, 1996) (see also Chapters 2 and 3). To make subcellular and cellular biomarkers more useful in assessing the impact of environmental pollution requires fundamental research. However, since the gill is the prime organ for exposure to waterborne chemicals it would be relevant to include this organ in *in vivo* biomonitoring including biomarker responses in fish.

4.2.3.2 Sensory epithelia

There are indications that Cd in water causes morphological changes in the olfactory rosette of fish (Hernadi, 1993). Similarly, Cu has been shown to affect the ability of rainbow trout to respond to chemical cues in water (Saucier *et al.*, 1991; Saucier & Astic, 1995). High concentrations of Cu (40 mg l^{-1}) was found to cause irreversible damage to the olfactory epithelium, whereas lower concentrations (20 mg l^{-1}) caused a reversible insensitivity. Such effects are more probable in freshwater systems than in the marine environment as the bioavailable levels of Cu will be lower in seawater than in freshwater.

The results of some studies suggest that the presence of xenobiotics, in this case an organophosphate pesticide, in water could have serious implications for spawning in salmonid species as male fish may not receive necessary pheromone cues from females (Moore & Waring, 1996). In this study, milt maturation in male fish was delayed and levels of circulating steroid hormones reduced. Similar scenarios could be envisaged for some marine fish species with complex spawning behaviour.

4.2.3.3 Liver and other visceral organs

The liver performs many essential body functions including regulation of metabolism, synthesis of plasma proteins, energy storage, storage of certain vitamins and trace metals, and transformation and excretion of steroids and xenobiotics. Studies of fish liver related to data obtained from field experiments mostly concern its functions in the detoxification of xenobiotics, specific responses to planar compounds (i.e. AhR-related induction of EROD activity) and its susceptibility to cellular lesions visible for morphological examinations. Other biomarker responses are frequently detected in fish livers from field studies including for example responses in antioxidant defence systems, metal homeostasis, stress proteins, DNA damage, etc. (see Chapters 2 and 3). In field monitoring of pollutant effects biomarkers measure fundamental acclimation responses that eventually may evolve into adverse effects. Therefore many biomarkers aid in indicating organ dysfunction by acting as early indicators of toxicant exposure, subsequent abnormal function and information about cause and effect relationships.

There is a long list of subcellular biomarkers that indicate or have the potential to indicate liver dysfunction in fish. By using a set of selected biomarker responses the assessment is strengthened. For example, induction of EROD may not indicate dysfunction but in conjunction with the presence of DNA-damage and lipid peroxidation the liver function can no longer be regarded as normal. It should at least be followed by additional studies. The induction of EROD may be regarded as the normal acclimation to exposure to PAH and planar dioxin-like compounds. However, should inhibition of the EROD activity caused by certain metals (for example Cd) and high levels of planar PCBs and chlorinated dioxins be regarded as a dysfunction of the liver? Such monitoring studies require chemical residue analyses, and/or that fish from contaminated sites are subject to intensified studies, e.g. taken to the laboratory for controlled exposure experiment. Each biomarker response needs careful characterisation and rigorous evaluation as indicator of liver dysfunction.

The spleen is involved in immune responses and is a storage organ for both red and white blood cells. In addition, it appears that the spleen in some fish species may have more diverse functions. In brown bullhead, islands of well-differentiated hepatocytes were observed in the spleen. The presence of hepatocytes appeared to be a normal phenomenon, but their role (if any) is unknown (Spitsbergen & Wolfe, 1995). Most studies on effects of xenobiotics on spleen concentrate on immune functions. In some species, there is a tendency for macrophage aggregation (see section 4.4). The immune functions of the spleen have been shown to be affected by various classes of xenobiotics, e.g. pulp mill effluents (BKME) (Barker *et al.*, 1994; Couillard & Hodson, 1996), polycyclic aromatic hydrocarbons (Hart *et al.*, 1998) and pesticides (Hart *et al.*, 1997).

For most fish species, there are two morphologically distinct tissues referred to as kidney. The head kidney is a very complex organ with a mixture of tissues, including renal, hematopoietic, immune and endocrine. In most fish species, the trunk kidney mainly consists of renal tissue, active in the secretion of divalent ions. There have been various studies concerning the immune function of cells derived from the head kidney. Similarly, there have been a large number of studies concerning the endocrine tissue in the head kidney, the interrenal tissue, as this is the tissue that will release cortisol into the bloodstream in response to stress (Robertson *et al.*, 1987) (see section 4.2.3.4). In contrast, there is a dearth of studies concerning the effects of xenobiotics on hematopoietic or renal functions of the kidney. Relatively few studies are available about subcellular and cellular biomarker responses in fish kidney. EROD and MT induction have been indicated but the responses appear to be less sensitive than in the liver (Pesonen *et al.*, 1987; George *et al.*, 1996). There are also relatively few studies on fish kidney using morphological techniques. There are two possible explanations: either the kidney is generally less sensitive to effects from pollutant exposure or less attention has been paid to kidney abnormalities.

4.2.3.4 Endocrine organs

The interest and concern about so-called endocrine disrupting substances has resulted in an increasing awareness of hormone-related disturbances in aquatic organisms. Effects on endocrine parameters related to reproduction in fish are mainly treated in Chapter 5. Measurements of plasma concentrations of hormones can give information about processes in maintaining homeostasis including compensatory mechanisms related to xenobiotics impact on biochemical and physiological processes. Deviation in hormone levels may also reflect aberration in synthesis, secretion, metabolism and/or clearance of hormones.

There is limited knowledge of how diseases of endocrine organs affect the health of fish. Thyroid hyperplasia have been reported in salmons species in the Great Lakes (reviewed by Leatherland & Farbridge, 1992). It is not evident that this hypertrophy of the thyroid can be linked to specific chemical contaminants. There are some indications that xenobiotics may interact with thyroxine metabolism (Ricard *et al.*, 1998). Hyperplasia has been reported of pancreatic cells producing insulin in winter flounder (*P. flesus*) caught at a very polluted site (Gardner & Yevich, 1988). Also, affected levels of plasma levels of blood glucose have been interpreted as alterations in pancreatic function (Larsson *et al.*, 1985). Altered levels of plasma retinol have been observed in flounders exposed to polyaromatic hydrocarbons (Besselink *et al.*, 1998).

The most extensively studied hormones related to stress in fish are the corticoid hormones. These hormones are formed in the interrenal tissue and regulated via the hypothalamic-pituitary-interrenal axis, which can be stimulated by many factors including chemical stimulation by a variety of contaminants, and biotic factors such as diet, temperature, photoperiod, social stress and sex (Mazeaud *et al.*, 1977; Donaldson, 1981). Contaminants have also been reported to cause decreases in corticoid hormone levels. Cortisol is the dominant corticosteroid, and regulates for example metabolic activity in many organs and has important functions in osmoregulation and immunosuppression. Stimulation of the hypothalamic-pituitary-interrenal-axis results in a rapid release of cortisol into the plasma. This occurs during handling stress, which may mask any effects related to the contaminant.

The exposure of fish to environmentally relevant concentrations of alkylphenols and/or their etoxylates has been shown to cause oestrogenic responses, and to modify growth in fish (Ashfield *et al.*, 1998 and reference therein). It is not known if alkylphenols and alkylphenol ethoxylates affect growth by an oestrogenic effect or if other mechanisms of action are involved.

Studies with pulp mill effluents have revealed evidence for endocrine effects in fish (see also later sections). In the middle of the 1980s fish caught near pulp mills frequently showed physiological disturbances including reduced plasma sex hormone levels, reduced gonad growth and delayed sexual maturity, suggesting exposure to endocrine disrupting substances. Although these reproductive disturbances and other fish health disturbances during the 1990s have been less frequently observed, reproduction disorders, impaired growth and immune defence are considered as a major remaining effect of pulp mill effluents (Sandström *et al.*, 1997). Investigations of possible effects of endocrine disrupting substances by measuring the oestradiol-inducible protein vitellogenin in wild male and female perch caught near pulp mills seem to indicate a weak antioestrogenic response in the female and possibly a weak oestrogenic response in the male fish (Larsson *et al.*, 1999). The chemicals causing these effects are not known, but suspected compounds that may interact with the endocrine systems in fish are plant sterols (Howell & Denton, 1989; Tremblay *et al.*, 1995; Lehtinen *et al.*, 1999).

Sex differentiation in fish can possibly be disturbed by endocrine disruption. In most fish larvae sex differentiation takes place around hatching. In eelpout, a viviparous species, Larsson *et al.* (1999) showed that the undifferentiated gonads differentiate within 3–4 weeks in the unborn fish embryos. Near a pulp mill effluent outlet in the Baltic Sea more male than female embryos were found (Larsson *et al.*, 1999). Sex ratio in fish embryo affects recruitment of fish and thus affects the performance of the fish population.

Altered endocrine functions related to reproduction, development, immune defence and growth are likely to affect the performance of fish populations. Survival of fish in the natural environment is influenced by, for example, the body size and gonad growth for successful reproduction. Growth rate is closely related to survival because small fish generally compete less successfully for food. In order to facilitate linking of responses at different lower to higher levels of organisation, parallel studies at subindividual, individual and population levels are needed.

4.2.3.5 Blood

Chemistry of blood plasma, blood cells and blood cell counts offer many vital responses to measure toxicity in fish. Blood is relatively easy to sample and many effect parameters are easy and fast to analyse. The list of effect parameters in blood can be made very long. Blood indexes including ion and enzyme levels, red and white blood cell counts and haemoglobin content are commonly used in various fish studies. Many metals (Larsson *et al.*, 1985), chlorinated hydrocarbons (Haux & Larsson, 1979) and complex mixtures for example pulp mill effluents (Andersson *et al.*, 1988) are known to affect for example plasma ion concentrations and white and red blood cell numbers. These methods are relatively inexpensive, fairly sensitive and provide good indices of pathological changes in ion regulatory and blood forming tissues (Larsson *et al.*, 1985).

Since strict ion regulation is necessary for fish to maintain water and ion homeostasis, disturbances in ion regulation induced by pollutants are manifested by altered plasma ion concentration. Many blood parameters are sensitive to stress including handling and sampling stress. To overcome such problems standardised procedure must be employed for fishing, handling and sampling (Larsson *et al.*, 1985; Förlin *et al.*, 1986).

Blood parameters such as ion concentrations and red blood cell counts are sensitive to general stress and therefore not often included in health studies of fish. However, with appropriate and standardised methods and trained personnel these difficulties can be treated. As with many biomarkers there is a general lack of mechanistic linkage to important parameters such as growth and reproduction, but extensive field studies such as studies of pulp mill effluents in fish clearly demonstrate relationships between these vital plasma variables and fish population performances (Sandström *et al.*, 1997). Therefore, in order to facilitate linking of responses at different lower to higher levels of organisation, parallel studies at subindividual, individual and population levels are needed.

In humans, liver dysfunction is often accompanied by increased bleeding because the dysfunction results in malproduction of blood coagulation factors in the liver. In fish, blood clotting has been studied in conjunction with stress which increase blood clotting time (Smit & Schoonbee, 1988; Ruis & Bayne, 1997) and exposure to certain pesticides and PAH seems to increase blood clotting time (Zbanyszek & Smith, 1984; Sing & Srivastava, 1992). Blood clotting mechanisms in fish need further studies. It would be valuable with biomarkers indicating deviation in bleeding performance, including blood clotting time, with the potential to indicate liver dysfunction in fish.

4.2.3.6 Nervous tissue

Acetylcholinesterase inhibition in fish has for many years been known to be a good biomarker for exposure to certain insecticides, i.e. organophosphates and carbamates (Holland *et al.*, 1967; Mayer *et al.*, 1992; Nemcsok, 1994). This biomarker has also been used in field monitoring, e.g. Kirby *et al.* (2000). The function of acetylcholinesterase is to cleave acetylcholine into choline and acetic acid. Acetylcholine is a neurotransmitter and is thus essential for normal neural functioning of sensory, integrative and muscular systems. In fish, inhibition of the enzyme affects, for example, respiration, feeding and swimming (Wildish & Lister 1973; Post & Leisure, 1974; Klaverkamp *et al.*, 1977).

4.3 Pathological abnormalities

Histological methods remain the primary tools used for the evaluation of pathological changes in tissues, although much information can be gained from tissue imprints and smears for detecting cellular changes at the light microscope level. Cytology, in the broadest sense to include studies on isolated cells, is also a powerful tool for the visualisation of cellular injury using biochemical techniques (Moore, 1992). The ultrastructural analysis of cells and tissues provides essential information on the pathological changes occurring in a variety of organelles which can be related to both biochemical changes at the cellular level and to tissue pathology (Klaunig *et al.*, 1979; Kohler, 1989). Historically, the interpretation of histological changes has depended to a great extent on the experience and opinion of the pathologist examining the material. Consequently, there has been increasing emphasis on the standardisation of both the diagnosis of lesions detected and in assessments of their relative importance and significance (Reimschuessel *et al.*, 1992; Bernet *et al.*, 1998).

This section provides an overall assessment of the principal pathologies reported in fish exposed to xenobiotics (i.e. contaminants) with emphasis on data obtained from field studies. Each organ system is dealt with separately and a general description of gross and microscopic pathology provided. It should be stressed that the understanding of the normal ultrastructural and light microscopic anatomy of the cells and tissues is fundamental to the correct interpretation of pathological changes in whichever organism is under investigation. There are a number of general texts providing excellent descriptions of ultrastructural pathology of cellular organelles in tissues and organs exposed to a variety of toxic compounds (Slauson & Cooper, 1982; Majno & Joris, 1996). These subcellular changes are often similar in a variety of different tissues and cell types and are highlighted where they have been associated specifically with contaminant exposure. Tissue lesions tend to be more specific to the organ involved and depending on the severity may directly affect function and consequently the general well-being of the fish.

General tissue responses to contaminants are well summarised by Couch (1975), Meyers and Hendricks (1982), Hinton and Lauren (1990) and Hinton *et al.* (1992). For the purposes of this review, lesions known to be induced by infectious agents are excluded unless relevant for differential diagnosis purposes. Nevertheless, it is important to recognise that it can be very difficult to discriminate between histopathological changes induced by xenobiotics and those caused by pathogens. In particular, where these cannot be visualised directly, in the case of most viral infections, or where they may be present at very low levels or even absent. However, idiopathic lesions consisting of, or incorporating inflammatory cells, most often represent the host response to foreign objects or pathogens.

4.3.1 Integument

The use of gross pathology of fish is well established as an integral part of the suite of 'biomarkers' used in biological effects of contaminants monitoring programmes in Europe. Gross pathology recorded according to International Council for the Exploration of the Sea (ICES) guidelines includes acute and healing ulcers, lymphocystis, epidermal hyperplasia/ papilloma and the presence of liver nodules (ICES, 1996). Ulcerations are likely to result from a variety of causes, including physical trauma and subsequent infection with bacteria and other opportunistic organisms. Lymphocystis disease is known to be caused by an iridovirus. Infected cells become greatly hypertrophied, forming clusters of nodules on the surface of the fish (Bucke *et al.*, 1983). Occasionally, lymphocystis nodules occur on the gills and in internal organs. At present, there does not appear to be a strong link between xenobiotic exposure and lymphocystis disease in flatfish (Möller, 1990), but there are seasonal and annual variations (Riersen & Fugelli, 1984). Epidermal hyperplasia and papilloma have been recorded from many fish species from contaminated and relatively clean environments (Cross, 1986; Baumann *et al.*, 1987; Hayes *et al.*, 1990; Bowser *et al.*,

1991; Poulet *et al.*, 1994). It is not yet clear whether infectious agents are involved in the development of epidermal papillomas, but viral particles have been recorded from these lesions in some fish species (see Baumann, 1992; Grizzle & Goodwin, 1998).

Cellular changes in the integument predominantly affect the epithelial layers of the skin and if the stimulant persists, more severe lesions may occur affecting deeper tissues, including the underlying musculature (Bucke *et al.*, 1983; Lindesjöö & Thulin, 1994). Additional pathologies of the integument include cellular hypertrophy, necrosis, erosion and sloughing of the epidermis (including fin rot) (Haensly *et al.*, 1982), and proliferation of mucous cells or changes in their relative abundance. Hyperplasia of pigment cells (melanocytes and iridophores) leading to hyperpigmentation has been recorded in some flatfish species, such as the dab (*Limanda limanda*), from the North Sea. The occurrence of this condition has been increasingly noted. Histological examination of affected tissues has so far failed to demonstrate evidence of pathogen involvement and the aetiology of the condition remains unknown (Lang & Feist, unpublished). With the exception of papillomas, cutaneous neoplasms, including carcinomas, are relatively rare. However, Baumann (1992) suggested that chromatophoromas in certain fish species may be of use in environmental monitoring. Further, Kranz (1989) identified relationships between ulcers of the skin and melanomacrophage (MMC) centres in the spleen.

There are no known direct links between xenobiotic exposure and pathologies of the integument, although exposure to xenobiotics could affect the progress of skin and integument pathologies. If the ability of the immune system of a fish to combat pathogens is compromised, e.g. through effects of xenobiotics on either humoral or cellular immune system components (see section 4.2), one obvious consequence could be slower healing of skin ulcerations. Similarly, effects of xenobiotics on blood clotting could delay healing processes in integument.

4.3.2 Gills

Gill epithelium is continuously exposed to the environment and is one of the main routes for the uptake of soluble xenobiotics. The use of morphological techniques in the evaluation of adaptive responses in the gill has proved to be a powerful tool. The basic structure of the gill is well described for several fish species and the range of pathological changes in this organ is also comprehensively reported in the literature (Mallatt, 1985; Perry & Laurent, 1993). Cellular damage includes hypertrophy of the epithelium and chloride cells, necrosis and hyperplasia (Solangi & Overstreet, 1982; Daoust *et al.*, 1984; Khan & Kiceniuk, 1984; Stoker *et al.*, 1985; Khan *et al.*, 1994), as well as cartilage dysplasia which appears to be a common response to irritants (Spies *et al.*, 1996). In addition, mucous cell proliferation and excess mucous production are also frequently present (Haensly *et al.*, 1982). Haaparanta *et al.* (1997) reported chloride cell proliferation as a significant pathological alteration in roach *Rutilus rutilus* from polluted lakes in Finland. Epithelial hyperplasia, if continued, usually results in lamellar fusion and/or distal clubbing of the secondary lamellae. The use of scanning electron microscopy (SEM) is often used to visualise this category of gill pathology (Lindesjöö & Thulin, 1994) and is essential for the assessment of changes to the surface structure of the epithelium. Other important lesions are those associated with disturbances of blood flow, including vascular congestion, aneurysms, thrombi, telangiectiasis and constriction of blood sinuses. Although there are a variety of pathological changes that can be observed in gills exposed to contaminants, several of these also arise if there is even a slight delay in fixation. Special care must be taken in sampling of field collected material where fish may have been in a trawl net for many minutes before landing, and thereafter there could be a delay before fish are placed in holding tanks prior to post mortem. The stresses involved during these operations could quite possibly produce one or more of the changes indicated above, in particular separation and hypertrophy of the respiratory epithelium (Speare & Ferguson, 1989).

In freshwater fish it is well known that exposure to metals, especially Cd, Cu and Zn, can lead to increased amount of chloride cells and there is also evidence that Cd affects the cytoskeleton of the chloride cells (Devos *et al.*, 1998). Other studies, e.g. Karan *et al.* (1998), indicate that Cu will also cause morphological and physiological changes in gills. It is not known whether marine fish will be exposed to sufficiently high levels of metals to elicit such responses. In one study, Grinwis *et al.* (1998) showed that TBT-exposure would cause gill lesions in flounder (*Platichthys flesus*). Field exposure to pulp and paper mill effluent caused hyperplasia in the gills of shorthorn sculpin (*Myoxocephalus scorpius*) (Barker *et al.*, 1994).

It is clear that any damage to the respiratory epithelium compromises the respiratory ability of the host. Minimal or moderate damage, although not likely to result in mortality directly, may adversely affect the performance of the fish, specifically in feeding and for avoiding predation. Locally severe or extensive damage may allow a route of access for pathogens or directly result in mortality.

4.3.3 Sensory epithelia

In teleosts these consist of the lining of the nares and the lateral line. There is limited published work on the effects of xenobiotics (such as heavy metals and petroleum hydrocarbons) on these structures but similar cellular changes to those already alluded to above are produced in these epithelia (Gardner, 1975; Hawkes, 1980), in particular, severe oedema and spongiosis, necrosis, epithelial sloughing and metaplasia (Gardner, 1975). The sensory epithelium of the lateral line would seem to offer some promise for the detection of early toxicopathic injury. In addition, the sampling of this organ would also provide data on the skin and underlying tissues of the fish since these would also be present in the tissue section. Since damage to sensory organs is likely to affect behaviour (Solangi & Overstreet, 1982), the potential of xenobiotics to induce pathological and potentially behavioural changes at the population level should be seriously considered and investigated.

There are indications that Hg accumulate in the inner ear of fish and could affect balance and co-ordination (Skak & Baatrup, 1993). Various studies indicate that the olfactory epithelium of fish will accumulate various contaminants, especially metals (Gottofrey & Tjälve, 1991; Tjälve *et al.*, 1986). Metal contaminants, especially Cu, have been shown to cause apoptosis of olfactory epithelial cells (Julliard *et al.*, 1996). Decreased sense of taste could have serious implications for many demersal fish species that mainly use chemical cues to find food, or need such cues for appropriate reproductive behaviour.

4.3.4 Visceral organs

4.3.4.1 Liver

The liver is the main organ for the detoxification of xenobiotics and several categories of hepatocellular pathology are now regarded as reliable biomarkers of toxic injury and representative of a biological end-point of contaminant exposure. Consequently, the liver has attracted the most attention as a target organ for biological effects monitoring programmes in both Europe and the USA (Pierce *et al.*, 1978; Bucke & Feist, 1984, 1993; Bucke *et al.*, 1984; Malins *et al.*, 1984; Murchelano & Wolke, 1985; Rhodes *et al.*, 1987; Kranz & Dethlefsen, 1990; Myers *et al.*, 1990, 1992, 1994, 1998a,b; Murchelano & Wolke, 1991; Vethaak *et al.*, 1992; Bucke, 1993; Moore & Stegeman, 1994; Vethaak & Jol, 1996; Vethaak & Wester, 1996). Of particular importance, several fish species exhibit the presence of macroscopic liver nodules or tumours that are easily visible and can be recorded in the field. Guidelines for the sampling and recording of these and other external diseases have been published by ICES (1996) and detailed recommendations on the diagnosis and reporting of histological liver lesions have been the subject of an ICES Special Meeting (ICES, 1997). There is relatively little data on toxicopathic hepatic lesions of non-flatfish species from European coastal waters (Kranz & Peters, 1985) but there is a wealth of information on non-flatfish species from North American and Canadian waters (Smith *et al.*, 1979; Solangi & Overstreet, 1982; Khan & Kicenuik, 1984; Malins *et al.*, 1984; Stoker *et al.*, 1985; Baumann *et al.*, 1987, 1991; Hayes *et al.*, 1990; Vogelbein *et al.*, 1990; Spies *et al.*, 1996; Stehr *et al.*, 1998). However, most attention has been given to the occurrence of toxicopathic lesions, including neoplasms in marine and estuarine flatfish. The diagnostic criteria for the description of the histological features of these lesions have largely been derived and adapted from models developed from mammalian studies (Frith & Ward, 1980; Bannasch, 1986). The currently accepted system has been verified by a number of studies investigating the development of hepatic neoplasia in several fish species (for example Hinton *et al.*, 1988; Köhler, 1990a,b; Vethaak *et al.*, 1996; Moore *et al.*, 1997; Stehr *et al.*, 1998).

Hepatic lesions in flatfish have generally been categorised into several distinct groups (Myers *et al.*, 1987) and it has been possible to rank them according to their relative importance as indicators of contaminant exposure. Of most importance are those that have been recognised in experimental hepatocarcinogenesis studies. This group includes:

- Unique degenerative lesions including hydropic degeneration of biliary epithelial cells and hepatocellular and nuclear polymorphism
- Foci of cellular alteration (FCA), including basophilic, eosinophilic, clear cell and vacuolated foci recognisable by their staining reaction with haematoxylin and eosin technique
- Benign neoplasms, including hepatocellular adenoma, cholangioma arising from the bile ducts and hemangioma arising from blood vessels and capillaries in particular
- Malignant neoplasms, including hepatocellular carcinoma cholangiocarcinoma and hemangiosarcoma.

A second group of lesions includes certain non-neoplastic proliferative lesions (hepatocellular regeneration, bile duct hyperplasia and hepatic fibrosis) as well as general or non-specific degenerative lesions including focal or diffuse cellular necrosis, the presence of hyaline inclusion bodies and increased apoptosis. Finally, a third group of lesions comprises storage conditions and inflammatory changes. This final group is regarded as the least significant in terms of relevance as indicators of contaminant exposure, and provides more information on the general health status and condition of the fish. There are some indications that macrophage aggregates in liver relate to environmental contamination, specifically BKME from pulp and paper mills (Couillard & Hodson, 1996).

Without doubt, analysis of gross and microscopic liver lesions provides a sensitive technique for the evaluation of contaminant exposure. Ultrastructural changes including abnormalities of the lysosomal system and endoplasmic reticulum as well as proliferation of peroxisomes also appear to be good indicators of contaminant exposure (Köhler, 1990a; Braunbeck & Völkl, 1991; Köhler *et al.*, 1992). However, the presence of liver pathology can only provide evidence of previous exposure. The development of lesions may take months (or years) depending on factors as yet not fully understood. It is therefore important to collect age data from individuals exhibiting gross hepatic lesions, from those sampled for routine histopathological analysis and also from subsets of the population sampled. This data is essential in order to determine relationships between the occurrence of different lesion categories in the same species and of the age distribution of fish exhibiting toxicopathic lesions in the general population. There is also a continuing research need for the development of more sensitive techniques to detect changes in the enzyme activities and of DNA damage in hepatocytes as well as ultrastructural changes (Myers *et al.*, 1998a; Winzer & Köhler, 1998). The application of these techniques in conjunction with histopathological analyses is important in order to determine the relationships between the expression of marker enzymes, adduct formation and the presence of microscopic and gross lesions. However, it must be noted that these relationships may differ between fish species (Husøy *et al.*, 1996).

4.3.4.2 Spleen

This organ is of major importance in haematopoiesis and antigen trapping. Macrophages remove particulate material from the ellipsoids to the melanomacrophage centres (MMC). Several workers have advocated the use of these structures as biomarkers for exposure to xenobiotics (Bucke *et al.*, 1984; Kranz, 1989; Bucke *et al.*, 1992; Khan *et al.*, 1994; Khan, 1995). It is well recognised that MMCs vary both in number and size between fish species and individuals within a species. In general terms, they may at best provide an overall assessment of general health status based on the current state of knowledge (Wolke *et al.*, 1981; Blazer *et al.*, 1987; Wolke, 1992). However, the very fact that MMCs can be formed relatively rapidly following a wide variety of environmental and biological insults can be of benefit for the detection of adverse environmental change (Bucke & Dixon, 1992). In addition, the function of MMCs as storage units for unwanted materials which have not been excreted by other means can be exploited for the assessment of contamination by heavy metals since these become trapped in these centres. X-ray microanalysis can be used for the specific localisation and quantification of such deposits (Pulsford *et al.*, 1992). Histological and ultrastructural indicators of contaminant exposure are also well expressed in the spleenic tissues. Spazier *et al.* (1992) provided clear evidence of ultrastructural pathology in the spleen of eels (*Anguilla anguilla*), exposed to a chemical spill in the river Rhine. Principal alterations included loss of cell surface structures such as pseudopodia and organelle damage, including mitochondrial swelling and increased numbers of secondary lysosomes containing membranous material, presumably representing degraded organelles. There are few studies that link spleen pathologies to exposure to xenobiotics.

4.3.4.3 Kidney

The major lesion categories of particular importance in the present context affect the excretory elements. Hydropic vacuolation, presence of proteinaceous droplets and necrosis of the tubule epithelia have all been recognised in fish exposed to hydrocarbons (Haensly *et al.*, 1982; Spies *et al.*, 1996). In addition, pathological changes of the glomerulus were reported, including dilation of Bowman's space, hypercellularity and fibrosis of the glomerular tuft and basement membrane thickening. As in the spleen, MMCs are prominent structures in the kidney and increased numbers may be used as non-specific indicators of stress.

Renal lesions by themselves can provide evidence of toxic insult but it is in combination with the observation of pathological change in other organs that they give stronger clues of xenobiotic impact (Rhodes *et al.*, 1987). The evaluation of other organs is important since different fish species respond in different ways and exhibit varied susceptibility in the induction of tissue and organ pathology.

4.3.5 Skeletal muscle

Skeletal muscle has been little studied with respect to contaminant exposure. The main pathological change is degeneration and necrosis of the myofibrils (Haensley *et al.*, 1982; Khan & Kicenuik, 1984). Clearly, significant damage to the musculature will adversely affect the performance of the fish for feeding, spawning migrations and predator avoidance.

4.3.6 The skeleton

The skeletal development of fish has been shown to be sensitive to contaminant exposure. Some studies have shown effects from metal exposure (Bengtsson *et al.*, 1975, 1988) and from exposure to effluents from pulp and paper mills in the Baltic (Bengtsson, 1991; Lindesjöö *et al.*, 1994). The effects are thought to be caused by perturbation of Ca-metabolism and calcification processes. It is not clear what the ecological significance is, although at least some effects cannot be altogether deleterious as adult specimens with skeleton abnormalities have been collected in the wild.

4.3.7 Endocrine organs

Among the most important of the endocrine organs are the pituitary, thyroid, adrenals (interrenal and suprarenals), Corpus of Stannius and the endocrine pancreas. In each of these organs necrosis, hypertrophy and hyperplasia have been reported (Gardner, 1975). In addition, degranulation of pituitary cells has also been observed in eels exposed to DDT

(Ball & Baker, 1969). Although there is a dearth of information on the pathology of endocrine organs, the utility of using the pituitary in particular as a biomarker for contaminant effects has been proposed (Couch, 1984). Despite practical difficulties in examining this organ, its importance as a central controlling organ suggests that more effort should be given to determine whether significant pathological changes in endocrine organs occur in wild fish populations exposed to xenobiotics.

4.3.8 Nervous tissue

Pathology of the brain and spinal cord caused by xenobiotics has been little studied (Meyers & Hendricks, 1982). Expected lesions would include hyperaemia, cellular hyperplasia, necrosis and possibly hydropic vacuolation of nerve tissue cells. Investigations including the pituitary would provide the opportunity to study the brain and spinal cord.

4.3.9 Gastro-intestinal tract

The gastro-intestinal tract is one of the main routes for the uptake of xenobiotics. The principal lesion type reported is hydropic degeneration of the digestive gland (Haensly *et al.*, 1982). Other pathological changes that might be expected include proliferation of mucous cells, hyperaemia, atrophy and metaplasia. Some studies have indicated that high levels of some metals in diet may cause increased apoptosis of intestinal cells (Berntssen *et al.*, 1999). However, based on current information, tissues of the gastro-intestinal tract do not seem to exhibit lesions which may be of value for biological effects monitoring.

4.3.10 Gonads

Pathology of the testis and ovary relating to endocrine disrupters is considered in Chapter 5. Little data is available on the pathological effects of xenobiotics on these organs and some studies were unable to detect specific histopathological changes after contaminant exposure. However, disturbances in the pattern of ovarian follicle development (including atresia), and inflammation were noted in some studies (Stott *et al.*, 1983; Johnson *et al.*, 1988; Spies *et al.*, 1996). However, Khan & Kiceniuk (1984) demonstrated that in cod *Gadus morhua* chronically exposed to crude oil, the synchronous development of the testis was disturbed and that multinucleate giant cells, thought to be involved in the removal of cellular debris, were present prematurely in the seminiferous tubules. The application of a suite of immunohistochemical techniques may provide valuable evidence of damage in exposed fish. In particular, development of techniques to evaluate the viability of developing sperm and oocytes will be required to ascertain reproductive potential.

Lesions affecting the gonad clearly have the potential to affect reproductive success in individuals. Overall recruitment to the population could be reduced if sufficient numbers of fish are affected.

4.3.11 Eyes

Little data is available on toxicopathic lesions in the teleost eye. Enlargement and softening of the lens and haemorrhaging of the anterior chamber have been reported. Hargis and

Zwerner (1990) described the occurrence of lens cataracts in sciaenid fish from the Elizabeth River, Virginia, USA, and in fish exposed experimentally to the polynuclear aromatic hydrocarbon (PAH) contaminated Elizabeth River sediments. Ocular lesions do not appear to be common in marine or estuarine fish species inhabiting contaminated sites, although Payne *et al.* (1978) reported degeneration of lens fibres in cunner (*Tautogolabrus adsperus*) exposed to petroleum, and similar changes have been noted in experiments using rainbow trout (Hawkes, 1977).

4.4 Larval and embryological development

The larvae of most, and the eggs of some, marine fish species are planktonic and will thus be exposed to xenobiotics in the water or the surface microlayer (Hardy *et al.*, 1987). There are observations of aberrations in larval development of commercial fish species from both European and American coastal waters (see reviews by von Westernhagen *et al.*, 1988; Longwell *et al.*, 1992; Dethlefsen *et al.*, 1996).

Fish larvae may be exposed to xenobiotics while still in the egg, through water after hatching, from yolk or from food consumed during larval development. In addition, mutagenic effects in eggs could also be expressed during larval development. In studies from the North Sea there are indications that xenobiotic exposure through yolk may be at least part of the reason for the observed effects (von Westernhagen *et al.*, 1987, 1989). Larval aberrations may also derive from natural factors, however, including poor quality of eggs, low food reserves in the embryo, and predation, as well as hostile environmental conditions such as physical disturbance, low dissolved oxygen and variable salinity (see Rosenthal & Alderdice (1976), Wiegand *et al.* (1989) and Purceli *et al.* (1990)). Observations from the North Sea suggest that higher water temperature may increase the number of deformed larvae (Dethlefsen *et al.*, 1996). In addition, it is known that survival of larvae (and presumably development) will be affected by food availability immediately after the yolk sac stage.

There are two established situations of high larval mortality: the M74 syndrome in the Baltic and the EMS syndrome in the North American Great Lakes. In both syndromes, xenobiotics may contribute to reduced larval survival, but no clear link has been established (see section 4.6).

4.4.1 Early development in fish

Fish eggs have a large amount of yolk and a protective membrane, the chorion, which is composed of a polysaccharide and proteinaceous material. The chorion becomes completely toughened after fertilisation and acts as a physical and possibly a chemical barrier to the influx of chemicals (Tesoriero, 1977). After fertilisation the cytoplasm of the egg cell becomes segregated from the yolk and forms a blastodisc. The blastodisc further subdivides during cleavage to form the blastoderm, which later forms the body of the fish embryo. Towards the end of cleavage the blastomeres spread, which is followed by the process of epiboly, during which the primary germ layers of the embryo are established and the embryonic axis is defined. As a result of cell movements the embryonic shield develops, within which the primary organs of the embryo are formed including the neural tube, the notochord and the somites. In general, most mature fish are highly fecund, and embryo and larval survival is naturally low (<5%). The embryo larval stage lasts for several weeks in most fish species.

4.4.2 Methods

Most commonly, pelagic eggs or larvae are collected in plankton nets in surface waters by trawling at low speed. Such larvae will have undergone a 'natural' exposure. Alternatively, ripe male and female of the species in question can be collected and brought to spawn onboard ship or in the laboratory. Eggs or embryos may then be exposed or allowed to develop with no post-fertilisation exposure. A second alternative is to use eggs from hatcheries in the afflicted areas, as is done in studies of the M74 syndrome. Eggs from some fish species, e.g. salmonids and gadiids, are sufficiently large to allow microinjection of contaminants following fertilisation. Various methods can be used to assess effects on embryos or larvae. The simplest is to assess hatching success, which needs a low-magnification binocular microscope. Larval aberrations can be scored using a similar binocular microscope. In some studies, biochemical and cytological techniques have been used to assess specific processes in the larvae, e.g. metallothionein (George *et al.*, 1996) or cytochrome P4501A activity (Goksøyr *et al.*, 1991).

4.4.3 Mechanisms

The observed larval deformities within a fish species are surprisingly similar whether the effect is due to natural factors such as temperature or xenobiotics. Weis and Weis (1991) identified four categories for developmental abnormalities in fish larvae: effects on morphogenetic movements, tissue interactions, growth and degeneration. For most purposes, it is not really necessary to distinguish between different categories as they all will result in developmental aberration. In addition to those four categories, sex differentiation is a highly relevant parameter. Sex differentiation can be affected by exposure of larvae to oestrogens or androgens during a specific period of development.

From the above, it will be apparent that a large number of xenobiotics released into the aquatic environment can interfere with larvae development. The skeletal, circulatory and optical systems and rates of development to specific stages appear to be very sensitive. Rosenthal and Alderdice (1976) concluded that the most sensitive stages would be the gonadal tissue, the early embryo, and the stage of larval transition between endogenous and exogenous food sources.

The precise details of the mechanisms by which contaminants influence the developing embryo or larvae are not known. This is compounded by the fact that there appears to be a total similarity in morphological defects observed for embryos exposed to the major classes of contaminants (i.e. heavy metals, chlorinated hydrocarbons, petroleum hydrocarbons). The reviews by Rosenthal and Alderdice (1976), von Westernhagen (1988) and Weis and Weis (1989) clearly show that notochord abnormalities, crano-facial defects, brain and eye defects, cardiovascular defects and spinal abnormalities may be induced in developing embryos exposed to a variety of contaminants. This led Rosenthal and Alderdice (1976) to suggest that an embryo responds to toxic insults with a generalised 'stress' response. This may be clearly seen in the study by Bodammer (1993): the initial treatment of the egg and continued exposure of the embryo to cadmium can:

(1) Modify the permeability of the egg membrane prior to and after fertilisation
(2) Disrupt gastrulation and axiation during the mid to later stages of embryogenesis
(3) Retard the growth, development, and organogenesis
(4) Reduce embryonic heart rate
(5) Reduce or modify embryo motility
(6) Decrease the activity of several biosynthetic enzymes in late-stage embryos
(7) Disrupt normal osteogenesis, resulting in skeletal abnormalities
(8) Reduce yolk-sac size via osmotic effects on perivitilline fluid, resulting from cadmium affected membranes
(9) Result in premature or delayed hatching.

A further important and complicating factor, which must also be considered, is the possibility of stage-dependent sensitivity to contaminants (Marty *et al.*, 1990).

4.4.4 Experimental studies

Embryos and larvae may be used in screening tests for aquatic toxicity testing to derive maximum acceptable toxicant concentrations. The most common end-points are hatching success, early larval survival and growth (Dave & Xiu, 1991; Hutchinson *et al.*, 1994; Matta *et al.*, 1997). Other studies include observations on the occurrence of developmental abnormalities in embryos and larvae (Call *et al.*, 1985; Henry *et al.*, 1997; Olivieri & Cooper, 1997) or larval behaviour (Olivieri & Cooper, 1997). It is generally accepted that embryos and larvae are very sensitive to contaminants and as a result whole life cycle toxicity tests have often been replaced by early life-stage tests. Much of the data that has been reported in the literature was derived from experiments carried out on embryos exposed after fertilisation for short time periods to environmentally unrealistic concentrations of contaminants. As noted above, embryos in the natural environment could be affected by xenobiotics in at least four additional ways, through genetic damage, via the yolk synthesised during oogenesis by exposed females, during the brief period between the shedding of the gametes and formation of the chorion and through feeding. For organic contaminants there is a high correlation between maternal transfer during oogenesis and the lipid content of the egg (Nimi, 1983). There are few studies of the impact of contaminants on gametes prior to fertilisation. There are some examples, however: reduced sperm motility was observed in trout after exposure to Hg (McIntyre, 1973) and there was reduced hatch in trout eggs when sperm from male trout exposed to Hg were used to fertilise the eggs (Birge *et al.*, 1979).

4.4.5 Field studies

In addition to the embryo abnormalities identified in field studies by Perry *et al.* (1999), Cameron *et al.* (1992, 1996) and Cameron & von Westerhagen (1997) (see Chapter 3, section 3.3.10.5), early life mortality have been observed in Baltic salmon and in fish from the

Great Lakes. In the Baltic Sea it has also been indicated that sea trout may be affected (Landergren *et al.*, 1999). Also, downstream of pulp mills impaired larvae development has been indicated (Sandström, 1994).

There are few examples in the literature on field-related effects of contaminants causing larval abnormalities and poor recruitment. However, the examples that do exist provide convincing evidence that this is an important area for further investigation. Longwell & Hughes (1981) studying the mackerel (*Scomber scombrus*) in the New York Bight, showed significantly lower egg viability in highly impacted Bight areas (disposal sites) than in areas offshore and that the health of the embryos correlated with contaminant concentrations. Several investigators have worked on the sea surface microlayer (Cross *et al.*, 1987; Hardy *et al.*, 1987; Kocan *et al.*, 1987). These studies have demonstrated that sea surface waters from contaminated areas produce significantly higher morphological or chromosome abnormalities than waters from control areas.

4.4.6 Links between cellular effects and larval development

While there are few studies that link processes in the larvae themselves to increased level of aberrations, there are some studies that suggest cellular processes in the female affect the offspring. In one study, Monosson *et al.*, (1994) found that larval survival was affected by PCB exposure, possibly through disruption of gonadal maturation. The treatment did not affect plasma levels of sex steroid hormones or yolk precursor protein, vitellogenin. Black *et al.* (1998) found that embryo and larval survival was inversely related to hepatic cytochrome P4501A activity in female *Fundulus heteroclitus* exposed to PCB under field conditions. While cytochrome P4501A activity in the larvae themselves have been found to relate to PAH exposure (Goksøyr *et al.*, 1991), no clear link has been observed between such induction and survival of the larvae (Serigstad, pers.comm.).

4.5 Case studies

While there are numerous observations of how xenobiotics affect fish health, there are only a few cases in which an attempt has been made to link such effects to population or community impacts. Such knowledge would only become available following a major effort over many years. Relevant studies include monitoring of pulp mill effluents in Sweden and Canada, of metals from mining activities in Canada (Munkittrick *et al.*, 1991), of PAHs in Puget Sound, USA (Collier *et al.*, 1992; Landahl *et al.*, 1997; Myers *et al.*, 1998b), and the work on EMS (early mortality syndrome) and the M74 syndrome in the Great Lakes and the Baltic, respectively. Below, a brief review is given for two cases, one concerning the effects of pulp mill effluents on fish health and fish populations, the other concerning the M74 syndrome. The two cases represent different viewpoints; whereas there is a clear link between contaminant stressors and the observed effects for pulp mills, it is still not clear whether xenobiotics play a role in the development of the M74 syndrome. Common to both is the clear impact on fish populations and the link from health effects to population impacts.

4.5.1 Pulp mill effluent

Biological investigations have revealed many disturbances in the fish community and on the health status of fish in the receiving water of pulp mills effluents. Below is summarised the important work performed in receiving waters of pulp mills along the Swedish east coast (Baltic Sea and Gulf of Bothnia) (Andersson *et al.*, 1987, 1988; Bengtsson *et al.*, 1988; Härdig *et al.*, 1988; Karås *et al.*, 1991; Lindesjöö & Thulin, 1992, 1994; Balk *et al.*, 1993; Lindesjöö *et al.*, 1994; Förlin *et al.*, 1995; Sandström, 1995; Sandström *et al.*, 1997; Larsson *et al.*, 2003).

The main strategy used in the studies of adverse effects of pulp mill effluents was to use biochemical, physiological and morphological biomarkers covering effects on different levels of biological organisation from cell to individual, to detect disturbances at an early stage in selected stationary species (Larsson *et al.*, 1985; Förlin *et al.*, 1986; Haux & Förlin 1988; Adams *et al.*, 1989). To strengthen the cause and effects relationship, controlled laboratory exposure experiments were run in parallel to field studies. From the controlled exposure experiments a map of typical responses caused by pulp mill effluents was established. The same parameters were then studied in a discharge gradient. Where possible, the results were compared to observed effects in investigations on ecological end-points like growth, and abundance of fry and adults, reproductive capacity, recruitment and community structure.

The usefulness of biochemical and physiological variables as health indicators in fish exposed to pulp mill effluents was thoroughly tested in Sweden within the research project Environment/Cellulose. The studies included both long-term laboratory investigations and monitoring in the receiving waters of different pulp mills. The main field studies performed in perch from the receiving waters of the Norrsundet pulp mill in 1984/1985 revealed significant effects on several fundamental biochemical and physiological functions (Andersson *et al.*, 1988). Typical symptoms were reduced gonad growth, liver enlargement and very strong induction of hepatic ethoxyresorufin-O-deethylase (EROD) activity. Furthermore, elevated levels of ascorbic acid in the liver and abnormal carbohydrate metabolism pointed towards pronounced metabolic disorders. Marked effects on the white blood cell (WBC) patterns indicated a suppressed immune defence. In addition, the exposed fish showed a stimulated red blood cell (RBC) production and a disturbed ion balance. The toxic effects were dose-dependent, with the most pronounced effects at the two innermost capture sites (2 and 4.5 km; degree of waste water dilution 166 and 330 times, respectively), but many serious disturbances (for example EROD induction, reduced gonad growth, and altered red and white blood cell pattern) could be detected also in perch caught 8–10 km from the pulp mill (degree of waste water dilution >1100 times).

Parallel morphological investigations in the same area showed increased prevalence of fin erosion (Lindesjöö & Thulin, 1994) and skeletal deformities including deformation of jaw bones (Lindesjöö & Thulin, 1992) and abnormalities of gill cover bone (Lindesjöö *et al.*, 1994). In addition, parallel ecological studies in the same area showed delayed sexual maturity and inhibited gonad growth, markedly impaired fish fry production, growth disturbances, increased mortality and low abundance of perch as well as other species of fish (Neuman & Karås, 1988; Sandström & Thoresson, 1988; Sandström *et al.*, 1988; Karås *et al.*, 1991). This wide approach generated effect data in fish from subcellular to community levels.

Biomarker studies later during the 1980s on perch outside other Swedish pulp mills showed effect patterns which to a great extent were similar to those found in perch in the receiving water of Norrsundet (Andersson *et al.*, 1988; Förlin *et al.*, 1991, 1995; Balk *et al.*, 1993). In addition, many of the same responses were subsequently observed at a number of North American pulp mills (Hodson *et al.*, 1992; Munkittrick *et al.*, 1992; Servos *et al.*, 1992). These studies seem to indicate that effluents from conventional pulp production process with chlorine bleaching, but usually without secondary treatment, caused several disturbances which affected the health status and possibly also the reproduction and survival of the fish in the receiving water.

Since the mid-1980s, a rapid development and introduction of new techniques for bleaching of kraft pulp (elemental chlorine free, ECF; total chlorine free, TCF, respectively) in combination with other alterations in several other process changes and cleaning measures in the production of pulp, have also led to a considerable decrease of chlorinated organic and other materials discharge from Swedish pulp mills to the receiving waters (Sandström *et al.*, 1997). At the Norrsundet pulp mill at the Bay of Bothnia, the successive modifications and introduction of new process techniques and effluent treatments started in 1983 and 1984 when the introduction of a new production line, oxygen prebleaching and new washing techniques reduced AOX (a measure of chlorinated organic materials) by more than 40% and also reduced discharge of organic materials (COD reduction about 30%). The use of chlorine was successively decreased during the following years and ceased entirely in April 1992, leading to a low AOX level in the effluents after 1993. In addition, a secondary treatment system, installed in 1992, led to the reduction of several more components.

In order to investigate if the improved discharge situation was accompanied by a positive alteration of the health status in fish, the recovery studies were performed outside the Norrsundet pulp mill in 1988, 1990, 1993, 1995 (Larsson *et al.*, 2003) and in 1997 (Larsson *et al.*, 2000). Compared to the situation in 1984/85, considerably lower toxic responses in fish were found at all sampling events. Both the number and degree of biomarker responses were reduced. The previously observed reduced levels in blood plasma sex steroids (van der Kraak *et al.*, 1992) were not observed after 1990. Inhibited gonad growth was sometimes observed and together with observation of delayed sexual maturity, smaller perch embryos with higher prevalence of malformations and increased larval mortality (Sandström, 1995) suggest that the exposure from the pulp mill still had an effect on reproduction and recruitment of the perch population in water areas closest to the pulp mill.

In the earlier studies at Norrsundet, marked effects were observed on EROD activity, WBC, RBC and carbohydrate metabolism. All these variables showed a relatively rapid recovery from the pronounced responses noted in 1984/85 (Larsson *et al.*, 2003). This reflects the positive reduction or elimination of several environmental stressors previously present in the effluents. Although the induction of EROD activity has markedly decreased, it is noteworthy that the EROD activity at all sampling occasions has shown a dose-response relationship following the pollution gradient. This suggests the pulp mill discharge to be the most likely source of EROD inducer(s). It has been suggested that retene, a rather degradable compound found in pulp mill effluents, could be a major EROD inducing compound (Fragoso *et al.*, 1998; Ronisz & Förlin, 1998; Billard *et al.*, 1999).

Previous studies on effects of pulp mill effluents in fish clearly indicate endocrine disrupting responses by the effluents, including reduced gonad growth, reduced sex steroid levels, reduced growth and impaired reproduction and recruitment (Andersson *et al.*, 1988; Sandström *et al.*, 1988; McMaster *et al.*, 1991; van der Kraak *et al.*, 1992; Gagnon *et al.*, 1994; Munkittrick *et al.*, 1994; Sandström, 1995; Sandström *et al.*, 1997). Masculinising effects of pulp and paper mill effluents include gonopodium formation of female poecilids (Howell *et al.*, 1980; Cody & Burtone, 1997), testicular growth and enlarged eyes in eels (Caruso *et al.*, 1988), spawning warts in female fish (Munkittrick *et al.*, 1999) and more male embryos in viviparous eelpout close to a pulp mill (Larsson *et al.*, 1999). Endocrine disruption in fish near pulp mills seems not to have disappeared after introduction of chlorine free bleaching or secondary treatment. Suggested responsible candidate compounds include plant sterols and/or other compounds endogenous to the wood raw material.

The repeated fish investigations in the receiving waters at Norrsundet pulp mill clearly show that a very positive but not complete recovery has occurred as a result of the internal process modification and improved treatment of the waste water. Most individual organism effects previously observed in exposed fish have either disappeared or show a very limited response. Additionally, investigations of the fish community in the receiving water outside Norrsundet pulp mill (Sandström, 1995) showed a much improved situation except for a disturbed recruitment and reduced abundance of fish.

The remaining effects, such as delayed sexual maturity, indication of some genotoxicity, slight EROD induction, possibly reduced vitellogenin production and male biased sex ratio in the water area closest to the discharge point, suggest that the fish populations remain exposed to effluents containing substances with a toxic potential.

4.5.2 The M74 syndrome

The M74 syndrome is characterised by a diet-related deficiency of vitamin B1 (thiamine) in mature female salmon, eggs and fry, which shows many similarities to EMS (early mortality syndrome) in salmonids from the Great Lakes (see reviews by Bengtsson *et al.*, 1999; Fitzsimons, 1995). The syndrome was initially observed at the beginning of the 1970s, hence the name, and has had a fluctuating impact on the recruitment of natural populations of Swedish salmon (*Salmo salar*) since then. Aspects of the M74 syndrome were addressed in a Swedish research programme, FiRe, the results of which have been summarised (Anonymous, 1999).

4.5.2.1 The Baltic salmon

The salmon in the Baltic is the same species as the Atlantic salmon (*Salmo salar*), but is genetically distinct from other salmon populations in Europe. As with Atlantic salmon elsewhere, the Baltic salmon generally returns to spawn in the same river from which it was hatched. Following smoltification and migration to the Baltic, the salmon generally stays for up to 4 years before maturing and returning to spawn. It appears probable that the Baltic salmon does not migrate out of the Baltic during this period. The Baltic is a brackish-water system with a narrower selection of food-items than marine habitats. Salmon in the Baltic appear to feed on whatever is available at the time, ranging from sprat and herring to stickleback and crustaceans. Although feeding habits differ between areas of the Baltic, the Baltic salmon appear to feed less on crustaceans than their Atlantic counterparts. The availability of specific food items is presumed to be one important factor in determining the extent of M74 in any given year.

4.5.2.2 A history of the M74 syndrome

The percentage of afflicted spawning female salmon has varied from 0 (in 1981–83) to nearly 80% (in 1993). Since spawning salmon are captured in the wild each year from salmon rivers, there is a good record of the extent of M74 since 1974. This syndrome only affects salmon that live and feed in the Baltic. Two-thirds of the Swedish salmon rivers are regulated and the appropriate habitats for salmon spawning are thus largely destroyed. To assure the conservation of river-specific genetic material, mature salmon are captured each year, stripped and the eggs fertilised artificially. The resulting fry are raised to smolts, which are then released. This practice differs from that most commonly used in Finland, where river-specific brood stock are kept continuously. Without the Swedish rearing programme using wild salmon, it is probable that even such a dramatic effect as the M74 syndrome could have gone unnoticed, at least until the salmon population of many afflicted areas had crashed beyond recovery. The M74 syndrome does not affect all salmon populations in the Baltic. Salmon in the Gulf of Riga are not afflicted.

4.5.2.3 Possible causes for M74

The symptoms of the disease occur in newly hatched yolk-sac fry during the phase when the yolk sac is resorbed before the fry start to feed. The fry initially show hyperactivity followed by loss of co-ordination, hyperpigmentation, precipitates in yolk sac apathy and exophthalmia. Death occurs usually a few days after the first external symptoms. The symptoms are female-specific; that is, the M74 affects offspring of certain females with total mortality. As indicated above, it is by now well-established that the immediate cause for M74 in a fish is a lack of thiamine (vitamin B1) and female salmon have been routinely treated with thiamine since 1994. Treatment with thiamine has been shown to protect both adult and fry from M74. It is by no means entirely clear why Baltic salmon are deficient in thiamine in some years and not in others. Possible causal factors include food availability and selection, exposure to xenobiotics, infectious agents, endogenous levels of antioxidants, activity of antioxidant enzymes and low genetic diversity. Each of the factors mentioned may interact and could in its turn be affected by environmental influences such as eutrophication, acid precipitation and overfishing of salmon or other Baltic species.

In both EMS and M74 the diet appears to be an important parameter. Thiamine cannot be synthesised by higher organisms and must therefore be taken in through diet (or water) or produced by endosymbiontic microorganisms (Cooray *et al.*, 1999). Shifts in the diet of salmonids from crustaceans and fish species low in thiaminase to thiaminase-rich fish species appear to be important factors in the development of the diseases (Fitzsimons *et al.*, 1999; Karlsson *et al.*, 1999). In the Baltic, different diets could also explain why some salmon populations are not affected.

The Great Lakes and the Baltic are among the most contaminated partially enclosed water systems in the world. While there does not appear to be a direct link between organohalogen residues in the tissue of spawning salmon and the development of M74, it is not possible to rule out interactions with other factors or the presence of xenobiotics not presently known (Asplund *et al.*, 1999).

There is no evidence that M74 is related to infection of adult salmon by virus, bacteria or fungi (Cooray *et al.*, 1999). The natural bacterial microflora of salmon intestine was able to produce vitamins and this microflora was disturbed following formalin treatment. It remains to be clarified whether bacterial symbionts in salmon intestine could synthesise thiamine and whether this ability is affected by environmental factors (Cooray *et al.*, 1999).

There appears to be a link between the antioxidant defence system and the development of M74 in developing larvae (Lundström *et al.*, 1999). The levels of antioxidants such as astaxanthin and carotenoids are depressed in the liver of M74 fry compared to healthy fry (Pettersson & Lignell, 1999). The pattern for enzymes involved in antioxidant responses was not as clear: catalase was depressed in 3-week old M74-fry, whereas the activities of both glutathione peroxidase and glutathione reductase were increased in M74-fry compared to healthy fry (Lundström *et al.*, 1999). In addition to the antioxidant defence mechanisms, metabolising enzymes were depressed in M74-fry. At present, it is not known whether changes in various markers for cellular health are part of the cause for thiamine deficiency or whether they are attempts from the cell to compensate.

At the current state of knowledge, it is known that the M74 syndrome is directly caused by thiamine deficiency and that diet is important. Many components in the causal chain are still unresolved, however, including the influence of xenobiotics, the relevance of gastrointestinal symbionts and the involvement of cellular processes relating to antioxidant defence and metabolising enzymes. The outbreaks of M74 appear to have a cyclic pattern, possibly related to large-scale changes in the Baltic ecosystem. Co-ordinated studies are required to resolve the mechanisms behind the outbreak of M74. An integrated approach is required including studies at molecular level to the ecosystem level.

4.6 Conclusions

While there can be no doubt that a compromised immune system is detrimental to the individual, there are no clear links to effects on populations. Research is needed on links between changes in cellular markers, immune parameters and pathological changes. Such links can only be made through chronic mesocosm or possibly field studies in which a broad selection of markers is analysed at regular time intervals.

There is a large volume of studies concerning the effects of xenobiotics on the normal function of fish tissues. In laboratory studies, there is well-documented evidence that xenobiotics accumulate in specific tissues. By far the largest amount of information is available for the liver, but these results are possibly also the ones most difficult to interpret in terms of health or survival. However, in contrast to other sections of this chapter, there is some knowledge of links between cellular processes and observed physiological aberrations in the liver.

While there are more or less clear links between xenobiotic exposure and some pathological effects, e.g. PAHs leading to neoplasms or cancers of the liver, there is surprisingly limited knowledge of the extent to which contaminants cause pathological changes in fish tissues. More effort should be put into clarifying effects on some organs that appear to be particularly sensitive to contaminants, e.g. olfactory epithelium, gills and nervous tissue. Disruption to the functioning of each of these tissues may have significant impacts on the fish at a physiological level (see Chapter 3) and consequently on the future survival of populations.

Perhaps one of the most important levels of organisation in which to determine pathological effects is in relation to larval aberrations. It is here that immediate impacts on a population would be likely to be determined. However, it is not possible at present to link larval aberrations to population effects. Two views prevail: one view argues that mortality of fish larvae is so high normally that it probably does not matter if a few more die as a result of pollution insult. However, it could be argued that an increased toll of a few percent would have dramatic consequences for the relevant stocks. The importance of such effects will obviously vary between species. In addition, until recently it has also not been clear how or whether a changed sex ratio would affect marine fish populations (see Chapter 7).

4.7 References

Adams, S.M., K.L. Shepard, M.S. Greely Jr, B.D. Jimenez, M.G. Ryon, L.R. Shugart & J.F. McCarthy (1989) The use of bioindicators for assessing the effects of pollutant stress on fish. *Marine Environmental Research*, **28**, 459–464.

Ainsworth, A.J., C. Dexiang & P.R. Waterstrat (1991) Changes in peripheral blood leukocyte percentages and function of neutrophils in stressed channel catfish. *Journal of Aquatic Animal Health*, **3**, 41–47.

Albergoni, V. & A. Viola (1995) Effects of cadmium on catfish, *Ictalurus melas*, humoral immune response. *Fish and Shellfish Immunology*, **5**, 89–95.

Anderson, D.P. (1990) Immunological indicators: effects of environmental stress on immune protection and disease outbreaks. *American Fisheries Society Symposium*, **8**, 38–50.

Anderson, M.J., M.R. Miller & D.E. Hinton (1996) *In vitro* modulation of 17-β-estradiol-induced vitellogenin synthesis: Effects of cytochrome P4501A1 inducing compounds on rainbow trout (*Oncorhynchus mykiss*) liver cells. *Aquatic Toxicology*, **34** (4), 327–350.

Anderson, R.S. & L.L. Brubacher (1993) Inhibition by pentachlorophenol of production of reactive-oxygen intermediates by medaka phagocytic blood cells. *Marine Environmental Research*, **35**, 125–129.

Andersson, T., B.E. Bengtsson, L. Förlin, J. Härdig & Å. Larsson (1987) Long-term effects of bleached kraft pulp mill effluents on carbohydrate metabolism and xenobiotic biotransformation enzymes in fish. *Ecotoxicology and Environmental Safety*, **13**, 53–60.

Andersson, T., L. Förlin, J. Härdig & Å. Larsson (1988) Physiological disturbances in fish living in coastal water polluted with bleached kraft mill effluents. *Canadian Journal of Fisheries and Aquatic Sciences*, **45**, 1525–1536.

Anonymous (1999) *Ambio*, **28**, 1–109.

Arkoosh, M.R., E. Clemons, M. Myers & E. Casillas (1994) Suppression of B-cell mediated immunity in juvenile chinook salmon (*Oncorhynchus tshawytscha*) after exposure to either a polycyclic aromatic hydrocarbon or to polychlorinated biphenyls. *Immunopharmacology and Immunotoxicology*, **16**, 293–314.

Arkoosh, M.R., E. Clemons, P. Huffman, H.R. Sanborn, E. Casillas & J.E. Stein (1996) Leukoproliferative response of splenocytes from English sole (*Pleuronectes vetulus*) exposed to chemical contaminants. *Environmental Toxicology and Chemistry*, **15**, 1154–1162.

Ashfield, L.A., T.G. Pottinger & J.P. Sumpter (1998) Exposure of female juvenile rainbow trout to alkylphenolic compounds results in modifications to growth and ovosomatic index. *Environmental Toxicology and Chemistry*, **17** (4), 679–686.

Asplund, L., M. Athanasiadou, A. Sjodin, Å. Bergman & H. Börjeson (1999) Organohalogen substances in muscle, egg and blood from healthy Baltic salmon (*Salmo salar*) and Baltic salmon that produced offspring with the M74 syndrome. *Ambio*, **28**, 67–76.

Balk, L., T. Andersson, L. Förlin, M. Söderström & Å. Larsson (1993) Indications of regional and large-scale biological effects caused by bleached pulp mill effluents. *Chemosphere*, **27**, 631–650.

Ball, J.N. & B.I. Baker (1969) The pituitary gland: Anatomy and histophysiology. In: (eds Hoar, W.S. & D.J. Randall) *Fish Physiology*. Vol. 2. Academic Press, New York, pp. 1–110.

Bang, J.D., J.W. Kim, S.D. Lee, S. Park, S.G. Chun, C.S. Jeong & J.W. Park (1996) Humoral immune response of flounder to *Edwardsiella tarda*: The presence of various sizes of immunoglobulins in flounder. *Diseases of Aquatic Organisms*, **26**, 197–203.

Bannasch, P. (1986) Pre-neoplastic lesions as end-points in carcinogenicity testing. I. Hepatic preneoplasia. *Carcinogenesis.*, **7**, 689–695.

Barker, D.E., R.A. Khan, E.M. Lee, R.G. Hooper & K. Ryan (1994) Anomalies in sculpin (*Myoxocephalus* spp.) sampled near a pulp and paper mill. *Archives of Environmental Contamination and Toxicology*, **26**, 491–496.

Baumann, P.C. (1992) Methodological considerations for conducting tumor surveys of fishes. *Journal of Aquatic Ecosystem Health*, **1**, 127–133.

Baumann, P.C., W.D. Smith & W.K. Parland (1987) Tumor frequencies and contaminant concentrations in brown bullheads from a industrialized river and a recreational lake. *Transactions of the American Fisheries Society*, **116**, 79–86.

Baumann, P.C., M.J. Mac, S.B. Smith & J.C. Harshbarger (1991) Tumor frequencies in walleye (*Stizostedion vitreum*) and brown bullhead (*Ictalurus nebulosus*) and sediment contaminants in tributaries of the Laurentian Great Lakes. *Canadian Journal of Fisheries and Aquatic Sciences*, **48** (9), 1804–1810.

Bengtsson, Å. (1991) Effects of bleached pulp mill effluents on vertebral defects in fourhorn sculpin (*Myoxocephalus quadricornis* L.) in the Gulf of Bothnia. *Archiv fur Hydrobiologie*, **121**, 373–384.

Bengtsson, Å., B.-E. Bengtsson & G. Lithner (1988) Vertebral defects in fourhorn sculpin, *Myxocephalus quadricornis* L., exposed to heavy metal pollution in the Gulf of Bothnia. *Journal of Fish Biology*, **33**, 517–529.

Bengtsson, B.E., C.H. Carlin, Å. Larsson & O. Svanberg (1975) Vertebral damage in minnows, *Phoxinus phoxinus* L. exposed to cadmium. *Ambio*, **4**, 166–168.

Bengtsson, B.E., C. Hill, Å. Bergman, I. Brandt, N. Johansson, C. Magnhagen, A. Sodergren & J. Thulin (1999) Reproductive disturbances in Baltic Fish: A synopsis of the FiRe project. *Ambio*, **28**, 2–8.

Bernet, D., H. Schmidt, W. Meier, P. Burkhardt-Holm & T. Wahli (1998) Histopathology in fish: proposal for a protocol to assess aquatic pollution. *Journal of Fish Diseases*, **22** (1), 25–34.

Berntssen, M.H.G., K. Hylland, B.S.E. Wendelaar & A. Maage (1999) Toxic levels of dietary copper in Atlantic salmon (*Salmo salar* L.) parr. *Aquatic Toxicology*, **46**, 87–99.

Besselink, H.T., E.M.T.E. Flipsen, M.L. Eggens & *et al.* (1998) Alterations in plasma and hepatic retinoid levels in flounder (*Platichthys flesus*) after chronic exposure to contaminated harbour site sludge in a mesocosm study. *Aquatic Toxicology*, **42** (4), 271–285.

Betoulle, S., D. Troutaud, N. Khan & P. Deschaux (1995) Antibody response, cortisol and prolactin levels in rainbow trout. *Comptes Rendus de l'Academie des Sciences Serie III – Sciences de la Vie-Life Sciences*, **318**, 677–681.

Billard, S.M., K. Querbach & P.V. Hodson (1999) Toxicity of retene to early life stages of two freshwater fish species. *Environmental Toxicology and Chemistry*, **18** (9), 2070–2077.

Birge, W.J., J.A. Black, A.G. Westerman & J.E. Hudson (1979) The effects of mercury on reproduction of fish and amphibians. In: (ed. Nriagu, J.) *The Biogeochemistry of mercury in the Environment*. Elsevier, Amsterdam, Holland, 629 pp.

Black, D.E., R. Gutjahr-Gobell, R.J. Pruell, B. Bergen, L. Mills & A.E. McElroy (1998) Reproduction and polychlorinated biphenyls in *Fundulus heteroclitus* (Linnaeus) from New Bedford Harbour, Massachusetts, USA. *Environmental Toxicology and Chemistry*, **17**, 1405–1414.

Blazer, V.S., R.E. Wolke, J. Brown & C.A. Powell (1987) Piscine macrophage aggregate parameters as health monitors: effects of age, sex, relative weight, season and site quality in largemouth bass (*Micropterus salmoides*). *Aquatic Toxicology*, **10**, 199–215.

Blom, J.H. & K. Dabrowski (1996) Ascorbic acid metabolism in fish: Is there a maternal effect on the progeny? *Aquaculture*, **147**, 215–224.

Bodammer, J.E. (1993) The teratological and pathological effects on embryonic and larval fishes exposed as embryos: a brief review. *American Fisheries Society Symposium*, **14**, 77–84.

Bowser, P.R., M.J. Wolfe, J. Reimer & B.S. Shane (1991) Epizootic papillomas in brown bullheads *Ictalurus nebulosus* from Silver Stream reservoir, New York. *Diseases of Aquatic Organisms*, **11**, 117–127.

Braunbeck, T. & A. Völkl (1991) Induction of biotransmformation in the liver of eel (*Anguilla anguilla* L.) by sublethal exposure to Dinitro-o-cresol: an ultrastructural and biochemical study. *Ecotoxicology and Environmental Safety*, **21**, 109–127.

Bremner, I. (1991) Nutritional and physiological significance of metallothionein. In: (eds Riordan, J.F. & B.L. Vallee) *Metallobiochemistry, Part B: metallothionein and related molecules*. Academic Press, London, pp. 25–35.

Bucke, D. (1993) Aquatic pollution: effects on the health of fish and shellfish. *Parasitology*, **106**, S25–S37.

Bucke, D. & P.F. Dixon (1992) *Serological and histopathological studies on fish exposed* in vitro *to contaminated harbour sludge. International Council for the Exploration of the Sea (ICES)*, Copenhagen, E:35.

Bucke, D. & S.W. Feist (1984) *Histological changes in livers of dab* Limanda limanda *L. International Council for the Exploration of the Sea (ICES)*, Copenhagen, F:8.

Bucke, D. & S.W. Feist (1993) Histopathological changes in the livers of dab, *Limanda limanda* (L.). *Journal of Fish Diseases*, **16**, 281–296.

Bucke, D., S.W. Feist, M.G. Norton & M.S. Rolfe (1983) A histopathological report of some epidermal anomalies of Dover sole, *Solea solea* L., and other flatfish species in coastal waters off south-east England. *Journal of Fish Biology*, **23**, 565–578.

Bucke, D., B. Waterman & S. Feist (1984) Histological variations of hepato-sphenic organs from the North Sea dab, *Limanda limanda*. *Journal of Fish Diseases*, **7** (4), 255–268.

Bucke, D., A.D. Vethaak & T. Lang (1992) Quantitative assessment of melanomacrophage centres (MMCs) in dab *Limanda limanda* along a pollution transect in the German Bight. *Marine Ecology Progress Series*, **91**, 193–196.

Call, D.J., L.T. Brooke, M.L. Knuth, S.H. Poirier & M.D. Hoglund (1985) Fish subchronic toxicity prediction model for industrial organic chemicals that produce narcosis. *Environmental Toxicology and Chemistry*, **4**, 335–341.

Cameron, P. & H. von Westernhagen (1997) Malformation rates in embryos of North sea fishes in 1991 and 1992. *Marine Pollution Bulletin*, **34**, 129–134.

Cameron, P., J. Berg, V. Dethlefsen & H. Vonwesternhagen (1992) Developmental defects in pelagic embryos of several flatfish species in the southern North-Sea. *Netherlands Journal of Sea Research*, **29** (1–3), 239–256.

Cameron, P., J. Berg & H. Von Westernhagen (1996) Biological effects monitoring of the North Sea employing fish embryological data. *Environmental Monitoring and Assessment*, **40** (2), 107–124.

Caruso, J.H., R.D. Suttkus & G.E. Gunning (1988) Abnormal expression of secondary sex characteristics in a population of *Anguilla rostrata* (Pisces: Anguillidae) from a dark coloured Florida stream. *Copeia*, **4**, 1077–1079.

Churchich, J.E., G. Scholz & F. Kwok (1988) Modulation of the catalytic activity of pyridoxal kinase by metallothionein. *Biochem. Int.*, **17** (3), 395–403.

Cierszko, A. & K. Dabrowksi (1995) Sperm quality and ascorbic acid concentration in rainbow trout semen are affected by dietary vitamin-c: An across-season study. *Biology of Reproduction*, **52** (5), 982–988.

Clemons, E., M.R. Arkoosh & E. Casillas (1999) Enhanced superoxide anion production in activated peritoneal macrophages from English sole (*Pleuronectes vetulus*) exposed to polycyclic aromatic compounds. *Marine Environmental Research*, **47** (1), 71–87.

Cody, R.P. & S.A. Burtone (1997) Masculinization of mosquitofish as an indicator of exposure to kraft mill effluent. *Bulletin of Environmental Contamination and Toxicology*, **58**, 429–436.

Collier, T.K., J.E. Stein, H.R. Sanborn, T. Hom, M.S. Myers & U. Varanasi (1992) Field studies of reproductive success and bioindicators of maternal contaminant exposure in english sole (parophrys-vetulus). *Science of the Total Environment*, **116** (1–2), 169–185.

Cooray, R., M. Holmberg, A. Hellstrom, J. Hardig, R. Mattsson, A. Gunnarsson, H. Börjeson, J. Lindeberg & B. Morein (1999) Screening for microorganisms associated with M74 disease syndrome in sea-run Baltic salmon (*Salmo salar*). *Ambio*, **28**, 77–81.

Couch, J.A. (1975) Histopathological effects of pesticides and related chemicals on the livers of fishes. In: (eds Ribelin, W.K. & G. Migaki) *Pathology of fishes*. University of Wisconsin Press, Madison, pp. 559–584.

Couch, J.A. (1984) Histopathology and enlargement of the pituitary of a teleost exposed to the herbicide trifluralin. *Journal of Fish Diseases*, **7**, 157–163.

Couillard, C.M. & P.V. Hodson (1996) Pigmented macrophage aggregates: A toxic response in fish exposed to bleached-kraft mill effluent? *Environmental Toxicology and Chemistry*, **15**, 1844–1854.

Cross, J.N. (1986) Epidermal tumours in *Microstomus pacificus* (Pleuronectidae) collected near a municipal wastewater outfall in the coastal waters off Los Angeles (1971–1983). *Calif. Fish and Game*, **72** (2), 68–77.

Cross, J.N., J.T. Hardy, J.E. Hose, G.P. Hershelman, L.D. Antrim, R.W. Gossett & E.A. Crecelius (1987) Contaminant concentrations and toxicity of sea-surface microlayer near Los Angeles, California. *Marine Environmental Research*, **23** (4), 307–323.

Cutcomp, L.K., H.H. Yap, D. Desaiah & R.B. Koch (1972) The sensitivity of fish ATPases to polychlorinated biphenyls. *Environmental Health Perspectives*, **1**, 165–168.

Daoust, P.Y., G. Wobeser & J.D. Newstead (1984) Acute pathological effects of inorganic mercury and copper in gills of rainbow trout. *Vet. Pathol.*, **21**, 93–101.

Dave, G. & R.Q. Xiu (1991) Toxicity of mercury, copper, nickel, lead, and cobalt to embryos and larvae of zebrafish, *Brachydanio rerio. Archives of Environmental Contamination and Toxicology*, **21** (1), 126–134.

Demers, N.E. & C.J. Bayne (1997) The immediate effects of stress on hormones and plasma lysozyme in rainbow trout. *Dev. Comp. Immunol.*, **21**, 363–373.

Desaiah, D. & R.B. Koch (1975) Inhibition of ATPase activity in Channel catfish brain by kepone and its reduction products. *Bulletin of Environmental Contamination and Toxicology*, **13**, 153–158.

Dethlefsen, V., H. von Westernhagen & P. Cameron (1996) Malformations in North Sea pelagic fish embryos during the period 1984–1995. *ICES Journal of Marine Science.*, **53** (6), 1024–1035.

Devos, E., P. Devos & M. Cornet (1998) Effect of cadmium on the cytoskeleton and morphology of gill chloride cells in parr and smolt Atlantic salmon (*Salmo salar*). *Fish Physiology and Biochemistry*, **18**, 15–27.

Dexiang, C. & A.J. Ainsworth (1991) Effect of temperature on the immune system of channel catfish (*Ictalurus punctatus*) – II. Adaptation of anterior kidney phagocytes to 10°C. *Comparative Biochemistry and Physiology*, **100A**, 913–918.

Donaldson, E.M. (1981) The piturary-interrenal axis as an indicator of stress in fish. In: (ed. Pickering, A.D.) *Stress in Fish*. Academic Press, London, New York, pp. 11–47.

Ellis, A.E. (1981) Stress and the modulation of defence mechanisms in fish. In: (ed. Pickering, A.D.) *Stress in Fish*. Academic Press, London, pp. 147–169.

Evans, M.R., S.J. Larsen, G.H.M. Riekerk & K.G. Burnett (1997) Patterns of immune response to environmental bacteria in natural populations of the red drum, *Sciaenops ocellatus* (Linnaeus). *Journal of Experimental Marine Biology and Ecology*, **208**, 87–105.

Faisal, M. & R.J. Huggett (1993) Effects of polycyclic aromatic hydrocarbons on the lymphocyte mitogenic response in spot, *Leistomus xanthurus*. *Marine Environmental Research*, **35**, 121–124.

Fitzsimons, J.D. (1995) A critical-review of the effects of contaminants on early-life stage (ELS) mortality of Lake trout in the Great-Lakes. *Journal of Great Lakes Research*, **21** (S1), 267–276.

Fitzsimons, J.D., S.B. Brown, D.C. Honeyfield & J.G. Hnath (1999) A review of early mortality syndrome (EMS) in Great Lake salmonids: Relationship with thiamine deficiency. *Ambio*, **28**, 9–15.

Förlin, L., T. Andersson, C. Haux, P.E. Olsson & Å. Larsson (1986) Physiological methods in fish toxicology: Laboratory and field studies. In: (eds Nilsson & Holmgren) *Fish Physiology: Recent Advances*. Croom Helm, London, pp. 158–169.

Förlin, L., T. Andersson, L. Balk & Å. Larsson (1991) Biochemical and physiological effects of pulp mill effluents on fish. In: (ed. Södergren, A.) *Environmental Fate and Effects of Bleached Pulp Mill Effluents*. Swedish EPA Report (4031), Sweden, pp. 235–243.

Förlin, L., T. Andersson, L. Balk & Å. Larsson (1995) Biochemical and physiological effects of bleached pulp mill effluents in fish. *Ecotoxicology and Environmental Safety*, **30**, 164–170.

Fragoso, N.M., J.L. Parrott, M.E. Hahn & P.V. Hodson (1998) Chronic retene exposure causes sustained induction of CYP1A activity and protein in rainbow trout (*Oncorhynchus mykiss*). *Environmental Toxicology and Chemistry*, **17**, 2347–2353.

Frith, C.H. & J.M. Ward (1980) A morphologic classification of proliferative and neoplastic hepatic lesions in mice. *J. Environ. Pathol. Toxicol. Oncol.*, **3**, 329–351.

Gagnon, M.M., J.J. Dodson, P.V. Hodson, G.J. van der Kraak & J.H. Carey (1994) Seasonal effects of bleached kraft mill effluent on reproductive parameters of white sucker (*Catostomus commersoni*) populations of the St. Maurice river, Quebec, Canada. *Canadian Journal of Fisheries and Aquatic Sciences*, **51**, 337–347.

Gardner, G.R. (1975) Chemically induced lesions in estuarine or marine teleosts. In: (eds Ribelin, W.E. & G. Migakli) *The Pathology of Fishes*. University of Wisconsin Press, Madison, Wisconsin, pp. 657–693.

Gardner, G.R. & P.P. Yevich (1988) Comparative histopathological effects of chemically contaminated sediment on marine organisms. *Marine Environmental Research*, **24** (1–4), 311–316.

George, S.G., P.A. Hodgson, P. Tytler & K. Todd (1996) Inducibility of metallothionein mRNA expression and cadmium tolerance in larvae of a marine teleost, the turbot (*Scophthalmus maximus*). *Fund. Appl. Toxicol.*, **33** (1), 91–99.

Glynn, P.J. & A.L. Pulsford (1990) Isolation and characterisation of the serum immunoglobulin of the flounder, *Platichthys flesus*. *Journal of the Marine Biological Association of the United Kingdom*, **70**, 429–440.

Goksøyr, A., T.S. Solberg & B. Serigstad (1991) Immunochemical detection of cytochrome-P4501A1 induction in cod larvae and juveniles exposed to a water-soluble fraction of North sea crude oil. *Marine Pollution Bulletin*, **22** (3), 122–127.

Gottofrey, J. & H. Tjälve (1991) Axonal transport of cadmium in the olfactory nerve of the pike. *Pharmacol. Toxicol.*, **69**, 242–252.

Grinwis, G.C.M., A. Boonstra, E.J. VandenBranhof, J.A.M.A. Dormans, M. Engelsma, R.V. Kuiper, H. vanLoveren, P.W. Wester, M.A. Vaal, A.D. Vethaak & J.G. Vos (1998) Short-term toxicity of bis(tri-n-butyltin)oxide in flounder (*Platichthys flesus*): Pathology and immune function. *Aquatic Toxicology*, **42**, 15–36.

Grizzle, J.M. & A.E. Goodwin (1998) Neoplasms and related lesions. In: (eds Leatherland, J.F. & P.T.K. Woo) *Fish Diseases and Disorders*. Vol. 2: Non-infectious disorders, pp. 37–104.

Guha, G., K. Dutta & M. Das (1993) Vitamin C as antioxidant factors in DDT induced haemotoxicity in *Clarius batrachus*. *Proc. Zool. Soc., Calcutta*, **46**, 11–15.

Haaparanta, A., E.T. Valtonen & R.W. Hoffmann (1997) Gill anomalies of perch and roach from four lakes differing in water quality. *Journal of Fish Biology*, **50**, 575–591.

Haensly, W.E., J.M. Neff, J.R. Sharp, A.C. Morris, M.F. Bedgood & P.D. Boem (1982) Histopathology of *Pleuronectes platessa* L. from Aber Benoit, Brittany, France: long term effects of the Amoco Cadiz oil spill. *Journal of Fish Diseases*, **5**, 365–391.

Halliwell, B. & J.M.C. Gutteridge (1989) *Free radicals in biology and medicine*, 2nd ed. Clarendon Press, Oxford.

Hardie, L.J., T.C. Fletcher & C.J. Secombes (1990) The effect of vitamin E on the immune response of the Atlantic salmon (*Salmo salar* L.). *Aquaculture*, **87**, 1–13.

Hardie, L.J., T.C. Fletcher & C.J. Secombes (1991) The effect of dietary vitamin C on the immune response of the Atlantic salmon (*Salmo salar* L.). *Aquaculture*, **95**, 201–214.

Hardie, L.J., T.C. Fletcher & C.J. Secombes (1994) Effect of temperature on macrophage activation and the production of macrophage activating factor by rainbow trout (*Oncorhynchus mykiss*) leucocytes. *Dev. Comp. Immunol.*, **18** (1), 57–66.

Härdig, J., T. Andersson, L. Förlin, B.E. Bengtsson & Å. Larsson (1998) Longterm effects of pulp bleached mill effluents on hematology and ion balance in fish. *Toxicol. Environ. Safety*, **15**, 96–106.

Hardy, J., S. Kiesser, L. Antrim, A. Stubin, R. Kocan & J. Strand (1987) The sea-surface microlayer of Puget Sound, Washington, USA. I. Toxic effects on fish eggs and larvae. *Marine Environmental Research*, **23**, 227–250.

Hargis, W.J. & D. Zwerner (1990) Some effects of sediment-borne contaminants on development and cytomorphology of teleost eye-lens epithelial cells and their derivatives. In: (eds Moore, M.N. & J.J. Stegeman) *Responses of marine organisms to pollutants*. **28** (1–4), pp. 399–405.

Hart, L.J., S.A. Smith, J. Robertson & S.D. Holladay (1997) Exposure of tilapian fish to the pesticide lindane results in hypocellularity of the primary hematopoietic organ (pronephros) and the spleen without altering activity of phagocytic cells in these organs. *Toxicol.*, **118**, 211–221.

Hart, L.J., S.A. Smith, B.J. Smith, J. Robertson, E.G. Besteman & S.D. Holladay (1998) Subacute immunotoxic effects of the polycyclic aromatic hydrocarbon 7,12-dimethylbenzanthracene (DMBA) on spleen and pronephros leukocytic cell counts and phagocytic cell activity in tilapia (*Oreochromis niloticus*). *Aquatic Toxicology*, **41**, 17–29.

Haux, C. & L. Förlin (1988) Biochemical methods for detecting effects of contaminants on fish. *Ambio*, **6**, 376–380.

Haux, C. & Å. Larsson (1979) Effects of DDT on blood plasma electrolytes in the flounder, *Platichthys flesus* L., in hypotonic brackish water. *Ambio*, **8**, 171–173.

Haux, C., Å. Larsson, G. Lithner & M.L. Sjöbeck (1986) A field study of physiological effects on fish in lead-contaminated lakes. *Environmental Toxicology and Chemistry*, **5**, 283–288.

Hawkes, J.W. (1977) The effects of petroleum hydrocarbon exposure on the structure of fish tissues. In: (ed. Wolf, D.A.) *Fate and Effects of Petroleum Hydrocarbons in Marine Ecosystems and Organsims*. Pergamon Press, New York, pp. 115–128.

Hawkes, J.W. (1980) The effects of xenobiotics on fish tissues: morphological studies. *Fed. Proc.*, **39**, 3230–3236.

Hayes, M.A., I.R. Smith, T.H. Rushmore, T.L. Crane, C. Thorn, T.E. Kocal & H.W. Ferguson (1990) Pathogenisis of skin and liver neoplasms in white suckers from industrially polluted areas in Lake Ontario. *Science of the Total Environment*, **94**, 105–123.

Haynes, L. & E.C. McKinney (1991) Shark spontaneous cytotoxicity: characterization of the regulatory cell. *Dev. Comp. Immunol.*, **15**, 123–134.

Henry, T.R., J.M. Spitsbergen, M.W. Hornung, C.C. Abnet & R.E. Peterson (1997) Early life stage toxicity of 2,3,7,8-tetrachlorodibenzo-p-dioxin in zebrafish (*Danio rerio*). *Toxicology and Applied Pharmacology*, **142**, 56–68.

Hernadi, L. (1993) Fine structural characterization of the olfactory epithelium and its response to divalent cations cadmium in the fish *Alburnus alburnus* (Teleostei, Cyprinidae): A scanning and transmission electron microscopy study. *Neurobiology*, **1**, 11–31.

Hinton, D.E. & D.J. Lauren (1990) Integrative histopathological approaches to detecting effects of environmental stressors on fishes. *American Fisheries Society Symposium*, **8**, 51–66.

Hinton, D.E., J.A. Couch, S.J. Teh & L.A. Courtney (1988) Cytological changes during progression of neoplasia in selected fish species. *Aquatic Toxicology*, **11**, 77–112.

Hinton, D.E., P.C. Baumann, G.R. Gardner, W.E. Hawkins, J.D. Hendricks, R. Murchelano & M.S. Okihiro (1992) Histopathologic markers. In: (eds Ward, C.H., B.T. Walton & T.W. LaPoint) *Biomarkers: Biochemical, Physiological and Histological Markers of Anthropogenic Stress*. Lewis Publishers, Boca Raton, pp. 155–209.

Hodson, P.V. (1976) δ-amino levulinic acid dehydratase activity of fish blood as an indicator of a harmful exposure to lead. *Journal of the Fisheries Research Board of Canada*, **33**, 268–271.

Hodson, P.V., M. McWirther, K. Ralph, B. Gray, D. Thivierge, J. Carey, G. van der Kraak, D. Wittle & M.C. Levesque (1992) Effects of bleached kraft mill effluent on fish in the St. Maurice river, Quebec. *Environmental Toxicology and Chemistry*, **11**, 1635–1651.

Holland, H.T., D.L. Coppage & P.A. Butler (1967) Use of fish brain acetylcholinesterase to monitor pollution by organophosphorus pesticides. *Bulletin of Environmental Contamination and Toxicology*, **2**, 156–162.

Houlihan, D.F., M.J. Costello, C.J. Secombes, R. Stagg & J. Brechin (1994) Effects of sewage sludge exposure on growth, feeding and protein synthesis of dab (*Limanda limanda* (L.)). *Marine Environmental Research*, **37** (4), 1–1.

Howell, W.M. & T.E. Denton (1989) Gonopodial morphogenesis in female mosquitofish, *Gambusia affinis affinis*, masculinized by exposure to degradation products from plant sterols. *Environmental Biology of Fishes*, **24**, 43–51.

Howell, W.M., D.A. Black & S.A. Bortone (1980) Abnormal expression of secondary sex characters in a population of mosquitofish, *Gambusia affinis holbrooki*: evidence for environmentally-induced masculinization. *Copeia*, **1980**, 676–681.

Huggett, R.J., Kimerle, R.A., Mehrle, P.M. & Bergman, H.L. (eds) (1992) Biomarkers – biochemical, physiological, and histological markers of anthropogenic stress. SETAC special publication series, Lewis Publ. 335 pp.

Husøy, A.M., M.S. Myers, M.L. Willis, T.K. Collier, M. Celander & A. Goksøyr (1994) Immunohistochemical localization of CYP1A and CYP3A-like isozymes in hepatic and extrahepatic tissues of Atlantic cod (*Gadus morhua*), a marine fish. *Toxicology and Applied Pharmacology*, **129**, 292–308.

Husøy, A.M., M.S. Myers & A. Goksøyr (1996) Cellular localisation of cytochrome P450 (CYP1A) induction and histology in Atlantic cod (*Gadus morhua* L.) and European flounder (*Platichthys flesus*) after environmental exposure to contaminants by caging in Sørfjorden, Norway. *Aquatic Toxicology*, **36** (1–2), 53–74.

Hutchinson, T.H., T.D. Williams & G.J. Eales (1994) Toxicity of cadmium, hexavalent chromium and copper to marine fish larvae (*Cyprinodon variegatus*) and copepods (*Tisbe battagliai*). *Marine Environmental Research*, **38** (4), 275–290.

ICES (1996) *ICES Techniques in Marine Environmental Science, No.19. Common diseases and parasites of fish in the North Atlantic: Training guide for identification.* ICES Palaegade 2–4, DK-1261, Copenhagen K, Denmark. ISSN 0903-2606.

ICES (1997) Report of the Special Meeting on the Use of Liver Pathology of Flatfish for Monitoring Biological Effects of Contaminants. Weymouth, UK. 22–25 October, 1996.

Inoue, M., S. Satoh, M. Maita, V. Kiron & N. Okamoto (1998) Recovery from derangement of natural killer-like activity of leukocytes due to Zn or Mn deficiency in rainbow trout, *Oncorhynchus mykiss* (Walbaum), by the oral administration of these elements. *Journal of Fish Diseases*, **21**, 233–236.

Johnson, L.L., E. Casillas, T. Collier, B.B. McCain & U. Varanasi (1988) Contaminant effects on ovarian development in English sole (*Parophrys vetulus*) from Puget Sound, Washington. *Canadian Journal of Fisheries and Aquatic Sciences*, **45**, 2133–2146.

Julliard, A.K., D. Saucier & L. Astic (1996) Time-course of apoptosis in the olfactory epithelium of rainbow trout exposed to a low copper level. *Tissue & Cell*, **28**, 367–377.

Kaattari, S.L. & J.D. Piganelli (1996) The specific immune system: humoral defense. In: (eds Iwama, G.K. & S. Nakanishi) *The fish immune system – organism, pathogen, and environment.* Academic Press, San Diego, pp. 207–254.

Karan, V., S. Vitorovic, V. Tutundzic & V. Poleksic (1998) Functional enzymes activity and gill histology of carp after copper sulfate exposure and recovery. *Ecotoxicology and Environmental Safety*, **40**, 49–55.

Karås, P., E. Neuman & O. Sandström (1991) Effects of pulp mill effluents on the population dynamics of perch, *Perca fluviatilis*. *Canadian Journal of Fisheries and Aquatic Sciences*, **48**, 28–34.

Karlsson, L., E. Ikonene, A. Mitans & S. Hansson (1999) The diet of salmon (*Salmo salar*) in the Baltic Sea and connections with the M74 syndrome. *Ambio*, **28**, 37–42.

Karrow, N.A., H.J. Boermans, D.G. Dixon, A. Hontela, K.R. Solomon, J.J. Whyte & N.C. Bols (1999) Characterizing the immunotoxicity of creosote to rainbow trout (*Oncorhynchus mykiss*): a mesocosm study. *Aquatic Toxicology*, **45**, 223–239.

Khan, R.A. (1995) Histopathology in winter flounder, *Pleuronectes americanus*, following chronic exposure to crude oil. *Bulletin of Environmental Contamination and Toxicology*, **54**, 297–301.

Khan, R.A. & J. Kiceniuk (1984) Histopathological effects of crude oil on Atlantic cod following exposure. *Canadian Journal of Zoology*, **62** (10), 2038–2043.

Khan, R.A., D.E. Barker, R. Hooper, E.M. Lee, K. Ryan & K. Nag (1994) Histopathology in winter flounder (*Pleuronectes americanus*) living adjacent to a pulp and paper mill. *Archives of Environmental Contamination and Toxicology*, **26**, 95–102.

Kirby, M.F., S. Morris, M. Hurst, S.J. Kirby, P. Neall, T. Tylor, & A. Fagg (2000) The use of Cholinesterase activity in flounder (*Platichthys flesus*) muscle tissue as a biomarker of neurotoxic contamination in UK estuaries. *Mar. Pollut. Bull.*, **40**, 780–791.

Klaunig, J.E., M.M. Lipsky, B.F. Trump & D.E. Hinton (1979) Biochemical and ultrastructural changes in teleost liver following subacute exposure to PCB. *J. Environ. Pathol. Toxicol.*, **2**, 953–963.

Klaverkamp, J.F., M. Duangsawadsi, W.A. MacDonald & H.S. Majewski (1977) An evaluation of fenitrothion toxicity in four life stages of rainbow trout, *Salmo gairdneri*. In: (eds Mayer, F.L. &

J.L. Hamelink) *Aquatic Toxicology and Hazard Evaluation*. ASTM STP 634, ASTM Philadelphia, pp. 231–240.

Kocan, R.M., H. von Westernhagen, M.L. Laandolt & G. Furstenburg (1987) Toxicity of sea surface microlayer: effects of hexane extracts on Baltic herring and Atlantic cod embryos. *Marine Environmental Research*, **23**, 291–305.

Köhler, A. (1989) Regeneration of contaminant-induced liver lesions in flounder-experimental studies towards the identification of cause-effect relationships. *Aquatic Toxicology*, **14**, 203–232.

Köhler, A. (1990a) Identification of contaminant-induced cellular and subcellular lesions in the liver of flounder (*Platichthys flesus* L.) caught at differently polluted estuaries. *Aquatic Toxicology*, **16**, 271–294.

Köhler, A. (1990b) *Cellular responses in fish liver as indicators for toxic effects of environmental pollution*. International Council for Exploration of the Sea (ICES), Copenhagen, Denmark. pp. E:30.

Köhler, A., H. Deisemann & B. Lauritzen (1992) Histological and cytochemical indices of toxic injury in the liver of dab *Limanda limanda*. *Marine Ecology Progress Series*, **91** (1/3), 141–153.

Kranz, H. (1989) Changes in splenic melano-macrophage centers of dab *Limanda limanda* during and after infection with ulcer disease. *Diseases of Aquatic Organisms*, **6**, 167–174.

Kranz, H. & V. Dethlefsen (1990) Liver anomalies in dab *Limanda limanda* from the southern North Sea with special consideration given to neoplastic lesions. *Diseases of Aquatic Organisms*, **9**, 171–185.

Kranz, H. & N. Peters (1985) Pathological conditions in the liver of ruffe, *Gymnocephalus cernua* (L.) from the Elbe estuary. *Journal of Fish Diseases*, **8**, 13–24.

Landahl, J.T., L.L. Johnson, J.E. Stein, T.K. Collier & U. Varanasi (1997) Approaches for determining effects of pollution on fish populations of Puget Sound. *Transactions of the American Fisheries Society*, **126** (3), 519–535.

Landergren, P., L. Vallin, L. Westin, P. Amcoff, H. Börjeson & B. Ragnarsson (1999) Reproductive failure in Baltic Sea trout (*Salmo trutta*) compared with the M74 syndrome in Baltic salmon (*Salmo salar*). *Ambio*, **28**, 87–91.

Lanno, R.P., S.J. Slinger & J.W. Hilton (1985) Effect of ascorbic acid on dietary copper toxicity in rainbow trout (*Salmo gairdneri* Richardson). *Aquaculture*, **49**, 269–287.

Larsson, Å., C. Haux & M.L. Sjöbeck (1985) Fish physiology and metal pollution: Results and experiences from laboratory and field studies. *Ecotoxicology and Environmental Safety*, **9**, 250–281.

Larsson, Å., L. Förlin, E. Lindesjöö & O. Sandström (2003) Monitoring of individual organisms responses in fish populations exposed to pulp mill effluents. In: (ed. Stuthridge, T.R.) *Environmental Fate and Effects of Pulp and Paper Mill Effluents*. SETAC Press, Pensacola, Florida, in press.

Larsson, D.G.J., H. Hällman, S.J. Hyllner & L. Förlin (1999) *More male embryos near a pulp mill*. Poster presentation at the 6th International Symposium on Reproductive Physiology of Fish, Bergen, Norway, 4–9 July 1999.

Larsson D.G.J., H. Hällman & L. Förlin (2000) More male fish embryos near a pulp mill. *Environ. Toxicol. Chem.*, **19** (12), 2911–2917.

Leatherland, J.F. & K.J. Farbridge (1992) Chronic fasting reduces the response of the thyroid to growth hormone and TSH, and alters the growth hormone related changes in hepatic 5′-monodeiodinase activity in rainbow trout *Oncorhynchus mykiss*. *General and Comparative Endocrinology*, **87** (3), 342–353.

Lehtinen, K.J., K. Mattson, J. Tana, C. Engström, O. Lerche & J. Hemming (1999) Effects of wood-related sterols on the reproduction, egg survival, and offspring of brown trout (*Salmo trutta lacustris*). *Ecotoxicology and Environmental Safety*, **42**, 40–49.

Lindesjöö, E. & J. Thulin (1987) Fin erosion of perch (*Perca fluviatilis*) in a pulp mill effluent. *Bull. Eur. Assoc. Fish Pathol.*, **7**, 11–14.

Lindesjöö, E. & J. Thulin (1992) A skeletal deformity of northern pike (*Esox lucius*) related to pulp mill effluents. *Canadian Journal of Fisheries and Aquatic Sciences*, **49**, 166–172.

Lindesjöö, E. & J. Thulin (1994) Histopathology of skin and gills of fish in pulp mill effluents. *Diseases of Aquatic Organisms*, **18**, 81–93.

Lindesjöö, E., J. Thulin, B.E. Bengtsson & U. Tjarnlund (1994) Abnormalities of a gill cover bone, the operculum, in perch *Perca fluviatilis* from a pulp mill effluent area. *Aquatic Toxicology*, **28**, 189–207.

Longwell, A.C. & J.B. Hughes (1981) Cytologic, cytogenic, and embryological state of Atlantic mackerel eggs from surface waters of the New York Bight in relation to pollution. *Rapports et Procès-verbaux de Réunions du Conseil International pour l'Exploration de la Mer*, **178**, 76–78.

Longwell, A.C., S. Chang, A. Hebert, J.B. Hughes & D. Perry (1992) Pollution and developmental abnormalities of Atlantic fishes. *Environmental Biology of Fishes*, **35** (1), 1–21.

Lundström, J., B. Carney, P. Amcoff, A. Pettersson, H. Börjeson, L. Förlin & L. Norrgren (1999) Antioxidative systems, detoxifying enzymes and thiamine levels in Baltic salmon (*Salmo salar*) that develop M74. *Ambio*, **28**, 24–29.

Majno, G. & I. Joris (1996) *Cells, tissues and disease: principles of general pathology*. Blackwell Science, Oxford.

Malins, D.C., B.B. McCain, D.W. Brown, S.L. Chan, M.S. Myers, J.T. Landahl, P.G. Prohaska, A.J. Friedman, L.D. Rhodes, D.G. Burrows, W.D. Gronlund & H.O. Hodgins (1984) Chemical pollutants in sediments and diseases of bottom-dwelling fish in Puget Sound, Washington. *Environ. Sci. Technol.*, **18** (9), 705–713.

Mallatt, J. (1985) Fish gill structural changes induced by toxicants and other irritants: a statistical review. *Canadian Journal of Fisheries and Aquatic Sciences*, **42** (4), 630–648.

Manning, M.J. & S. Nakanishi (1996) The specific immume system: cellular defenses. In: (eds Iwama, G.K. & S. Nakanishi) *The Fish Immune System – organism, pathogen, and environment*. Academic Press, San Diego, pp. 160–206.

Marty, G.D., J.M. Nunez, D.J. Lauren & D.E. Hinton (1990) Age-dependent changes in toxicity of N-nitroso compounds to Japanese medaka (*Oryziass latipes*) embryos. *Aquatic Toxicology*, **17**, 45–62.

Matta, M.B., C. Cairncross & R.M. Kocan (1997) Effect of a polychlorinated biphenyl metabolite on early life stage survival of two species of trout. *Bulletin of Environmental Contamination and Toxicology*, **59**, 146–151.

Mayer, F.L., D.J. Versteeg, M.J. McKee, L.C. Folmar, R.L. Graney, D.C. McCume & B.A. Rattner (1992) Physiological and nonspecific biomarkers. In: (eds Hugget, R.J., R.A. Kimerle, P.M. Mehrle & H.L. Bergman) *Biomarkers: Biochemical, Physiological, and Histological Markers of Anthropogenic Stress*. SETAC Special Publications Series, Lewis Publishers, pp. 5–85.

Mazeaud, M.M., F. Mazeaud & E.M. Donaldson (1977) Primary and secondary effects of stress in fish: some new data with a general review. *Transactions of the American Fisheries Society*, **106**, 201–212.

McIntyre, A.D. (1973) Toxicity of methylmercury for steelhead trout sperm. *Bulletin of Environmental Contamination and Toxicology*, **9**, 98.

McMaster, M.E., G.J. van der Kraak, C.B. Portt, K.R. Munkittrick, P.K. Sibley, I.R. Smith & D.G. Dixon (1991) Changes in hepatic mixed function oxygenase (MFO) activity, plasma steroid levels and age at maturity of a white sucker (*Catostomus commersoni*) population exposed to bleached kraft pulp mill effluent. *Aquatic Toxicology*, **21**, 199–218.

Meyers, T.R. & J.D. Hendricks (1982) A summary of tissue lesions in aquatic animals induced by controlled exposures to environmental contaminants, chemotherapeutic agents, and potential carcinogens. *Marine Fisheries Review*, **44** (12), 1–17.

Möller, H. (1990) Association between diseases of flounder (*Platichthys flesus*) and environmental conditions in the Elbe estuary, FRG. *J. Cons. int. Explor. Mer*, **46**, 187–199.

Monosson, E., W.J. Fleming & C.V. Sullivan (1994) Effects of the planar PCB 3,3′,4,4′-tetrachlorobiphenyl (TCB) on the ovarian development, plasma levels of sex steroid hormones and vitellogenin, and progeny survival in the white perch (*Morone americana*). *Aquatic Toxicology*, **29**, 1–19.

Moore, A. & C.P. Waring (1996) Sublethal effects of the pesticide Diazinon on the olfactory function in mature male Atlantic salmon parr. *Journal of Fish Biology*, **48**, 758–775.

Moore, M.J. & J.J. Stegeman (1994) Hepatic neoplasms in winter flounder *Pleuronectes americanus* from Boston Harbor, Massachusetts, USA. *Diseases of Aquatic Organisms*, **20**, 33–48.

Moore, M.J., R.M. Smolowitz & J.J. Stegeman (1997) Stages of hydropic vacuolation in the liver of winter flounder (*Pleuronectes americanus*) from a chemically contaminated site. *Diseases of Aquatic Organisms*, **31**, 19–28.

Moore, M.N. (1992) Pollutant-induced cell injury in fish liver: use of fluorescent molecular probes in live hepatocytes. *Marine Environmental Research*, **34**, 25–31.

Munkittrick, K.R., P.A. Miller, D.R. Barton & D.G. Dixon (1991) Altered performance of white sucker populations in the Manitouwadge chain of lakes Ontario, Canada is associated with changes in benthic macroinvertebrate communities as a result of copper and zinc contamination. *Ecotoxicology and Environmental Safety*, **21**, 318–326.

Munkittrick, K.R., G.J. van der Kaak, M.E. McMaster & C.B. Portt (1992) Response of hepatic MFO activity and plasma sex steroids to secondary treatment of bleached kraft pulp mill effluent and mill shutdown. *Environmental Toxicology and Chemistry*, **11**, 1427–1439.

Munkittrick, K.R., G.J. van der Kraak, M.E. McMaster, C.B. Portt, M.R. van den Heuvel & M.R. Servos (1994) Survey of receiving water environmental impacts associated with discharges from pulp mills. 2. Gonad size, liver size, hepatic EROD activity and plasma sex steroid levels in white sucker. *Environmental Toxicology and Chemistry*, **13**, 1089–1101.

Munkittrick, K.R., M.E. McMaster, M.R. Servos & G.J. van der Kraak (1999) Changes in secondary sex characteristics and gonadal size in fish exposed to pulp mill effluents over the period of mill modernization. In: *Proceedings from 6th International Symposium on Reproductive Physiology of Fish*, Bergen, Norway, July 4–9.

Murchelano, R.A. & R.E. Wolke (1985) Epizootic carcinoma in the winter flounder, *Pseudopleuronectes americanus*. *Science*, **228**, 587–589.

Murchelano, R.A. & R.E. Wolke (1991) Neoplasms and non-neoplastic liver lesions in winter flounder, *Pseudopleuronectes americanus*, from Boston Harbor, Massachusetts. *Environmental Health Perspectives*, **90**, 17–26.

Myers, M.S., L.D. Rhodes & B.B. McCain (1987) Pathologic anatomy and patterns of occurrence of hepatic neoplasms, putative preneoplastic lesions and other idiopathic hepatic conditions in English sole (*Parophrys vetulus*) from Puget Sound, Washington, USA. *J. Natl. Cancer. Inst.*, **78** (2), 333–363.

Myers, M.S., J.T. Landahl, M.N. Krahn, L.L. Johnson & B.B. McCain (1990) Overview of studies on liver carcinogenesis in English sole from Puget Sound; evidence for a xenobiotic chemical etiology I: Pathology and Epizootiology. *Science of the Total Environment*, **94**, 33–50.

Myers, M.S., O.P. Olson, L.L. Johnson, C.S. Stehr, T. Hom & U. Varanasi (1992) Hepatic lesions other than neoplasms in subadult flatfish from Puget Sound, Washington: Relationships with indices of contaminant exposure. *Marine Environmental Research*, **34**, 45–51.

Myers, M.S., C.M. Stehr, O.P. Olson, L.L. Johnson, B.B. McCain, S.-L. Chan & U. Varanasi (1994) Relationships between toxicopathic hepatic lesions and exposure to chemical contaminants in English sole (*Pleuronectes vetulus*), starry flounder (*Platichthys stellatus*), and white craoker

(*Genyonemus lineatus*) from selected marine sites on the Pacific Coast, USA. *Environmental Health Perspectives*, **102** (2), 200–215.

Myers, M.S., B.L. French, W.L. Reichert, M.L. Willis, B.F. Anulacion, T.K. Collier & J.E. Stein (1998a) Reductions in CYP1A expression and hydrophobic DNA adducts in liver neoplasms of English sole (*Pleuronectes vetulus*): further support for the 'resistant hepatocyte' model of hepatocarcinogenesis. *Marine Environmental Research*, **46** (1–5), 197–202.

Myers, M.S., L.L. Johnson, T. Hom, T.K. Collier, J.E. Stein & U. Varanasi (1998b) Toxicopathic lesions in subadult English sole (*Pleuronectes vetulus*) from Puget Sound, Washington, USA: relationships with other biomarkers of contaminant exposure. *Marine Environmental Research*, **45** (1), 47–67.

Nemcsok, J. (1994) Enzymatic determinations as tools for the detection of water pollution in fishes. In: (eds Salanki, J., D. Jeffrey & G.M. Hughes) *Biological Monitoring of the Environment: A Manual of Methods*. CAB International, Wallingford, UK.

Neuman, E. & P. Karås (1988) Effects of pulp mill effluent on a Baltic coastal fish community. *Water Science and Technology*, **20**, 95–106.

Nimi, A.J. (1983) Biological and toxicological effects of environmental contaminants in fish and their eggs. *Canadian Journal of Fisheries and Aquatic Sciences*, **40**, 306.

Olivieri, C.E. & K.R. Cooper (1997) Toxicity of 2,3,7,8-tetrachlorodibenxo-p-dioxin (TCDD) in embryos and larvae of the fathead minnow (*Pimephales promelas*). *Chemosphere*, **34**, 1139–1150.

Ourth, D.D. & V.D. Ratts (1991) Bactericidal complement activity and concentration of immunoglobulin M, transferrin, and protein at different ages of channel catfish. *Journal of Aquatic Animal Health*, **3**, 274–280.

Palace, V.P., J.F. Klaverkamp, W.L. Lockhart, D.A. Metner, D.C.G. Muir & S.B. Brown (1996) Mixed-function oxidase enzyme activity and oxidative stress in lake trout (*Salvelinus namaycush*) exposed to 3,3′,4,4′,5-pentachlorobiphenyl (PCB-126). *Environmental Toxicology and Chemistry*, **15**, 955–960.

Payne, J.F., I. Martins *et al.* (1978) Crankcase oils: are they a major mutagenic burden in the aquatic environment. *Science*, **200**, 329–330.

Perry, D.M., J.B. Hughes & A.T. Hebert (1991) Sublethal abnormalities in embryos of winter flounder, pseudopleuronectes-americanus, from long-island sound. *Estuaries*, **14** (3), 306–317.

Perry, S.F. & P. Laurent (1993) Environmental effects on gill structure and function. In: (eds Rankin, J.C. & F.B. Jensen) *Fish Ecophysiology*. Chapman & Hall, Fish & Fisheries Sci., London.

Pesonen, M., M. Celander, L. Förlin & T. Andersson (1987) Comparison of xenobiotic biotransformation enzymes in kidney and liver of rainbow trout (*Salmo gairdneri*). *Toxicology and Applied Pharmacology*, **91**, 75–84.

Pettersson, A. & A. Lignell (1999) Astaxanthin deficiency in eggs and fry of Baltic salmon (*Salmo salr*) with the M74 syndrome. *Ambio*, **28**, 43–47.

Pettey, C.L. & E.C. McKinney (1988) Induction of cell-mediated cytotoxicity by shark 19s IgM. *Cell Immunol.*, **111**, 28–38.

Pierce, K.V., B.B. McCain & S.R. Wellings (1978) Pathology of hepatomas and other liver abnormalities in English sole (*Parophrys vetulus*) from the Duwamish river estuary, Seattle, Washington. *J. Natl. Cancer. Inst.*, **60** (6), 1445–1453.

Post, G. & R.A. Leisure (1974) Sublethal effect of malathion to three salmonid species. *Bulletin of Environmental Contamination and Toxicology*, **12**, 312–319.

Poulet, F.M., M.J. Wolfe & J.M. Spitsbergen (1994) Naturally occurring orocutaneous papillomas and carcinomas of brown bullheads (*Ictalurus nebulosus*) in New York State. *Vet. Pathol.*, **31** (1), 8–18.

Pulsford, A.L., K.P. Ryan & J.A. Nott (1992) Metals and metalomacrophages in flounder, *Platichthys flesus*, spleen and kidney. *Journal of the Marine Biological Association of the United Kingdom*, **72**, 483–498.

Purceli, J.E., D. Grosse & J.J. Grover (1990) Mass abundance of abnormal Pacific herring larvae at a spawning ground in British Columbia. *Transactions of the American Fisheries Society*, **119**, 463–469.

Rana, S.V.S., R. Singh & S. Verma (1995) Mercury-induced lipid peroxidation in the liver, kidney, brain and gills of a freshwater fish, *Channa punctatus*. *Japanese Journal of Ichthyology*, **42**, 255–259.

Reimschuessel, R., R.O. Bennett & M.M. Lipsky (1992) A classification system for histological lesions. *Journal of Aquatic Animal Health*, **4**, 135–143.

Rhodes, L.D., M.S. Myers, W.D. Gronlund & B.B. McCain (1987) Epizootic characteristics of hepatic and renal lesions in English sole, *Parophrys vetulus*, from Puget Sound. *Journal of Fish Biology*, **31**, 395–407.

Ricard, A.C., C. Daniel, P. Anderson & A. Hontela (1998) Effects of subchronic exposure to cadmium chloride on endocrine and metabolic functions in rainbow trout *Oncorhynchus mykiss*. *Archives of Environmental Contamination and Toxicology*, **34**, 377–381.

Rice, C.D., D.H. Kergosien & S.M. Adams (1996) Innate immune function as a bioindicator of pollution stress in fish. *Ecotoxicology and Environmental Safety*, **33**, 186–192.

Riersen, L.O. & K. Fugelli (1984) Annual variation in lymphocystis infection frequency in flounder, *Platichthys flesus* (L.). *Journal of Fish Biology*, **24**, 187–191.

Robertson, L., P. Thomas, C.R. Arnold & J.M. Trant (1987) Plasma cortisol and secondary stress responses of red drum to handling, transport, rearing density, and a disease outbreak. *Prog. Fish Culturist*, **49**, 1–12.

Rodríguez-Ariza, A., J. Alhama, F.M. Diaz-Mendes & J. Lopez-Barea (1999) Content of 8-oxodG in chromosomal DNA of *Sparus aurata* fish as biomarker of oxidative stress and environmental pollution. *Mutat. Res-Gen Tox. En.*, **438** (2), 97–107.

Ronisz, D. & L. Förlin (1998) Interaction of isosafrol, β-naphthoflavone and other CYP1A1 inducers in liver of rainbow trout (*Oncorhynchus mykiss*) and eelpout (*Zoarces viviparus*). *Comparative Biochemistry and Physiology*, **121** (1–3), 289–296.

Rosenthal, H. & D.F. Alderdice (1976) Sublethal effects of environmental stressors, natural and pollutional, on marine fish eggs and larvae. *Journal of the Fisheries Research Board of Canada*, **33**, 2047–2065.

Rowley, A.F., J. Knight, E.P. Lloyd, J.W. Holland & P.J. Vickers (1995) Eicosanoids and their role in immune modulation in fish: A brief overview. *Fish and Shellfish Immunology*, **5**, 549–567.

Ruis, M.A.W. & C.J. Bayne (1997) Effects of acute stress on blood clotting and yeast killing by phagocytes of rainbow trout. *Journal of Aquatic Animal Health*, **9**, 190–195.

Røed, K.H., K. Fjalestad, H.J. Larsen & L. Midthjel (1992) Genetic variation in haemolytic activity in Atlantic salmon (*Salmo salar* L.). *Journal of Fish Biology*, **40**, 739–750.

Sandnes, K., Y. Ulgenes, O.R. Braekkan & F. Utne (1984) The effect of ascorbic acid supplementation in broodstock feed on reproduction of rainbow trout (*Salmo gairdneri*). *Aquaculture*, **43** (1–3), 167–177.

Sandström, O. (1994) Incomplete recovery in a coastal fish community exposed to effluent from a modernized Swedish bleached kraft mill. *Canadian Journal of Fisheries and Aquatic Sciences*, **51** (10), 2195–2202.

Sandström, O. (1995) Incomplete recovery in a coastal fish community exposed to effluents from a modernized Swedish bleached kraft mill. *Canadian Journal of Fisheries and Aquatic Sciences*, **51**, 2195–2202.

Sandström, O. & G. Thoresson (1988) Mortality in perch population in a Baltic pulp mill effluent area. *Marine Pollution Bulletin*, **19** (11), 564–567.

Sandström, O., E. Neuman & P. Karås (1988) Effects of bleached pulp mill effluent on growth and gonad function in Baltic coastal fish. *Water Science and Technology*, **20**, 107–118.

Sandström, O., L. Förlin, O. Grahn, L. Landner & Å. Larsson (1997) *Environmental impact of pulp and paper mill effluents*. Swedish EPA Report no 4785, Stockholm, Sweden.

Saucier, D. & L. Astic (1995) Morpho-functional alterations in the olfactory system of rainbow trout (*Oncorhynchus mykiss*) and possible acclimation in response to long-lasting exposure to low copper levels. *Comparative Biochemistry and Physiology A.*, **112**, 273–284.

Saucier, D., L. Astic & P. Rioux (1991) The effects of early chronic exposure to sublethal copper on the olfactory discrimination of rainbow trout, *Oncorhynchus mykiss. Environmental Biology of Fishes*, **30** (3), 345–351.

Scarpa, J., D.M. Gatlin III & D.H. Lewis (1992) Effects of dietary zinc and calcium on select immune functions of channel catfish. *Journal of Aquatic Animal Health*, **4**, 24–31.

Schneider, R. & H. Ambrosius (1987) The influence of environmental temperature on the lymphocyte populations in carp (*Cyprinus carpio* L.). *Biomed. Biochim. Acta*, **46**, 135–141.

Schreck, C.B. (1996) Immunomodulation: endogenous factors. In: (eds Iwama, G.K. & S. Nakanishi) *The Fish Immune System – organism, pathogen, and environment*. Academic Press, San Diego, pp. 311–338.

Secombes, C.J., T.C. Fletcher, J.A. O'Flynn, M.J. Costello, R. Stagg & D.F. Houlihan (1991) Immunocompetence as a measure of the biological effects of sewage sludge pollution in fish. *Comparative Biochemistry and Physiology*, **100C** (1/2), 133–136.

Secombes, C.J., T.C. Fletcher, A. White, M.J. Costello, R. Stagg & D.F. Houlihan (1992) Effects of sewage sludge on immune responses in the dab, *Limanda limanda* (L.). *Aquatic Toxicology*, **23**, 217–230.

Secombes, C.J., A. White, T.C. Fletcher, R. Stagg & D.F. Houlihan (1995) Immune parameters of plaice, *Pleuronectes platessa*, L. along a sewage sludge gradient in the Firth of Clyde, Scotland. *Ecotoxicology*, **4**, 329–340.

Secombes, C.J., L.J. Hardie & G. Daniels (1996) Cytokines in fish: an update. *Fish and Shellfish Immunology*, **6**, 291–304.

Servos, M.R., J.H. Carey, M.L. Ferguson, G.J. van der Kraak, H. Ferguson, J. Parrot, K. Gorman & R. Cowling (1992) Impact of a modern bleached kraft mill with secondary treatment on white suckers. *Water Pollution Research Journal of Canada*, **27**, 423–437.

Sheldon, W.M., Jr. & V.S. Blazer (1991) Influence of dietary lipid and temperature on bactericidal activity on channel catfish macrophages. *Journal of Aquatic Animal Health*, **3**, 87–93.

Sing, N.N. & A.K. Srivastava (1992) Blood dyscrasia in the freshwater Indian catfish *Heteropneustes fossilis* after acute exposure to a sublethal concentration of propuxur. *Acta Hydrobiologica*, **34**, 189–195.

Siolman, A.K., K. Jauncey & R.J. Roberts (1986) The effect of dietary ascorbic acid supplementation on hatchability, survival rate and fry performance in *Oreochromis mossambicus* (Peters). *Aquaculture*, **59**, 197–208.

Skak, C. & E. Baatrup (1993) Quantitative and histochemical demonstration of mercury deposits in the inner ear of trout, *Salmo trutta*, exposed to dietary methylmercury and dissolved mercuric chloride. *Aquatic Toxicology*, **25**, 55–70.

Slauson, D.O. & B.J. Copper (1982) *Mechanisms of disease: A textbook of comparative general pathology*. Williams and Wilkins, Baltimore/London, 420 pp.

Smit, G.L. & H.J. Schoonbee (1988) Blood coagulation factors in the freshwater fish *Oreochromis mossambicus*. *Journal of Fish Biology*, **32**, 673–678.

Smith, C.E., T.H. Peck, R.J. Klauda & J.B. McLaren (1979) Hepatomas in Atlantic tomcod *Microgadus tomcod* (Walbaum) collected in the Hudson River estuary in New York. *Journal of Fish Diseases*, **2**, 313–319.

Solangi, M.A. & R.M. Overstreet (1982) Histopathological changes in two estuarine fishes, *Menidia beryllina* (Cope) and Trinectes maculatus (Bloch and Schneider), exposed to crude oil and its water-soluble fractions. *Journal of Fish Diseases*, **5**, 13–35.

Spazier, E., V. Storch & T. Braunbeck (1992) Cytopathology of spleen in eel *Anguilla anguilla* exposed to chemical spill in the Rhine River. *Diseases of Aquatic Organisms*, **14**, 1–22.

Speare, D.J. & H.W. Ferguson (1989) Fixation artefacts in rainbow trout (*Salmo gairdneri*) gills: A morphometric evaluation. *Canadian Journal of Fisheries and Aquatic Sciences*, **46**, 780–785.

Spies, R.B., J.J. Stegeman, D.E. Hinton, B. Woodin, R. Smolowitz, M. Okihiro & D. Shea (1996) Biomarkers of hydrocarbon exposure and sublethal effects in embiotocid fishes from a natural petroleum seep in the Santa Barbara Channel. *Aquatic Toxicology*, **34**, 195–219.

Spitsbergen, J.M. & M.J. Wolfe (1995) Hepatocyte clusters in the spleen: A normal feature of some populations of brown bullheads in New York State. *Toxicologic Pathology*, **23**, 726–730.

Stehr, C.M., L.L. Johnson & M.S. Myers (1998) Hydropic vacuolation in the liver of three species of fish from the US West Coast: lesion description and risk assessment associated with contaminant exposure. *Diseases of Aquatic Organisms*, **32**, 119–135.

Stoker, P.W., J.R. Larsen, G.M. Booth & M.L. Lee (1985) Pathology of gill and liver tissues from two generations of fishes exposed to two coal-derived materials. *Journal of Fish Biology*, **27**, 31–46.

Stott, G.G., W. Haensly, J. Neff & J. Sharp (1983) Histopathologic survey of ovaries of plaice, *Pleuronectes platessa* L., from AberWarc'h and Aber Benoit, Brittany, France oil spills. *Journal of Fish Diseases*, **6** (5), 429–437.

Sunyer, J.O., E. Gomez, V. Navarro, J. Quesada & L. Tort (1995) Physiological responses and depression of humoral components of the immune system in gilthead sea bream (*Sparus aurata*) following daily acute stress. *Canadian Journal of Fisheries and Aquatic Sciences*, **52**, 2339–2346.

Suzuki, K.T., H. Sunaga, Y. Yamane & Y. Aoki (1991) Binding of cadmium to alcohol dehydrogenase in the liver before induction of metallothionein. *Research Communications in Chemical Pathology and Pharmacology*, **74**, 223–236.

Szalai, A.J., J.E. Bly & L.W. Clem (1994) Changes in serum concentrations of channel catfish (*Ictalurus punctatus* Rafinesque) phosphorylcholine-reactive protein (PRP) in response to inflammatory agents, low temperature shock and infection by the fungus *Saprolegnia* sp. *Fish and Shellfish Immunology*, **4**, 323–336.

Tahir, A. & C.J. Secombes (1995) The effects of diesel oil-based drilling mud extracts on immune responses of rainbow trout. *Archives of Environmental Contamination and Toxicology*, **29**, 27–32.

Tahir, A., T.C. Fletcher, D.F. Houlihan & C.J. Secombes (1993) Effect of short-term exposure to oil-contaminated sediments on the immune response of dab, *Limanda limanda* (L.). *Aquatic Toxicology*, **27**, 71–82.

Tatner, M. (1996) Natural changes in the immune system of fish. In: (eds Iwama, G.K. & S. Nakanishi) *The Fish Immune System – organism, pathogen, and environment*. Academic Press, San Diego, pp. 255–287.

Tesoriero, J.V. (1977) Formation of the chorion (*zona pellucida*) in the teleost, *Oryzias latipes*. II Polysaccharide cytochemistry of early oogenesis. *Journal of Histochemistry and Cytochemistry*, **25**, 1376.

Thiele, D.J. (1992) Metal-regulated transcription in eukaryotes. *Nucleic Acids Research*, **20**, 1183–1191.

Thomas, P. (1987) Influence of some environmental variables on the ascorbic acid status of striped mullet, *Mugil cephalus* (L.), tissue effects of exposure to oil. *Journal of Fish Biology*, **30** (4), 485–494.

Thomas, P. & J.M. Neff (1984) Effects of pollutants and other environmental variables on the ascorbic acid content of fish tissue. *Marine Environmental Research*, **14**, 489–491.

Thomas, P. & H.W. Wofford (1993) Effects of cadmium and Aroclor 1254 on lipid peroxidation, glutathione peroxidase activity, and selected antioxidants in Atlantic croaker tissues. *Aquatic Toxicology*, **27**, 159–178.

Thompson, I., A. White, T.C. Fletcher, D.F. Houlihan & C.J. Secombes (1993) The effect of stress on the immune response of Atlantic salmon (*Salmo salar* L.) fed diets containing different amounts of vitamin C. *Aquaculture*, **114**, 1–18.

Thuvander, A. (1989) Cadmium exposure of rainbow trout, *Salmo gairdneri* Richardson: effects on immune functions. *Journal of Fish Biology*, **35**, 521–529.

Tjälve, H., J. Gottofrey & I. Björklund (1986) Tissue disposition of 109 Cd 2+ in the brown trout *Salmo trutta* studied by autoradiography and impulse counting. *Toxicology and Environmental Chemistry*, **12**, 31–45.

Tremblay, L., X. Yao & G.J. van der Kraak (1995) Interaction of the environmental estrogen nonylphenol and β-sitosterol with liver estrogen receptors in fish. In: *Proceedings from 5th International Symposium on Reproductive Physiology of Fish*. Austin, Texas, USA, 2–8 July, 202 pp.

van der Kraak, G.J., K.R. Munkittrick, M.E. McMaster, C.B. Portt & J.P. Chang (1992) Exposure to bleached kraft mill effluent disrupts the pituitary-gonadal axis of white sucker at multiple sites. *Toxicology and Applied Pharmacology*, **115**, 224–233.

Vethaak, A.D. & J.G. Jol (1996) Diseases of flounder (*Platichthys flesus* L.) in Dutch coastal and estuarine waters, with particular reference to environmental stress factors. I. Epizootiology of gross lesions. *Diseases of Aquatic Organisms*, **26**, 81–97.

Vethaak, A.D. & P.W. Wester (1996) Diseases of flounder *Platichthys flesus* in Dutch coastal and estuarine waters, with particular reference to environmental stress factors. II. Liver histopathology. *Diseases of Aquatic Organisms*, **26**, 99–116.

Vethaak, A.D., D. Bucke, T. Lang, P.W. Wester, J. Jol & M. Carr (1992) Fish disease monitoring along a pollution transect: a case study using dab *Limanda limanda* in the German Bight. *Marine Ecology Progress Series*, **91**, 173–192.

Vethaak, A.D., J.G. Jol, A. Meijboom, M.L. Eggens, T. Aprheinallt, P.W. Wester, T. VandeZande, A. Bergman, N. Dankers, F. Ariese, R.A. Baan, J.M. Everts, A. Opperhuizen & J.M. Marquenie (1996) Skin and liver diseases induced in flounder (*Platichthys flesus*) after long-term exposure to contaminated sediments in large-scale mesocosms. *Environmental Health Perspectives*, **104**, 1218–1229.

Vogelbein, W.K., J.W. Fournie, P.A. Van Veld & R.T. Huggett (1990) Hepatic neoplasms in the Mummichog *Fundulus heteroclitus* from a Creosote-contaminated site. *Cancer Research*, **50**, 5978–5986.

von Westernhagen, H. (1988) Sublethal effects on fish eggs and larvae. In: (eds Hoar, W.S. & D.J. Randall) *Fish Physiology. Vol. XI. The physiology of developing fish*. Academic Press, New York, pp. 253–346.

von Westernhagen, H., V. Dethlefsen, P. Cameron & D. Janssen (1987) Chlorinated hydrocarbon residues in gondads of fish and effects on reproduction. *Sarsia*, **72**, 419–422.

von Westernhagen, H., V. Dethlefsen, P. Cameron, J. Berg & G. Fürstenberg (1988) Developmental defects in pelagic fish embryos from the western Baltic. *Helgoländer Meeresuntersuchungen*, **42**, 13–36.

von Westernhagen, H., P. Cameron, V. Dethlefsen & D. Janssen (1989) Chlorinated hydrocarbons in North Sea whiting (*Merlangius merlangus*), and effects on reproduction. 1. tissue burden and hatching success. *Helgoländer Meeresuntersuchungen*, **43**, 45–60.

Watson, T.A. & F.W.H. Beamish (1980) Effects of zinc on branchial ATPase activity in vivo in rainbow trout, *Salmo gairdneri*. *Comparative Biochemistry and Physiology*, **66C**, 77–82.

Weis, J.S. & P. Weis (1989) Effects of environmental pollutants on early fish development. *CRC Critical Reviews in Aquatic Sciences*, **1**, 45–73.

Weis, P. & J.S. Weis (1991) The developmental toxicity of metals and metalloids in fish. In: (eds Michael, C., C. Newman & A.W. McIntosh) *Metal Ecotoxicology*. CRC Lewis publishers, Florida, pp. 145–169.

White, A. & T.C. Fletcher (1983) Serum concentrations of C-reactive protein and serum amyoid P component in plaice (*Pleuronectes platessa* L.) in relation to season and injected polysaccharide. *Comparative Biochemistry and Physiology*, **74B**, 453–458.

Wiegand, M.D., J.M. Hately, C.L. Kitchen & L.G. Buchanan (1989) Induction of developmental abnormalities in larval goldfish *Carassius auratus* L., under cool incubation conditions. *Journal of Fish Biology*, **35**, 85–95.

Wildish, D.J. & N.A. Lister (1973) Biological effects of fenitrothion in the diet of brook trout. *Bulletin of Environmental Contamination and Toxicology*, **10**, 333–339.

Winkelhake, J.L., M.J. Vodicnik & J.L. Taylor (1983) Induction in rainbow trout of an acute phase (C-reactive) protein by chemicals of environmental concern. *Comparative Biochemistry and Physiology*, **74C**, 55–58.

Winzer, K. & A. Köhler (1998) Aldehyde dehydrogenase and glucose-6-phosphate dehydrogenase as markers for enzyme-altered foci in the liver of dab (*Limanda limanda* L.). *Marine Environmental Research*, **46** (1–5), 215–219.

Wolke, R.E. (1992) Piscine macrophage aggregates: a review. *Annual Review of Fish Diseases*, 91–108.

Wolke, R.E., C.J. George & V.S. Blazer (1981) Pigmented macrophage accumulations (MMC; PMB): possible monitors of fish health. In: (ed. Hargis, J.) *Parasitology and Pathology of Marine Organisms of the World Ocean*. NOAA Tech. Rep. NMFS 25, USA, pp. 93–97.

Yamamoto, Y., T. Ishii, M. Sato & S. Ikeda (1977) Effects of dietary ascorbic acid on the accumulation of copper in carp. *Bulletin of the Japanese Society of Scientific Fisheries*, **43**, 989–993.

Yamamoto, Y., K. Hayama & S. Ikeda (1981) Effects of dietary ascorbic acid on the poisoning in rainbow trout. *Bulletin of the Japanese Society of Scientific Fisheries*, **47**, 1085–1089.

Yano, T. (1996) The non-specific immune system: humoral defenses. In: (eds Iwama, G.K. & S. Nakanishi) *The Fish Immune System – organism, pathogen, and environment*. Academic Press, San Diego, pp. 106–159.

Zbanyszek, R. & L.S. Smith (1984) The effect of water-soluble aromatic hydrocarbons on some hematological parameters of rainbow trout, *Salmo gairdneri*, during acute exposure. *Journal of Fish Biology*, **24**, 545–552.

Chapter 5

Molecular/Cellular Processes and the Impact on Reproduction

A. Goksøyr, A. Arukwe, J. Larsson, M.P. Cajaraville, L. Hauser, B.M. Nilsen, D. Lowe and P. Matthiessen

5.1 Endocrine disruption

5.1.1 General aspects

The terms environmental oestrogens, endocrine disrupters, endocrine modulators, ecoestrogens, environmental hormones, xenoestrogens, hormone-related toxicants, and phytoestrogens all have one thing in common: they describe synthetic chemicals and natural plant or animal compounds that may affect the endocrine system (the biochemical messengers or communication systems of glands, hormones and cellular receptors that control the body's internal functions) of various organisms. Many of the effects caused by these substances have been associated with developmental, reproductive and other health problems in wildlife and laboratory animals. There is also some evidence suggesting that these compounds may be affecting humans in similar ways (Toppari *et al.*, 1996).

The exact mode of action of all endocrine disrupters is not fully understood, but it is known that these compounds can alter hormonal functions by several means (Fig. 5.1a,b). They can:

(1) Mimic or partly mimic the natural hormones by binding to hormone receptors or influencing cell signalling pathways
(2) Block, prevent and alter hormonal binding to hormone receptors or influencing cell signalling pathways
(3) Alter the production and breakdown of natural hormones
(4) Modify the production and function of hormone receptors (see Colborn & Clement, 1992).

5.1.2 Oestrogenic and antioestrogenic effects

Undesired effects on reproduction of oestrogens and antioestrogens are the most widely studied examples of endocrine disruption, although they have only been recognised as a pollution problem in the aquatic environment for less than a decade (Shore *et al.*, 1988, 1993; Purdom *et al.*, 1994). The subject has been reviewed by Sumpter (1995), Wiese and Kelce (1997), Arukwe and Goksøyr (1998) and Matthiessen and Sumpter (1998).

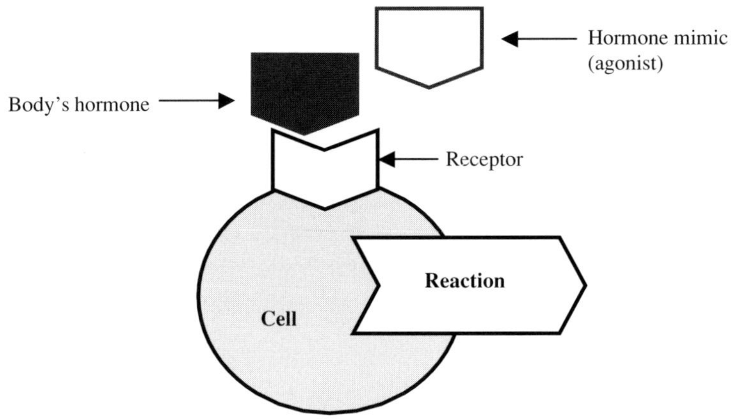

Fig. 5.1a Interaction of endocrine-disrupters with the body's signalling system: normal hormones. A normal hormone activates a receptor, either at the cell surface (as shown), or in cytosol, and initiates a cellular response. A hormone mimic (agonist) can initiate the same response by binding to the receptor.

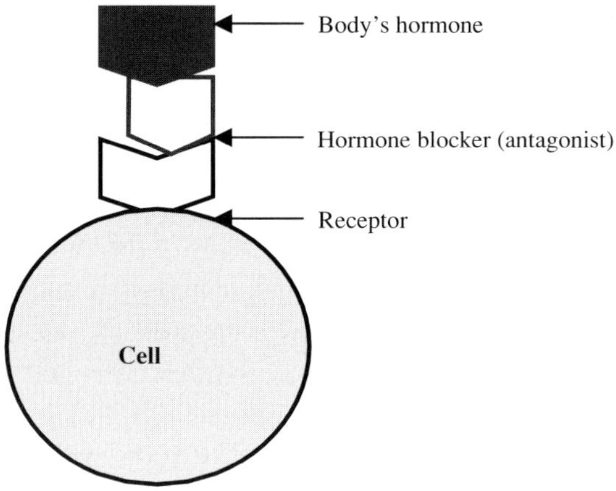

Fig. 5.1b Interaction of endocrine-disrupters with the body's signalling system: antagonist. An antagonist can interfere with the signal from the body hormones and block the cellular response.

5.1.2.1 Mechanisms

Oestrogenic and antioestrogenic effects in fish occur by definition via interactions of ligands with oestrogen receptor (ER) protein molecules.

In normal female fish, oestradiol hormone (17β-oestradiol, E_2) is produced by the ovary under the influence of pituitary-derived gonadotropin hormones (GTH I & II; also known as follicle-stimulating hormone, FSH, and luteinising hormone, LH, respectively (Fig. 5.2). The GTHs are peptide hormones synthesised by the pituitary in response to other hormones released by the hypothalamus. This organ is sensitive to environmental cues such as day length and temperature, thereby helping to synchronise the reproductive cycle with the appropriate season. E_2's mode of action is to interact agonistically with ER in the liver, oviduct, testes and brain. It binds reversibly with nuclear ER protein molecule in their active

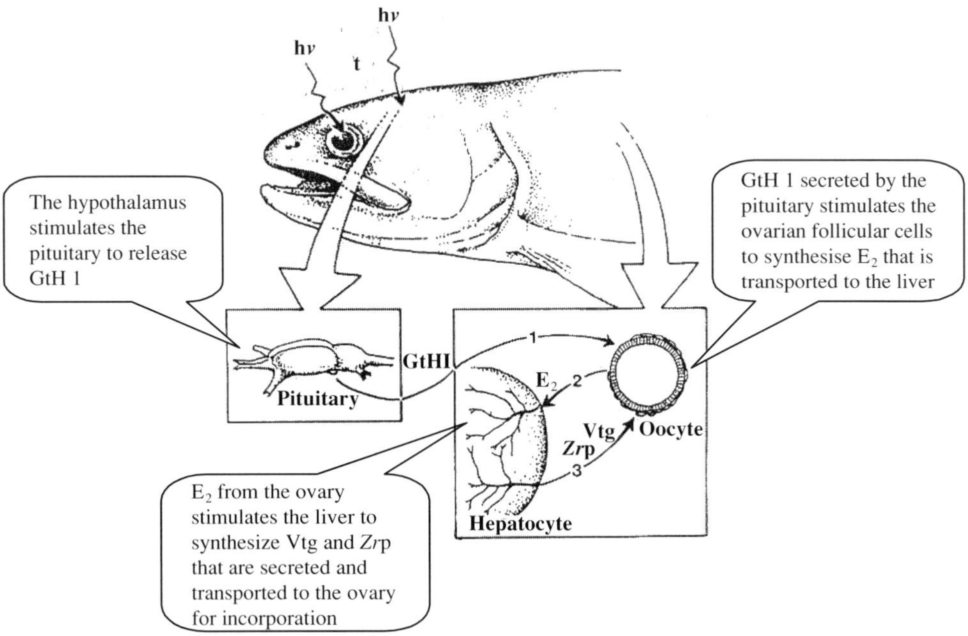

Fig. 5.2 A model for vitellogenin (Vtg) and eggshell *zona radiata* protein (*Zr*p) synthesis in salmon. In response to photoperiodic stimuli (hν), gonadotopin I (GtH I) is secreted from the pituitary and transported in the blood to the ovary where it induces synthesis of oestradion-17β (E_2; 1), E_2 is transported by the blood (2) to the liver (hepatocytes) where it induces the synthesis of Vtg and *Zr*p. The synthesised Vtg and *Zr*p are transported to the ovary for deposition/incorporation in the growing oocyte (3). Modified from Oppen-Berntsen, 1990.

'pocket', forming ER dimers which in turn bind with oestrogen response elements (EREs) in the chromosomal DNA to form an initiation complex (Brown, 1994; Campbell *et al.*, 1994; Beato *et al.*, 1995).

These complexes subsequently promote assembly of the polymerases needed to transcribe the genes coding for the synthesis of reproductively important proteins, for example, in the liver, the phospholipoglycoprotein yolk precursor vitellogenin (VTG) (Lazier & MacKay, 1993) and the eggshell zona radiata proteins (*Zr*p) (Hyllner *et al.*, 1991; Oppen-Berntsen *et al.*, 1992). Synthesis of these proteins is followed by a cascade of events leading up to full reproductive development and receptivity. Oestrogen hormone also causes fish hepatocytes to produce more ER (Lazier & MacKay, 1993), regulates fish gonadotropin secretion by an interesting feedback loop (Querat *et al.*, 1991) and can entirely feminise the male phenotype through exposure of fish larvae (Piferrer & Donaldson, 1989). E_2 may also affect the calcium balance of fish, with consequences to scale and bone resorption, and bone formation, as observed in rainbow trout (*Oncorhynchus mykiss*; Persson *et al.*, 1997).

Normal reproduction in fish can be disrupted by oestrogens and their mimics through the synthesis of abnormal numbers of oestrogen receptors during embryonic and larval development or later (Nimrod & Benson, 1997), through the stimulation of the oestrogen receptors and response elements in immature or adult females at inappropriate times, or through the stimulation of receptors and response elements in immature or adult males. In the latter case, ER and response elements are present in normal males, but generally are not stimulated due to the almost complete absence of endogenous oestradiol. Synthesis of both zona radiata protein and vitellogenin has been observed in male fish exposed to xenoestrogens (Sumpter, 1995; Arukwe *et al.*, 1997a), as has reduced testicular growth (Jobling *et al.*, 1996; Christiansen *et al.*, 1998).

True oestrogenic hormones (e.g. 17β-oestradiol, oestrone, oestriol) and their synthetic analogues (e.g. 17α-ethinyloestradiol; diethylstilbestrol) are able to act as full agonists of the ER, and the latter substances may be even more potent than the natural hormones, but most oestrogenic environmental contaminants are only partial ('weak') agonists and are hundreds to tens of thousands of times less potent than the natural hormones. This is apparently due to the fact that they only partly resemble the three-dimensional structure of true oestrogens and therefore only fit very imperfectly into the receptor pocket. The shape of the resulting dimer will also presumably align imperfectly with EREs. In either case, relatively few oestrogen-mimicking molecules actually trigger the cascade of oestrogenic effects which normally results from oestrogenic hormone exposure, and so the overall effect is muted. Further information on the possible effects of oestrogens in the aquatic environment can be found in Stahlschmidt-Allner *et al.* (1997) and useful reviews of the effects of pollutants on fish reproduction and reproductive behaviour can be found in Kime (1995) and Jones and Reynolds (1997).

However, antioestrogenic environmental contaminants can act as weak antagonists at the ER, binding irreversibly with it but not producing dimers which can interact with the EREs. They thus block the action of natural oestrogenic agonists, and the more receptors which are occupied, the weaker the action of the hormone will be. An example of an oestrogen antagonist with therapeutic value is the drug tamoxifen which is used to treat or prevent breast cancer by blocking the potentiating effect of natural oestrogens.

Two different ER forms have been identified in mammals (Kuiper *et al.*, 1996; Kuiper & Gustafsson, 1997), and possibly more forms occur in some fish species (Tchoudakova *et al.*, 1999). A testicular ER has a higher affinity than the liver ER for oestrogens and xenoestrogens (Loomis & Thomas, 1999). Different tissues may therefore have different susceptibilities to chemical disruption.

There have been reports of some endocrine effects not mediated through ER binding and activation. For example, β-hexachlorocyclohexane (β-HCH) produced oestrogen-like responses (cell division and growth) at levels found in human breast cancer tissue (Steinmetz *et al.*, 1996). Compounds, which do not bind ER (Coosen & van Velson, 1989; Steinmetz *et al.*, 1996), may promote DNA transcription and thus produce oestrogenic responses, by passing signals through a highway of hormone and non-hormone response elements that turn genes on (Steinmetz *et al.*, 1996). For example, *p,p′*-DDT, at or below levels found in human breast fat tissue, bypassed the oestrogen receptor and stimulated a complex mixture of cell signalling proteins (growth factor receptors) and processes that eventually led to cell division (Shen & Novak, 1997). Also, some superficially similar feminisation effects may be caused through antiandrogenic mechanisms whereby the masculinising action of testosterone or 11-ketotestosterone is blocked at the androgen receptor by androgen antagonists such as *p,p′*-DDE (e.g. Kelce *et al.*, 1995; see section 5.1.3.1). These results suggest that the same chemical and/or its metabolites can influence the endocrine system in more than one way. If understood, these complex modes of action may be able to answer the questions of how different molecules impact the endocrine system and how pollutants may induce or promote the development of some types of tumours. Although not yet studied, it is possible that these responses observed using mammalian cell systems may also occur in fish.

A complicating factor is that weak oestrogen mimics are also by their nature often able to act as weak oestrogen antagonists, some molecules reacting agonistically with the oestrogen receptor and some binding irreversibly with it. Little is known about the factors which determine the final phenotypic outcome.

5.1.2.2 *Contaminants*

Many oestrogenic contaminants are produced for specific purposes and are used in pesticides, plastics, electrical transformers and other products (Fig. 5.3; for review see Caldwell, 1985; Ahlborg *et al.*, 1995). Other substances are generated as by-products

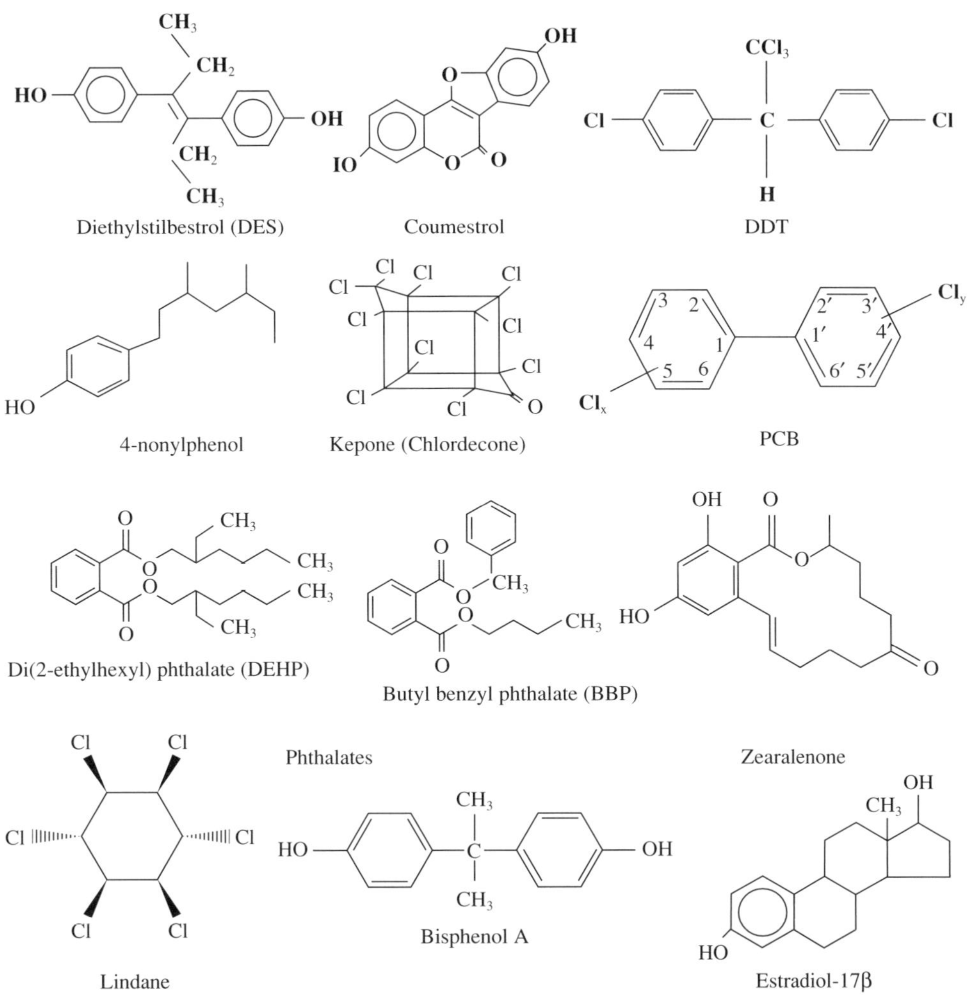

Fig. 5.3 Structure of selected endocrine modulating chemicals and the natural oestrogen, oestradiol-17β. Reproduced with permission from Arukwe, 1998.

during manufacturing or are breakdown products of some other chemical, and some, like 17α-ethinyloestradiol and diethylstilbestrol (DES), are drugs. However, natural compounds capable of producing oestrogenic responses, such as the phyto-oestrogens and myco-oestrogens which occur in a variety of plants and fungi, have also been studied in fish (Pelissero *et al.*, 1991, 1993; Arukwe *et al.*, 1999; Celius *et al.*, 1999). Regardless of the source or original intended use, substantial amounts of these chemicals end up in the aquatic environment due to physico-chemical, hydrologic and atmospheric processes (Barrie *et al.*, 1992; Guardans & Gimeno, 1994; Ayotte *et al.*, 1995; Bjerregaard, 1996).

Many of the oestrogenic and antioestrogenic environmental contaminants that have been reported in the literature have been detected in *in vitro* assays of various types based on the human or rodent oestrogen receptor (e.g. naked ER assays; breast cancer cell proliferation assays such as MCF-7; yeast assays with transfected human ER genes). Some have also been detected in *in vitro* assays which measure VTG induction in fish hepatocytes. In both cases, some caution should be used when extrapolating to fish *in vivo*. Although it is known that steroid hormone structure and function have been highly conserved during the evolution of the vertebrates such that the endocrine systems of fish and mammals retain great similarity, there may still be important differences between species. Another reason to be cautious when extrapolating from *in vitro* tests to living fish is that most assays possess little if any metabolic competence, and will therefore miss oestrogenic effects caused by metabolites or endocrine disruption caused by interferences with steroid metabolism. Finally, the various endocrine pathways in intact animals are extremely complex and interdependent (e.g. Cyr & Eales, 1996), so effects in an *in vitro* assay will not necessarily be replicated *in vivo*, as is true for the opposite case.

These caveats should therefore be kept in mind when considering the list of environmental chemicals with suspected oestrogenic and antioestrogenic action shown in Table 5.1.

A selection of the compounds referred to in Table 5.1 are described in more detail here:

- *Diethylstilbestrol (DES)*: A pharmaceutical oestrogen banned from use in the 1970s.
- *Coumestrol*: A phyto-oestrogen which is a natural plant compound with some oestrogenic properties.
- o,p′-*DDT*: A synthetic pesticide constituting between 10 and 25% of technical DDT. DDT is banned in many countries but is still used extensively in equatorial countries to control mosquitoes and malaria.
- *Alkylphenols (4-nonylphenol: NP; Octylphenol: OP; etc.)*: Breakdown products of detergents that are widely used in household products, in agricultural and industrial applications, and in plastics manufacturing. Nonylphenols are found in natural water bodies, sewage sludge and river sediments.
- *Kepone (chlordecone)*: A synthetic pesticide banned in the USA.
- *Bisphenol A*: Bisphenol A is used in the production of epoxy resins and polycarbonate plastics. These plastics are used in many food and drink packaging applications, whilst the resins are commonly used as lacquers to coat metal products such as food cans, bottle tops and water supply pipes (ENDS, 1995). Some polymers used in dental treatment contain Bisphenol A.
- *Lindane*: Lindane is used on many crops in the UK, including cereals, cabbages, apples, pears, tomatoes and strawberries (Maynard, 1995).

Table 5.1 Environmental contaminants with known or suspected oestrogenic or anti-oestrogenic action in fish.

Contaminant	Assay system	Reference
(A) Oestrogenic action		
equol	Sturgeon (*in vivo*)	Pelissero *et al.* (1991)
β-sitosterol	Rainbow trout (*in vivo*)	Tremblay & van der Kraak (1999)
Daidzein	Sturgeon (*in vivo*)	Pelissero *et al.* (1991)
genistein	Sturgeon (*in vivo*)	Pelissero *et al.* (1991)
biochanin A	Sturgeon (*in vivo*)	Pelissero *et al.* (1991)
formononetin	Sturgeon (*in vivo*)	Pelissero *et al.* (1991)
coumestrol	Sturgeon (*in vivo*)	Pelissero *et al.* (1991)
zearalenone	Salmon (*in vivo*)	Arukwe *et al.* (1999)
	Salmon (*in vitro*)	Celius *et al.* (1999)
zearalenol	Salmon (*in vivo*)	Arukwe *et al.* (1999)
	Salmon (*in vitro*)	Celius *et al.* (1999)
o,p′-DDT	Mosquitofish (*in vivo*)	Denison *et al.* (1981)
	Croaker (*in vivo*)	Khan & Thomas (1998)
	Salmon (*in vitro*)	Celius & Walther (1998)
o,p′-DDE	Trout (*in vivo*)	Donohoe & Curtis (1996)
β-HCH, g-HCH	Guppy, medaka (*in vivo*)	Wester (1991)
	Salmon (*in vivo*)	Arukwe *et al.* (2000)
	Salmon (*in vitro*)	Celius *et al.* (1999)
1-hydroxychlordene	Rainbow trout FHVSA	White *et al.* (1994)
4-tert-butylphenol	Rainbow trout FHVSA	Jobling & Sumpter (1993)
4-tert-octylphenol	Rainbow trout FHVSA	Jobling & Sumpter (1993)
4-tert-pentylphenol	Carp (*in vivo*)	Gimeno *et al.* (1996)
4-nonylphenol	Rainbow trout (*in vivo*)	Jobling *et al.* (1996), Christiansen *et al.* (1998)
	Salmon (*in vivo*)	Arukwe *et al.* (1997a,b)
	Salmon (*in vitro*)	Celius *et al.* (1999)
4-nonylphenol-diethoxylate	Rainbow trout FHVSA	White *et al.* (1994)
4-nonylphenoxy-carboxylic acid	Rainbow trout FHVSA	White *et al.* (1994)
3-trifluoromethyl-4-nitrophenol	Rainbow trout FHVSA	Hewitt *et al.* (1998)
di-n-butylphthalate	Rainbow trout ZR-75	Jobling *et al.* (1995)
butylbenzylphthalate	Rainbow trout ZR-75	Jobling *et al.* (1995)
	Rainbow trout (*in vivo*)	Christiansen *et al.* (1998)
toxaphene	Rainbow trout (*in vivo*)	Donohoe & Curtis (1996)
methoxychlor	Channel catfish (*in vivo*)	Schlenk *et al.* (1998)
ethinylestradiol	Rainbow trout (*in vivo*)	Jobling *et al.* (1996); Larsson *et al.* (1999); Christiansen *et al.* (1998)
diethylstilbestrol	Rainbow trout (*in vivo*)	Christiansen *et al.* (1998)
bisphenol A	Salmon (*in vivo*)	Arukwe *et al.* (2000)
	Rainbow trout (*in vivo*)	Christiansen *et al.* (1998)
		Lindholst *et al.* (2000)
tert-butyl-hydroxyanisole	Rainbow trout ZR-75	Scholz *et al.* (1997)
(B) Anti-oestrogenic action		
PCBs		Jansen *et al.* (1993); Krishnan and Safe (1993); Safe (1995)
Aroclor 1254	Rainbow trout (*in vivo*)	Chen *et al.* (1986)
	Croakers (*in vivo*)	Thomas (1990)
3,3′,4,4′-TCB	White perch (*in vivo*)	Monosson *et al.* (1994)
TCDD	Rainbow trout (*in vitro*)	Anderson *et al.* (1996a)
PCDF	Rainbow trout (in vitro)	Anderson *et al.* (1996a)
TCDF	Rainbow trout (*in vitro*)	Anderson *et al.* (1996a)
Benzo[a]pyrene	Croakers (*in vivo*)	Thomas (1990); Safe (1995)
3-methylcholanthrene		Safe (1995)
indole[3,3b]carbazole		Safe (1995)

N.B. triCDF = 6-t-butyl-1,3,8-trichlorodibenzofuran

- *Atrazine*: Atrazine is the most frequently detected pesticide in UK drinking water, with 28% of drinking water samples taken in 1990 exceeding the EC limit of 0.1 $\mu g\ l^{-1}$ (ENDS, 1995), which is particularly worrying in view of its possible involvement in breast cancer.
- *Polychlorinated biphenyls (PCBs), polychlorinated dibenzo-p-dioxins (PCDDs) and polychlorinated dibenzofurans (PCDFs)*: PCBs have been used since 1929 in a variety of applications, including as heat transfer fluids in large transformers and as dielectric fluids in capacitors. Though their use has now ceased, they are still present in many older electrical installations. A typical PCB is made up from a mixture of congeners, each having different numbers and positions of the chlorine. PCDDs (or 'dioxins') and PCDFs are different from the other chemicals described here because they are not manufactured deliberately. They can be produced during incineration, paper manufacture, and in the production of chlorinated aromatics such as 2,4,5-trichlorophenol (an intermediate in the manufacture of the herbicide 2,4,5-T). Like PCBs, many congeners of PCDD exist, but one of the most studied is 2,3,7,8-tetrachlorodibenzodioxin (TCDD). It appears that the main oestrogenic effect of PCBs may be due to their hydroxylated metabolites, which are produced when the body attempts to break them down. Additionally, some PCB congeners may be antioestrogenic in fish (Anderson *et al.*, 1996a,b). Those metabolites with a para-hydroxylation on one of the rings are particularly effective at mimicking oestradiol (McKinney & Waller, 1994), though others are also oestrogenic (Soto *et al.*, 1995). Although many PCDDs are known to be toxic and carcinogenic, PCDDs appear to be antioestrogenic (Astroff & Safe, 1990; Safe *et al.*, 1991; Safe, 1995; Safe & Krishnan, 1995).
- *Kraft pulp mill effluents*: β-sitosterol in bleached kraft mill effluent, which can be present at concentrations above 1 $mg\ l^{-1}$, has been shown to alter the reproductive status of fish, possibly through weakly oestrogenic action, although its degradation products are thought to be masculinising (xenoandrogens), and there also seems to be interference with the pituitary-gonadal axis (Howell *et al.*, 1980; Bortone & Drysdale, 1981; Bortone *et al.*, 1989; Drysdale & Bortone, 1989; Howell & Denton, 1989; van der Kraak *et al.*, 1992; Bortone & Davies, 1994; Munkittrick *et al.*, 1994; Gagnon *et al.*, 1995; MacLatchy & van der Kraak, 1995; Mellanen *et al.*, 1996; Tremblay & van der Kraak, 1999). Stigmastanol has also been implicated in these endocrine disrupting effects of pulp mill effluents.
- *Polycyclic aromatic hydrocarbons (PAHs)*: PAHs are naturally occurring hydrocarbons, but are also generated in combustion processes and smelterworks, where they constitute a major pollution problem. Some PAH compounds (e.g. benzo[a]pyrene) are strongly carcinogenic, and many are effective inducers of the CYP1A system through binding to the Ah-receptor. These compounds are also antioestrogenic, possibly through the same receptor cross-talk mechanisms as planar PCBs and PCDDs.

There appears to be little structural similarity among documented oestrogenic and/or antioestrogenic chemicals (Fig. 5.3), although enhanced oestrogenicity of *para*-substituted phenolic and halogenated aromatic hydrocarbons and structural rigidity has been noted (Jordan *et al.*, 1985). Since it is impossible to empirically establish the sensitivity, susceptibility and resistance of every species to each type of compound with regard to disturbances in processes important for successful reproduction, current research approaches have assumed the concept of structure activity relationships (SARs).

The principle of SARs is an assumption that the properties and behaviour of chemicals are directly derived from their molecular structural characteristics. Basically, SARs describe the chemical and/or biological properties of a series of chemicals relative to their molecular structure and/or other physico-chemical properties (Bradbury, 1994; Kaiser, 1997). Two conceptual approaches have been used in studying SARs:

(1) Correlative approaches that relate variation in molecular structure, assessed quantitatively by molecular descriptors, within a congeneric series of compounds to variance in a toxicological property
(2) Pattern recognition approaches that attempt to identify common stereoelectronic characteristics among structures that elicit similar toxicological activity (Bradbury, 1995).

Both approaches require a clear definition of chemicals or biological end-points of concern, and a set of mechanistically-based assumptions regarding the process in question, in addition to the identification of a common mode of action. Despite efforts to construct quantitative structure-activity relationships (QSARs) for oestrogenic substances (see Tattersfield *et al.*, 1997), we are not yet in a position to predict weak oestrogenic action with any confidence.

5.1.2.3 Immediate consequences

As described above, the immediate consequences of xenoestrogen exposure in juvenile or adult male fish, or in females outside the breeding season, are to induce the inappropriate synthesis inter alia of zona radiata protein and vitellogenin. In particular, induction of male vitellogenesis seems to be widespread in the aquatic environment near sewage and industrial discharges, at least in Europe, the USA and Japan (Folmar *et al.*, 1996; Harries *et al.*, 1996, 1997; Lye *et al.*, 1997; Allen *et al.*, 1999a,b; Larsson *et al.*, 1999; Hashimoto *et al.*, 2000). These substances have no function in males, while in females they may already be produced maximally.

The direct consequences of VTG and *Zr*p synthesis in males are poorly understood, but can include reduced calcium in the scales and skeleton, liver hypertrophy, and kidney damage (Herman & Kincaid, 1988). They also represent a substantial waste of energy to the male fish, and thus their production will almost inevitably reduce their reproductive fitness. In females, the effects of xenoestrogen exposure may be less serious, although there have been reports of premature maturation in female flatfish which could be due to oestrogen exposure switching on the inappropriate production of GTH, causing unseasonably early development of the ovary (Johnson *et al.*, 1997). Also, it has been hypothesised that unscheduled *Zr*p synthesis in females may give oocytes with abnormal eggshells, resulting in lower egg quality or problems with hatching and survival (Arukwe & Goksøyr, 1998). Together with these changes, it is to be expected that testicular development will slow down (Fig. 5.4; Jobling *et al.*, 1996; Harries *et al.*, 1997; Christiansen *et al.*, 1998). Furthermore, Nagler & Cyr (1997) have shown that male flatfish exposed to potentially oestrogenic

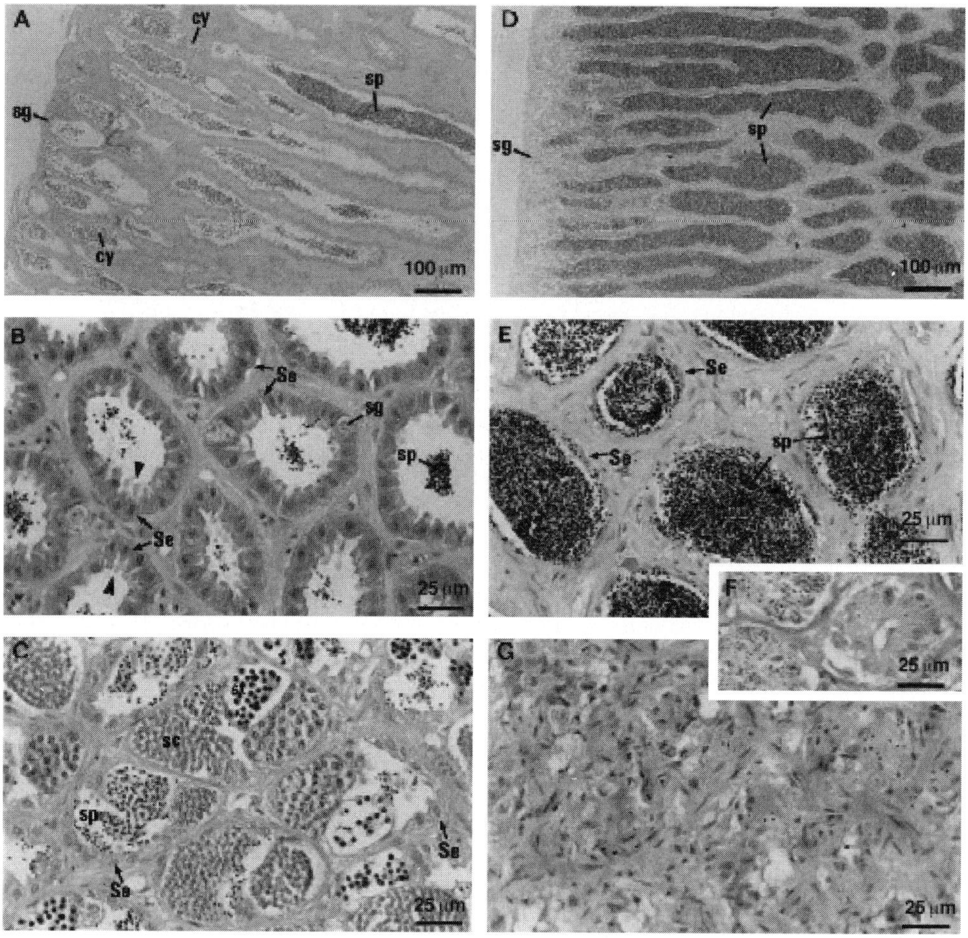

Fig. 5.4 Light micrographs of the testis of male eelpout (*Zoarces viviparous*). The light micrographs taken from an experiment performed in June 1995 and May 1996, show the seminiferous lobules in control fish (A-C) and in nonylphenol- or oestradiol-treated fish (D-G). The seminiferous lobules were filled with spermatozoa (and/or late spermatoza) and the testis showed a reduction in mass (D). The sertoli cells lining the seminiferous lobule walls were very squamous (E). **A:** Longitudinal section (Control, June); **B:** Transverse section (Control, June); **C:** Transverse section (Control, May); **D:** Longitudinal section (nonylphenol 100 $\mu g\ g^{-1}$, June); **E:** Transverse section (nonylphenol 100 $\mu g\ g^{-1}$, May); **F:** Transverse section (nonylphenol 100 $\mu g\ g^{-1}$, May); **G:** Transverse section (estradiol, May). cy: spermatocysts; sp: spermatozoa; sg: spermatogonium; Se: sertoli cell; sc: spermatocytes. Arrowheads in B indicate possible secretory material from sertoli cells. Reprinted from Christiansen *et al.*, 1998 with permission from The Company of Biologists Ltd.

sediments have greatly reduced fertilisation success, i.e. the quality of their sperm has been impaired, perhaps through delayed or abnormal testicular development. There is no data on oestrogenic effects on sperm density, but by analogy with mammals, this effect is to be expected. Secondary sexual characteristics and body markings in males may also not develop properly, leading to abnormal or absent reproductive behaviour (Jones & Reynolds, 1997). This has been observed in cichlid fish in areas where the oestrogen-mimic endosulfan has been sprayed from the air for tsetse fly control (Douthwaite *et al.*, 1981, 1983; Matthiessen & Logan, 1984).

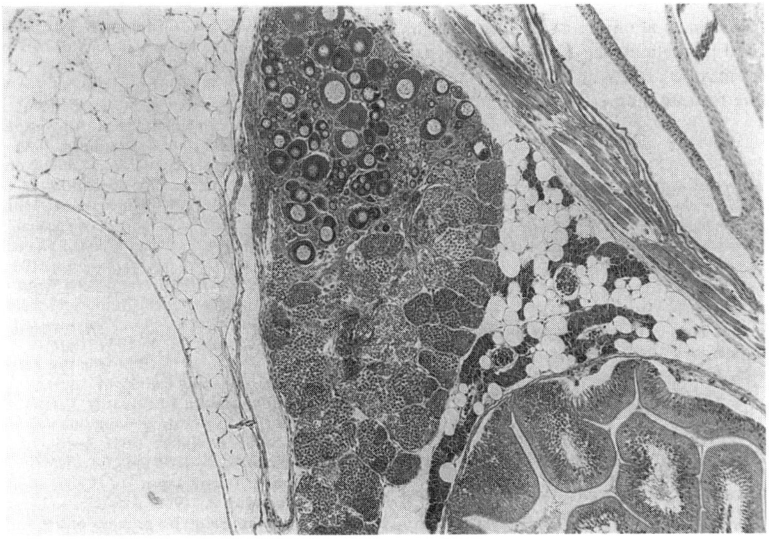

Fig. 5.5 Testis-ova in the gonad of male medaka exposed to 100 mg/l nominal concentration of NP from 1 day posthatch to 3 months. The testicular tissue is in the anterior (right-hand) part of the gonad, and ovarian tissue is in the posterior part of the gonad (40x magnification, H&E staining) (Gray & Metcalfe, 1997). Reprinted with permission from *Environmental Toxicology and Chemistry*, 1997. Copyright Society of Environmental Toxicology and Chemistry (SETAC), Pensacola, Florida, USA.

As shown by Piferrer and Donaldson (1989), exposure of male fish larvae to oestrogenic hormones during the sensitive part of gonadal development can completely feminise the phenotype. This results in apparently normal females although they remain genetically male. It was reported by Jobling *et al.* (1996) that exposure of rainbow trout to four different alkylphenolic chemicals that induced Vtg synthesis, at the same time inhibited testicular growth. In a three-month exposure study using medaka (*Oryzias latipes*) and β-HCH, Wester and Canton (1986) observed the development of testis-ova in males and induced vitellogenesis in either sex, demonstrating oestrogenic effects of this compound.

A similar response was observed when medaka was exposed to NP in a more recent study (Fig. 5.5; Gray & Metcalfe, 1997). In a study using juvenile salmon and di(2-ethylhexyl)phthalate, phenotypic feminisation was observed without concurrent Vtg induction (Norrgren *et al.*, 1999), suggesting that this compound's action may not be via the oestrogen receptor.

Larval carp (*Cyprinus carpio*) exposed to 4-tert-pentylphenol (TPP) and E2 for 60 days showed the development of oviduct in almost all male carp, in addition to reduction in the number of primordial germ cells (PGCs) in fish gonads at day 40 post-exposure (Gimeno *et al.*, 1996).

Most studies of endocrine disrupting chemicals (EDCs) have focused on their activational modifications on adult model systems. Usually, activational effects occur as transitory actions during adulthood (Phoenix *et al.*, 1959). Another event occurring early in an individual's lifetime that induces permanent effects is referred to as an organisational effect and has received little attention. Both organisation and activation have been useful concepts in explaining the role of hormones in the differentiation of vertebrate sexual dimorphism, be it morphological, physiological or behavioural (Arnold & Breedlove, 1985; Guillette *et al.*, 1995).

For example, male specific hormones (androgens) can organise embryos by stimulating the development of the male reproductive duct systems and external genitals in mammals (see section 5.1.3). Androgens can also activate growth and secretary activity of glands associated with the male reproductive tract. Given the pivotal role played by hormones in sexual development and reproduction, it is obvious that the organisational and activational concepts are also central to the role of environmental EDCs.

If the complete environmental impact of EDCs is to be assessed, their organisational and reorganisational effects on embryos should be a major concern. In this respect, it is important to note that many organisational modifications do not become apparent until later in life.

The immediate consequences of exposure of fish to oestrogens and their mimics may thus be profound, altering the overt sexuality of the animal in many ways and thereby damaging its ability to reproduce normally.

5.1.3 Androgenic and antiandrogenic effects

Androgens are a group of hormones or compounds that primarily influence the growth and development of the male reproductive system, while an antiandrogen is a drug or compound that blocks the activity of an androgen hormone. The main androgen hormone in fish is 11-ketotestosterone. The major actions of androgens include:

(1) Regulation of gonadotropin releasing hormone (GnRH) pulse generator
(2) Spermatogenesis
(3) Expression of normal male behavioural patterns (however, in fish it is not clear which role oestradiol and dihydrotestosterone (DHT, the active form of testosterone in mammals) play in this regard)
(4) Normal function of male accessory sex gland
(5) Other non-reproductive effects including immune function, bone metabolism and muscle development (especially in mammals).

A biomarker of androgenic action in fish is being developed under the UK research programme known as EDMAR (Endocrine Disruption in the Marine Environment (see DEFRA, 2002), using the euryhaline stickleback, *Gasterosteus aculeatus*, in which the males build a tubular nest of plant matter which they glue together with a protein (spiggin – named after the Swedish word for stickleback, *spigg*) secreted by the kidneys and stored in the urinary bladder. It is known that the production of spiggin is under the control of the androgenic hormone 11-ketotestosterone (Jakobsson *et al.*, 1999), and can be induced in females through exposure to exogenous androgens, in an analogous way to the induction of vitellogenin in male fish through oestrogen exposure. The effect, along with increased tubular epithelial height, has been reported in 11-ketotestosterone exposed sticklebacks, but its environmental significance has not been established yet (see section 5.1.5). As yet unpublished preliminary data from the UK (P. Matthiessen, pers. comm., 2001) shows that some estuarine sewage discharges contain measurable amounts of natural androgens, but that these have only small effects on spiggin levels in caged female sticklebacks.

5.1.3.1 Mechanisms

The mechanisms of action of environmental androgens are not well understood however; as discussed section 5.1.1 (see Fig. 5.1a,b), chemical compounds that function as androgens, mimic or partly mimic the natural androgens (the male sex hormone) by binding to androgen receptors (AR) and influencing cell signalling pathways. Alternatively, they can block, prevent and alter androgen binding to AR and interfere with cell signalling pathways. Chemicals that block or antagonise androgens are labelled antiandrogens. As with ER, two distinct nuclear AR forms have been identified in fish tissues (Sperry & Thomas, 1999), showing different affinity for xenobiotic antiandrogens.

5.1.3.2 Contaminants

The limited available data on the discharge of androgen agonists in sewage to the aquatic environment suggests that these largely consist of natural androgenic steroids related to testosterone (P. Matthiessen, pers. comm., 2001). However, it is important to note that some masculinising effects in the environment, e.g. TBT-induced imposex in molluscs, are caused most probably by an effect on hormone synthesis (aromatase inhibition) or excretion rather than by direct androgenic action (see below). Furthermore, the masculinising effects of the plant-derived substances (e.g. β-sitosterol) in pulpmill effluents appear to be due to a range of effects on the pituitary-gonadal axis, as well as some oestrogen receptor-mediated action and alterations in cholesterol availability (van der Kraak *et al.*, 1998; Tremblay & van der Kraak, 1999).

5.1.3.3 Immediate consequences

The immediate consequences of fish exposure to environmental androgens and/or antiandrogens are not very well understood. However, several studies have reported the masculinisation of female poeciliids (mosquitofish) from streams recipient of kraft mill effluents (KME) (Howell *et al.*, 1980; Bortone *et al.*, 1989; Drysdale & Bortone, 1989; Davis & Bortone, 1992). The masculinisation features include the development of a modified anal fin (gonopodium) on pregnant females. The development of a gonopodium was concomitant with male mating behaviours, and a hermaphroditic condition (vitellogenic oocytes and cysts of spermatids and spermatozoa). The consequences of masculinisation for individual fish is not known, since masculinisation does not continue after removal from KME and masculinised females (at early stages) gave birth to non-masculinised viable offspring in the aquaria.

Presently, there is no information on the fate of masculinised poeciliid populations subjected to continued exposures to KME androgens (Davis & Bortone, 1992). Male-biased sex ratios of embryos from the viviparous eelpout (*Zoarces viviparus*) have been identified in fish living in the effluent gradient from a large Swedish pulp mill (Fig. 5.6). However, the underlying mechanisms behind the overrepresentation of male embryos remain to be established (Larsson *et al.*, 1999, 2000). Nevertheless, there is a potential danger that the observed masculinisation may represent stages of intersexuality in a progression toward hermaphroditism or sex reversal. The apparent and long-term ecological consequences may

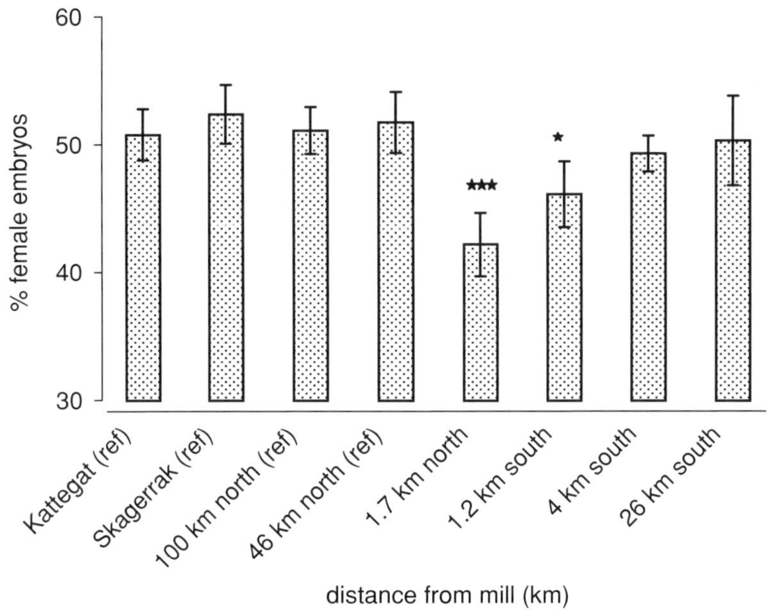

Fig. 5.6 Embryonic sex ratios ± SEM of the viviparous eelpout (*Zoarces viviparus*) sampled in 1998 at four reference sites and four sites near a pulp mill outfall on the Swedish Baltic coast. Asterisks refer to significant differences from the four pooled reference sites ($p = 0.0005$ and 0.042, respectively) (Larsson *et al.*, 2000). Reprinted with permission from *Environmental Toxicology and Chemistry*, 2000. Copyright Society of Environmental Toxicology and Chemistry (SETAC), Pensacola, Florida, USA.

be the eradication of the masculinised populations, since high mortality rates have been observed among masculinised fish in the laboratory (Davis & Bortone, 1992). Masculinisation has been induced experimentally using androgens, and masculinisation of mosquitofish in the laboratory is related to microbially degraded phytosterol components.

5.1.4 Effects on hormone synthesis, metabolism and regulation

5.1.4.1 Mechanisms

As mentioned above, endocrine disruption may occur through several pathways, not only by compounds mimicking natural hormones. Endocrine response pathways are very complex and involve synthesis, release and transport of signalling molecules (e.g. hormone) to target cells and interactions with cellular membrane or intracellular receptors. The formation of hormone receptor complexes results in the induction of a signalling cascade or directly interacts with specific genomic sequences and modulates gene transcription and translation, leading to an altered biochemical or physiological response. Homeostasis in cells is maintained through these complex signalling pathways, and unscheduled modulation of any of these processes can lead to development of endocrine disruption or toxicity at various levels.

The biosynthesis of steroid hormones is catalysed by a series of enzymatic steps involving a number of different steroid hydroxylases, reductases and other enzymes introducing

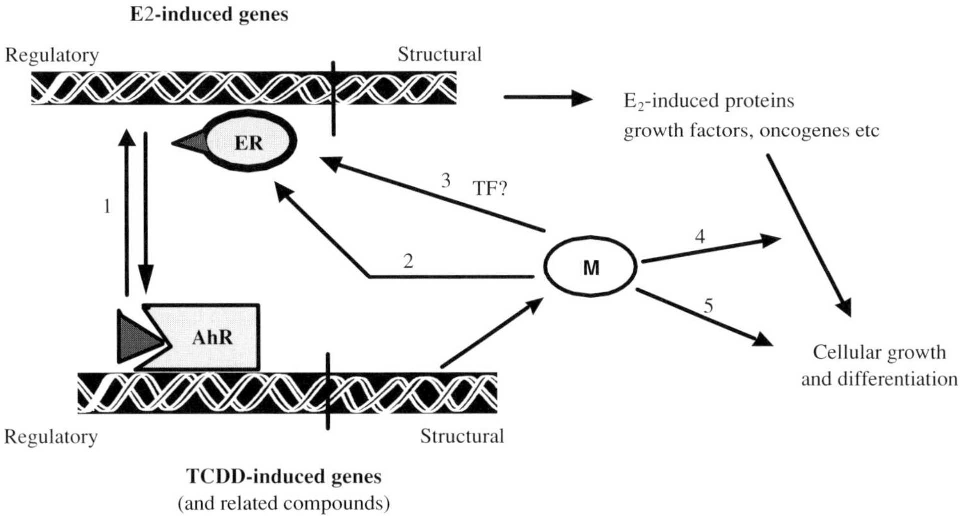

Fig. 5.7 Proposed mechanisms of action of TCDD and related compounds as antioestrogens. The AhR or ER complex may directly inhibit oestrogen- or TCDD-induced genes (1), TCDD or related compound may induce modulatory protein(s) which degrade the nuclear oestrogen receptor (2) directly inhibit oestrogen-induced gene transcription (3), inhibit the action of oestrogen-induced growth factors (4), or exhibit other antimitogenic activities (5). M: modulatory proteins; TF: transcription factors. Modified from Safe *et al.*, 1991.

modifications from the parent cholesterol molecule. These steps take place in several different tissues and cell types that are all possible target cells for an endocrine disrupter.

Several EDCs may act via cross-talk between the aryl hydrocarbon receptor (AhR) and ER (Fig. 5.7). Several presumed antioestrogens (dioxins, polyaromatic hydrocarbons (PAHs), and planar polychlorinated biphenyls, PCBs) are strong AhR agonists that also are capable of inducing the cytochrome P450 1A (CYP1A) system (Chapter 2). In many ER regulated genes there are imperfect response elements for the AhR (also called xenobiotic response elements, XREs, or dioxin response elements, DREs), and studies have shown that the binding of this ligand-AhR complex may suppress oestrogen responses and thereby elicit the so-called antioestrogenic effects (see review by Safe, 1995).

Another type of apparent antioestrogenic action concerns interference with the synthesis of oestradiol from testosterone. This synthetic pathway is mediated by the enzyme aromatase which is a member of the cytochrome P450 (CYP) superfamily. Inhibition of aromatase (CYP19) leads to the build-up of testosterone and the reduced synthesis of oestradiol. Such effects can lead to masculinisation of females. For example, tributyltin causes the masculinisation of the oviduct in neogastropod molluscs known as imposex most probably by inhibiting aromatase, and other aromatase inhibitors can also cause masculinisation in female fish (see Matthiessen & Gibbs, 1998, for a review of this effect).

Since the early 1990s, it has become clear that many of the enzymes involved in the metabolism of steroids as well as other signal compounds are regulated by a series of transcription factors called nuclear receptors, of which the steroid hormone receptors are only members of a larger cast (Mangelsdorf *et al.*, 1995; Mangelsdorf & Evans, 1995). The nuclear receptor superfamily is comprised of over 150 different proteins that share an involvement in mediating a complex array of extracellular signals into transcriptional responses. Many, but not all, of these proteins bind directly to signalling molecules, which, because of their small lipophilic character, can easily enter the target cell. In contrast to membrane-bound receptors, the nuclear receptors are intracellular and function to control the activity of target genes directly. The ligands for these receptors are chemically diverse and include vitamin D, thyroid hormone, retinoids, prostanoids, pregnanes, fatty acids, and possibly a number of other still undiscovered compounds (hence the term 'orphan receptor'). This opens a number of sites for endocrine disrupting contaminants to exert their effects.

While the steroid receptors function as homodimers, most of the other nuclear receptors act as heterodimers with the retinoid X receptor (RXR) (Mangelsdorf & Evans, 1995), binding to specific response elements for each heterodimer pair (hormone response elements, HREs). These receptors include the peroxisome proliferator-activated receptor (PPAR) (Issemann & Green, 1990), the liver expressed receptor (LXR), and the pregnane X receptor (PXR) (Kliewer *et al.*, 1998), which among other genes regulate CYP4 genes, CYP7 genes, and CYP3A genes, respectively, the latter two being especially important in steroid metabolism.

Some of the receptors may be upregulated by the presence of ligand, giving a positive feedback for the hormone mimicking signal. This has been shown in fish for ER (Yadetie *et al.*, 1999), and AhR. More receptor means more binding sites for the activating compound. It is possible that the levels of other nuclear receptors may also be regulated in a similar manner.

Another site of action for EDCs may be the plasma membrane steroid receptors. In Atlantic croaker and spotted sea trout, the maturation-inducing steroid (17,20β,21-trihydroxy4-pregnen-3-one, 20β-S) receptor has been characterised in the ovaries, testes and sperm. Both *o,p′*-DDD and Kepone (chlordecone) have been shown to antagonise 20β-S-induced final maturation of croaker oocytes, indicating that they may be competing with the steroid for binding to the ovarian 20β-S membrane receptor (Ghosh & Thomas, 1995). Later, a number of xenoestrogens (Kepone, *o,p′*-DDE, 2′,4′,6′-PCB-4-OH and the mycotoxin zearalenone) were shown to be able to compete with 20β-S for binding to the croaker sperm membrane 20β-S receptor (Thomas *et al.*, 1998). In addition, Kepone was shown to inhibit sperm motility in a direct assay (Thomas *et al.*, 1998).

The levels of circulating hormones are controlled by releasing factors, e.g. the release of GTH hormones from the pituitary controlled by gonadotropin releasing hormone (GnRH) produced in the hypothalamus. Virtually no studies exist on such effects in fish, although it is highly possible that many endocrine disrupting chemicals may act at this level. Studies have shown that oestrogens such as NP and E2 can induce mRNA levels of GTHII (LH) in the pituitary of female fish (Yadetie & Male, 2002). Similarly, Khan and Thomas (1998) observed increased release of GTHII in plasma of *o,p′*-DDT and E2 treated Atlantic croaker. The mechanism behind this effect is not known. It could either be a direct effect on the gonadotrops in the pituitary or an indirect effect by altering GnRH synthesis and/or release in the anterior hypothalamus. An indication that cyclic AMP pathways were involved, when cadmium caused increases in gonadotropin release in Atlantic croaker, has been observed (Thomas, 1999).

Apparently, many endocrine disrupters, e.g. alkylphenols, may have their effects at multiple sites of the pituitary-gonadal-liver axis. Arukwe *et al.* (1997b) showed that the

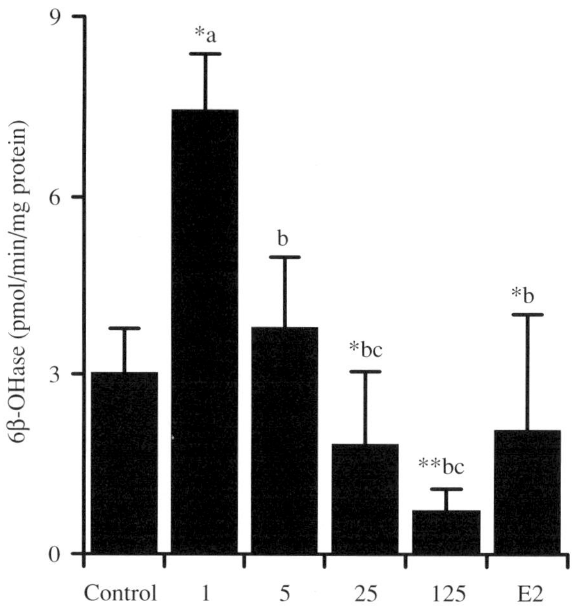

Fig. 5.8 Effects of 4-nonylphenol on live microsomal metabolism of [4-^{14}C] progesterone. The activity of 6β-hydroxylase (Ohase) is expressed as pmol progesterone metabolite formed per mg microsomal protein per minute in liver microsomes of control, 4-nonylphenol- (NP; single *i.p.* injection at 1-, 5-, 25- and 125 mg/kg body weight) and oestradiol–17 – (E_2) treated juvenile salmon. Data are given as mean values ± SD. *Significantly different from control ($p < 0.05$); **Significantly different from control ($p < 0.01$); Different letters indicate significant differences between treatment groups ($p < 0.05$) (Arukwe *et al.*, 1997b). Reprinted with permission from *Environmental Toxicology and Chemistry*, 1997. Copyright Society of Environmental Toxicology and Chemistry (SETAC), Pensacola, Florida, USA.

xenoestrogen 4-nonylphenol at low doses was able to induce some steroid hydroxylases in salmon liver, whereas the same activities were suppressed at higher doses (Fig. 5.8). Accordingly, plasma 17β-oestradiol was lower in the low-dose treated group compared with both control and high-dose treated fish. This biphasic effect suggests that endocrine modulators may have different effects depending on the dose encountered by the target tissue. Another complicating factor is that different effects may be observed depending on the stage of the reproductive cycle (Thomas, 1999).

5.1.4.2 Contaminants

Dioxin and related Ah receptor agonists modulate P450 gene expression and related activities, and this can result in tissue-specific changes in hormone levels. In rats treated with TCDD, adrenal 21-hydroxylase (CYP21A), testicular 17-hydroxylase (CYP17A) and 17,20-lyase activities were decreased. In the same species, testicular steroidogenesis and testosterone synthesis was also decreased, and this was associated with decreased mobilisation of cholesterol as a substrate for P450scc (CYP11A) (reviewed by Safe, 1995). Several reports indicate that the testis is a target for TCDD-induced toxicity, and the effects include alterations in Leydig cell number and function, testicular hypoplasia and impairment of spermatogenesis. At least in part, these effects may result from TCDD-induced decrease of testicular androgen levels (Safe, 1995). TCDD also affects pituitary-adrenal function and alters the effect of adrenocorticotropic hormone, and decreases serum melatonin levels in rats.

Through molecular and cellular transformation studies using gene constructs with specific HREs, reporter assay systems and receptor complexes, ligand activation studies have shown that RXR can be activated by the insecticide methoprene (Harmon *et al.*, 1995). In addition, PXR can be activated by synthetic steroids such as pregnenolone 16a-carbonitrile (Kliewer *et al.*, 1998), as well as some highly-chlorinated non-planar PCBs and pesticides (Schuetz *et al.*, 1998). Whether the same effects can occur in fish has not been studied yet.

5.1.4.3 Immediate consequences

The immediate consequence of effects at this level is a disturbance or modulation of the levels of circulating hormones, or hormone receptors in the cell, and of cellular and physiological processes mediated by these endocrine signals. Whether these disturbances may translate into reproductive toxicity, developmental toxicity or other types of toxicity, will depend on a number of factors, including the dose of contaminant, co-exposure with other interacting compounds, the timing in relation to gonad development and maturation, temperature, and season. Very little is known about how such modulations may transfer into higher level effects, although some studies suggest that they do. These findings, reviewed below, must therefore be considered in the light that the mechanistic relationships remain speculative.

Johnson *et al.* (1988) and Casillas *et al.* (1991) have reported the effects of environmental pollutants on ovarian development in English sole (*Parophrys vetulus*) from Puget Sound, Washington, USA. One significant finding of these authors was that female English sole from sites heavily contaminated with PCBs and PAHs had lower plasma oestradiol levels and were significantly less likely to undergo gonadal recrudescence than females from the less contaminated sites. Collier *et al.* (1998a,b) have also reported precocious juvenile sexual maturation and inhibited gonadal development in female flatfish from the Hylebos Waterway, in central Puget Sound, known to be severely contaminated by a variety of organic and inorganic contaminants. Current work in Puget Sound with male English sole is showing that fish from some contaminated locations are showing vitellogenin induction (Lomax *et al.*, 2001), indicating that multiple mechanisms of endocrine disruption are in operation. This is also the case for male flounder (*Platichthys flesus*) in UK estuaries which as well as showing elevated oestradiol titres (Scott *et al.*, 2000), simultaneously show VTG induction and ovotestis (Allen *et al.*, 1999a,b).

Elsewhere, altered ovarian development in plaice exposed to crude oil as a result of the *Amoco Cadiz* oil spill near Brittany, France, has been reported by Stott *et al.* (1983). Furthermore, the grounding of the tanker *Exxon Valdez* in 1989 that spilled 42 000 000 litres of crude oil into the Prince William Sound in Alaska, has resulted in severe effects on the reproductive success of pink salmon (*Oncorhynchus gorbuscha*) (Wertheimer & Celewycz, 1996) and Pacific herring (*Clupea pallasi*) (Hose *et al.*, 1996; Kocan *et al.*, 1996; Norcross *et al.*, 1996). Parameters used in evaluating Pacific herring reproductive success include egg and larval mortality, morphological deformities, cytogenetic abnormalities and premature hatch. Significant correlations were found between these effects and crude oil exposure.

There are also reports of reduced viable hatch in the Baltic flounder (*Platichthys flesus*) and Baltic herring (*Clupea harengus*) in correlation with elevated PCB concentrations in the eggs (von Westernhagen *et al.*, 1981; Hansen *et al.*, 1985). High egg mortality of Lake Geneva charr (*Salvelinus alpinus*) in correlation with elevated PCB and DDT in charr eggs (Monod, 1985), and reduced fertilisation success and viable hatch in female starry flounder (*Platichthys stellatus*) from contaminated areas of San Francisco Bay have also been reported (Spies & Rice, 1988).

In studies conducted by Dethlefsen, Cameron and co-workers between 1984 and 1995 on developmental disturbances in eggs of dab (*Limanda limanda*), whiting (*Merlangius merlangus*), cod (*Gadus morhua*), flounder and plaice (*Pleuronectes platessa*) in the Southern North Sea, high incidences of embryo malformations were observed in coastal waters known to receive high pollution loads (Cameron *et al.*, 1996; Dethlefsen *et al.*, 1996; Cameron & von Westernhagen, 1997). Developmental defects were evaluated using deviation of life-stage morphology from normal morphological differentiation. Common defects recorded by these authors include blister proliferation in early and late embryos, failure to close the blastopore and deformation of the notochord. However, significant correlations were only found for malformations of dab and concentrations of *p,p′*-DDE residues.

5.1.5 Methodology

At present, the available methodology for detecting endocrine disruption effects of chemicals is limited, and methods for specifically detecting (anti-) androgenic and (anti-) thyroid mediated effects are particularly inadequate (Zacharewski, 1998). The methods available for detecting oestrogenic and antioestrogenic effects are also rather primitive and not standardised internationally (Tattersfield *et al.*, 1997). Figure 5.9 shows a review of existing *in vitro* and *in vivo* screening methods for (anti-) estrogenic and (anti-) androgenic chemicals in wildlife (Ankley *et al.*, 1998), and a more detailed review of *in vitro* techniques for assessing estrogenic substances (Zacharewski, 1997) has been published.

The most relevant *in vitro* techniques for fish are those which observe the production of vitellogenin or zona radiata protein by fish hepatocytes (e.g. Jobling & Sumpter, 1993; Celius *et al.*, 1999). This method may possibly be used for detecting both oestrogen receptor (ER) agonists and antagonists, the latter by measuring decreased responsiveness to E_2. However, other *in vitro* methods such as oestrogen receptor binding assays (e.g. Thomas & Smith, 1993) and the recombinant yeast screen (YES) (e.g. Routledge & Sumpter, 1996) have also been widely used. Other *in vitro* receptor systems, e.g. utilising the androgen receptor or steroid membrane receptors, may be anticipated in the near future to cover other aspects of receptor-mediated endocrine disruption.

In general, *in vitro* models are ideal test systems for studying the actual mechanisms of endocrine disrupting chemicals, but they have clear limitations and may be misleading as predictive tools in risk assessment (Ankley *et al.*, 1998). These systems lack generally the metabolic capacity of the whole organ or *in vivo* situation, making it difficult to detect effects when metabolites are more active than the parent compound. In addition, the aspect of bioaccumulation may be lost in these generally short-term tests. It is therefore widely

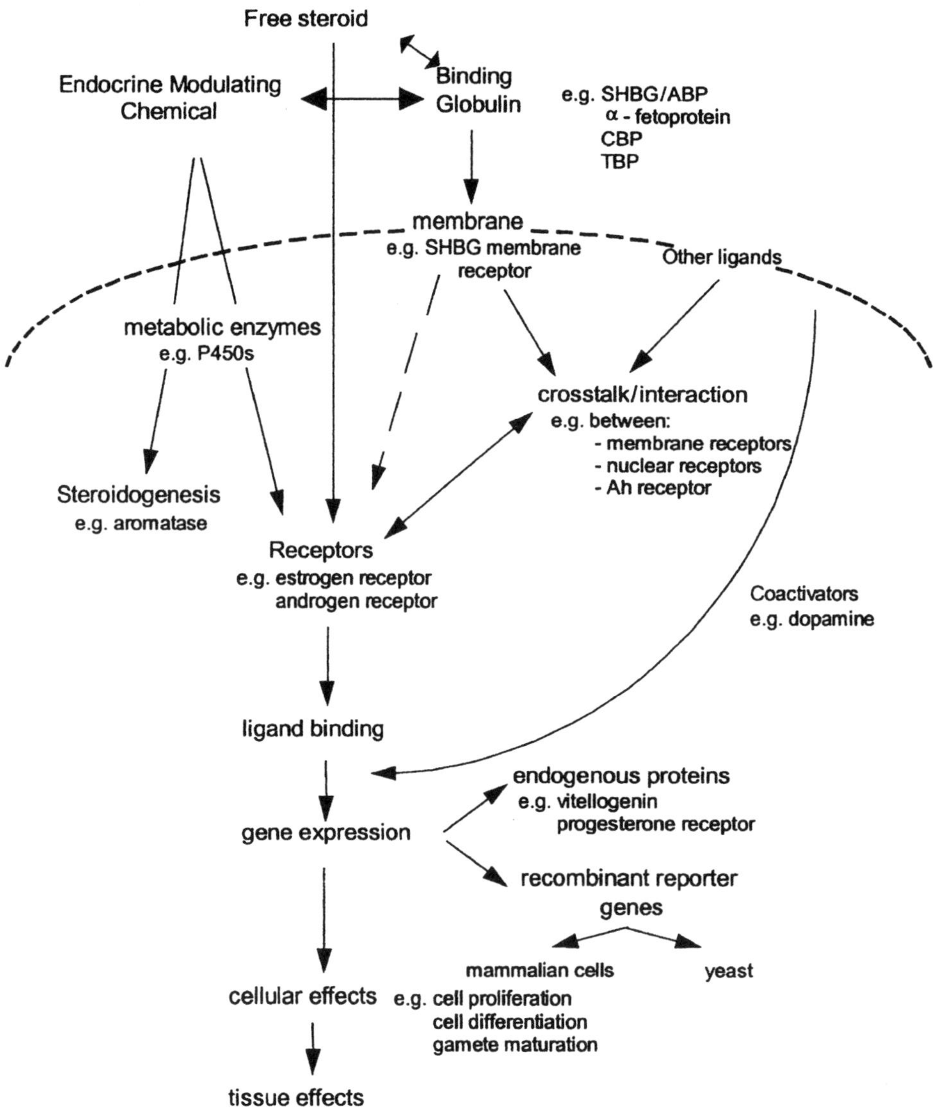

Fig. 5.9 Potential targets and mechanisms of action of endocrine modulators. Endocrine modulators may elicit adverse effects through a number of different mechanisms such as interaction with binding globulins, inhibition of steroidogenic enzymes and/or binding receptors. Crosstalk occurs when mechanisms interact to elicit unique responses that may modulate endogenous expression. The figure illustrates possible *in vivo* and *in vitro* assays that can be developed (or are developed) to identify and assess endocrine modulators, provided that the mechanism of action is known (Zacharewski *et al*., 1997). Reprinted with permission from *Environmental Toxicology and Chemistry*, 1997. Copyright Society of Environmental Toxicology and Chemistry (SETAC), Pensacola, Florida, USA.

recognised that *in vivo* data should be used whenever possible. One approach is to use an all-male strain of fish like that employed by Gimeno *et al.* (1996), and expose them during gametogenesis, using feminisation of the gonoduct and testis as a quick end-point. Normal strains of fish can also be used (e.g. Gray & Metcalfe, 1997), but then they have to be reared to adulthood before effects on sexual development can be assessed (because the sex of each individual is not known in advance).

Measurement of vitellogenesis or zona radiata protein synthesis in juvenile or adult males (or in juvenile females) is a simple way of determining whether a substance or effluent to which they are exposed is oestrogenic. However, not enough is yet known about the relative sensitivity of VTG or *Zr*p induction and more overtly reproductive end-points. The preferred methods for measuring VTG and *Zr*p induction involve analysis of plasma samples by immunoassays such as ELISA, RIA or western blotting with antibodies to VTG or *Zr*p. Since VTG structure differs among species it may be necessary to develop species-specific antibodies, although studies have shown that it is possible to develop antibodies that can be used for a number of species while maintaining acceptable sensitivity (e.g. Heppell *et al.*, 1995; Tyler *et al.*, 1996, 1999a; Nilsen *et al.*, 1998). VTG and *Zr*p induction assays need to be developed into standardised test systems with interlaboratory calibration and ring testing (OECD, 1997). Monoclonal and polyclonal antibodies for fish VTG have been developed for this purpose (e.g. Tyler *et al.*, 1996, 1999a; Nilsen *et al.*, 1998).

A number of international fora (e.g. the EMWAT Workshop – Tattersfield *et al.*, 1997) have made it clear that, due to the complexity of the endocrine system, procedures employed for screening chemicals for endocrine disrupting properties must use *in vivo* tests if they are to avoid high rates of false positives and false negatives. Furthermore, it is inconceivable that risk assessments of EDCs could be conducted solely with *in vitro* data, due to the need to account for route of exposure, rate of uptake, transfer to the site of toxic action, metabolism and excretion. This does not, of course, mean that *in vitro* tests have no part to play in hazard assessment schemes for EDCs, but at the present state of knowledge they will have to take a secondary role, being used primarily to assist interpretation of modes of action, etc.

At present, there are no internationally agreed *in vivo* fish tests which are suitable for assessing the hazards of EDCs, although several (e.g. OECD, 1992, 1999) have end-points such as hatching success and growth rate which could well be interfered with via endocrine processes (but also by non-endocrine routes). The Organisation of Economic Co-operation and Development (OECD) has recognised this gap, and it convened an OECD Expert Consultation on the issue in October 1998 (OECD, 1999). This meeting accepted the need for both fish-based screening procedures (which would provide information about a chemical's basic endocrine disrupting properties), and fish testing proper, the latter being suitable for environmental risk assessments. In the first case, the objective is to inform priorities for testing and to throw light on mechanisms of action, whereas the latter procedures will fully characterise the effects to be expected under realistic exposure scenarios.

The Expert Consultation concentrated on three major types of endocrine disruption: oestrogenic/anti-oestrogenic effects, androgenic/anti-androgenic effects, and thyroid hormone interference, although in practice most progress was made with the first two categories. The meeting agreed on a tiered fish testing approach for these EDCs (Table 5.2), as follows.

For Tier 1, the Expert Consultation identified four possible candidates:

- An enhancement of the 3 week OECD TG204 (Prolonged fish toxicity test) or TG215 (Fish juvenile growth test) to include such end-points as sexual differentiation, vitellogenin

Table 5.2 Tiered fish testing approach for endocrine disrupting chemicals (EDCs) (OECD, 1999).

Screening Tier 1	Test(s) up to 8 weeks in duration (but preferably <3 weeks), aimed at identifying EDC activity *in vivo*, setting priorities for subsequent testing, and investigating mechanisms of action
Testing Tier 2	Extended developmental and reproductive end-points aimed at identification and characterisation of endocrine disrupting effects. These data could be usd for ecological risk assessment
Testing Tier 3	A confirmatory test based on full life-cycle end-points, which could also be used for ecological risk assessment

induction, measurement of sex steroid titres, gonadosomatic index (GSI), gonad morphology and secondary sexual characteristics
- A 3 week fish gonadal recrudescence assay developed in the USA (EDSTAC, 1998) which uses adult fathead minnows (*Pimephales promelas*) in the sexually regressed state, and measures gonadal maturation and related end-points (GSI, secondary sexual characteristics, plasma steroids and vitellogenin)
- A 3 week adult fish terminal reproductive assay developed by the USEPA NHEERL at Duluth, which also involves adult fathead minnows, and measures survival and growth, vitellogenin induction, sex steroids, gonadal pathology and GSI, secondary sexual characteristics, gamete production, and embryo viability
- An 8 week fish sex reversal assay being developed by Sumitomo under the SPEED programme (SPEED, 1998) which uses the medaka (*Oryzias latipes*), exposing newly hatched fry for 2–4 weeks and observing gonad morphology and secondary sexual characteristics after a further 4 weeks.

For Tier 2, the meeting recommended two possible candidates:

- An enhancement of OECD TG210 (fish early life-stage toxicity test), to include exposure during sexual differentiation and early gametogenesis, histopathology at end of sexual differentiation, sex ratio, fecundity of resulting adults (possibly), and biomarkers such as vitellogenin and sex steroid titres
- A partial life cycle, terminal reproductive test under development by Schering under the EMSG/ECETOC programme, which starts with sexually mature adult fathead minnows exposed for 4 weeks, and measures time to first spawning, spawning frequency, number of eggs per batch, number of eggs per female, number of fertilised eggs, hatching success of the F1 generation, and various biomarkers.

It is possible that either test may be needed for application in different situations, depending on existing knowledge about modes of action, etc.

For Tier 3, the meeting recommended development of a single candidate, the USEPA fish whole life cycle test (USEPA, undated) which uses fathead minnows; end-points recommended included embryo hatching and viability (F0 and F1), larval survival, growth and development (F0 and F1), time to maturity (F0), secondary sexual characteristics (F0), sex ratio (F0 and F1), egg production (F0), spawning frequency (F0), fertilisation success

(F0 and F1), gonad histopathology (F0 and F1), gamete maturation (F0) and various biomarkers including vitellogenin, steroids, and steroid metabolic enzymes (F0 and F1).

The main problem with the latter test, however, is that it is extremely difficult to bring it to a successful conclusion, and hence the costs are very high.

More recently, the OECD Validation Management Group, which is developing guidelines for ecotoxicological testing of endocrine disrupters (VMG-eco), has decided to validate a 2-week screening procedure which can be used with a variety of fish species, and which focuses on the measurement of VTG, gonad histology, and gonadosomatic index and other changes in gross morphology such as secondary sexual characteristics, as primary end-points (OECD, 2001).

Overall, the OECD proposals appear sensible, although it is clear that much research is still required before a definitive set of test guidelines can be fully validated. In particular, it will be essential to devise tests which are capable of linking apical end-points such as reproductive success, with biomarkers of endocrine disruption (e.g. vitellogenin induction in males), so that the integrated output of such tests is both diagnostic of causes or mechanisms, and capable of providing information in support of environmental risk assessments.

The overwhelming majority of research which has been conducted on endocrine disruption in fish has concerned the effects of oestrogens on reproduction (Matthiessen & Sumpter, 1998). Almost nothing is known about other types of receptor-mediated effects in fish, one of which is the agonistic action of androgens and their mimics.

However, as noted earlier, in an attempt to redress this situation, a biomarker of androgenic action in fish has been developed under the UK research programme known as EDMAR (Endocrine Disruption in the Marine Environment) (DEFRA, 2002). The males of the euryhaline stickleback (*Gasterosteus aculeatus*), produce a glue protein and it was suspected that it could be induced in females.

This predicted phenomenon has now been demonstrated in the laboratory (Katsiadaki *et al.*, 2000). It has been shown that exposure for 2 weeks of both male and female stickleback to methyltestosterone (3–500 $\mu g\ l^{-1}$) or 11-ketotestosterone (10–20 $\mu g\ l^{-1}$) in the ambient water produces a large increase in the epithelial cell height of the tubules in the secondary proximal kidney segment. This hypertrophy results from the interaction of androgenic hormone with the androgen receptor, which in turn triggers the spiggin gene to produce the mRNA which controls spiggin production. Kidney tubule hypertrophy itself may be useful as a biomarker of androgen exposure, but it is thought that direct measurement of spiggin is more sensitive and quicker. Katsiadaki *et al.* (2000) have therefore isolated spiggin from stickleback nests and bladders, purified it using SDS-PAGE (to give a single band of 203 kDa) (Jakobsson *et al.*, 1999), used the pure protein to induce antibody in rabbits, and finally developed an ELISA assay with this antibody. More recently, spiggin induction in caged sticklebacks has been used in the UK to survey for the presence of androgens in estuaries, but little activity has been detected to date (P. Matthiessen, pers. comm., 2001).

The spiggin ELISA is still being tested in the laboratory, but early results indicate that this assay will be a useful androgen biomarker in female stickleback kidneys. As with the vitellogenin assay, it has obvious value as a biomarker of EDC exposure, but it remains to be seen whether its induction in females also has adverse effects on their physiology, or whether it is associated with undesirable reproductive impacts such as masculinised behaviour or intersex ovaries.

The final test of whether a substance can interfere with reproduction in fish is to deploy an early life-stage test (e.g. OECD, 1992) or life cycle test (USEPA, 1986), but these are very expensive and are not necessarily sensitive to or diagnostic of oestrogens or other endocrine disrupters as they stand. However, Tyler *et al.* (1999a) presented an *in vivo* testing system for early life-stages of fathead minnow, based on the carp Vtg ELISA. The study showed that fathead minnow are sensitive to oestrogens and are able to synthesise Vtg very early in development.

5.2 Other types of reproductive interferences

5.2.1 Protein/membrane damage in gonads

Studies by Lowe and Pipe (1986, 1987) demonstrated and quantified germ cell damage in mussels, and were able to show an increase in degenerating gametes in response to experimental exposure to diesel oil emulsions. Similarly, Widdows *et al.* (1982) in a multidisciplinary study of contaminant impacts on mussels also demonstrated a significant increase in degenerating gametes in mussels exposed to the water accommodated fraction of North Sea crude oil. In a study of metal ion distribution in mussels, Lowe and Moore (1979) showed that excess zinc is excreted in the eggs of mussels; however, in males it is cleared via the kidneys. Whilst this observation does not specifically demonstrate damage, excess zinc is known to be toxic and therefore the prognosis for zinc laden eggs is not good.

Chlorinated hydrocarbons have been shown to accumulate in ovarian tissues of whiting and exert a negative effect on embryo development as well as on the production of normal early life-stages (von Westernhagen *et al.*, 1989). The ovarian cycle was also shown to be negatively affected in flounder (*Platichthys flesus*) exposed to pollutants (Hansen *et al.*, 1985). Dethlefsen (1977) demonstrated that DDT had a direct effect on the developing eggs of cod. Also investigating the consequences of contaminants exposure in Swedish cod, Swedmark and Granmo (1981) demonstrated direct correlations between increased mortality of eggs and larvae, reduced hatching frequency and larval viability and abnormal development and contaminant concentrations. Similarly, viable hatch was significantly reduced in Baltic herring where ovarian levels of PCBs and DDT were greater than 120 and 18 mg kg^{-1} respectively (Hansen *et al.*, 1985).

Studies in the USA have investigated the consequences for reproductive success and larval viability of parental exposure to contaminants in striped bass and starry flounder (Whipple *et al.*, 1981; Westin *et al.*, 1985). The studies all showed significant negative correlations between the determinands of effect and concentrations of contaminants in eggs. By contrast, studies by Couillard *et al.* (1997) on migrating eels (*Anguilla rostrata*) in the St Lawrence River, Canada, were unable to detect any significant impact of contaminant exposure on reproductive processes in terms of rate of maturation or oocyte size. Ionising radiation has also been shown to cause damage to reproductive tissues of the oyster *Crassostrea gigas* (Mix & Sparks, 1971). In conclusion, contaminant exposure has been shown to damage ovarian tissues in a range of finfish and shellfish species resulting in enhanced egg mortality and reduced hatching frequency. Whilst not specifically stated, much of the damage associated with eggs undoubtedly relates to the fact that they are extremely rich in lipids and many organic contaminants are lipophilic and will therefore concentrate in the eggs. The yolk granules within eggs are a specialised form of lysosome, therefore the probability is that lysosomal damage will also occur as a consequence of the contaminants in the eggs which will affect their development.

5.2.2 Spermatotoxic effects

Whilst an extensive literature is available on the impact of contaminants on oocyte development and eggs, the impact on the development and viability of male germ cells has attracted much less attention (however, see section 5.1.4.1). Furthermore, many of the studies that have been undertaken on sperm relate to cryopreservation procedures and their applications for aquaculture. Notwithstanding this situation, studies by Kiceniuk and Khan (1987) demonstrated that the rate of gametogenesis was slower in male cod exposed to oil fractions during summer-autumn as compared to a control group; furthermore, spermiation was delayed in cod treated during winter-spring. Studies on American plaice (*Hippoglossides platessoides*), demonstrated a significant reduction (50%) in the number of larvae hatched from uncontaminated eggs fertilised by males that had been exposed to contaminated sediments (Cyr & Nagler, 1996). In the eelpout (*Zoarces vivparus*), nonylphenol exposure resulted in significant degeneration of testis lobules (Christiansen *et al.*, 1998). van Look and Kime (1999) reported that of several heavy metals tested, mercury was the most toxic to goldfish and rainbow trout sperm motility, studied by computer-assisted sperm analysis. However, it may be of some significance for the sensitivity of male germinal tissues to contaminant damage, that studies by von Westernhagen *et al.* (1989) were unable to show any correlation between contamination of testes tissues in North Sea whiting and total or viable hatch. In conclusion, studies on sperm are very limited as compared to studies undertaken on eggs, so the true cost of the effects of contaminants exposure to the success of reproductive processes is difficult to define. The indication from studies that have been undertaken is that, like eggs, the effect is detrimental and results in reduced hatching success even when the crosses involved non-contaminant exposed eggs.

5.2.3 Effects of peroxisome proliferators on reproduction

Some peroxisome proliferators such as phthalate esters produce adverse effects on reproduction and development of sensitive species (Treinen *et al.*, 1990; IPCS, 1992; Grasso *et al.*, 1993; Laskey & Berman, 1993; Davis *et al.*, 1994a,b; Eagon *et al.*, 1994; Wine *et al.*, 1997). Phthalate esters cause testicular atrophy in rodents and one of their targets are Sertoli cells, at least *in vitro*. They also act as germ cell toxicants; for instance, MEHP causes changes in testicular germ cell apoptosis, thus affecting the normal balance between germ cell proliferation and apoptosis in the seminiferous epithelium (Roberts *et al.*, 1997).

Phthalate esters are weakly oestrogenic (Jobling *et al.*, 1995), possibly through their interaction with steroidogenesis. Thus, these peroxisome proliferators increase release of oestradiol from rat Leydig cells and increase aromatase expression (Liu *et al.*, 1996a,b), increase expression of steroid hormone receptors (androgen and oestrogen) and reduce sex steroid metabolism (Eagon *et al.*, 1994, 1996). The finding that the PPAR:RXR heterodimer competes with ER for the ERE (Keller *et al.*, 1995) constitutes a possible explanation of these effects.

In goldfish ovarian follicles, clofibrate decreased hCG-stimulated production of testosterone (Mercure & van der Kraak, 1995), possibly indicating a role of ovarian peroxisomes in steroidogenic processes, perhaps supplying cholesterol to steroidogenic enzymes.

5.3 Higher level consequences of reproductive damage

5.3.1 Altered sex ratios

It is clear from section 5.1 that one of the main consequences of exposure to oestrogens (and probably also to androgens) during the sensitive part of gonadogenesis (about 10 days either side of hatching in some species) (Sumpter, 1995) is that genotypic males may be completely feminised, developing apparently normal ovaries and oviducts and reproducing normally when adult (although with all-male offspring). Exposure of juvenile or adult males cannot cause these effects as far as we know. The effects of antioestrogens are less well-known. The all-embracing feminisation caused by oestrogen exposure of larvae is impossible to detect without sophisticated genetic analysis, but it implies that the sex ratios of adult populations may become skewed (at least in the short term) towards females.

The implications of skewed sex ratios for the reproductive capabilities and recruitment capacity of a fish population might be profound. However, such effects have rarely been observed in wild fish populations and are very difficult to distinguish from effects caused by a range of natural stressors, including climate and fishing pressure (e.g. Lang *et al.*, 1995). Differential mortality of one sex can be very misleading. Studies of sex ratios among embryos of the viviparous eelpout (*Zoarces viviparus*) appear, however, to be a practical model system. In this species, the preceding mortality is small and known, and problems with sex differences in growth or behaviour that often affect estimations of adult sex ratios are avoided. The normal sex ratio of the species was close to 50% females at four reference sites along the Swedish coast. Near a large pulp mill there were significantly more males in the broods (58%), again approaching 50% females further from the effluent tube (Larsson *et al.*, 1999, 2000). Whether effects of oestrogenic substances on fish sex ratios are more widespread than currently suspected remains to be seen.

5.3.2 Intersex

A much more common condition caused by early exposure of fish larvae to oestrogenic substances is intersex, which in males usually takes the form of ovotestis. The subject has been reviewed by Bortone and Davis (1994), particularly with respect to the masculinisation of females caused by pulpmill effluents. Ovotestis is a partial feminisation in which oocytes may appear in otherwise normal testes. More rarely, one entire testis becomes an ovary, and the other remains fully male. Sometimes, fully developed eggs are formed, but the organ can still be recognised as a testis. However, little is known about the implications of this condition for reproductive functionality. Ovotestis can be induced in the laboratory by exposing fish larvae to weak oestrogens like nonylphenol (Gray & Metcalfe, 1997), and has also been observed at prevalences ranging from 20% to 100% in wild fish populations exposed to oestrogenic effluents (Jobling *et al.*, 1998; Allen *et al.*, 1999a). However, caution must be exercised when making field observations because some fish species undergo natural sex reversal at certain stages of their life history. Pathogens can also induce intersex conditions. Among roach (*Rutilus rutilus*) in the Baltic Sea, individuals with both testicular and ovarian tissue can be found. The observed changes were not linked to a specific pollutant source, but were probably caused by a microsporidian parasite (*Pleistophora mirandellae*).

5.3.3 Life cycle strategies

Since each fish species lives under a specific set of ecological conditions, it has a specific strategy (note that the use of the term 'strategy' in this regard does not imply that it is regarded as a conscious decision by the animal), with special anatomical, behavioural, physiological and energetic adaptations for reproduction (Moyle & Cech, 1988). The reproductive strategies of fishes are often reflected in the differences between the sexes. The onset of sexual maturity represents a critical transition in the life of any fish individual. Before, the allocation of time and resources is related predominantly to growth and survival. After, there is a potential conflict between the allocation of time and resources to reproduction or to survival and growth.

The reproductive cycles of fishes are closely tied to environmental changes, particularly seasonal changes in light and temperature. These two factors are often most important because they can act, directly or through sense organs, on the glands that produce hormones, which in turn produce the appropriate physiological or behavioural responses. Thus, fish (and indeed all animals), have adopted different life history strategies as a means of solving the problem of successful reproduction in a fluctuating environment (Thorpe, 1994, 1989).

Fish species are products of several hundred million years of evolution, and as such have adopted several different life cycle strategies. The maturation decision is annual (in most temperate species) and depends on some genetically determined performance threshold, and the maturation processes will continue if this performance exceeds a set point at this critical time. The maturation decision is based on some critical decisions:

(1) At what age?
(2) Where and how often?
(3) How much surplus energy to invest?
(4) How many eggs and how large?
(5) Guard them, or leave them?

These questions involve the concept of iteroparity, 'bet-hedging' thereby decreasing the risk of reproduction in an uncertain environment; and semelparity, producing all the offspring at the same time. The adaptation of iteroparity and semelparity assumes that juvenile mortality is high and low, respectively.

It can be speculated that since life cycle strategies are products of several million years of evolution, xenobiotic-induced reproductive disturbances of individual fish in a given population might affect life cycle strategies and result in serious ecological consequences in the longer term. However, no knowledge of such effects exists in fish.

5.3.4 Reduced recruitment

Reduction in fecundity may be caused by several mechanisms, such as impaired gonadal development, reduced spawning ability, and reduction in egg number and egg weight, but also by high mortality of early life history stages (eggs, larvae, juveniles) (section 5.1.4.3). In species with size-dependent fecundity, such as many fish and invertebrates, changes in growth rate due to xenobiotics may also affect total egg production. However, despite this multitude of mechanisms, the ample evidence for reduced individual fecundity and the conceptual logic of predictions on population effects, causal links between xenobiotic damage of individuals and population recruitment are difficult to demonstrate, mainly because unknown density-dependent factors may compensate for losses in early life history stages. For example, despite a reduction of total egg production of English sole (*Pleuronectes vetulus*) in polluted areas of the Puget Sound by about 30% (Collier *et al.*, 1998a,b), population models suggest that declines in individual fecundity would decrease the population growth rate only if density-dependent mortality is weak or moderate (Landahl *et al.*, 1997). Indeed, density-dependent effects may even sustain a constant population growth rate despite 60% acute mortality, as suggested by life-table experiments on pea aphids (*Acyrthosiphon pisum*) (Walthall & Stark, 1997). The investigation of density-dependent factors in recruitment dynamics is thus probably the most important task to predict population level effects of pollutants from histological or other biomarkers (Boreman, 1997).

Density-dependent mechanisms maintaining abundance by compensating for reduction in recruitment may provide enough time for a population to adapt to pollutants. Such adaptations have been shown in several aquatic species, and may significantly alter the susceptibility to xenobiotics. In a laboratory population of *Chironomus riparius* (Diptera), for example, LC_{50} values were 13–250 times lower than in a field population, suggesting different selection pressures in the laboratory and the field (Hoffman & Fisher, 1994). In mosquitofish, genotypes tolerant to heavy metal pollution identified by allozyme electrophoresis were more common in polluted environments, demonstrating the selective advantage of such genotypes at the population level (Guttman, 1994; Newman & Jagoe, 1998).

Despite the lack of conclusive evidence of causal links between toxicological effects on individuals and the response of populations, there is ample evidence of reduced abundance in wild populations most probably caused by pollution-related reductions in recruitment. Such effects were observed in, for example, brown trout (*Salmo trutta*) (Kubecka & Matena, 1991), stoneloach, bullhead (*Cottus cottus*) and minnow (*Phoxinus phoxinus*) (Bagge & Hakari, 1992), salmon (Hesthagen *et al.*, 1995), striped bass and American shad (Weisberg *et al.*, 1996), and whole fish communities have suffered from the effects of pollution (Lyons *et al.*, 1998). Reduction in abundances of more sensitive species has been used in species diversity indices to estimate the effect of pollutants on fish communities (Paller *et al.*, 1996). It is thus neither the effect of pollution on individual fish, nor their consequences for entire populations or communities that need investigation, but the links between the two, possibly leading to biomarkers for imminent population collapse.

The effects of pollution at the population level are comparable in magnitude to the effects of fishing pressure (Landahl *et al.*, 1997), and considerable research has been done attempting to apply well-developed fishery models to predict the population effects of pollution (Griswold, 1997). However, most models are extremely sensitive to changes in survival estimates of eggs and larvae, which are often difficult to estimate due to complex spatial and temporal dynamics of ichthyoplankton and to sampling problems (Horst, 1977), but may be more affected by pollution than adult fish (Rose *et al.*, 1993). Furthermore, while most models are capable of incorporating density-dependent factors, in practice they have not (Boreman, 1997), and may thus be of limited value to predict long-term effects of reduced recruitment to wild populations. They are, however, useful in identifying important factors and the kind of data needed for improved analysis.

The importance of density-dependent factors for population responses to pollution outlined above points to the necessity to consider whole ecosystems and the ecological relationships within ecosystems, and several ecologists have suggested that the 'single species' approach is inadequate in ecotoxicology (Cairns, 1983; Kareiva *et al.*, 1996). Indeed, species interactions may produce surprising outcomes when several species are exposed to a pollutant. For example, a phytoplankton-*Daphnia* system exposed to malathion, a toxicant inhibiting *Daphnia* growth, achieved higher *Daphnia* densities than the control system, probably because the slowed growth of *Daphnia* caused an increase in phytoplankton densities, which subsequently more than outweighed the inhibitory effect of malathion on the *Daphnia* population (Taub *et al.* (1988) cited in deAngelis, 1996).

Clearly, this difference between treatment and control does not reflect the ecological equilibrium, but it does demonstrate that short-term population responses may sometimes be unexpected. Even long-term effects may be difficult to predict: another phytoplankton-*Daphnia* system exposed to cadmium did not show the expected reduction in *Daphnia* biomass, but instead their complete elimination, probably caused by the extreme density of phytoplankton inhibiting *Daphnia* growth (Borgmann *et al.*, 1989). The authors note that in the field the elimination of *Daphnia* would probably have led to the establishment of another zooplankton species, which would thus appear to benefit from pollution. Indeed, both empirical and theoretical studies have suggested that such indirect effects of pollution may be as important as direct toxic effects on populations (Talmage & Walton, 1991; deAngelis, 1996). Unfortunately, such indirect effects are not easy to predict, and computer simulations suggest that often even the direction of change in abundance (increase or decrease) after pollution exposure may be unexpected (Yodzis, 1988). It is thus apparent that much basic ecological research is needed, including mesocosm studies and computer modelling (deAngelis, 1996).

5.3.5 Reproductive behaviour

The effects of pollution on the reproductive behaviour of fish have been reviewed by Jones and Reynolds (1997). While there is comparatively little work on behavioural changes due to pollution, there is increasing interest in the issue, both because of the possibility of using behaviour as a biomarker for sublethal doses of xenobiotics, and because of effects on the affected species themselves. Most studies focused on either male courtship or parental care, which may be closely related to reproductive output, though the link has only been demonstrated in a few studies where reduction in parental care caused death of offspring (Ryabov, 1985; Breitburg, 1992; Lorenz & Taylor, 1992). The masculinised females of mosquitofish exposed to KME (see section 5.1.3), displayed clear male sexual behaviour when placed in aquaria with either males or females (Howell *et al.*, 1980). In a study with male goldfish (*Carassius auratus*) exposed to E_2 in physiological concentrations via food or water, sexual behaviour was almost totally inhibited (Bjerselius *et al.*, 1999). A test of reproductive performance of fathead minnow is currently being developed at Brunel University (Tyler *et al.*, 1999b).

5.4 References

Ahlborg, U.G., L. Lipworth, L. Titus-Ernstoff, C.-C. Hsieh, A. Hanberg, J. Baron, D. Trichopoulos & H.-O. Adami (1995) Organochlorine compounds in relation to breast cancer, endometrial cancer, and endometriosis: An assessment of the biological and epidomiological evidence. *Critical Reviews in Toxicology*, **25**, 463–531.

Allen, Y., A.P. Scott, P. Matthiessen, S. Haworth, J.E. Thain & S. Feist (1999a) Survey of oestrogenic activity in United Kingdom estuarine and coastal waters and its effects on gonadal development of the flounder *Platichthys flesus*. *Environmental Toxicology and Chemistry*, **18** (8), 1791–1800.

Allen, Y., P. Matthiessen, A.P. Scott, S. Haworth, S. Feist, & J.E. Thain (1999b) The extent of oestrogenic contamination in the UK marine environment – further surveys of flounder. *Science of the Total Environment* **233**, 5–20.

Anderson, M.J., M.R. Miller & D.E. Hinton (1996a) *In vitro* modulation of 17-β-estradiol-induced vitellogenin synthesis: Effects of cytochrome P4501A1 inducing compounds on rainbow trout (*Oncorhynchus mykiss*) liver cells. *Aquatic Toxicology*, **34** (4), 327–350.

Anderson, M.K., H. Olsen, F. Matsumura & D.E. Hinton (1996b) *In vivo* modulation of 17β-estradiol-induced vitellogenin synthesis and estrogen receptor in rainbow trout (*Oncorhynchus mykiss*) liver cells by b-naphthoflavone. *Toxicology and Applied Pharmacology*, **137**, 210–218.

Ankley, G., E. Mihaich, R. Stahl, D. Tillitt, T. Colborn, S. McMAster, R. Miller, J. Bantle, P. Campbell, N. Denslow, R. Dickerson, L. Folmar, M. Fry, J. Giesy, L.E. Gray, P. Guiney, T. Hutchinson, S. Kennedy, V. Kramer, G. LeBlanc, M. Mayes, A. Nimrod, R. Patino, R. Peterson, R. Purdy, R. Ringer, P. Thomas, L. Touart, G. van der Kraak & T. Zacharewski (1998) Overview of a workshop on screening methods for detecting potential (anti-) estrogenic/androgenic chemicals in wildlife. *Environmental Toxicology and Chemistry*, **17**, 68–87.

Arnold, A.P. & S.M. Breedlove (1985) Organizational and activational effects of sex steroids on brain and behaviour: a reanalysis. *Hormones and Behaviour*, **10**, 469–498.

Arukwe, A. (1998) *Xenobiotic modulation of fish endocrine systems: molecular and biochemical studies of the estrogen- and Ah-receptor pathways in Atlantic salmon (Salmo salar)*. PhD Dissertation, University of Bergen, Norway.

Arukwe, A. & A. Goksøyr (1998) Xenobiotics, xenoestrogens and reproduction disturbances in fish. *Sarsia*, **83** (3), 225–241.

Arukwe, A., F.R. Knudsen & A. Goksøyr (1997a) Fish *zona radiata* (eggshell) protein: a sensitive biomarker for environmental estrogens. *Environmental Health Perspectives*, **105**, 418–422.

Arukwe, A., L. Forlin & A. Goksøyr (1997b) Xenobiotic and steroid biotransformation enzymes in Atlantic salmon (*Salmo salar*) liver treated with an estrogenic compound, 4-nonylphenol. *Environmental Toxicology and Chemistry*, **16** (12), 2576–2583.

Arukwe, A., T. Grotmol, T.B. Haugen, F.R. Knudsen & A. Goksøyr (1999) A fish model for assessing the *in vivo* estrogenic potency of the mycotoxin zearalenone and its metabolites. *Science of the Total Environment*, **236**, 153–161.

Arukwe, A., T. Celius, B.T. Walther & A. Goksøyr (2000) Effects of xenoestrogen treatment on *zona radiata* protein and vitellogenin expression in Atlantic salmon (*Salmo salar*). *Aquatic Toxicology*, **49** (3), 159–170.

Astroff, B. & S. Safe (1990) 2,3,7,8-Tetrachlorodibenzo-p-dioxin as an antiestrogen: effect on rat uterine peroxidase activity. *Biochemical Pharmacology*, **39**, 485–488.

Ayotte, P., E. Dewailly, S. Bruneau, H. Careau & A. Vezina (1995) Arctic air pollution and human health: what effects should be expected. *Science of the Total Environment*, **161**, 529–537.

Bagge, P. & L. Hakkari (1992) Effects of paper mill effluents on the fish fauna of stony shores of Lake Paijanne. *Hydrobiologia*, **243**, 413–420.

Barrie, L.A., D. Gregor, B. Hargrave, R. Lake, D. Muir, R. Shearer, B. Tracey & T. Bidleman (1992) Arctic contaminants: sources, occurrence and pathways. *Science of the Total Environment*, **122**, 1–74.

Beato, M., P. Herrlich & G. Schütz (1995) Steroid hormone receptors: many actors in search of a plot. *Cell*, **83**, 851–857.

Bjerregaard, P. (1996) Cardiovascular disease and environmental pollutants: the Arctic aspect. *Arct. Med. Res.*, **55** (Suppl. 1), 25–31.

Bjerselius, R., K. Lundstedt-Enkel, I. Mayer & K.H. Olsén (1999) Estrogen in food or water severely affect the male goldfish (*Carassius auratus*) behaviour (Abstract PP-161). 6th International Symposium on the Reproductive Physiology of Fish, Bergen, Norway, 4–9 July 1999.

Boreman, J. (1997) Methods for comparing the impacts of pollution and fishing on fish populations. *Transactions of the American Fisheries Society*, **126** (3), 506–513.

Borgmann, U., D.S. Cherry & J. Cairns (1989) Effect of cadmiun on a stable, large volume, laboratory ecosystem containing Daphnia and phytoplankton. *Canadian Journal of Fisheries and Aquatic Sciences*, **46**, 399–405.

Bortone, S.A. & D.T. Drysdale (1981) Additional evidence for environmentally-induced intersexuality in Poeciliid fishes. *Assoc. Southeastern Biologists Bulletin*, **28**, 67.

Bortone, S.A. & W.P. Davis (1994) Fish intersexuality as an indicator of environmental stress. *Bioscience*, **44**, 165–172.

Bortone, S.A., W.P. Davis & C.M. Bundrick (1989) Morphological and behavioural characters in mosquito fish as potential bioindicators of exposure to kraft mill effluent. *Bulletin of Environmental Contamination and Toxicology*, **43**, 370–377.

Bradbury, S.P. (1994) Predicting modes of toxic action from chemical structure: An overview. *SAR and QSAR Environmental Research*, **2**, 89–104.

Bradbury, S.P. (1995) Quantitative structure activity relationships and ecological risk assessment: An overview of predictive aquatic research. *Toxicology Letters*, **79**, 229–237.

Breitburg, D.L. (1992) Episodic hypoxia in Chesapeake Bay: interacting effects of recruitment, behavior, and physical disturbance. *Ecological Monographs*, **62**, 525–546.

Brown, M. (1994) Estrogen receptor molecular biology. *Breast Cancer*, **8**, 101–111.

Cairns, J. (1983) Are single species tests alone adequate for estimating hazard? *Hydrobiologia*, **100**, 45–57.

Caldwell, J. (1985) *Conjugation mechanisms of xenobiotic metabolism: Mammalian aspects*. Paper presented at the Xenobiotic Conjugation Chemistry, Miami Beach, Florida.

Cameron, P. & H. von Westernhagen (1997) Malformation rates in embryos of North sea fishes in 1991 and 1992. *Marine Pollution Bulletin*, **34**, 129–134.

Cameron, P., J. Berg & H. von Westernhagen (1996) Biological effects monitoring of the North Sea employing fish embryological data. *Environmental Monitoring and Assessment*, **40** (2), 107–124.

Campbell, P.M., T.G. Pottinger & J.P. Sumpter (1994) Changes in the affinity of estrogen and androgen receptors accompany changes in receptor abundance in brown and rainbow trout. *General and Comparative Endocrinology*, **94**, 329–340.

Casillas, E., D. Misitano, L.L. Johnson, L.D. Rhodes, T.K. Collier, J.E. Stein, B.B. McCain & U. Varanasi (1991) Inducibility of spawning and reproductive success of female english sole (*Parophrys vetulus*) from urban and non-urban areas of Puget Sound, Washington. *Marine Environmental Research*, **31** (2), 99–122.

Celius, T. & B.T. Walther (1998) Oogenesis in Atlantic salmon (*Salmo salar*) occurs by zonagenesis preceeding vitellogenesis *in vivo* and *in vitro*. *Journal of Endocrinology*, **158** (2), 259–266.

Celius, T., T.B. Haugen, T. Grotomol & B.T. Walther (1999) A sensitive zonagenetic assay for rapid *in vitro* assessment of estrogenic potency of xenobiotics and mycotoxins. *Environmental Health Perspectives*, **107**, 63–68.

Christiansen, T., B. Korsgaard & A. Jespersen (1998) Effects of nonylphenol and 17β-oestradiol on vitellogenin synthesis, testicular structure and cytology in male eelpout *Zoarces viviparus*. *Journal of Experimental Biology*, **201** (2), 179–192.

Colborn, T. & C. Clement (1992) *Chemically-induced alterations in sexual and functional development: The wildlife/human connection*. Vol. XXI. Princeton Scientific Publishing, Princeton, NJ.

Collier, T.K., L.L. Johnson, M.S. Myers, C.M. Stehr, M.M. Krahn & J.E. Stein (1998a) *Fish Injury in the Hylebos Waterway of Commencement Bay, Washington*. NOAA, U.S.A. May 1998, Technical Memorandum NMFS-NWFSC-36.

Collier, T.K., L.L. Johnson, C.M. Stehr, M.S. Myers & J.E. Stein (1998b) A comprehensive assessment of the impacts of contaminants on fish from an urban waterway. *Marine Environmental Research*, **46**, 243–247.

Coosen, R. & F.L. van Velson (1989) Effects of the β-isomer of hexacholocyclohexane on estrogen-sensitive human mammary tumor cells. *Toxicology and Applied Pharmacology*, **101**, 310–318.

Couillard, C.M., P.V. Hodson & M. Castonguay (1997) Correlations between pathological changes and chemical contamination in American eels, *Anguilla rostrata*, from the St Lawrence River. *Canadian Journal of Fisheries and Aquatic Sciences*, **54**, 1916–1927.

Cyr, D.G. & J.G. Eales (1996) Interrelationships between thyroidal and reproductive endocrine systems in fish. *Reviews in Fish Biology*, **6**, 165–200.

Cyr, D.G. & J.J. Nagler (1996) Effects of environmental contaminants on the male gamete of American plaice. ICES CM. 1996/Q:1.

Davis, B.J., R.R. Maronpot & J.J. Heindel (1994a) Di-(2-ethylhexyl) phthalate suppresses estradiol and ovulation in cycling rats. *Toxicology and Applied Pharmacology*, **128** (2), 216–223.

Davis, B.J., R. Weaver, L.J. Gaines & J.J. Heindel (1994b) Mono-(2-ethylhexyl) phthalate suppresses estradiol production independent of FSH-cAMP stimulation in rat granulosa cells. *Toxicology and Applied Pharmacology*, **128** (2), 224–228.

Davis, W.P. & S.A. Bortone (1992) Effects of kraft mill effluents on the sexuality of fishes: An environmental early warning. In: (eds Colborn, T. & C. Clement) *Chemically-induced alterations in sexual and functional development: The wildlife/human connection*. Vol. XXI. (Advances in modern environmental toxicology.) Princeton Scientific Publishing, Princeton NJ, pp. 113–127.

deAngelis, D.L. (1996) Indirect effects: concepts and approaches from ecological theory. In: (eds Baird, D.J., L. Maltby, P.W. Greig-Smith & P.E.T. Douben) *Ecotoxicology: Ecological Dimensions*. (Series eds Depledge, M.H. & B. Sanders. Ecotoxicology.) Chapman & Hall, London, pp. 25–42.

DEFRA (2002) *Endocrine Disruption in the Marine Environment Programme*. The EDMAR Secretariat, Department for Environment Food and Rural Affairs, 3/E6 Ashdown House, 123 Victoria Street, London SW1E 6DE.

Denison, M.S., J.E. Chambers & J.D. Yarbrough (1981) Persistent vitellogenin-like protein and binding of DDT in the serum or insecticide resistant mosquitofish (*Gambusia affinis*). *Comparative Biochemistry & Physiology*, **69C**, 109–112.

Dethlefsen, V. (1977) The influence of DDT and DDE on the embryogenesis and mortality of larvae of cod (*Gadus morhua* L.). *Ber. Dt. wiss. Komm Meerresforsch*, **25**, 115–148.

Dethlefsen, V., H. von Westernhagen & P. Cameron (1996) Malformations in North Sea pelagic fish embryos during the period 1984–1995. *ICES Journal of Marine Science*, **53** (6), 1024–1035.

Donohoe, R.M. & L.R. Curtis (1996) Estrogenic activity of chlordecone, o,p′-DDT and o,p′-DDE in juvenile rainbow trout: induction of vitellogenesis and interaction with hepatic estrogen binding sites. *Aquatic Toxicology*, **36**, 31–52.

Douthwaite, R.J., P.J. Fox, P. Matthiessen & A. Russell-Smith (1981) *Environmental Impact of Aerosols of Endosulfan, Applied for Tsetse Fly Control in the Okavango Delta, Botswana*. Overseas Development Administration, London, 141pp (Final report of the Endosulfan Monitoring Project).

Douthwaite, R.J., P.J. Fox, P. Matthiessen & A. Russell-Smith (1983) Environmental impact of aerial spraying operations against tsetse fly in Botswana. In: *17th Meeting of the International Scientific Council for Trypanosomiasis Research and Control, Arusha, Tanzania 1981*. Organisation of African Unity/International Scientific Council for Trypanosomiasis Research & Control, Eleza Services, Nairobi, pp. 626–633. (OAU/ISTRC Publication No. 112.)

Drysdale, D.T. & S.A. Bortone (1989) Laboratory induction of inter-sexuality in the mosquitofish *Gambusia affinis*, using paper mill effluent. *Bulletin of Environmental Contamination and Toxicology*, **43**, 611–617.

Eagon, P.K., N. Chandar, M.J. Epley, M.S. Elm, E.P. Brady & K.N. Rao (1994) Di(2-ethylhexyl) phthalate-induced changes in liver estrogen metabolism and hyperplasia. *International Journal of Cancer*, **58** (5), 736–743.

Eagon, P.K., M.S. Elm, M.J. Epley, H. Shinozuka & K.N. Rao (1996) Sex steroid metabolism and receptor status in hepatic hyperplasia and cancer in rats. *Gastroenterology*, **110** (4), 1199–1207.

EDSTAC (1998) Endocrine Disruptor Screening and Testing Advisory Committee (EDSTAC). *Final Report*. http://www.epa.gov/scipoly/oscpendo/history/finalrpt.htm.

ENDS (1995) Public exposed to oestrogen risks from food cans. ENDS Report, **246**, 3 pp.

Folmar, L.C., N.D. Denslow, V. Rao, M. Chow, D.A. Crain, J. Enblom, J. Marcino & L.J. Guillette Jr. (1996) Vitellogenin induction and reduced serum testosterone concentrations in feral male carp (*Cyprinus carpio*) captured near a major metropolitan sewage treatment plant. *Environmental Health Perspectives*, **104** (10), 1096–1101.

Gagnon, M.M., D. Bussieres, J.J. Dodson & P.V. Hodson (1995) White sucker (*Catostomus commersoni*) growth and sexual maturation in pulp mill contaminated and reference rivers. *Environmental Toxicology and Chemistry*, **14**, 317–327.

Ghosh, S. & P. Thomas (1995) Antagonistic effects of xenobiotics on steroid-induced final maturation of Atlantic croaker oocytes *in vitro*. *Marine Environmental Research*, **39**, 159–163.

Gimeno, S., A. Gerritsen, T. Bowmer & H. Komen (1996) Feminization of male. *Nature*, **384**, 221–222.

Grasso, P., J.J. Heindel, C.J. Powell & L.E. Reichert Jr. (1993) Effects of mono(2-ethylhexyl) phthalate, a testicular toxicant, on follicle-stimulating hormone binding to membranes from cultured rat Sertoli cells. *Biology of Reproduction*, **48** (3), 454–459.

Gray, M.A. & C.D. Metcalfe (1997) Induction of testis-ova in Japanese medaka (*Oryzias latipes*) exposed to *p*-nonylphenol. *Environmental Toxicology and Chemistry*, **16**, 1082–1086.

Griswold, B.L. (1997) Fisheries and pollution. *Transactions of the American Fisheries Society*, **126**, 504–505.

Guardans, R. & B.S. Gimeno (1994) Long distance transportation of atmospheric pollutants and its effects on ecosystems. *Microbiology*, **10**, 145–158.

Guillette, J.L., D.A. Crain, A.A. Rooney & D.B. Pickford (1995) Organization versus activation: The role of endocrine-disrupting contaminants (EDCs) during embryonic development in wildlife. *Environmental Health Perspectives*, **103** (Suppl. 7), 157–164.

Guttman, S.I. (1994) Population genetic structure and ecotoxicology. *Environmental Health Perspectives*, **102** (12), 97–100.

Hansen, P.D., H. von Westernhagen & H. Rosenthal (1985) Chlorinated hydrocarbons and hatching success in baltic herring spring spawners. *Marine Environmental Research*, **15**, 59–76.

Harmon, M.A., M.F. Boehm, R.A. Heyman & D.J. Mangelsdorf (1995) Activation of mammalian retinoid X receptors by the insect growth regulator methoprene. *Proceedings of the National Academy of Sciences of the USA*, **92**, 6157–6160.

Harries, J.E., D.A. Sheahan, S. Jobling, P. Matthiessen, P. Neall, E. Routledge, R. Rycroft, J.P. Sumpter & T. Tylor (1996) A survey of estrogenic activity in United Kingdom inland waters. *Environmental Toxicology and Chemistry*, **15**, 1993–2002.

Harries, J.E., D.A. Sheahan, S. Jobling, P. Matthiessen, P. Neall, J.P. Sumpter, T. Tyler & N. Zaman (1997) Estrogenic activity in five United Kingdom rivers detected by measurement of vitellogenesis in caged male trout. *Environmental Toxicology and Chemistry*, **16**, 534–542.

Hashimoto, S., H. Bessho, A. Hara, M. Nakamura, T. Iguchi, & K. Fujita (2000) Elevated serum vitellogenin levels and gonadal abnormalities in wild male flounder (*Pleuronectes yokohamae*) from Tokyo Bay, Japan. *Marine Environmental Research*, **49**, 37–53.

Heppell, S.A., N.D. Denslow, L.C. Folmar & C.V. Sullivan (1995) Universal assay of vitellogenin as a biomarker for environmental estrogens. *Environmental Health Perspectives*, **103** (Suppl. 7), 9–15.

Herman, R.L. & H.L. Kincaid (1988) Pathological effects of orally administered estradiol to rainbow trout. *Aquaculture*, **72**, 165–172.

Hesthagen, T., H.M. Berger, B.M. Larsen & R. Saksgard (1995) Monitoring fish stocks in relation to acidification in Norwegian watersheds. *Water Air and Soil Pollution*, **85** (2), 641–646.

Hewitt, L.M., L. Tremblay, G.J. van der Kraak, K.R. Solomon & M.R. Servos (1998) Identification of the lampricide 3-trifluoromethyl-4-nitrophenol as an agonist for the rainbow trout estrogen receptor. *Environmental Toxicology and Chemistry*, **17**, 425–432.

Hoffman, E.R. & S. Fisher (1994) Comparison of a field and laboratory-derived population of *Chironomus riparius* (Diptera, Chironomidae) – biochemical and fitness evidence for population divergence. *Journal of Economic Entomology*, **87** (2), 320–325.

Horst, T.J. (1977) Use of the Leslie matrix for assessing environmental impact with an example for a fish population. *Transactions of the American Fisheries Society*, **106**, 253–257.

Hose, J.E., M.D. McGurk, G.D. Marty, D.E. Hinton, E.D. Brown & T.T. Baker (1996) Sublethal effects of the *Exxon Valdez* oil spill on herring embryos and larvae: morphological, cytogenetic and histopathological assessments, 1989–1991. *Canadian Journal of Fisheries and Aquatic Sciences*, **53** (10), 2355–2365.

Howell, W.M. & T.E. Denton (1989) Gonopodial morphogenesis in female mosquitofish, *Gambusia affinis affinis*, masculinized by exposure to degradation products from plant sterols. *Environmental Biology of Fishes*, **24**, 43–51.

Howell, W.M., D.A. Black & S.A. Bortone (1980) Abnormal expression of secondary sex characters in a population of mosquitofish, *Gambusia affinis holbrooki*: evidence for environmentally-induced masculinization. *Copeia*, **1980**, 676–681.

Hyllner, D.J., D.O. Oppen-Bernsten, J.V. Helvik, B.T. Walther & C. Haux (1991) Oestradiol-17β induces major vitelline envelope proteins in both sexes in teleosts. *Journal of Endocrinology*, **131**, 229–236.

IPCS (1992) *International Programme on Chemical Safety*. World Health Organization, Geneva, 141pp (Environmental Health Criteria, 131).

Issemann, I. & S. Green (1990) Activation of a member of the steroid hormone receptor superfamily by peroxisome proliferators. *Nature*, **347**, 645–650.

Jakobsson, S., B. Borg, C. Haux & S.J. Hyllner (1999) An 11-ketotestosterone induced kidney-secreted protein: the nest building glue from male three-spined stickleback, *Gasterosteus aculeatus*. *Fish Physiology and Biochemistry*, **20** (1), 79–85.

Jobling, S.J. & J.P. Sumpter (1993) Detergent components in sewage effluent are weakly oestrogenic to fish: an *in vitro* study using rainbow trout hepatocytes. *Aquatic Toxicology*, **27**, 361–372.

Jobling, S., T. Reynolds, R. White, M.G. Parker & J.P. Sumpter (1995) A variety of environmentally persistent chemicals, including some phthalate plasticisers, are weakly estrogenic. *Environmental Health Perspectives*, **103**, 582–587.

Jobling, S., D. Sheahan, J.A. Osborne, P. Matthiessen & J.P. Sumpter (1996) Inhibition of testicular growth in rainbow trout (*Oncorhynchus mykiss*) exposed to estrogenic alkylphenolic chemicals. *Environmental Toxicology and Chemistry*, **15** (2), 194–202.

Jobling, S., C.R. Tyler, M. Nolan & J. Sumpter (1998) *The identification of oestrogenic effects in wild fish*. Environment Agency, UK (R&D Technical Report W119).

Johnson, L.L., E. Casillas, T. Collier, B.B. McCain & U. Varanasi (1988) Contaminant effects on ovarian development in English sole (*Parophrys vetulus*) from Puget Sound, Washington. *Canadian Journal of Fisheries and Aquatic Sciences*, **45**, 2133–2146.

Johnson, L.L., S.Y. Sol, G.M. Ylitalo, T. Hom, B. French, O.P. Olson & T.K. Collier (1997) *Precocious sexual maturation and other reproductive anomalies in English sole from an urban waterway*. International Council for the Exploration of the Sea, Copenhagen, ICES CM 1997/U:07, 15 pp.

Jones, J.C. & J.D. Reynolds (1997) Effects of pollution on reproductive behaviour of fishes. *Reviews in Fish Biology*, **7** (4), 463–491.

Jordan, V.C., S. Mittal, B. Gosden, R. Koch & M.E. Lieberman (1985) Structure-activity relationships of estrogens. *Environmental Health Perspectives*, **61**, 97–110.

Kaiser, J. (1997) Synergy paper questioned at toxicology meeting. *Science*, **275**, 1879.

Kareiva, P., J. Stark & U. Wennergren (1996) Using demographic theory, community ecology and spatial models to illuminate ecotoxicology. In: (eds Baird, D.J., L. Maltby, P.W. Greig-Smith & P.E.T. Douben) *ECOtoxicology: Ecological Dimensions*. (Series eds Depledge, M.H. & B. Sanders. Ecotoxicology.) Chapman & Hall, London, pp. 13–24.

Katsiadaki, I., A.P. Scott & P. Matthiessen (2000) The use of the three-spined stickleback as a potential biomarker for androgenic xenobiotics. In: (eds Norber, B., O.S. Kjesbu, G.L. Taranger, E. Andersson & S.O. Stefa) *Proceedings of the 6th International Symposium on the Reproductive Physiology of Fish*. Institute of Marine Research & University of Bergen, Norway, 4–9 July 1999, pp. 359–361.

Kelce, W.R., C.R. Stone, S.C. Laws, L.E. Gray, J.A. Kemppainen & E.M. Wilson (1995) Persistent DDT metabolite *p,p'*-DDE is a potent androgen receptor antagonist. *Nature*, 375, 581–585.

Keller, H., F. Givel, M. Perroud & W. Wahli (1995) Signaling cross-talk between peroxisome proliferator-activated receptor retinoid X receptor and estrogen receptro through estrogen response elements. *Molecular Endocrinology*, **9** (7), 794–804.

Khan, I.A. & P. Thomas (1998) Estradiol-17β and o,p'-DDT stimulate gonadotropin release in Atlantic croaker. *Marine Environmental Research*, **46**, 149–152.

Kiceniuk, J.W. & R.A. Khan (1987) Effects of petroleum hydrocarbons on Atlantic cod, *Gadus morhua* following chronic exposure. *Canadian Journal of Zoology*, **65**, 490–494.

Kime, D.E. (1995) The effects of pollution on reproduction in fish. *Reviews in Fish Biology*, **5** (1), 52–95.

Kliewer, S.A., J.T. Moore, L. Wade, J.L. Staudinger, M.A. Watson, S.A. Jones, D.D. McKee, B.B. Oliver, T.M. Willson, R.H. Zetterström, T. Perlmann & J.M. Lehmann (1998) An orphan nuclear receptor activated by pregnanes defines a novel steroid signaling pathway. *Cell*, **92**, 1–20.

Kocan, R.M., J.E. Hose, E.D. Brown & T.T. Baker (1996) Pacific herring (*Clupea pallasi*) embryo sensitivity to Prudhoe Bay petroleum hydrocarbons: laboratory evaluation and in situ exposure at oiled an unoiled sites in Prince William Sound. *Canadian Journal of Fisheries and Aquatic Sciences*, **53** (10), 2366–2375.

Kubecka, J. & J. Matena (1991) Downstream regeneration of the fish populations of 3 polluted trout streams in southern Bohemia. *Ekologia CSFR*, **10** (4), 389–404.

Kuiper, G.G.J.M. & J.A. Gustafsson (1997) The novel estrogen receptor-beta subtype: Potential role in the cell- and promoter-specific actions of estrogens and anti-estrogens. *FEBS Letters*, **410** (1), 87–90.

Kuiper, G.G.J.M., E. Enmark, M. Pelto Huikko, S. Nilsson & J.A. Gustafsson (1996) Cloning of a novel estrogen receptor expressed in rat prostate and ovary. *Proceedings of the National Academy of Sciences of the USA*, **93** (12), 5925–5930.

Landahl, J.T., L.L. Johnson, T.K. Collier, J.E. Stein & U. Varanasi (1997) Marine pollution and fish population parameters: English sole (*Pleuronectes vetulus*) in Puget Sound, WA. *Transactions of the American Fisheries Society*, **126**, 519–535.

Lang, T., U. Damm & V. Dethlefsen (1995) *Changes in the sex ratio of North Sea dab (Limanda limanda) in the period 1981–1995*. International Council for the Exploration of the Sea, Copenhagen, ICES CM 1995/G:25, 11 pp.

Larsson, D.G.J., M. Adolfsson Erici, J. Parkkonen, M. Pettersson, A.H. Berg, P.E. Olsson & L. Förlin (1999) Ethinyloestradiol – an undesired fish contraceptive? *Aquatic Toxicology*, **45**, 91–97.

Larsson, D.G.J., H. Hällman, S.J. Hyllner & L. Förlin (2000) More male embryos near a pulp mill. Poster presentation at the 6th International Symposium on Reproductive Physiology of Fish, Bergen, Norway, July 4–9 1999. *Environmental Toxicology and Chemistry*, **19** (129), 2911–2917.

Laskey, J.W. & E. Berman (1993) Steroidogenic assessment using ovary culture in cycling rats: effects of bis(2-diethylhexyl) phthalate on ovarian steroid production. *Reproductive Toxicology*, **7** (1), 25–33.

Lazier, C.B. & M.E. MacKay (1993) Vitellogenin gene expression in teleost fish. In: (eds Hochachka, P.W. & T.P. Mommsen) *Biochemistry and Molecular Biology of Fishes*. Vol. 2. Elsevier Science, New York, pp. 391–405.

Liu, R.C.M., C. Hahn & M.E. Hurtt (1996a) The direct effect of hepatic peroxisome proliferators on rat Leydig cell function *in vitro*. *Fundamental and Applied Toxicology*, **30** (1), 102–108.

Liu, R.C.M., M.E. Hurtt, J.C. Cook & L.B. Biegel (1996b) Effect of the peroxisome proliferator, ammonium Perfluorooctanoate (C8), on hepatic aromatase activity in adult male Crl:CD BR (CD) rats. *Fundamental and Applied Toxicology.*, **30** (2), 220–228.

Lomax, D.P., L.L. Johnson, W.T. Roubal, J.E. West, S.M. O'Neill & T.K. Collier (2001) Abnormal production of vitellogenin in marine fish from urban embayments in Puget Sound, Washington, USA. Poster paper presented to the 11th International Symposium on Pollutant Responses in Marine Organisms (PRIMO), Plymouth, UK, 10–13 July 2001.

Loomis, A.K. & P. Thomas (1999) Binding characteristics of estrogen receptor (ER) in Atlantic croaker (*Micropogonias undulatus*) testis: Different affinity for estrogens and xenobiotics from that of hepatic ER. *Biology of Reproduction*, **61** (1), 51–60.

Lorenz, J.J. & D.H. Taylor (1992) The effects of low pH as a chemical stressor on the ability of convict cichlids to raise their young. *Copeia*, **1992**, 832–839.

Lowe, D.M. & M.N. Moore (1979) The cytochemical distributions of zinc (ZN 11) and iron (Fe 111) in the common mussel, *Mytilus edulis*, and their relationship with lysosomes. *Journal of the Marine Biological Association of the United Kingdom*, **59**, 851–858.

Lowe, D.M. & R.K. Pipe (1986) Hydrocarbon exposure in mussels: a quantitative study of the responses in the reproductive and nutrient storage cell systems. *Aquatic Toxicology*, **8**, 265–272.

Lowe, D.M. & R.K. Pipe (1987) Mortality and quantitative aspects of storage cell utilization in mussels, *Mytilus edulis*, following exposure to diesel oil hydrocarbons. *Marine Environmental Research*, **22**, 243–251.

Lye, C.M., C.L.J. Frid, M.E. Gill & D. McCormick (1997) Abnormalities in the reproductive health of flounder *Platichthys flesus* exposed to effluent from a sewage treatment works. *Marine Pollution Bulletin*, **34** (1), 34–41.

Lyons, J., G. Gonzalez Hernandez, E. Soto Galera & M. Guzman Arroyo (1998) Decline of freshwater fishes and fisheries in selected drainages of west-central Mexico. *Fisheries*, **23** (4), 10–18.

MacLatchy, D.L. & G.J. van der Kraak (1995) The phytoestrogen β-sitosterol alters the reproductive endocrine status of goldfish. *Toxicology and Applied Pharmacology*, **134**, 305–312.

Mangelsdorf, D.J. & R.M. Evans (1995) The RXR heterodimers and orphan receptors. *Cell*, **83**, 841–850.

Mangelsdorf, D.J., C. Thummel, M. Beato, P. Herrlich, G. Schütz, K. Umesono, B. Blumberg, P. Kastner, M. Mark, P. Chambon & R.M. Evan (1995) The nuclear receptor superfamily: the second decade. *Cell*, **83**, 835–839.

Matthiessen, P. & P.E. Gibbs (1998) Critical appraisal of the evidence for tributyltin-mediated endocrine disruption in molluscs. *Environmental Toxicology and Chemistry*, **17**, 37–43.

Matthiessen, P. & J.W.M. Logan (1984) Low concentration effects of endosulfan insecticide on reproductive behavior in the tropical cichlid fish *Sarotherodon mossambicus*. *Bulletin of Environmental Contamination and Toxicology*, **33**, 575–583.

Matthiessen, P. & J.P. Sumpter (1998) Effects of estrogenic substances in the aquatic environment. In: (eds Braunbeck, E.T., D.E. Hinton & B. Streit) *Fish Ecotoxicology*. Birkhäuser Verlag, Basel, pp. 319–335.

Maynard, R. (1995) Sperm alert. *Living Earth*, **188**, 8–9.

McKinney, J.D. & C.L. Waller (1994) Polychlorinated biphenyls as hormonally active structural analogues. *Environmental Health Perspectives*, **102**, 290–297.

Mellanen, P., T. Petanen, S. Lehtim, G. Bylund, B. Holmbom, E. Mannila, A. Oikari & R. Santti (1996) Wood-derived estrogens – studies *in vitro* with breast cancer cell lines and *in vivo* in trout. *Toxicology and Applied Pharmacology*, **136**, 381–388.

Mercure, F. & G.J. van der Kraak (1995) Evidence of peroxisomal involvement in ovarian steroidogenesis in teleosts. In: (eds Goetz, F.W. & P. Thomas) *Proceedings of the Fifth International Symposium on the Reproductive Biology of Fish*. Fish Symposium 95, Austin, 321 pp.

Mix, M.C. & A.K. Sparks (1971) Repair of digestive tubule tissue of the pacific oyster, *Crassostrea gigas*, damaged by ionizing radiation. *Journal of Invertebrate Pathology*, **17** (2), 172–177.

Monod, G. (1985) Egg mortality of Lake Geneva Charr (*Salvelinus alpinus* L.) contaminated by PCB and DDT derivatives. *Bulletin of Environmental Contamination and Toxicology*, **35**, 531–536.

Moyle, P.B. & J.J. Cech Jr. (1988) *Fishes: An introduction to ichthyology*, 2nd ed. Princeton Hall, Englewood Cliffs, New York, 559 pp.

Munkittrick, K.R., G.J. van der Kraak, M.E. McMaster, C.B. Portt, M.R. van den Heuvel & M.R. Servos (1994) Survey of receiving water environmental impacts associated with discharges from pulp mills. 2. Gonad size, liver size, hepatic EROD activity and plasma sex steroid levels in white sucker. *Environmental Toxicology and Chemistry*, **13**, 1089–1101.

Nagler, J.J. & D.G. Cyr (1997) Exposure of male American plaice (*Hippoglossoides platessoides*) to contaminated marine sediments decreases the hatching success of their progeny. *Environmental Toxicology and Chemistry*, **16** (8), 1733–1738.

Newman, M.C. & R.H. Jagoe (1998) Allozymes reflect the population-level effect of mercury: simulations of the mosquitofish (*Gambusia holbrooki* Girard) GPI-2 response. *Ecotoxicology*, **7** (3), 141–150.

Nilsen, B.M., K. Berg, A. Arukwe & A. Goksøyr (1998) Monoclonal and polyclonal antibodies against fish vitellogenin for use in pollution monitoring. *Marine Environmental Research*, **46** (1–5), 153–157.

Nimrod, A.C. & W.H. Benson (1997) Xenobiotic interaction with and alteration of channel catfish estrogen receptor. *Toxicology and Applied Pharmacology*, **147** (2), 381–390.

Norcross, B.L., J.E. Hose, M. Frandsen & E.D. Brown (1996) Distribution, abundance, morphological conditions, and cytogenetic abnormalities of larval herring in Prince William Sound, Alaska, following the *Exxon Valdez* oil spill. *Canadian Journal of Fisheries and Aquatic Sciences*, **53** (10), 2376–2387.

Norrgren, L., A. Blom, P.L. Andersson, H. Börjesson, D.G.J. Larsson & P.E. Olsson (1999) Effects of potential xenoestrogens (DEHP, nonylphenol and PCB) on sexual differentiation in juvenile Atlantic salmon (*Salmo salar*). *Aquatic Ecosystem Health and Management*, **2** (3), 311–318.

OECD (1992) *OECD Guideline for Testing of Chemicals. Fish, Early-life Stage Toxicity Test, OECD 210*. Organisation of Economic Cooperation and Development, Paris, 18 pp.

OECD (1999) *Detailed Review Document on Classification Systems for Reproductive Toxicity in OECD Member Countries*. Organisation of Economic Cooperation and Development, Paris. OECD Series on Testing and Assessment, Number 15, 24 pp.

OECD (2001) First meeting of the Validation Management Group on Ecotoxicity Test Methods for Endocrine Disrupters – VMG-eco1 – 28–29 March 2001, Paris. *Final report of the meeting*. ENV/JM/TG/EDTA(2001)7, OECD, Paris, 25 pp.

Oppen-Berntsen, D.O. (1990) *Oogenesis and hatching in teleostean fishes with special reference to eggshell proteins*. PhD Dissertation, University of Bergen, Norway.

Oppen-Berntsen, D.O., E. Gram-Jensen & B.T. Walther (1992) Zona radiata proteins are synthesized by rainbow trout (*Oncorhynchus mykiss*) hepatocytes in response to oestradiol-17β. *Journal of Endocrinology*, **135** (2), 293–302.

Paller, M.H., M.J.M. Reichert & J.M. Dean (1996) Use of fish communities to assess environmental impacts in South Carolina coastal plain streams. *Transactions of the American Fisheries Society*, **125** (5), 633–644.

Pelissero, C., B. Bennetau, P. Babin, F. Le Menn & L. Dunogues (1991) The estrogenic activity of certain phytoestrogens in the Siberian sturgeon, *Acipenser baeri*. *Journal of Steroid Biochemistry and Molecular Biology*, **38** (3), 293–299.

Pelissero, C., G. Flouriot, J.L. Foucher, B. Bennetau, J. Dunogues, F.L. Gac & J.P. Sumpter (1993) Vitellogenin synthesis in cultured Hepatocytes; an *in vitro* test for the estrogenic potency of chemicals. *Journal of Steroid Biochemistry and Molecular Biology*, **44**, 263–272.

Persson, P., S.H. Johannsson, Y. Takagi & B.T. Bjornsson (1997) Estradiol-17β and nutritional status affect calcium balance, scale and bone resorption, and bone formation in rainbow trout, *Oncorhynchus mykiss*. *Journal of Comparative Physiology*, **167** (7), 468–473.

Phoenix, C.H., R.W. Goy, A.A. Gerall & W.C. Young (1959) Organizing action of prenatally administered testosterone propionate on the tissues mediating mating behaviour in the female guinea pig. *Endocrinology*, **65**, 369–382.

Piferrer, F. & E.M. Donaldson (1989) Gonadal differentiation in coho salmon, *Oncorhynchus kisutch*, after a single treatment with androgen or estrogen at different stages during ontogenesis. *Aquaculture*, **77**, 251–262.

Purdom, C.E., P.A. Hardiman, V.J. Bye, N.C. Eno, C.R. Tyler & J.P. Sumpter (1994) Estrogenic effects of effluents from sewage treatment works. *Journal of Chemical Ecology*, **8**, 275–285.

Querat, B., A. Hardy & Y.A. Fontaine (1991) Regulation of gonadotropin (GTH-2) a and b-subunit mRNAs by oestradiol and testosterone in the European eel. *Journal of Molecular Endocrinology*, **7**, 81–86.

Roberts, R.A., D.W. Nebert, J.A. Hickman, J.H. Richburg & T.L. Goldsworthy (1997) Symposium overview. Perturbation of the mitosis/apoptosis balance: a fundamental mechanism in toxicology. *Fundamental and Applied Toxicology*, **38**, 107–115.

Rose, K.A., E.D. Cowan, E. Houde & C.C. Coutant (1993) Individual based modelling of environmental quality effects on early life stages of fishes: a case study using striped bass. In: (ed. Fuiman, L.A.) *Water quality and the early life history stages of fishes* (American Fisheries Society Symposium). American Fisheries Society, Bethesda, Maryland, pp. 125–145.

Routledge, E.J. & J.P. Sumpter (1996) Estrogenic activity of surfactants and some of their degradation products assessed using a recombinant yeast screen. *Environmental Contamination and Toxicology*, **15**, 241–248.

Ryabov, I.N. (1985) The behaviour of *Gasterosteus aculeatus* L. in the zone of action of warm waste water. *Behaviour*, **93**, 56.

Safe, S. (1995) Modulation of gene expression and endocrine response pathways by 2,3,7,8-tetrachlorodibenzo-p-dioxin and related compounds. *Pharmacology and Therapeutics*, **67**, 247–281.

Safe, S. & V. Krishnan (1995) Cellular and molecular biology of aryl hydrocarbon (Ah) receptor-mediated gene expression. *Archives of Toxicology*, **17**, 99–115.

Safe, S., B. Astroff, M. Harris, T. Zacherewski, R. Dickerson, M. Romkes & L. Biegel (1991) 2,3,7,8-tetrachlorodibenzo-p-dioxin (TCDD) and related compounds as antiestrogens: characterisation and mechanism of action. *Pharmacology and Toxicology*, **69**, 400–409.

Schlenk, D., D.M. Stresser, J. Rimoldi, L. Arcand, J. McCants, A.C. Nimrod & W.H. Benson (1998) Biotransformation and estrogenic activity of methoxychlor and its metabolites in channel catfish (*Ictalurus punctatus*). *Marine Environmental Research*, **46**, 159–162.

Schuetz, J.D. (1998) Environmental xenobiotics and the antihormones cyproterone acetate and spironolactone use the nuclear hormone pregnenolone X receptor to activate the CYP3A23 hormone response element. *Molecular Pharmacology*, **54** (6), 1113–1117.

Scott, A.P., Stewart, C., Allen, Y. and Matthiessen, P. (2000) 17β-oestradiol in male flatfish. In: (eds Norberg, B., O.S. Kjesbu, G.L. Taranger, E. Andersson & S.O. Stefansson) *Proceedings of the 6th International Symposium on the Reproductive Physiology of Fish*. Institute of Marine Research and University of Bergen, 4–9 July 1999, 382 pp.

Shen, K. & R.F. Novak (1997) DDT stimulates c-erbB2, c-met and STATS tyrosine phosphorylation, Crb2-Sos association, MAPK phosphorylation and proliferation of human breast epithelial cells. *Biochemical and Biophysical Research Communications*, **231**, 17–21.

Shore, L.S., M. Shemesh & R. Cohen (1988) The role of oestradiol and oestrone in chicken manure silage in hyperoestrogenism in cattle. *Australian Veterinary Journal*, **65**, 67.

Shore, L.S., M. Gurevitz & M. Shemesh (1993) Estrogen as an environmental pollutant. *Bulletin of Environmental Contamination and Toxicology*, **51**, 361–366.

Soto, A.M., C. Sonnenschein, K.L. Chung, M.F. Fernandez, N. Olea & F.O. Serrano (1995) The E-screen assay as a tool to identify estrogens: an update on estrogenic environmental pollutants. *Environmental Health Perspectives*, **103** (Suppl. 7), 113–122.

SPEED (1998) *Strategic Programs on Environmental Endocrine Disruptors '98*. Japan Environment Agency. http://www.eic.or.jp/eanet/e/end/sp98.html.

Sperry, T.S. & P. Thomas (1999) Characterization of two nuclear androgen receptors in Atlantic croaker: comparison of their biochemical properties and binding specificities. *Endocrinology*, **140** (4), 1602–1611.

Spies, R.B. & D.W. Rice (1988) Effects of organic contaminants on reproduction of the starry flounder (*Platichthys stellatus*) in San Francisco Bay. II. Reproductive success of fish captured in San Francisco Bay and spawned in the laboratory. *Marine Biology*, **98**, 191–200.

Stahlschmidt-Allner, P., B. Allner, J. Römbke & T. Knacker (1997) Endocrine disrupters in the aquatic environment. *Environmental Science and Pollution Research.*, **4** (3), 155–162.

Steinmetz, R., P.C.M. Young, A. Caperell-Grant, E.A. Gize, B.V. Madhukar, N. Ben-Jonathan & R.M. Bigsby (1996) Novel estrogenic action of the pesticide residue β-hexachlorocyclohexane in human breast cancer cells. *Cancer Research*, **56**, 5403–5409.

Stott, G.G., W. Haensly, J. Neff & J. Sharp (1983) Histopathologic survey of ovaries of plaice, *Pleuronectes platessa* L., from AberWarc'h and Aber Benoit, Brittany, France oil spills. *Journal of Fish Diseases*, **6** (5), 429–437.

Sumpter, J.P. (1995) Feminized responses in fish to environmental estrogens. *Toxicology Letters*, **82–83**, 737–742.

Swedmark, M. & A. Granmo (1981) Effects of mixtures of heavy metals and a surfactant on the development of cod (*Gadus morhua* L). *Rapports et Procès-verbaux de Réunions du Conseil International pour l'Exploration de la Mer*, **178**, 95–103.

Talmage, S.S. & B.T. Walton (1991) Small mammals as monitors of environmental contaminants. *Reviews of Environmental Contamination and Toxicology*, **119**, 47–145.

Tattersfield, L., P. Matthiessen, P. Campbell, N. Grandy & R. Länge (eds) (1997) *SETAC-Europe/OECD/EC Expert Workshop on Endocrine Modulators and Wildlife: Assessment and Testing*. Veldhoven, The Netherlands, 10–13 April 1997, Society of Environmental Toxicology and Chemistry, Brussels, 126 pp.

Tchoudakova, A., S. Pathak & G.V. Callard (1999) Molecular cloning of an estrogen receptor beta subtype from the goldfish, *Carassius auratus*. *General and Comparative Endocrinology*, **113** (3), 388–400.

Thomas, P. (1999) Nontraditional sites of endocrine disruption by chemicals on the hypothalamus-pituitary-gonadal axis: interactions with steroid membrane receptors, monoaminergic pathways and signal transduction. In: (ed. Naz, R.K.) *Endocrine Disruptors: Effects on Male and Female Reproductive Systems*. CRC Press, Boca Raton, pp. 3–38.

Thomas, P. & J. Smith (1993) Binding of xenobiotics to the estrogen receptor of spotted seatrout: A screening assay for potential estrogenic effects. *Marine Environmental Research*, **35**, 147–151.

Thomas, P., D. Breckenridge-Miller & C. Detweiler (1998) The teleost sperm membrane progestogen receptor: interactions with xenoestrogens. *Marine Environmental Research*, **46**, 163–167.

Thorpe, J.E. (1989) Developmental variation in salmonid populations. *Journal of Fish Biology*, **35** (Suppl. A), 295–303.

Thorpe, J.E. (1994) Reproductive strategies in Atlantic salmon, *Salmo salar* L. *Aquaculture and Fisheries Management*, **25**, 77–87.

Toppari, J., J.C. Larsen, P. Christiansen, A. Giwercman, P. Grandjean, L.J. Guillette Jr., B. Jégou, T.K. Jensen, P. Jouannet, N. Keiding, H. Leffers, J.A. McLachlan, O. Meyer, J. Müller, E. Rajpert-De Meyts, T. Scheike, R. Sharpe, J. Sumpter & N.E. Skakkerbæk (1996) Male reproductive health and environmental xenoestrogens. *Environmental Health Perspectives*, **104** (Suppl. 4), 741–803.

Treinen, K.A., W.C. Dodson & J.J. Heindel (1990) Inhibition of FSH-stimulated cAMP accumulation and progesterone production by mono(2-ethylhexyl) phthalate in rat granulosa cell cultures. *Toxicology and Applied Pharmacology*, **106** (2), 334–340.

Tremblay, L. & G. van der Kraak (1999) Comparison between the effects of the phytosterol beta-sitosterol and pulp and paper mill effluents on sexually immature rainbow trout. *Environmental Toxicology and Chemistry*, **18**, 329–336.

Tyler, C.R., B. van der Erden, S. Jobling, T. Panter & J.P. Sumpter (1996) Measurement of vitellogenin, a biomarker for exposure to oestrogenic chemicals, in a wide variety of cyprinid fish. *Journal of Comparative Physiology B-Biochemical Systemic and Environmental Physiology*, **166**, 418–426.

Tyler, C.R., R. van Aerle, T.H. Hutchinson, S. Maddix & H. Trip (1999a) An *in vivo* testing system for endocrine disruptors in fish early life stages using induction of vitellogenin. *Environmental Toxicology and Chemistry*, **18** (2), 337–347.

Tyler, C.R., T. Hutchinson, J. Harries, K. Thorpe, S. Maddix & J. Sumpter (1999b) Development of *in vivo* testing systems for endocrine disrupting chemicals in fish (Abstract 2j/001). 9th Annual Meeting of SETAC-Europe, Leipzig, Germany, 25–29 May 1999.

USEPA (1986) *Fish Life-Cycle Toxicity Tests*. US Environmental Protection Agency, Hazard Evaluation Division, Standard Evaluation Procedure, USA, EPA 540/9-86-137, 11 pp.

van der Kraak, G.J., K.R. Munkittrick, M.E. McMaster, C.B. Portt & J.P. Chang (1992) Exposure to bleached kraft mill effluent disrupts the pituitary-gonadal axis of white sucker at multiple sites. *Toxicology and Applied Pharmacology*, **115**, 224–233.

van der Kraak, G., K.R. Munkittrick, M.E. McMaster, & D.L. MacLatchy (1998) A comparison of bleached kraft mill effluent, 17β-estradiol, and β-sitosterol effects on reproductive function in fish. In (eds Kendall, R.J., R.L. Dickerson, J.P. Giesy, & W.A. Suk) *Principles and Processes for Evaluating Endocrine Disruption in Wildlife*. SETAC Technical Publication, SETAC Press, Pensacola, pp. 249–265.

van Look, K.J.W. & D.E. Kime (1999) Fish sperm motility as a monitor of reproductive disruption by heavy metals (Abstract PP-138). *6th International Symposium on Reproductive Physiology of Fish*, 4–9 July 1999, Bergen, Norway.

von Westernhagen, H., H. Rosenthalm, V. Dethlefsen, W. Ernst, U. Harms & P.-D. Hansen (1981) Bioaccumulating substances and reproductive success in Baltic flounder *Platichthys flesus*. *Aquatic Toxicology*, **1**, 85–99.

von Westernhagen, H., P. Cameron, V. Dethlefsen & D. Janssen (1989) Chlorinated hydrocarbons in North Sea whiting (*Merlangius merlangus*), and effects on reproduction. 1. tissue burden and hatching success. *Helgoländer Meeresuntersuchungen*, **43**, 45–60.

Walthall, W.K. & J.D. Stark (1997) A comparison of acute mortality and population growth rate as endpoint of toxicological effect. *Ecotoxicology and Environmental Safety*, **37**, 45–52.

Weisberg, S.B., P. Himchak, T. Baum, H.T. Wilson & R. Allen (1996) Temporal trends in abundance of fish in the tidal Delaware River. *Estuaries*, **19** (3), 723–729.

Wertheimer, A.C. & A.G. Celewycz (1996) Abundance and growth of juvenile pink salmon in oiled and non-oiled locations of western Prince William Sound after the *Exxon Valdez* oil spill. *American Fisheries Society Symposium*, **18**, 518–532.

Wester, P.W. (1991) Histopathological effects of environmental pollutants b-HCH and methyl mercury on reproductive organs of freshwater fish. *Comparative Biochemistry and Physiology*, **100C**, 237–239.

Wester, P.W. & J.H. Canton (1986) Histopathological study of *Oryzias latipes* (Medaka) after long-term b-hexachlorocyclohexane exposure. *Aquatic Toxicology*, **9**, 21–45.

Westin, D.T., C.E. Olney & B.A. Rogers (1985) Effects of parental and dietary organochlorines on survival and body burdens of striped bass larvae. *Transactions of the American Fisheries Society*, **114**, 125–136.

Whipple, J., M. Eldridge, P. Benville, M. Bowers, B. Harvis & N. Stapp (1981) The effects of inherent parental factors on gamete condition and variability in striped bass (*Morone saxatilis*). *Rapports et Procès-verbaux de Réunions du Conseil International pour l'Exploration de la Mer*, **178**, 93–94.

White, R., S. Jobling, S.A. Hoare, J.P. Sumpter & M.G. Parker (1994) Environmentally persistent alkylphenolic compounds are estrogenic. *Endocrinology*, **135**, 175–182.

Widdows, J., T. Bakke, B.L. Bayne, P. Donkin, D.R. Livingstone, D.M. Lowe, M.N. Moore, S.V. Evans & S.L. Moore (1982) Responses of *Mytilus edulis* on exposure to the water accommodated fraction of North Sea oil. *Marine Biology*, **67**, 15–31.

Wiese, T.E. & W.R. Kelce (1997) An introduction to environmental oestrogens. *Chemistry and Industry*, 18 August 1997, 648–653.

Wine, R.N., L.-H. Li, L. Hommel Barnes, D.K. Gulati & R.E. Chapin (1997) The reproductive toxicity of di-n-butylphthalate in a continuous breeding protocol in Sprague-Dawley rats. *Environmental Health Perspectives*, **105**, 102–107.

Yadetie, F. & R. Male (2002) Effects of 4-nonylphenol on gene expression of pituitary hormones in juvenile Atlantic salmon (*Salmo salar*). *Aquatic Toxicology*, **58** (1–2), 113–129.

Yadetie, F., A. Arukwe, A. Goksøyr & R. Male (1999) Induction of hepatic estrogen receptor in juvenile Atlantic salmon *in vivo* by the environmental estrogen, 4-nonylphenol. *Science of the Total Environment*, **233**, 201–210.

Yodzis, P. (1988) The indeterminacy of ecological interactions as perceived through perturbation experiments. *Ecology*, **69**, 508–515.

Zacharewski, T. (1997) *In vitro* bioassays for assessing estrogenic substances. *Environmental Science and Technology*, **31**, 613–623.

Zacharewski, T. (1998) Identification and assessment of endocrine disruptors: Limitations of *in vivo* and *in vitro* assays. *Environmental Health Perspectives*, **106** (Suppl. 2), 577–582.

Zacharewski, T., P. Campbell, E. Routledge, M.-C. Huet, R. Cooper, D. Kime & A. Wenzel (1997) Current approaches for the use of *in vitro* tests in identifying the hazards of endocrine modulating chemicals to wildlife. In: (eds Tattersfield, L., P. Matthiessen, P. Campbell, N. Grandy & R. Länge) *SETAC-Europe/OECD/EC Expert Workshop on Endocrine Modulators and Wildlife: Assessment and Testing (EMWAT)*. SETAC-Europe, Brussels, pp. 41–58.

Chapter 6
From the Individual to the Population and Community Responses to Pollution

M. Elliott, K.L. Hemingway, D. Krueger, R. Thiel, K. Hylland, A. Arukwe, L. Förlin and M. Sayer

6.1 Introduction

Changes in the marine system due to pollutants and other anthropogenic causes of change can be regarded as a set of attributes which in turn are used for the diagnosis of ecosystem pathology (Harding, 1992). These attributes refer to nutrients, productivity, abiotic zones, species diversity, size distribution, disease prevalence, biotic composition, and the bioaccumulation of contaminants. In turn, these can be reduced to seven indicators for general application: primary production, nutrients, species diversity, instability, disease prevalence, size spectrum and contaminants. The use of these indicators is included in turn within the increasingly adopted DPSIR approach (Elliott, 2002) which summarises the causes of change, the effects within the natural system and the successive policy responses to control those causes and effects. Overall 'Drivers' for change in the marine system include the activities responsible for using and transporting chemicals, e.g. shipping and industry. This in turn produces a set of precise 'Pressures', such as the discharge of particular chemicals either legally (after authorisation) or accidentally via spillages. The 'Status' of the physical, chemical and biological system then requires to be assessed, together with any 'Impacts' on that system due to the pressures. Finally, the 'Responses' made by man to those impacts will enable problems to be solved.

This chapter considers the status and impacts at the population and community level of fishes exposed to pollutants. However, it is reiterated here that in many cases the pressures act in tandem and there are few examples in which pollution is the only stressor on the system.

The links between subcellular and population responses to pollution and their repercussions can be assessed through a set of case studies giving the genetic response to pollutants, changes in the genetic composition and structure of the populations and the population shift in genetic structure via any stressors on the population. It is therefore necessary to consider pollution effects on single and mixed stocks and also to determine whether the physiological responses to pollutants are translated into population changes, i.e. single species stocks. Any such population changes will influence stocks and thus have socio-economic repercussions. In essence, these changes can be seen as a set of bottom-up drivers, especially changes at the genetic and cellular level, and a set of top-down responses, at the population, ecosystem and socio-economic levels. This chapter considers the latter top-down responses.

The chapter aims to cover the levels of pollution response in individual fish and their populations and these in relation to eventual changes in community structure. This incorporates the contamination and quality of individual fish, the effects of pollution leading to a reduction in habitat integrity and the concomitant effect on fish production and yield, and the removal of habitat as the result of pollution, such as the loss of feeding area or changes to the nature of the prey community. The latter in turn will lead to poor quality and tainting of the fish, or the perception of this by consumers, and thus to socio-economic repercussions. Although this chapter is concerned with pollution by xenobiotics, the introduction of organic pollutants, nutrients and organic matter and the often concomitant production of a water quality barrier, especially in estuaries and migrations routes, gives lessons for the responses of fish populations and communities to stressors (de Jonge & Elliott, 2001; Elliott & Hemingway, 2002).

The changes to individual quality, as the result of pollutants, reflect a set of responses by the individual fish, their populations or communities. For example, the bioaccumulation of pollutants and the possible biomagnification of them through food chains, can be determined through concentration factors. Such an analysis may show uptake of pollutants and pathways but not necessarily show harm. Even the evidence of detoxification mechanisms may not be termed pollution *sensu stricto* (as the reduction in fitness for survival), i.e. if an organism has the ability to detoxify a contaminant without there being a deleterious effect on its survival (or that of its progeny) then pollution would not have occurred. Similarly, contamination may be regarded merely as an increase in an introduced substance without there being a biological consequence.

The population response to pollutants may be described using an interpretation based on a Leslie-Matrix model which aims to quantify the age-specific survival and fecundity under different conditions, especially different exposure regimes. In such an approach, pollution requires to be assessed as a mechanism for removing individuals within a population; this is by definition the outcome if pollution is regarded as reducing the fitness for survival of a level of biological organisation (at the cellular, individual, population or ecosystem levels). However, as indicated throughout this book, pollutant response at the lower levels of biological organisation is often demonstrated experimentally. Hence, given the conditions under which pollutants occur, it is then necessary to extrapolate from an experimental and/or single species response to a field or multispecies situation. Similarly, where commercial fisheries' considerations are paramount, any pollutant response as a reduction in the number or quality of juveniles has to be translated to adult-equivalent stock and to the effects on breeding populations. In following such effects of pollutants on the stock, some of the effects will be density-dependent and others density-independent. For example, any change to growth and survival as the result of pollutant exposure will be confounded by all other environmental and biological factors affecting growth and survival. As such, whereas populations have the ability to withstand and absorb environmental change (i.e. 'population homeostasis'), such an ability, as the result of compensation mechanisms, has to be separated from pollution-induced changes.

Similarly, in complex and variable environments such as nearshore areas and estuaries, the inherent natural variability is likely to make any anthropogenic signal more difficult to detect. In addition, those variable environments will have an increased ability to absorb change such that effects do not occur at the higher biological levels (see below).

The higher level responses (at population, community and ecosystem levels) which can be attributed to either a single stressor or multiple stressors have been well-defined for invertebrates but less well for fishes (Elliott, 1994; Walker *et al.*, 1996; Elliott & Hemingway, 2002). Stress is defined in this context as the cumulative, quantifiable response to adverse environmental conditions or factors as the result of anthropogenic activities which results in a reduction of fitness to survive at any biological level of organisation (cellular, individual, population or community). In addition, certain extreme stressors, such as a loss of habitat (McLusky *et al.*, 1992) or overfishing (North Sea QSR and Inter-Governmental Ministerial Meeting, Bergen report (Svelle *et al.*, 1997)) have an overarching effect such that lesser stressors, such as pollution, are difficult to detect and quantify.

The loss and/or replacement of community members depending on their susceptibility to stress, has been well documented (e.g. Pearson & Rosenberg, 1978; Odum, 1985). Under both acute and chronic stress, larger species and stress-intolerant species will be replaced by smaller forms and tolerant species. With regard to marine invertebrates, this is regarded as producing a movement along the Pearson-Rosenberg continuum and there will be the decline of some species (termed k-strategists, species which are good competitors and long-lived, slow to reach maturity), perhaps to the level of removal from an area, and replacement by others. Chronic stress will lead to a selection of small/tolerant and perhaps opportunistic forms (r-strategists) and there may be a change in abundance until a new equilibrium is formed. In addition to r and k-strategists, Gray (1992) considers a further strategy: t-(tolerant) strategists have characteristics between the r and k-strategists. However, whereas these ideas have been developed extensively for marine and estuarine invertebrates, there are no readily available examples for fishes.

In communities exposed to chronic stress, the abundance of tolerant species may increase, together with the development of resistance as organisms induce the ability to detoxify or sequester pollutants. If the stress remains, then a new equilibrium will develop. Ecosystems where the stress is then removed or reduced will recover through recruitment, recolonisation and/or immigration, although the recovery stages may be transient until a stable system is regained. The ability of any system to withstand and tolerate such changes may be regarded as 'environmental homeostasis'. With regard to fishes, the effects of stress may be manifest at one or more of several levels of biological organisation and there are many diagnostic techniques to investigate and explain such changes (Whitfield & Elliott, 2002). Such levels cover the cell, individual, population, community and ecosystem, each of which has some ability to absorb and ameliorate environmental change. However, the speed of response by each level decreases with progression to the higher system levels, and the inherent complexity increases with the same progression (i.e. cell to ecosystem).

In order to link the scientific and socio-economic aspects of pollutant response, it is necessary to consider those pollutants which may affect quality or the perception of quality by the consumers of fish in relation to the health of those fish. Spoilage of fish may occur as the result of bioaccumulation and biomagnification of pollutants (i.e. as internal accumulation) or by tainting by pollutants (external contamination). In such cases, it is necessary to quantify pollutant uptake, using for example critical path analysis by identifying consumers who may obtain a critically high burden of contaminants by eating fish. Within each of these, it is necessary to determine which mechanisms affect fish quality or, in the minds of consumers, the perceptions of quality. It is difficult to quantify the latter, except by a tasting panel, and thus only anecdotal evidence may occur. For example, salmon migrating through industrialised estuaries may get tainted by hydrocarbons if water quality barriers do not prevent successful migration (Elliott & Hemingway, 2002). Aspects of fish quality and pollution effects on fishery economics are considered in more detail in Chapter 8.

6.2 Changes manifested in individuals

Inshore coastal and estuarine areas support many organisms that have important direct or indirect influences on productivity and which may be of ecosystem and/or commercial significance (e.g. Methven & Bajdik, 1994; Arico, 1995; Diaz & Rosenberg, 1995). These are also areas that are vulnerable to chronic or acute episodes of substantial natural and/or anthropogenic variations in water quality through principally the effects of land-associated run-off (e.g. Martin *et al.*, 1996; Sayer & Davenport, 1996; Hawkins *et al.*, 1999; Miller, 1999; Wells, 1999; Elliott & Hemingway, 2002). The magnitude and scope of biomonitoring and assessing risk in vulnerable coastal marine ecosystems is potentially too large to be practical (Wells, 1999), although human-induced change can be regarded as producing a set of symptoms of ecosystem pathology (Harding, 1992). Among others, those symptoms of ecosystem pathology include changes in the diversity and productivity of communities as well as changes in the accumulation of toxic and tainting materials. The uses and users of coasts and estuaries are often so diverse and widespread that it is difficult to demonstrate conclusively that one stressor is the cause of biological change. Consequently, at present, most ecotoxicological knowledge is derived from short-term exposure of a single species to high (often environmentally unrealistic) and uniform pollutant concentrations under standard physico-chemical conditions. Data so derived is largely inadequate in predicting medium and long-term ecological effects in the field, in which multispecies aggregations are being exposed to varying, low concentrations of pollutants in interacting and complex environments (Wu, 1999).

Because of the complexity of most marine ecosystems there is a requirement to determine the levels of biological organisation that provide the most sensitive yet robust method of assessing environmental health. It is therefore important to explore the mechanisms linking the different levels of biological organisation in an attempt to understand how toxicological responses at the individual level may be translated and manifest at other levels, especially the community level (Attrill & Depledge, 1997). Links between sublethal responses from the individual level will all involve related behavioural actions: the detection of the pollutant and if possible avoidance, the locomotion of avoidance and any concomitant alteration of predator-prey behaviour (e.g. Olla *et al.*, 1980; Blackstock, 1984). Some or all of these behaviours can be affected by the health of the organism and in turn these can affect the organism's health. The degree to which fitness may be impaired by altered physiological condition can be assessed using any of a suite of several bioassays which may indicate the links between responses in different aspects of a fish's biology (e.g. Elliott *et al.*, 1988) (Fig. 6.1). The significance of the links between those aspects may therefore determine the potential ecological consequences.

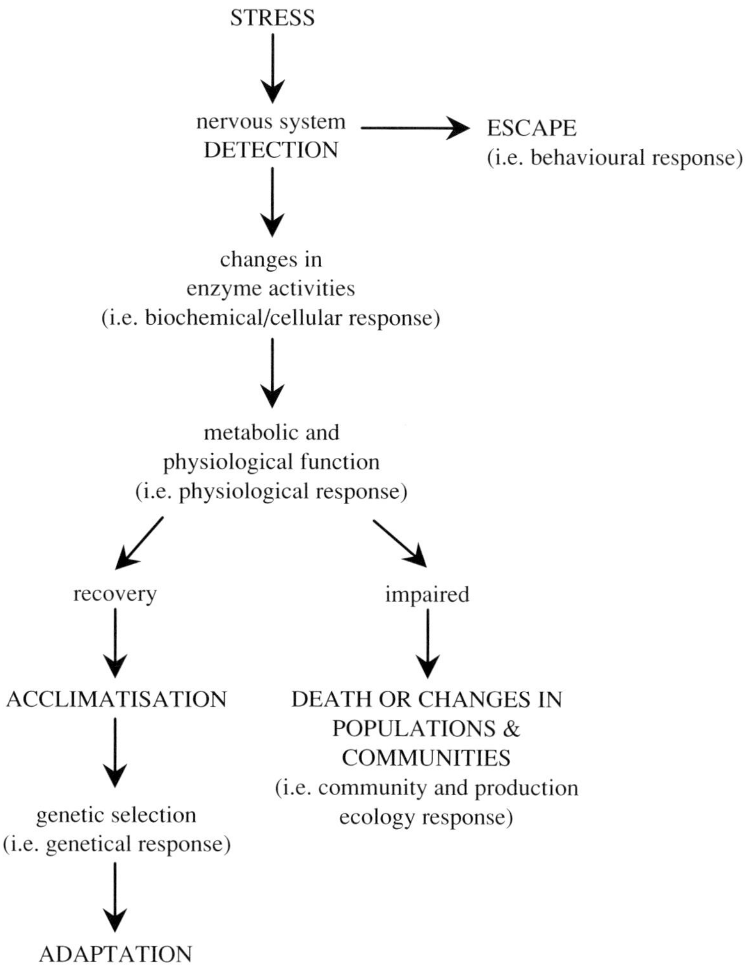

Fig. 6.1 Schematic sequence of the effects of environmental stress on estuarine animals. Modified from Blackstock, 1984.

Given the behaviour of organisms, the nature of the marine environment and the behaviour of pollutants both in the environment and in organisms, some marine species and trophic levels are more susceptible than others to environmental change (Stark, 1998). As a result of these relationships and characteristics, those species may be either proxy indicators of xenobiotic effect, or of importance because of their ecological relevance. For example, many pollutants have an affinity with sediments, especially fine-grained, organic-rich sediments. As such, they will accumulate in bottom deposits and undergo diagenesis within those sediments which may sequester the pollutants (Libes, 1992). As a result, any fish in intimate contact with those sediments will be exposed to high levels of contaminants and will subsequently show a selection of sublethal responses. In addition, the fish are likely to be feeding on prey within those sediments, thereby acting as a route for the uptake of pollutants. Bottom-dwelling flatfish such as flounder and plaice reflect these processes (e.g. Sulaiman *et al.*, 1991; Elliott *et al.*, 1998).

6.2.1 Bioaccumulation of contaminants in fish

As indicated throughout this book, bioaccumulation of materials by fishes should be considered as contamination rather than being pollution per se, and thus it requires a biological effect to be manifest before pollution is registered. Whereas some texts (e.g. Phillips & Rainbow, 1994; Elliott & Hemingway, 2002; Neff, 2002) provide greater detail of mechanisms of uptake and thus the reasons for bioaccumulation, it is necessary here to discuss bioaccumulation as a response at the individual level. The level of contaminants within an organism is the net result of the behaviour of that material in the environment and in the organism (including uptake, storage, sequestration and excretion), and of the routes of uptake and levels in the prey. As shown here, the likely effects of that bioaccumulation are either in the organism, the progeny, or predators of the fish. In relation to the latter, any increasing contamination along a food chain is regarded as biomagnification, a feature well-demonstrated in some higher trophic levels (Neff, 2002).

Biomagnification following trophic transfer to higher trophic levels such as fishes has been suggested for arsenic and mercury because of their high affinity to organic substances. Despite this, evidence for biomagnification is inconclusive and variable – whereas it occurs for some organic chemicals and organo-metals it has not been found for other components (Neff, 2002). As an example, in the Forth estuary (Scotland, UK), Elliott and Griffiths (1986) observed that for mercury (Hg) and within a more contaminated site, biomagnification occurred only along direct consumer routes where the consumer was a true estuarine resident or largely dependent on a single food source. This study, in assessing Hg contamination in all of the major components of the estuarine system, indicated both the role of the sedimentary components and the bioaccumulation in resident rather than migratory fish species (Fig. 6.2). Although biomagnification occurs only with some pollutants, in contaminated areas, all trophic levels bioaccumulate (Elliott & Hemingway, 2002; Neff, 2002). This is primarily due to direct uptake from the surrounding environment by direct absorption or via the food chain (Amiard *et al.*, 1980; Metayer *et al.*, 1980; Ferreira *et al.*, 1985).

The concentrations of metals in aquatic organisms vary because they reflect the net effect of two competing processes: uptake and depuration; they vary with the sex and size of an organism, the species under study, the season of sampling and the site located (Phillips & Rainbow, 1994; Elliott & Hemingway, 2002; Neff, 2002). This net result (degree of contamination) will reflect the ambient water concentrations and the external and internal processes of uptake, storage, detoxification and depuration. As such, both fishes and lower aquatic organisms have been used as sentinels in monitoring programmes (Amiard *et al.*, 1980; Elliott & Griffiths, 1986; Elliott *et al.*, 1988; Environment Agency, 1999; Köhler *et al.*, 1986; Lucas *et al.*, 1986; National Rivers Authority, 1993; Sauriau *et al.*, 1994).

In addition to the environmental concentrations, other factors affect the rate and processes of bioaccumulation of pollutants and these require to be considered when comparing different environments, species and areas (Phillips & Rainbow, 1994; Elliott & Hemingway, 2002; Neff, 2002):

- Physiological condition – the seasonal maxima of tissue concentrations of pollutants may occur just after spawning, partly as the result of a loss of condition but without a concomitant loss of metal content

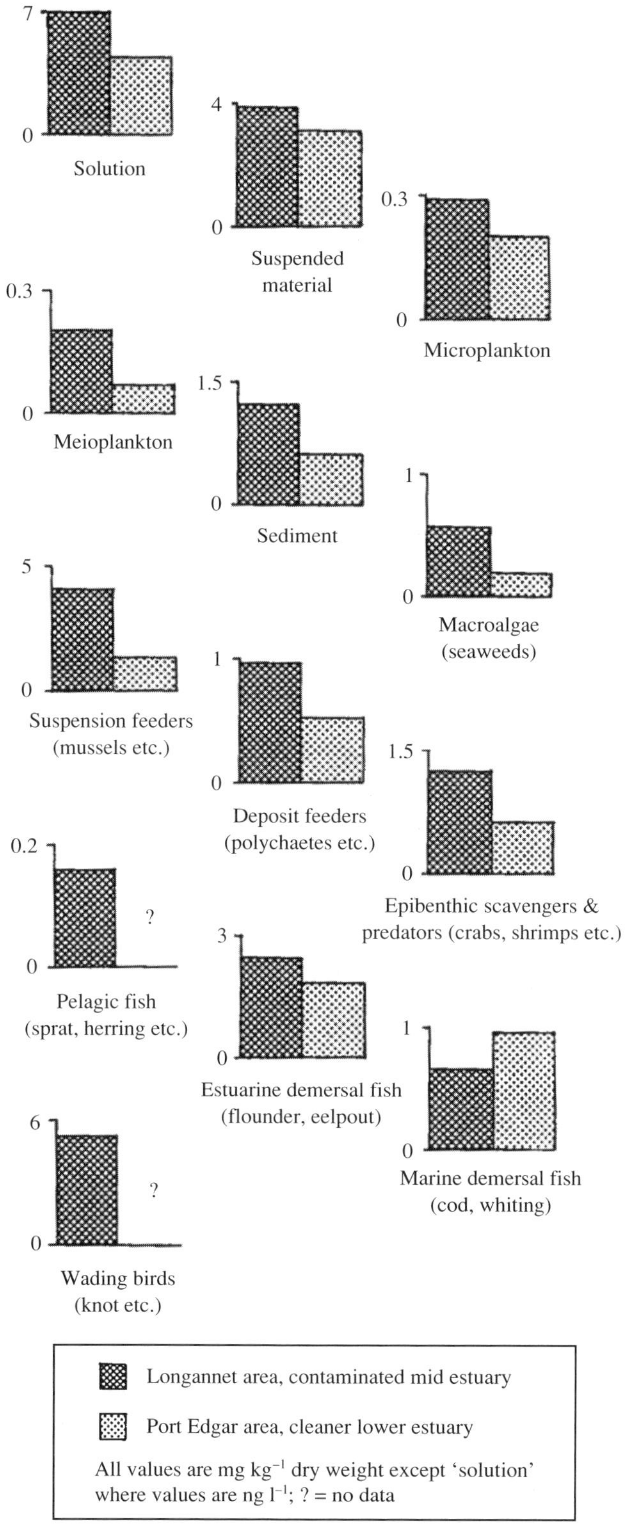

Fig. 6.2 The concentration of mercury in components of the Forth estuarine ecosystem, UK. Data derived from Elliott & Griffiths, 1986.

- Tissue – depending on the internal chemical transformations and storage processes, for example the liver is usually the main storage organ for heavy metals in fish
- Growth – low growth results in increasing tissue concentrations whereas high growth and sudden increases of flesh condition can decrease tissue concentrations
- Salinity – pollutant uptake rates increase as the salinity decreases depending on the osmoregulatory intake of dissolved pollutants
- Temperature – increases in temperature may cause an increase in the rate of accumulation depending on the concomitant physiological changes
- Age – pollutant concentration changes with age and size, although this differs with pollutant type
- Interactions between metals – interactions between pollutants may affect uptake depending on the effects of such linkages on the bioavailability of pollutants
- The chemical form and nature of binding of the bioavailability – stable forms and non-organic forms of pollutants will have a lowered uptake and availability to organisms.

As indicated throughout this book, although the accumulation of metals in the tissues of organisms is an effect per se of the inputs and concentrations of contaminants in the water and sediments, it is of greater relevance to determine the organisms reaction to the pollutants (through sublethal effects on biochemistry, pathology, genetics, behaviour) and also to assess the changes that pollutants cause in the communities (Gray, 1992). Toxic effects of metals occur when excretory, metabolic, storage, and detoxification mechanisms are no longer capable of matching uptake rates (Langston, 1990). However, information for sublethal effects is mainly derived from laboratory experiments which are static or semi-static and normally under conditions of stable (or discrete) salinity and temperature. Such stable conditions rarely occur in the marine and estuarine environment (Mance, 1987). As such, McLusky *et al.* (1986) believe that toxicity values estimated under laboratory fixed temperature and salinity regimes are inappropriate for evaluating the effect of such environmental factors in modifying the toxicity of metals to estuarine species.

Field studies attempting to determine sublethal effects of pollutants such as heavy metals are sometimes inconclusive. For example, Pohl (1990) attempted to relate skeletal deformities and heavy metal concentrations in juvenile smelts from the Elbe estuary (Germany). In this estuary, the spawning areas of the species are exposed to high heavy metal concentrations which might induce damage to the egg and embryonic stages. The levels of metals found in the muscle and liver of juvenile (group 0 and 1) smelts were considered to be low. The only indication of a relationship between skeletal deformation and metal concentrations was observed for the significantly higher lead and cadmium concentrations in liver tissue of malformed 0-group fish. The author indicates that hydrographical factors, the presence of other toxic organic and inorganic substances, and genetic factors might have influenced the occurrence of such skeletal deformities. A more conclusive field study is that of the inhibition of Na^+, K^+-ATPase activity of flounder by mercury contamination in the Forth estuary (Scotland, UK) (Stagg *et al.*, 1992). This enzyme plays a key role in branchial ion transport and the maintenance of osmotic and ionic homeostasis, but the inhibition of its activity occurs before considerable osmoregulatory dysfunction occurs and thus could be used as an early warning of damage to the osmoregulatory system in fish.

6.2.2 Link 1: Individual health to condition and growth

Subcellular biological effects at the population level of organisation normally display a long response time and when effects eventually occur it is often too late for effective counter measures to be taken (Goksøyr *et al.*, 1996). Prior to manifesting irreversible damage, pollutant exposure may lead to a range of sublethal effects such as depressed growth rate, elevated levels of infection and disease and decreased reproductive rate. A generally accepted concept in ecotoxicology is that these responses are preceded in time by effects at the molecular and cellular level in individuals (Parrett, 1998). A common approach is therefore to initially identify biochemical disturbances (usually by identifying the common biomarkers; see Chapters 2 and 3) and then examine for physiological and/or physical irregularities. However, interpreting the significance of environmental induction of biochemical disruption depends, in part, on a number of factors, such as the identity and concentration of the inducing contaminants to which a fish is exposed, as well as the identification of several natural abiotic and biotic variables such as water temperature, age, sex, dietary factors, reproductive status and geographical location (e.g. Burgeot *et al.*, 1994; Sayer *et al.*, 1995; Sleiderink & Boon, 1995). In a study of total growth and gonadal investment rates of goldsinny (*Ctenolabrus rupestris* L.) from different locations on the west coast of Scotland, Sayer *et al.* (1995) measured extreme differences in the indices of both sexes, which were attributed to differences in densities of a territorial fish. On the whole, similar studies on wide-scale geographical comparisons on natural growth and condition indices of other fish species are lacking, and if coincidence cannot be discounted with certainty, caution is needed in the interpretation of findings from surveys.

Given that most environmental stressors occur in tandem, especially in estuaries and nearshore areas, there are few examples linking the effects of a single stressor such as a pollutant with changes to individual health. Elliott and Griffiths (1986, 1988) and Elliott *et al.* (1988), using a field analysis, studied mercury and hydrocarbon contamination across all components of the Forth estuarine system, Scotland. As an example of the uptake of pollutants and their transfer across trophic levels, mercury was analysed in many components of the systems (Fig. 6.2) and indicated that the fish most at risk from uptake were those in contact with the sediments, or whose prey were in intimate contact with the sediments; this is a reflection of the affinity of pollutants for the sedimentary components. As a consequence, migratory fish species were less affected by the pollutant than the resident fish such as flounder, *Platichthys flesus*. Elliott and Hemingway (2002) give further details of pollution levels in fishes in estuarine systems.

Taking the above study further, and using those specimens analysed for mercury contamination, the pathology of a resident component of the estuarine fish community, the flounder *Platichthys flesus*, was analysed to determine any reduction of health as the result of contaminant exposure (Elliott *et al.*, 1988). The proportion of individuals showing morphological anomalies and disease differed with site within the industrialised Forth Estuary, Scotland, and that proportion increased with distance downstream despite the most industrialised and polluted areas being in the middle region of the estuary (Table 6.1). Overall, 18% of the individuals examined showed some anomaly although the latter ranged from minor blemishes to major skeletal deformities. However, the authors concluded that although the

Table 6.1 Pathological disorders* in *Platichthys flesus* (L.) (Elliott *et al.*, 1988).

Station♦	l	b	n	fr	ma	ed	s	lw	No. fish examined	% disorders
UE	3	—	1	1	—	—	—	—	54	9.3
LO	21	5	2	2	1	1	1	1	180	18.9
Ta	1	1	—	—	—	—	—	—	10	20.0
PE	6	7	1	1	—	—	—	—	68	22.1
Estuary§	1	—	—	1	—	—	—	—	13	—
Total No.	32	13	4	5	1	1	1	1	325	
%	55	22	7	9	2	2	2	2		17.8

*l, lesions; b, blemishes; n, nodules; fr, fin-rot; ma, mouth abcesses; ed, eye deformities; s, scoliosis; lw, lamprey wound.
♦UE, Upper estuary; LO, Longannet; Ta, Tancred; PE, Port Edgar.
§Fish not recorded by station.

individuals showed a deterioration in health, as manifested by morphological anomalies and disease, the large number of concurrent stressors in that system made it difficult to conclusively link the cause of pollutant exposure and the effects detected. Following this, Mathieson (1993), Mathieson *et al.* (1996) and van Egmond (1993) then used an experimental approach to link the causes and effects and detected sublethal responses in the resident fish species.

A major difficulty is encountered firstly in explaining the incidence of pathological anomalies, and secondly in attempting to relate those to the ambient and accumulated contaminant concentrations. Whilst contaminants may cause the anomaly, the source could be in the individual under study at the time of study or throughout its development or in a parent. However, the contaminant could also provide entry damage which in turn is colonised and exacerbated by microbial contamination. Disease agents, possibly induced by xenobiotics in fish populations, are thought to have effects on host population dynamics through enzootic or epizootic events. Enzootic disease can influence host abundance through long-term impacts on physiological processes which affect growth, reproduction and survival, whereas epizootic diseases generally affect population dynamics by reducing populations in short-term events (Arkoosh *et al.*, 1998), which if sufficiently large scale might result in stochastic processes causing extinction (Gulland, 1995).

Many of the biochemical, molecular and cellular effects of pollutant exposure have been positively correlated with other indicators of animal health including physiological indices (Chapter 4). Marine teleosts in full-strength seawater (or any strength seawater greater than the iso-osmotic relationship) incur the energetically demanding process of hypo-osmoregulation. An external challenge to the physiological maintenance of the fish will demand increased energy partitioning in favour of maintained hypo-osmoregulatory functions (Sayer & Reader, 1996). During chronic low-stress exposure these increased energy demands will gradually affect less immediate life-threatening processes such as growth and gonadal investment. Acute high-stress exposures will rapidly result in osmoregulatory loss and death (Sayer & Reader, 1996).

6.2.3 Link 2: Individual health to production and yield

It is considered that the production by an individual provides an integrative response to all environmental perturbations and that any depression in the parameters of production ecology, such as growth, energy budget, individual production (yield) and production to mean biomass ratio will occur as the result of stress (Elliott & McLusky, 1985). However, whereas changes in those parameters have been used for invertebrates (Elliott & McLusky, 1985) and for the community of estuarine fishes (Elliott & Taylor, 1989) in relation to general stressors, they have not been used for specific or single stressors such as concentrations of individual pollutants.

In general, biomarkers can be classified according to exposure and/or effect. Biomarkers of exposure indicate a general stress response or to more specific groups of contaminants, whereas a biomarker of effect indicates that an adverse effect is conferred on an individual. The biomarker of effect has more potential to indicate an effect of ecological significance as it indicates that an impairment of health has taken place or is likely to take place. The consequences of biomarker activity for biological fitness of fish populations has not yet been established, albeit that any disturbances at the physiological level may have serious adverse effects on the organism (Goksøyr *et al.*, 1994; Parrett, 1998). Increased knowledge about the relationship between 'early warning' biomarkers and more serious consequences at the population and community level is required (Parrett, 1998). It is arguable that the recruitment of individuals to a population, based on reproduction and survival of offspring, is the ecosystem parameter of greatest concern. Despite this, the increasing amount of research on pollutant responses in marine organisms rarely, if at all, focus on the translation of responses in the individual to population, community and ecosystem changes (e.g. see the studies reported in Goksøyr, 1998).

6.3 Changes manifested in populations

6.3.1 Reproductive success of individual affected by pollutants (linking to reproductive capacity of population)

The abundance of a population varies in response to changes in the probabilities of survival and reproductive success of individual fish and, as a precursor to this, the abundance of recruits is determined by egg production and egg and larval mortality (Wootton, 1990; Jennings *et al.*, 2001). The influence and effects of xenobiotics on reproduction may occur on a variety of levels such as development of juveniles, coupling, quantity of eggs produced, egg quality, hatching of embryos and development of larvae (Donaldson & Scherer, 1983). As described in previous chapters, the younger stages of fishes are of greatest susceptibility to the effects of pollutants although, as described below, the mortality of those early life-stages by natural or anthropogenic stressors does not necessarily translate to an effect at the higher population level.

The unprotected eggs and sperm, embryos and larvae are likely to show the effects of pollutants at levels much less than those required to produce an effect in later stages

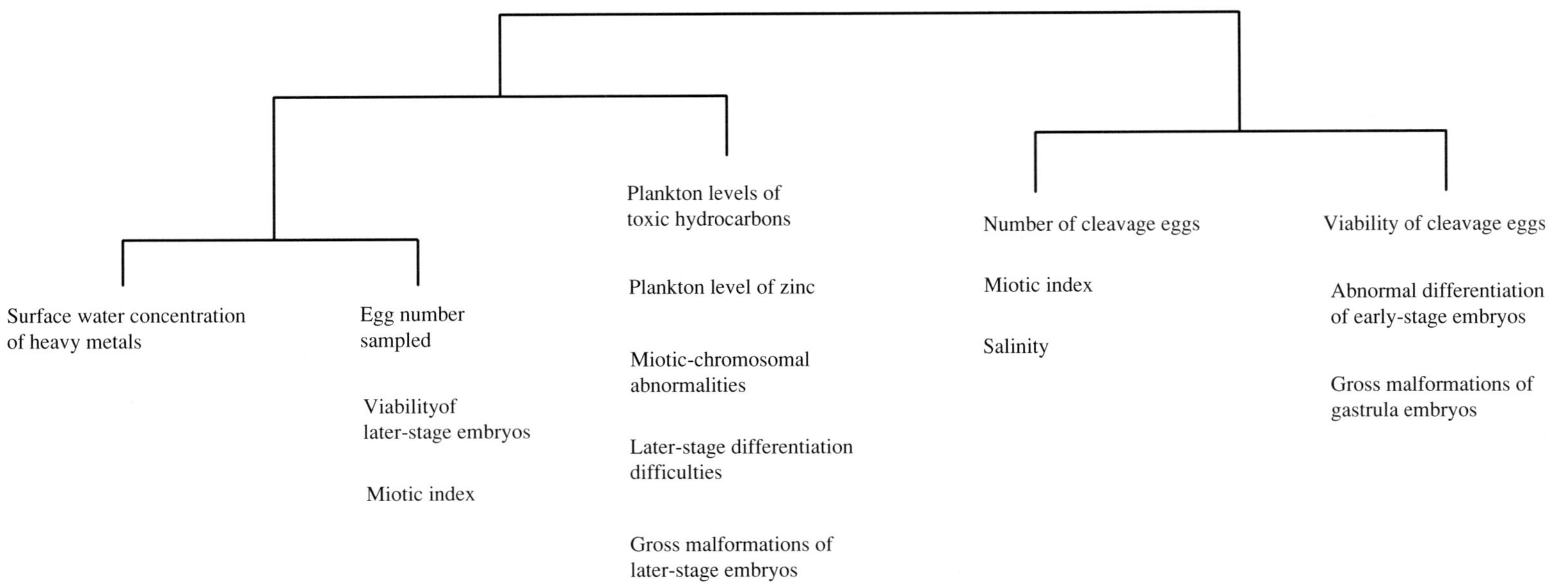

Fig. 6.3 Pollution effects on eggs and young stages, e.g. mackerel. Dendogram of signed correlations between Atlantic mackerel egg frequency, cytologic and embryonic measures of normality and abnormality of early-stage eggs, temperature, salinity, heavy metal, and toxic hydrocarbon levels. Correlation of a variable with other members of a cluster is higher than with variables not in the cluster. The level of joining of the clusters making the branches expresses the degree of relationships between cluster. The distance between is arbitrary. Modified from Longwell & Hughes, 1980.

(Longwell & Hughes, 1980). In a wide-ranging, empirical and correlative study, Longwell and Hughes (1980) assessed the incidence of reproductive and developmental abnormalities in relation to the presence of pollutants (Fig. 6.3). They assessed the pollutants in the water column and at the water-air interface, thus incorporating a knowledge of the behaviour of pollutants and measuring the concentrations at places likely to have an impact on pelagic eggs, embryos and larvae. They then used multivariate numerical techniques (cluster analyses) to relate these features to the number of eggs present, the viability of embryos, deformation, division and cleavage success and differentiation. While such correlative analysis does not imply cause and effect, it provides some evidence for the coincidence and processes of early stage damage due to pollutants. It is of note that this type of study has not been repeated more recently.

It is axiomatic that one of the most important xenobiotic effects on fish populations is the impairment of reproduction. Despite empirical studies such as that mentioned above, the widespread spawning and dispersal of reproductive stages dictate that it is difficult to link ambient pollution concentrations to reproduction and recruitment impairment. Secondly, spawning and reproductive problems/impairment cannot be related to recruitment success and population size. Finally, the degree of pollution response in the recruiting stages has not been related to the degree of contamination in adults.

Causal links between the impact of xenobiotics on reproductive success of an individual, and population responses, are difficult to detect, and it may not be a valid assumption that toxicant impacts can be generalised. Systems with similar characteristics may not respond to similar toxicant exposure in the same, or even similar ways. Additionally, detection becomes increasingly complex when there are sublethal or density-dependent factors impacting on survivors, and in such cases it is necessary to identify the point of impact (Munkittrick & Dixon, 1989). As indicated above, the dispersal nature of reproductive products, embryos and larvae prohibits the correlation of effects on those stages and ambient pollution concentrations. However, viviparous fish do give the opportunity to detect such cause and effect relationships. Reproduction by the viviparous blenny, *Zoarces viviparus*, in relation to mercury contamination was studied initially by Elliott and Griffiths (1986) and then in detail by Mathieson (1993). The former study indicated that there was no change in fecundity in polluted areas compared to cleaned areas, although the level of contamination in the young was related to that in the mother, i.e. females at a more polluted site had higher concentrations and produced more contaminated offspring than at a less-contaminated site (Table 6.2). However, there was no difference between the sites in the ratios of female to offspring content or concentration, suggesting that the same mechanism is operating at both sites irrespective of degree of contamination and that there is no greater control on mercury entering the brood at the more contaminated site. The authors hypothesised that, given the relatively sedentary nature of the species, as shown by a small home range, mothers in a polluted area would produce young with elevated contamination which in turn would produce other young which were contaminated. In addition, it appeared likely that there was no active control on the passage of the pollutant between generations.

Several species of wild teleost fish from the Baltic Sea show obvious signs of reproductive disorders, most notably salmon (*Salmo salar*), sea trout (*Salmo trutta*), cod (*Gadus morhua*) and burbot (*Lota lota*) (Bengtsson *et al.*, 1994). Although the reasons are not known, they are suspected to be related to the presence of anthropogenic substances. Within

Table 6.2 Significance tests of data (n = 10) relating to mercury contamination and fecundity of eelpout, *Zoarces viviparus* (Elliott & Griffiths, 1986).

Parameter		Port Edgar $\bar{x}$	Port Edgar S.D.	Longannet $\bar{x}$	Longannet S.D.	Test
Females analysed	size (cm)	19.6	3.44	19.2	3.01	n.s.
	weight (g)	36.4	16.9	36.7	18.4	n.s.
Mercury ($\mu g.g^{-1}$)	females	1.23	0.79	2.04	0.52	*
	brood	0.36	0.23	0.55	0.13	*
% Brood Hg content/ female Hg content		6.37	2.55	5.62	1.24	n.s.

n.s. = not significant
*denotes significant correlation when $p \leq 0.01$ (p = probability)

the marine field, however, these responses are complicated by the fact that disorders can be due both to levels of non-natural (xenobiotic) substances as well as non-natural levels (both elevated and reduced) of natural substances and conditions. Åkerman *et al.* (1996) found that reproductive success was dramatically lower in cod from the Baltic Sea compared with those from the relatively unpolluted Barents Sea, with the disturbances possibly being correlated to the female individuals rather than the males. These findings combined with earlier ones indicated that environmental pollution may be a factor involved in the observed disorders in the early development of cod in the Baltic Sea (Ericson *et al.*, 1996). As indicated above, the main difficulty with studies such as these is that the causal agent is non-specific and the effects may be the result of exposure to many stressors, both as contaminants and as unnatural levels of natural factors such as salinity, temperature and oxygen concentrations.

In July 1995, a workshop in Racine, Wisconsin, focused on studies which aimed to demonstrate the effects of contaminants on reproduction and development in fishes (Rolland *et al.*, 1997). In only eight cases did the participants feel confident that chemical contaminants could be linked with reproductive and/or developmental effects in wild fish populations. Approximately six cases were found to have an association or correlation between cause and effect, and in the majority of cases there was only some circumstantial evidence that contaminants may be involved. In some cases, laboratory induction of effects following exposure was observed, but these effects could not be extrapolated to population effects. Monosson (1997) indicates freshwater examples in which a decline in a lake trout (*Salvelinus namaycush*) population was probably related to contamination by dioxins and related compounds. This review indicates that there is a possible link in the decline of populations of English sole (*Parophrys vetulus*), Pacific herring (*Clupea pallasi*) and pink salmon (*Oncorhynchus gorbuscha*) and the presence of petrogenic hydrocarbons and sewage-related materials. In some cases, population size was affected although the role of contaminants was unclear. In these cases and in others where the population effects were unknown, sublethal effects on the reproductive physiology and biochemistry were observed.

With respect to the use of data in population assessments, Barnthouse *et al.* (1990) investigated three aspects of the use of toxicity test data for population-level risk assessment: the influence of life history characteristics on vulnerability to contaminant-induced stress; the importance of test data availability; and the influence of exploitation intensity. They quantified population-level effects of chronic contaminant exposure by coupling standard toxicity test data to matrix-type population models derived from long-term field studies of Chesapeake Bay striped bass (*Morone saxatilis*) and the Gulf of Mexico menhaden (*Brevoortia patronus*) populations. Regression analysis was used to quantify the uncertainty inherent in using test data ranging from life cycle tests to those produced by quantitative structure-activity relationships (QSARs) in order to estimate the effects of the availability and uptake of contaminants on the survival and reproduction parameters of the population models.

Barnthouse *et al.* (1990) found that due to differences in life history, menhaden and striped bass differ in terms of their capacity to sustain the same level of contaminant-induced mortality, and changes in exploitation intensity affect the responses of both populations to the same level of additional contaminant-induced mortality. The quantitative effects of both factors were, however, negligible compared to the uncertainty introduced by estimating long-term effects from short-term tests or QSARs, and the results suggest that consideration of life history may be important primarily for site-specific assessment, whereas for screening-level assessments, the substantial differences in uncertainty associated with different types of test data are of greater concern (Barnthouse *et al.*, 1990).

However, it is difficult to separate the pollutant from the non-pollutant factors which affect the reproductive capacity of a population, such as habitat alterations and other anthropogenic influences. For example, a reduction in reproduction due to the overfishing at sea of mature, commercial sizes or the prevention of spawning in estuaries by building barriers and restricting access to spawning grounds, will mask any reduction in reproductive fitness due to pollution. The importance of these factors on population dynamics is still unclear. Similarly, the relative strengths of these different stressors (habitat loss, overfishing, pollution) have not been determined although there may be mechanisms of determining relative loss of reproductive output.

An analogous system, which reflects the loss of juveniles through man-induced stress followed by population adjustment, is shown by impingement of juvenile fishes in power plant cooling systems (Turnpenny *et al.*, 1988). The loss of juveniles is compensated for such that there is little effect on the adult population equivalents, a feature which may be considered as 'population homeostasis'. Longwell & Hughes (1980) found empirical relationships between fecundity, embryo and larval integrity and contaminant levels following the analysis of field populations (see Fig. 6.3). Similarly, Luckenbach *et al.* (2001) found that the early life-stages of Brown trout (*Salmo trutta* f. *fario*) in differentially polluted streams had differences in the mortality rates, developmental rates, hatching period, proportion of malformations and growth rates. A stream polluted with organic chemicals and trace metals showed a retarded development, reduced growth rates and higher mortality rates. However, the differences were somewhat confounded by the overall differences in other physico-chemical and lino-chemical parameters.

More recently, studies have concentrated on the potential resistance by fishes exposed to organic pollutants, but have hypothesised that such a resistance will have potential costs and benefits to the populations. Nacci *et al.* (2002) compared populations of the estuarine species *Fundulus heteroclitus* (killifish) exposed to differing levels of PCBs in the field. Through differential metabolism of dioxin-like substances, the populations had a differential resistance which in turn may give populations protection following chronic exposure, and prevent the development of cancers.

6.3.2 Population models (e.g. Leslie matrix model)

An important aim of studies concerning the impact of xenobiotics and pollutants on fish population abundance is the development of predictive models for any given population. One such model is the Leslie matrix of age-specific survival rates and fecundities (Williamson, 1972). The analysis of such a model can focus on a number of indices, including long-term population size and the intrinsic rate of natural increase (*r*), reproductive value or potential, population resilience and risk of population extinction, and sensitivity of the dominant eigenvalue of the matrix to changes in model parameters (Caswell, 1989). The use of a Leslie matrix model provides a framework for further studies and also allows sensitivity analyses to be performed (Landahl *et al.*, 1997). Caswell (1989) and Usher (1972) give detailed descriptions of models such as the Leslie matrix, and Barnthouse (1993) summarises their use in assessing effects of toxicants on organism populations.

In order to construct a Leslie matrix model, it is extremely important to obtain accurate data on age-specific survival and reproductive rates for animals under various exposure conditions. In many previous studies of the impacts of contaminants on fish populations, the effects of single compounds based on water column toxicity have been used to estimate survival and reproductive impacts. Landahl *et al.* (1997) note that this method requires the extrapolation from tested species of fish to untested ones, and therefore, for more realistic model development it is preferable to base estimates of population-level toxicant impacts on data collected for the species of interest. As suggested above, the early study by Longwell and Hughes (1980) found changes as a result of pollution in reproduction and success of reproductive elements, but could not link these to changes at the population level.

Landahl *et al.* (1997) looked at approaches for determining effects of pollution on fish populations of Puget Sound (USA). The primary objectives of the study were to compare survival rates, reproduction rates and projected population growth rates of English sole (*Parophrys* (*Pleuronectes*) *vetulus*) from sites in Puget Sound with different levels of sediment contamination (Landahl & Johnson, 1993; Johnson & Landahl, 1994). Using field and laboratory data, they determined vital rates and other life history parameters in English sole subpopulations from urban and non-urban sites in Puget Sound, and used this information to estimate potential population level impacts of anthropogenic stressors, with a generalised Leslie matrix model.

Landahl *et al.* (1997) found that initial model projections indicated that contaminant effects, particularly in relation to reproductive capacity, could substantially reduce the intrinsic rate of increase (*r*) of English sole populations from contaminated sites in Puget Sound. Additionally, estimated reductions in *r* were comparable to reductions in *r* associated with fishing mortality rates between 15% and 30%, suggesting that contaminant impacts are a cause for concern. However, although the initial model provides insight into the potential effects of chemical contaminants on English sole populations, Landahl *et al.* (1997) note that several refinements are required in order to increase the model's ecological relevance. Firstly, better estimates of age-0 to age-3 survival obtained from field and laboratory studies with larval and juvenile English sole are critical as the model is extremely sensitive to these parameters. In addition, improved data is needed on both fishing and natural mortality for Puget Sound. Secondly, the model could be strengthened by a more careful analysis of site-specific growth rate and age at first sexual maturation, which would provide more accurate data on age-specific fecundity for fish from the sampling sites.

They concluded that the Leslie matrix model assumes a simple closed system within which local recruitment depends primarily on the reproductive output of adults residing at that site. Consequently, it is essential to consider the contribution of recruits from urban sites to the central Puget Sound English sole population, as well as the possibility that immigration into contaminated areas by offspring of fish from other sites could compensate for recruitment declines associated with contaminant exposure in localised areas. Without such information, the potential magnitude of pollution impacts on English sole abundance in central Puget Sound cannot be accurately assessed. However, Monosson (1997) suggested that the incorporation of field data into the Leslie matrix model predicts that effects of contaminants may exacerbate the effects of fishing pressure. In situations of equal fishing pressure, populations exposed to greater contaminant exposure were predicted to decline at a greater rate. This prediction, based on the use of the matrix model, was the result of pollutant-induced declines in fecundity which may, in turn, decrease population growth rate (assuming that there is no density-dependent compensation).

6.3.3 Reproductive capacity, survival, mortality to production and yield

6.3.3.1 Response-patterns of populations to reduced reproductive capacity

There is an extensive theory regarding the nature of spawning dynamics of fishes although the major drawback is linking that theory to the effects of pollutants. Aquatic ecosystems can only respond to changing conditions in a limited number of ways (Munkittrick & Dixon, 1989). A failure to reproduce at the maximum possible rate is regarded as a potential loss to the population (Moss *et al.*, 1982), due to the possibilities of reduced strength of the youngest year class and reduced recruitment. However, the ability of any population to absorb the loss of recruits without ultimate deterioration in the population is not known. Munkittrick and Dixon (1989) found that population responses to contaminants should be identical to any non-specific, density-independent stressor, despite the cause of the reduction. It is assumed that such density-independent (random) factors can cause population density to move away from a hypothetical equilibrium density, whilst density-dependent (regulatory) factors may lead to a return to it (Moss *et al.*, 1982).

Density-dependent population growth

Naturally, dispersal-spawning (as opposed to viviparous) fishes have high fecundities so the potential lifetime production of offspring is high. Although fish populations vary greatly in abundance (Cushing, 1982; Rothschild, 1986), increases in abundance are usually several orders of magnitude lower than the potential maximum increase, and even when a population is increasing, the fate of most zygotes is to die before growing to sexual maturity (Wootton, 1990). If the per capita death rate tends to increase with an increase in the population density (density-dependent mortality), or if the birth rate tends to decrease with an increase in population density (density-dependent natality), then there may be an equilibrium population density at which the birth and death rate balance each other. If the density

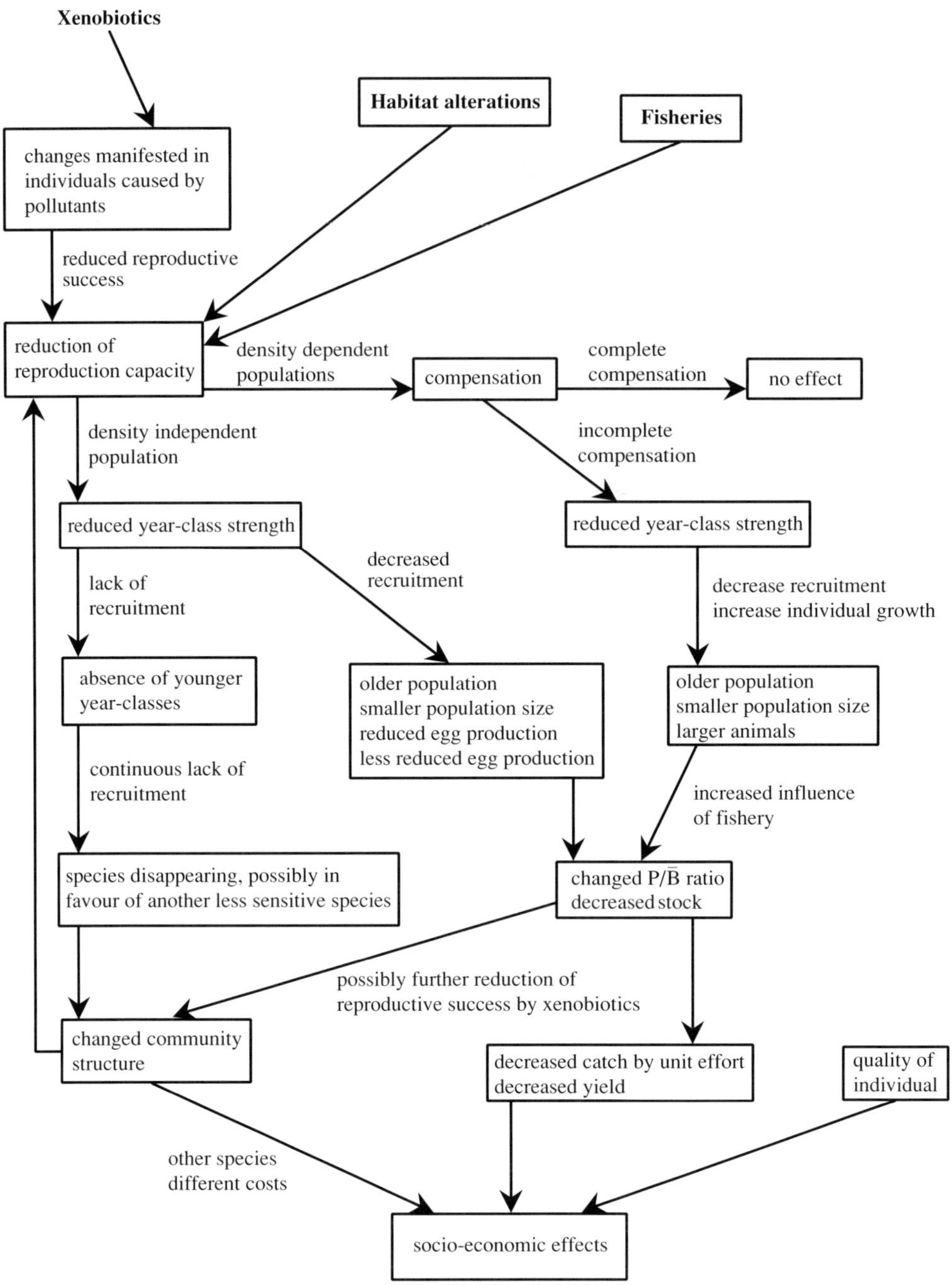

Fig. 6.4 Model to show potential population responses to reduced reproductive capacity as a result of pollution exposure.

drops below this equilibrium, the death rate will decrease (for example fewer fish are dying by lack of food) and/or the birth rate will increase (lower age at maturity), and therefore the population tends to return to the equilibrium. Changed emigration or immigration has the same effect as this.

Such density-dependent effects tend to buffer the direct effects of the factors which result in changes to mortality or fecundity. For fish in which fecundity and the risk of predation are functions of body size, density-dependent growth will also be an important demographic process (Wootton, 1990). Such inverse relationships between mortality factors are termed 'compensatory' (Moss *et al.*, 1982).

Density-independent population growth

The growth of a population is density-independent if the birth and death rates per individual do not depend on the population size (Hastings, 1997), for example, if the population is influenced by density-independent factors such as unnatural or fluctuating temperature and salinity or other stressors such as pollution. The abundance of the population will fluctuate through time, and density-dependent stressors do not occur. There is no situation in which the population can be thought of as regulated (Wootton, 1990).

6.3.3.2 Links between reproductive capacity, mortality rate, year-class strength and recruitment

The interlinkages between all aspects of population dynamics, recruitment success and survivorship require to be understood before changes due to pollutant exposure are superimposed on those linkages. Density-dependent and density-independent features are required to be determined and their links with survival and population maintenance need to be assessed in the light of pollution responses (Fig. 6.4).

Density-dependent populations

The increasing individual size of fishes with growth is matched by a decline in cohort abundance and the concomitant decline in the ability of the population to compensate for the effects of mortality, irrespective of whether that mortality is due to pollutants or natural factors. Although young stages or juveniles may undergo mortality, high levels of individual mortality of juveniles may not necessarily have a significant impact on adult population levels (Houde, 1989; Rose *et al.*, 1993). As such, mortality can be highly compensatory in fishery populations especially in the earliest life-stages (Jones *et al.*, 2002). Despite this, the pollution stress, either chronic or acute, could affect the population size if it occurs immediately after recruitment or in an area important as a nursery such as an estuary (Elliott & Hemingway, 2002; Jones *et al.*, 2002).

If density-dependent stressors are important in early life-stages of a fish population, then reduced reproductive capacity will be compensated for by mechanisms such as reduced juvenile mortality, faster individual growth, lower age at maturity and reduced competition for spawning area. In cases where a population is affected by a reduction of reproductive capacity, and yet still grows, reproduces and survives within the limits of a comparable reference population, it must be concluded that the reduction of reproductive capacity is completely compensated, i.e. a factor responsible for reduction of the year class strength is substituted by another. Although there are no examples with respect to pollution, if pollution is included merely as another stressor which may cause mortality but which is obscured by natural mortality, then the effect will not be detected.

Wootton (1990) noted that a density-dependent effect for the dynamics of the population may be present, but may be too weak to result in any significant regulation of the population. A strong density-dependent effect (fishery) can even destabilise a population by increasing the chance that very low densities are reached (den Boer, 1987).

With respect to *incomplete compensation*, as competition between adults would not initially change, growth characteristics of the adults may initially remain the same, although an increased growth rate in response to a decreased population size is generally characteristic (Munkittrick & Dixon, 1989). The fishery may have another impact on the fish stock as more younger fish can then be caught than would be the case without this compensation mechanism.

Density-independent populations

If the limits within which the compensatory mechanisms are able to operate are exceeded, or density-independent mechanisms are dominant, random larval mortality factors will be the same as in unstressed populations, and reduction of year-class strength is inevitable to an additional reduction in reproductive capacity. Rapport (1989) commented that in the Baltic Sea, many fish species are living at the limit of their distribution area. Population density and growth are mainly regulated by temperature and salinity such that the fish may be sensitive to other stressors. As a result, a number of fish populations in the Baltic Sea are good examples of density-independent populations.

The central difficulty in linking the above to the effects of pollution is that there are very limited empirical data to follow the sequence of pollutants affecting young stages through to population changes. As a way forward, Schaaf *et al.* (1987) developed a simulation model to estimate pollution effects on economically important estuarine-dependent fish populations. Available life history data on eight species (fourteen spatial-temporal stocks) is compiled, concentrating on age-specific rates of growth, survival and fecundity. Leslie matrix models of species population dynamics were used to predict pollutant impacts – mediated through changes in first year survival. Schaaf *et al.* (1987) found that on average, and without compensation, these modelled stocks respond to a one-time 50% reduction in first year survival by taking ten years to equilibrate at 88% of their pre-impact abundance. Synthesis of the data included a search for derived (standardised) population parameters to evaluate differences in susceptibility both among and within fish populations, to pollutant stresses. It was demonstrated that knowledge of a species' age-specific fecundity pattern provides additional predictive power of its response to pollution perturbation.

In the case of reduced reproductive capacity and/or increased juvenile and/or egg mortality, the mean age of the population will increase in response to a decreased population size, resulting in an older and weaker population, and possibly the absence of younger year classes. Munkittrick and Dixon (1989) called this a Type II response in contrast to the influence of fishery (adult removal), which leads to increased growth rate and fecundity, as well as an earlier age at maturation and decline of the mean age of the population (Type I response), multiple stressors (Type III response), limitation (Type IV response) and niche shift (Type V response). Type II changes have been reported by Beggs and Gunn (1986), Munkittrick and Leatherland (1984), Black *et al.* (1985) and Colby (1984).

As mortality factors may not be independent of each other (Moss *et al.*, 1982), random larval mortality factors may increase if the factors which affected reproductive success lead to an increase in larval mortality. An additional burden in this case leads to a further reduction in the strength of the younger year classes. If the factors result in complete removal of the first year class over the life span of a fish species, the population will become older and older and will eventually disappear, as seen in acidic waters since 1940. Ensenbach and Nagel (1997) forecasted the disappearance of their experimental population of zebrafish in the case of whole lifetime exposure to 100/40 $\mu g\ l^{-1}$ lindane as they observed no egg production. Hudd *et al.* (1986) found that reproductive failure of burbot and bream resulted from weak year classes, and Dutton *et al.* (1988) described an absence of younger year classes in white sucker populations collected from Hamell Lacke during the early 1980s. Similarly, Sandström *et al.* (1991) found decreased abundances of fish close to pulp mill effluents.

Until now, it has been extremely difficult to predict population fluctuations. Compensation limits are not known for different density-dependent populations, nor how close to the limit a population influenced by natural stressors may exist, despite the fact that the capacity to buffer anthropogenic influences depends on this basis. It is not possible to predict for each population, at what level of additional stressors compensation-processes are at their limit, and density-dependent population growth changes to density-independent.

6.3.3.3 Effects of changes in population structure on production, yield and the quantity of populations

Production and the parameters of production ecology of a species are a synthesis of population biomass, recruitment, growth and mortality, and as such are especially responsive to the health of a population in relation to environmental change (Elliott & McLusky, 1985; Mann & Penczak, 1986). Wootton (1990) and Elliott & Taylor (1989) noted that a feature of production in many populations is the high proportion contributed by the youngest age classes, which was first observed by Mann (1965), and Mann *et al.* (1972). This is a reflection of the high growth rates in the early life history stages, which are characterised by higher $P/\bar{B}$ ratios than the larger and older fish.

The change to older and weaker populations as a response of reduced reproductive capacity in density-independent populations is followed by a decrease of reproducing individuals, leading to reduced egg production and reproductive success of the survivor. Kwak and Waters (1997) found a significant linear inverse relationship between the annual $P/\bar{B}$ ratio and the number of population year-classes incorporating populations of each species of Minnesota salmonids. The weakness of population year classes during several years leads to decreased stock or in some cases disappearance of the whole species.

Iles (1994) reported that recruitment decreases with low stock size, leading to the assumption that reduced population size and changed population structure results in a further reduction to recruitment. This leads to lower stock size, especially if the stock is additionally reduced by fisheries, or reproductive success is reduced by the ongoing impact of pollution on spawners and offspring. As an example of this, Goodyear (1985) demonstrated that an increase in fishing mortality, or equivalent decrease in early life-stage survival caused by toxic effects of a pollutant, would cause similar stock declines. He found that if pollutant-induced mortality occurs after a period of high density-dependent mortality, the decline in yield would be more severe than that caused by an equivalent increase in fishing mortality.

Hudd *et al.* (1986) described loss of catches due to reproductive failure in response to anthropogenic acidification of spawning and nursery areas in the river Kyrönjoki, Finland. For burbot, conservative estimates described a loss of 50 tonnes from 1973 to 1982, and 70 tonnes for bream after the mass mortalities in 1970–73. In this case, they showed that the most important effect on the fish stocks and fisheries was the loss of reproduction. However, in response to the decline of one population, another perhaps less sensitive stock species may subsequently increase. Hildén *et al.* (1984) found changes to community structure in pelagic fish of the Baltic Sea in that over the last 50 years, the dominance appears to have changed from *Coregonus albula* (vendace) to the more tolerant species, smelt and herring.

The reduced catch of one species by unit of effort may result in a higher catch of another species. However, this may be more or less popular to both the fishery and the consumer, and an effect on the mixed fishery is expected through changes in diversity and yield.

6.4 Changes manifested in community response

It is of note that there are no readily available studies which attribute a loss of community, i.e. a reduction in species richness, to the presence of specific xenobiotics either singly or in combination. There are, however, many examples of the changes to species composition with other stressors such as organic enrichment, physical disruption of estuarine systems and overfishing (Blaber *et al.*, 2000; Elliott & Hemingway, 2002). For example, the estuaries of the Thames, the Mersey and the Clyde in the UK suffered large-scale reductions in their fish fauna as the result of organic pollution during the nineteenth and early twentieth centuries. While it is likely in these cases that other xenobiotic contaminants were also a contributory factor, it is not possible to indicate what proportion of the decline in the community was due to the different forms of pollution.

Despite the above, the community within any area will be a result of the differential tolerances of each member of that community to pollutant stress. In addition, interspecific relationships are suspected to be regulatory mechanisms of population dynamic processes (Thiel *et al.*, 1996). Changes in population size and structure caused by pollutants on one species will have an impact on the population of other species within the community. For example, Hildén (1997) found a decrease in smelt abundance (which are the most important prey to cod) in connection with an increase in cod abundance. In addition, the observed decrease in abundance of *Myoxocephalus scorpio*, *Myoxocephalus quadricornis*, *Zoarces viviparus* and *Perca fluviatilis* in the Baltic Sea are also thought to be a result of increasing population growth of cod. Changes of community structure in fish were detected by Karås (1989) in the shallow coastal waters of the Baltic, the most important spawning and nursery areas of some fish species. Despite the presence of pollutants from many sources, these changes were believed to be caused by eutrophication which in the southern Baltic provides the greatest stressor. Risk assessment has regularly utilised analysis at the community level as a tool for determining the health of an aquatic system (although to date, the majority of these studies have focused on invertebrates), and community structure itself is an expression of variation in the populations of the constituent species and the response of these populations to environmental stress (Attrill & Depledge, 1997). The loss of community diversity and stability in many damaged systems can lead to inefficient operation of the mechanisms regulating population size, resulting in fluctuations in total population size (Adams & Olver, 1977). As the result of such background knowledge, there are many examples of indices and other numerical methods which describe and present the state of communities exposed to anthropogenic change (Whitfield & Elliott, 2002). In many cases, these techniques describe either the change in species composition from that expected under unimpacted conditions, or the response of different guilds of fishes which have differential tolerances to anthropogenic change.

Maintenance of populations is dictated by the input of individuals within that population in terms of growth (biomass) and reproductive output. Both parameters can provide useful information on the health of a system. It is therefore important to explore the mechanisms linking the different levels of biological organisation to understand how individual toxicological responses may be expressed at the community level, and conversely what mechanisms are producing observed community structures in stressed systems (Attrill & Depledge, 1997).

Walker *et al.* (1996) outline four ways in which species within an ecosystem may be affected by the addition of pollutants. Firstly, the numbers of some species will decline, perhaps even to the extent of the species becoming locally extinct. Secondly, numbers may decline but level out lower than before and the population may persist at this level if the pollution continues (chronic pollution). Thirdly, population size may initially increase as if the pollution is chronic, resistance may evolve within the population, allowing population numbers eventually to increase to a new equilibrium. Finally, if the pollution is transient then the population may eventually recover, either rising from the level to which it was depressed by pollution or returning through immigration/recolonisation if pollution had rendered the population extinct.

However, pollution in this case often refers either to combinations of stressors or to organic inputs rather than xenobiotics. The changes to community structure as a result of pollution have been well studied in relation to marine invertebrate fauna (Elliott, 1994); there have been no similar syntheses with information related to marine and estuarine fishes. On a wider framework, Odum (1985) gives the features expected in stressed ecosystems (Table 6.3) and while many of these features can be seen with organic and nutrient pollution (de Jonge & Elliott, 2001), as yet this approach has not been attempted with xenobiotic stress.

6.4.1 *Effects on competition and behaviour*

It is necessary to understand the potential effects of xenobiotic bioaccumulation on fish behaviour and competition as they will influence the structure of a community. The concepts, basic knowledge and principles that underlie competition and behaviour in fish

Table 6.3 Trends expected in stressed ecosystems (Odum, 1985).

A. Energetics

1. Community respiration increases (H T Odum's 'pumping out' of disorder (Odum, 1967), or Prigogine's increase in the 'dissipative structure' (Prigogine *et al.*, 1972).
2. P/R (production/respiration) becomes unbalanced (< or > 1).
3. $P/\bar{B}$ and $R/\bar{B}$ (maintenance:biomass structure) ratios increase.
4. Importance of auxiliary energy increases (Margalef's (1975) exosomatic metabolism).
5. Exported or unused primary production increases.

B. Nutrient cycling

6. Nutrient turnover increases.
7. Horizontal transport increases and vertical cycling of nutrients decreases (cycling index decreases).
8. Nutrient loss increases (system becomes more 'leaky').

C. Community structure

9. Proportion of r-strategists increases.
10. Size of organisms decreases.
11. Lifespans of organisms or parts (leaves, for example) decrease.
12. Food chains shorten because of reduced energy flow at higher trophic levels and/or greater sensitivity of predators to stress.
13. Species diversity decreases and dominance increases; if original diversity is low, the reverse may occur; at the ecosystem level, redundancy of parallel processes theoretically declines.

D. General system-level trends

14. Ecosystem becomes more open (i.e. input and output environments become more important as internal cycling is reduced).
15. Autogenic successional trends reverse (succession reverts to earlier stages).
16. Efficiency of resource use decreases.
17. Parasitism and other negative interactions increase, and mutualism and other positive interactions decrease.
18. Functional properties (such as community metabolism) are more robust (homeostatic-resistant to stressors) than are species composition and other structural properties.

species should be considered. Competition takes place where resources are in short supply, and can take two forms as discussed earlier by Nicholson (1954) and Elton & Miller (1954).

Firstly, interference or contest competition occurs where access to a resource is denied to competitors by dominant individuals or species. Interference competition results in unequal access to resources, and subordinate individuals may not be able to participate in reproduction. Interference competition tends to couple numbers to available resources and often reduces population fluctuations. Many mechanisms involving behaviour and physiology are commonly involved in this kind of competition.

Secondly, exploitation or scramble competition is more involved in most of the biological world. This implies the direct use of a resource, reducing its availability to a competing individual or species simply because of consumption. Exploitation competition tends, at least in theory, to involve large fluctuations in density since the population may be built up based on a temporarily abundant resource, and then suffer a large collapse when the resource is exhausted.

There is a good conceptual understanding of the behavioural responses to pollutant exposure (Fig. 6.5) in which the main pathways relate to tolerance of the stressor or an

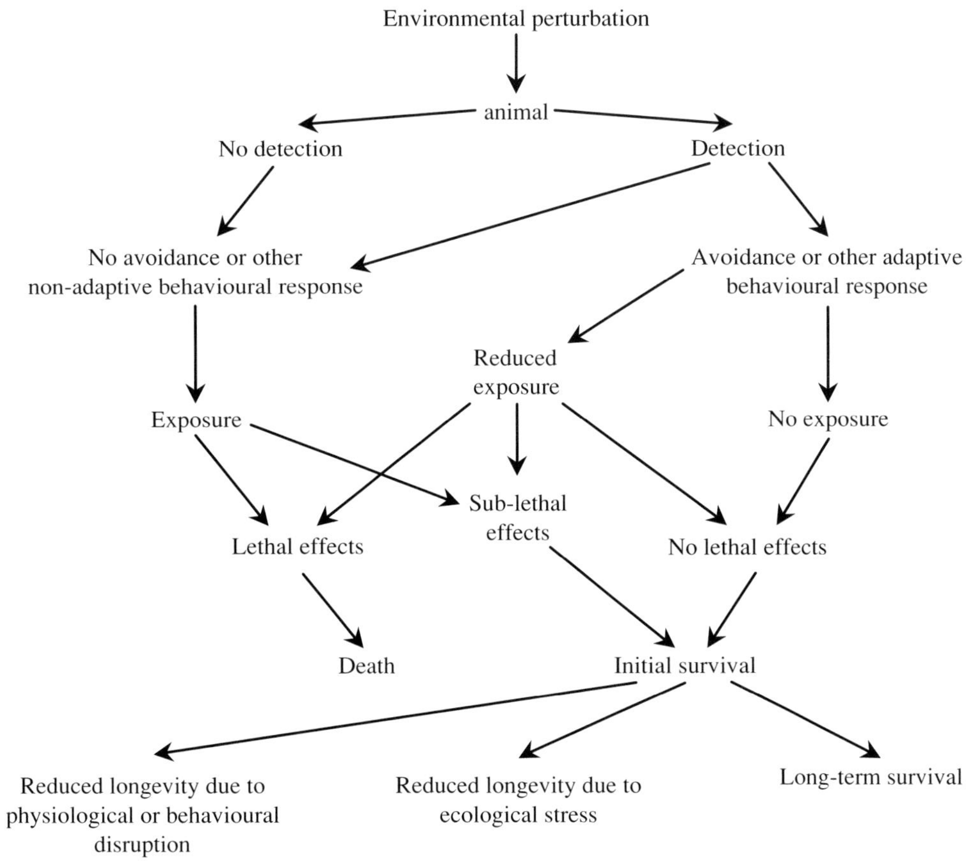

Fig. 6.5 Links between possible behavioural responses to environmental perturbations and their consequences. Modified from Olla *et al.*, 1980.

avoidance. Olla *et al.* (1980) suggest that in marine organisms these responses are shown by two sets of studies:

(1) Those which measure capabilities of organisms to mitigate effects of environmental perturbations (e.g. thermo-regulation in fishes exposed to thermal pollution)
(2) Those which measure departures from normal behaviour which, while not causing death immediately, may reduce probability for survival (e.g. avoidance of unfavourable conditions by plankton delaying metamorphosis).

Despite this, there are no studies which have quantified these links in fishes exposed to chemical (non-organic) pollutants.

The behavioural responses will relate to the resources commonly competed for by fishes, which include space, food and reproductive partners, hence the effects depend on whether pollution reduces any of these. Whereas organic pollution may remove the fishes' invertebrate prey or change its community structure, or may affect the quality of sediments and thus make the prey unpalatable, space could be lost by pollution affecting a habitat (Pearson

& Rosenberg, 1978; McLusky *et al.*, 1993; Elliott & Hemingway, 2002). Despite this, there are few available studies which, firstly, indicate that these effects can be caused by chemical pollutants, and, secondly, in turn affect the fish community.

In general, interspecific and intraspecific competition will occur when resources are limiting and where resource-partitioning is unable to prevent competition. The major determinant of the relative abundance of those resources, since exploitation competition is so common, is the number of individuals in a population. However, demographic properties can be affected by changes in the density of a given population of marine organisms and the competition within that species (i.e. intraspecific competition), and by changes in the density of other competing species (interspecific competition). Any reduction in the resource as a result of pollution will therefore increase competition and so organisms must anticipate environmental changes and respond appropriately (Jacobs, 1996). However, as indicated throughout this book, organisms modify their morphology, physiology and behaviour accordingly to stress exposure and thus reduce potential stress. Many organisms adjust their phenotype to maximise fitness in any given habitat (phenotypic plasticity) (West-Eberhard, 1989). Biological variables are, on a day-to-day basis, not always predictable, but can also be highly unpredictable with stressful consequences. Thus, in addition to predictable series of life history strategies within an organism's life cycle, there is growing evidence for an 'emergency life history stage' that is triggered by unpredictable events in the environment termed modifying factors or labile perturbation factors (Jacobs, 1996; Wingfield *et al.*, 1998). An example of labile perturbation factors can be pollution pressure which can trigger facultative behavioural and physiological responses that make up the emergency life history stage.

It has been noted that the anatomical and physiological characteristics of a fish are best described when explained in relation to their effects on its behaviour. For example, Lawrence and Poulter (1998) identified a clear biphasic swim response to copper in *Gammarus dubeni*, which appeared to be linked to respiration rate. In this case, the response observed in the amphipod was both time and concentration-dependent. Thus, the effect of pollutant exposure could be significant. Behaviour can be divided into several categories:

(1) Migratory behaviour
(2) Schooling behaviour
(3) Feeding behaviour
(4) Aggressive interaction (including competition)
(5) Resting behaviour
(6) Reproductive behaviour
(7) Interspecific interaction (including predator-prey, mimicry and symbiotic relationships).

Whereas the loss of resources will affect these behaviours, the latter are more difficult to relate to xenobiotics. Pollution due to organic enrichment, and thus the lowering of dissolved oxygen, may affect these behaviour patterns (Elliott & Hemingway, 2002).

The migratory behaviour of fish occurs on a seasonal to daily basis (Moyle & Cech, 1988). The use of migration to separate life history stages is characteristic not only of diadromous fishes such as salmon and eels but also of pelagic fishes such as herring and sardines, of many freshwater fishes, and of many marine benthic fishes (McDowall, 1988).

While the majority of fish migrations are related to reproduction and separation of life history stages, many are also in response to changing environmental conditions, particularly temperature, and the movements and abundance of food organisms.

It is not well understood how xenobiotic bioaccumulation will affect fish competition and behaviour discussed above. The alteration of fish physiology and behaviour as the result of chemical pollution is not well studied, thus predicting the effects of xenobiotic bioaccumulation on them is difficult. During emergencies in their natural environments, vertebrates initiate coping mechanisms that redirect behaviour away from non-essential activities and towards survival. For example, in passerine birds, evidence from field studies shows that this inhibition may not depend on the suppression of gonadal sex steroids, since during the breeding season they remain elevated despite activation of the stress response. In plainfin midshipman fish (*Porichthys notatus*), the pattern of androgen levels exhibited by reproductively active parental male may reflect a compromise between investment in paternal care versus courtship and/or territoriality (Knapp *et al.*, 1999). Given that competition and behaviour are coupled processes that are to a large extent controlled by the neuroendocrine system, xenobiotic bioaccumulation might have profound effects on fish. For example, during reproduction, fish show a wide range of behaviour patterns. Although such behaviour is highly adaptive and is strongly correlated with the overall ecology of each species and with morphological adaptations, a chemical influence that disrupts the sensory clues that induce such behaviour might result in reproductive failure. Such effects can be manifested through, but not limited to, the loss of breeding ground through competition, inability to guard or hide broods and nests (for brood hiders and guarders) and an inability to display appropriate mating rituals to attract mates.

6.4.2 Effects on mixed fishery – socio-economic changes

This chapter has considered the repercussions of pollutant exposure on the higher levels of biological organisation (populations, communities) but has emphasised the poor empirical evidence for any translation of effects on developmental stages, biochemical processes and individual health to those higher levels. Despite this, there are modelling studies which attempt to translate pollution responses to the higher level damage to fisheries, especially economic effects (Collins *et al.*, 1998). Studies such as those following the *Exxon Valdez* oil spill provide information for the reduction in the economic value of the fisheries and this can then be used in modelling for extrapolation to wider fisheries. Collins *et al.* (1998) produced a synoptic model (Fig. 6.6) and then used illustrative data to simulate the effects of chronic and acute pollution incidents to assess the economic changes likely. The model uses concentrations of pollutant in the waters and in individual fish and hence in the stock. It has to make assumptions regarding the mortality due to those contaminant levels as separated from the fishing mortality and acknowledges that the former may be minimal in relation to the latter. While this study is acknowledged to be an academic exercise, it provides the opportunity to test hypotheses and scenarios and thus it is considered here as a necessary and future development in the study of the effects of pollutants on fishes and fisheries (Chapter 8).

Above and beyond xenobiotic influences there are substantial problems in the context of management of the fish stocks. Many of the prime commercial species of the European

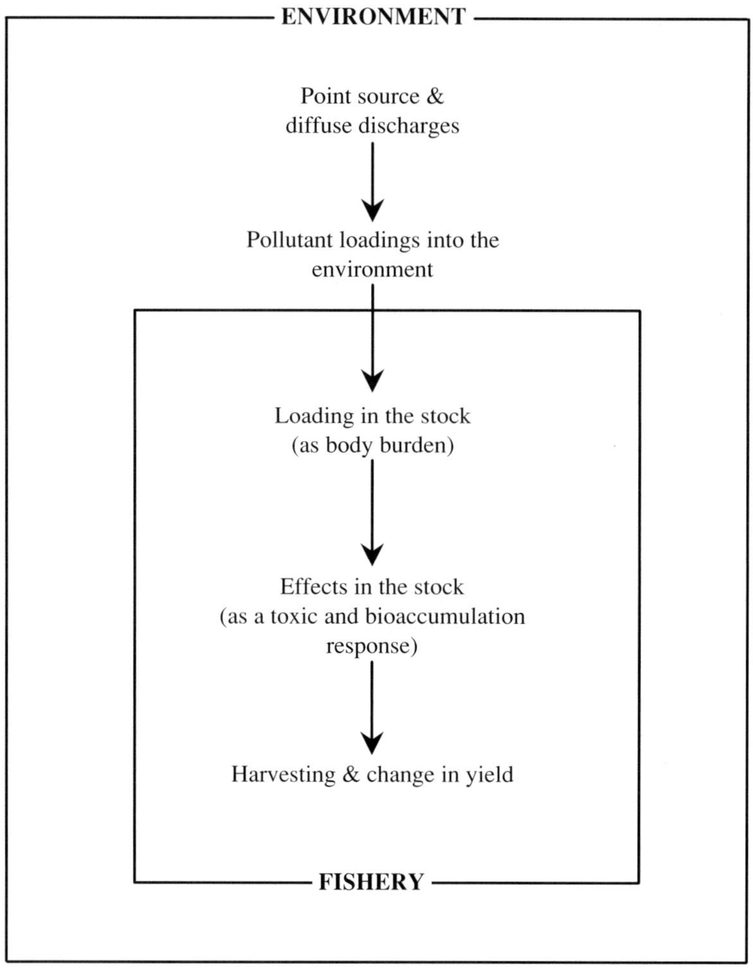

Fig. 6.6 A synoptic model of fishery-pollution interaction. Modified from Collins *et al.*, 1998.

Union are in decline as a result of adverse fisheries-dependent factors, notably excessive fishing effort. Indeed, those who are responsible for the design and implementation of the Common Fisheries Policy have been unsuccessful in halting this trend. The CFP revision, in 2002, is designed to take an ecosystemic approach to the management of the fish stocks and to achieve sustainability of the use of those stocks. While it is acknowledged that overfishing as a stressor will always dominate any effects due to pollutants, it is an aim to consider all stressors. Thus pollution effects at a multispecies level operate in a situation where the resource base is under threat from fisheries-dependent variables. The threat is posed by changes at a sublethal level as well as the occurrence of catastrophic incidents. In terms of a mixed fishery this might be manifested by differential die-off rates for individual species or differential changes in somatic quality. Whilst this would be of great importance in bio-scientific terms, it would be more difficult to ascribe specific socio-economic changes beyond those implicit in the loss or decline of individual fish species of commercial importance. A fuller treatment of these issues is given in Chapter 8.

6.5 References

Adams, G.F. & C.H. Olver (1977) Yield properties and structure of boreal percid communities in Ontario. *Journal of the Fisheries Research Board of Canada*, **34**, 1613–1625.

Åkerman, G., U. Tjärnlund, D. Broman, C. Näf, L. Westin & L. Balk (1996) Comparison of reproductive success of cod, *Gadus morhua*, from the Barents Sea and Baltic Sea. *Marine Environmental Research*, **42** (1–4), 139–144.

Amiard, J.C., C. Amiard-Triquet, C. Metayer & J. Marchand (1980) Etude du transfert de Cd, Pb, Cu et Zn dans les chaînes trophiques néritiques et estuariennes – I. Etat dans l'estuaire interne de la Loire (France) au cours de l'été 1978. *Water Research*, **14**, 665–673.

Arico, S. (1995) Report on international efforts in research monitoring and capacity building in the field of marine and coastal biological diversity. *Ocean Coastal Management*, **19**, 329–335.

Arkoosh, M.R., E. Casillas, E. Clemons, A.N. Kagley, R. Olson, P. Reno & J.E. Stein (1998) Effect of pollution on fish diseases: Potential impacts on salmonid populations. *Journal of Aquatic Animal Health*, **10** (2), 182–190.

Attrill, M.J. & M.H. Depledge (1997) Community and population indicators of ecosystem health: Targeting links between levels of biological organisation. *Aquatic Toxicology*, **38** (1–3), 183–197.

Barnthouse, L.W. (1993) Population-level effects. In: (ed. Sutter II, G.W.) *Ecological risk assessment*. Lewis Publishers, Boca Raton, Florida.

Barnthouse, L.W., G.W. Suter II & A.E. Rosen (1990) Risks of toxic contaminants to exploited fish populations: influence of life history, data uncertainty and exploitation intensity. *Environmental Toxicology and Chemistry*, **9**, 297–311.

Beggs, G.L. & J.M. Gunn (1986) Response of lake trout (*Salvelinus namaycus*) and brook trout (*S. fontinalis*) to surface water acidification in Ontario. *Water Air and Soil Pollution*, **30**, 711–717.

Bengtsson, B.E., A. Bergman, I. Brandt, C. Hill, N. Johansson, A. Södergren & J. Thulin (1994) *Reproductive disturbances in Baltic fish: Research Programme for the period 1994/95–1998/99*. Swedish Environmental Protection Agency, Report 4319, Sweden.

Blaber, S.J.M., J.-J. Albaret, Chong Ving Ching, D.P. Cyrus, J.W. Day, M. Elliott, D. Fonseca, J. Hoss, J. Orensanz, I.C. Potter & W. Silvert (2000) Effects of fishing on the structure and functioning of estuarine and nearshore ecosystems. *ICES Journal of Marine Science*, **57** (3), 590–602.

Black, J.J., E.D. Evans, J.C. Harshbarger & R.F. Ziegel (1985) Epizootic neoplasms in fishes from a lake polluted by copper mining wastes. *Journal of the National Cancer Institute*, **69**, 915–926.

Blackstock, J. (1984) Biochemical metabolic regulatory responses of marine invertebrates to natural environmental change and marine pollution. *Oceanography and Marine Biology*, Annual Review, **22**, 263–313.

Burgeot, T., G. Bocquené, G. Pingray, D. Godefroy, J. Legrand, J. Dimeet, F. Marco, F. Vincent, Y. Henocque, H.O. Jeanneret & F. Galgani (1994) Monitoring biological effects of contamination in marine fish along French coasts by measurement of ethoxyresorufin-O-deethylase activity. *Ecotoxicology and Environmental Safety*, **29**, 131–147.

Caswell, H. (1989) *Matrix population models: construction, analysis, and interpretation*. Sinauer Associates, Sunderland, Massachusetts.

Colby, P.J. (1984) Appraising the status of fisheries: rehabilitation techniques. In: (eds Cairns, V.W., P.V. Hodson & J.O. Nriagu) *Contaminant Effect on Fisheries. Advances in Environmental Science and Technology*, pp. 233–257.

Collins, A., M. Stapleton & D. Whitmarsh (1998) Fishery-pollution interactions: a modelling approach to explore the nature and incidence of economic damages. *Marine Pollution Bulletin*. **36** (3), 211–221.

Cushing, D.H. (1982) *Climate and Fisheries*. Academic Press, London.

de Jonge, V.N. & M. Elliott (2001) Eutrophication. In: (eds Steele, S., S. Thorpe & K. Turekian) *Encyclopedia of Ocean Sciences*, Vol. 2. Academic Press, London, pp. 852–870.

den Boer, P.J. (1987) Density dependence and the stabilization of animal numbers, 2. The pine looper. *Netherlands Journal of Zoology*, **37** (2), 220–237.

Diaz, R.J. & R. Rosenberg (1995) Marine benthic hypoxia: a review of its ecological effects and the behavioural responses of benthic macrofauna. *Oceanography and Marine Biology Annual Review*, **33**, 245–303.

Donaldson, E.M. & E. Scherer (1983) Methods to test and assess effects of chemicals on reproduction in fish. In: (eds Vouk, V.B. & P.J. Sheephan) *Methods for Assessing the Effects of Chemicals on Reproductive Functions*. Wiley and Sons, Chichester, pp. 365–405.

Dutton, M.D., H.S. Majewsky & J.F. Klaverkamp (1988) Biochemical stress indicators in fish from lakes near a metal smelter. In: *31st Conference, May 17, 1988*. International Association of Great Lakes Research, Hamilton, Ontario, pp. A-14.

Elliott, M. (1994) The analysis of macrobenthic community data. *Marine Pollution Bulletin*, **28** (2), 62–64.

Elliott, M. (2002) The role of the DPSIR approach and conceptual models in marine environmental management: an example for offshore wind power. *Marine Pollution Bulletin*, **44** (6), iii–vi.

Elliott, M. & A.H. Griffiths (1986) Mercury contamination in components of an estuarine ecosystem. *Water Science and Technology*, **18**, 161–170.

Elliott, M. & A.H. Griffiths (1988) Contamination and effects of hydrocarbons on the Forth ecosystem. *Proceedings of the Royal Society of Edinburgh*, **93B**, 327–342.

Elliott, M. & K.L. Hemingway (eds) (2002) *Fishes in Estuaries*. Blackwell Science, Oxford, UK, 636 pp.

Elliott, M. & D.S. McLusky (1985) Invertebrate production ecology in relation to estuarine management. In: (eds Wilson, J.G. & W. Halcrow) *Estuarine Management and Quality Assessment*. Plenum Press, New York, pp. 85–103.

Elliott, M. & C.J.L. Taylor (1989) The structure and functioning of an estuarine/marine fish community in the Forth Estuary, Scotland. *Proceedings of the 21st European Marine Biology Symposium*, Gdansk, September 1986. Polish Academy of Sciences – Institute of Oceanology, pp. 227–240.

Elliott, M., A.H. Griffiths & C.J.L. Taylor (1988) The role of fish studies in estuarine pollution assessment. *Journal of Fish Biology*, **33** (Suppl. A), 51–61.

Elliott, M., S. Nedwell, N.V. Jones, S.J. Read, N.D. Cutts & K.L. Hemingway (1998) *Volume II – Intertidal sand and mudflats & subtidal mobile sandbanks: an overview of dynamic and sensitivity characteristics for conservation management of marine SACs*. Institute of Estuarine & Coastal Studies, University of Hull. Report for and prepared by Scottish Association for Marine Science (SAMS) for the UK Marine SACs Project.

Elton, C. & R.S. Miller (1954) The ecological survey of animal communities with a practical system of classifying habitats by structural characters. *Journal of Ecology*, **42**, 460–496.

Ensenbach, U. & R. Nagel (1997) Toxicity of binary chemical mixtures: Effects on reproduction of zebrafish (*Brachydanio rerio*). *Archives of Environmental Contamination and Toxicology*, **32** (2), 204–210.

Environment Agency (1999) *Humber Estuary: State of the Environment 1998*. Environment Agency, Leeds, 41 pp.

Ericson, G., G. Åkerman, B. Liewenborg & L. Balk (1996) Comparison of DNA damage in the early life stages of cod, *Gadus morhua*, originating from the Barents Sea and Baltic Sea. *Marine Environmental Research*, **42** (1–4), 119–123.

Ferreira, A.M., C. Vale, O.G. Castro & C. Cortesao (1985) *Metals pesados e organoclorados em caranguejos da Ria de Aveiro*. Instituto Nacional de Investigaçao das Pescas, Lisboa, Portugal.

Goksøyr, A. (ed.) (1998) Pollutant Responses in Marine Organisms (PRIMO9). *Marine Environmental Research*, **46** (1–5), 607 pp.

Goksøyr, A., J. Beyer, A. Husøy, H.E. Larsen, K. Westerheim, S. Wilhelmsen & J. Klunsøyr (1994) Accumulation and effects of aromatic and chlorinated hydrocarbons in juvenile cod (*Gadus morhua*) caged in a polluted fjord. *Aquatic Toxicology*, **29**, 21–35.

Goksøyr, A., J. Beyer, E. Egaas, B.E. Grosvik, K. Hylland, M. Sandvik & J.U. Skaare (1996) Biomarker responses in flounder (*Platichthys flesus*) and their use in pollution monitoring. *Marine Pollution Bulletin*, **33** (1–6), 36–45.

Goodyear, C.P. (1985) Toxic materials, fishing, and environmental variations: simulated effects on striped bass population trends. *Transactions of the American Fisheries Society*, **114**, 107–113.

Gray, J.S. (1992) Biological and ecological effects of marine pollutants and their detection. *Marine Pollution Bulletin*, **25** (1–4), 48–50.

Gulland, F.M.D. (1995) The impact of infectious diseases on wild animal populations – a review. In: (eds Grenfell, B. & A.P. Dobson) *Ecology of infectious diseases in natural populations*. Cambridge University Press, Cambridge, UK, pp. 20–51.

Harding, L.E. (1992) Measures of Marine Environmental Quality. *Marine Pollution Bulletin*, **25** (1–4), 23–27.

Hastings, A. (1997) *Population Biology*. Springer, London, 220 pp.

Hawkins, S.J., J.R. Allen & S. Bray (1999) Restoration of temperate marine and coastal ecosystems: nudging nature. *Aquatic Conservation, Marine and Freshwater Ecosystems*, **9**, 23–46.

Hildén, M. (1997) Boundary conditions for the sustainable use of major fish stocks in the Baltic Sea. *Ecological Economics*, **20** (3), 209–220.

Hildén, M., H. Lehtonen & P. Böhling (1984) The decline of the Finnish vendace (*Coregonus albula*) (L.) catch and the dynamics of the fishery in the Bothnian Bay. *Aqua Fennica*, **14**, 33–47.

Houde, E.D. (1989) Subtleties and episodes in the early life of fishes. *Journal of Fish Biology*, **35** (Suppl. A): 29–38.

Hudd, R., M. Hildén & L. Urho (1986) The effects of anthropogenic acidification on the stocks and fisheries of bream and bubot in the sea area, influenced by the River Kyrönjoki in the Gulf of Bothnia. *Publ. Water Res. Institute*, **68**, 134–138. (Natural Board of Waters and Environment, Finland.)

Iles, T.C. (1994) A review of stock-recruitment relationship with reference to flatfish populations. *Netherlands Journal of Sea Research*, **32** (3–4), 399–420.

Jacobs, L.F. (1996) The economy of winter: phenotypic plasticity in behaviour and brain structure. *Biological Bulletin*, **191**, 92–100.

Jennings, S., M.J. Kaiser & J.D. Reynolds (2001) *Marine Fisheries Ecology*. Blackwell Science, Oxford, UK, 417 pp.

Johnson, L.L. & J.T. Landahl (1994) Chemical contaminants, liver-disease, and mortality-rates in English sole (*Pleuronectes vetulus*). *Ecological Applications*, **4** (1), 59–68.

Jones, R.F., D.M. Baltz & R.L. Allen (2002) Patterns of resource use by fishes and macroinvertebrates in Barataria Bay, Louisiana. *Marine Ecology Progress Series*, **237**, 271–289.

Karås, P. (1989) Some aspects of environmental disturbances in recruitment areas of Baltic fish populations. *Rapports et Procès-verbaux de Réunions du Conseil International pour l'Exploration de la Mer*, **190**, 193–197.

Knapp, R., J.C. Wingfield & A.H. Bass (1999) Steroid hormones and paternal care in the plainfin midshipman fish (*Porichthys notatus*). *Hormones and Behaviour*, **35**, 81–89.

Köhler, A., U. Harms & B. Luckas (1986) Accumulation of organochlorines and mercury in flounder – an approach to pollution assessments. *Helgoländer Meeresuntersuchungen*, **40**, 431–440.

Kwak, T.J. & T.F. Waters (1997) Trout production dynamics and quality in Minnesota streams. *Transactions of the American Fisheries Society*, **126** (1), 35–48.

Landahl, J.T. & L.L. Johnson (1993) Contaminant exposure and population growth of English sole in Puget Sound: the need for better early life-history data. In: (ed. Fuiman, L.A.) *Water Quality and the Early Life Stages of Fishes*. American Fisheries Society, Symposium 14, Bethesda, Maryland, pp. 117–123.

Landahl, J.T., L.L. Johnson, J.E. Stein, T.K. Collier & U. Varanasi (1997) Approaches for determining effects of pollution on fish populations of Puget Sound. *Transactions of the American Fisheries Society*, **126** (3), 519–535.

Langston, W.J. (1990) Toxic effects of metals and the incidence of metal pollution in marine ecosystems. In: (eds Furness, R.W. & P.S. Rainbow) *Heavy metals in the marine environment*. CRC Press Inc., Boca Raton, Florida, pp. 101–122.

Lawrence, A.J. & C. Poulter (1998) Development of a sub-lethal pollution bioassay using the estuarine amphipod *Gammarus duebeni*. *Water Research*, **32**, 569–578.

Libes, S.M. (1992) *An Introduction to Marine Biogeochemistry*. John Wiley & Son, New York.

Longwell, A.C. & J.B. Hughes (1980) Cytologic, cytogenetic, and developmental state of atlantic mackerel eggs from sea surface waters of the New York Bight, and prospects for biological effects monitoring with ichthyoplankton. *Rapports et Procés-verbaux des Réunions du Conseil International pour l'Exploration de la Mer*, **179**, 275–291.

Lucas, M.F., M.T. Caldeira, A. Hall, A.C. Duarte & C. Lima (1986) Distribution of mercury in the sediments and fishes of the lagoon of Aveiro, Portugal. *Water Science Technology*, **18**, 141–148.

Luckenbach, T., R. Triebskorn, E. Müller & A. Oberemm (2001) Toxicity of waters from two streams to early life stages of brown trout (*Salmo trutta* f. *fario* L.), tested under semi-field conditions. *Chemosphere*, **45** (4–5), 571–579.

Mance, G. (1987) *Pollution threat of heavy metals in aquatic environments*. Elsevier Applied Sciences, London, 372 pp.

Mann, K.H. (1965) Energy transformations by a population of fish in the River Thames. *Journal of Animal Ecology*, **34**, 253–275.

Mann, K.H., R.H. Britton, A. Kowakzewski, T.J. Lack, C.P. Mathews & I. McDonald (1972) Productivity and energy flow at all trophic levels in the River Thames, England. In: (eds Kajak, Z. & A. Hillbricht-Ilkowska) *Productivity Problems of Freshwaters*. PWN, Warsaw, pp. 579–596.

Mann, R.H.K. & T. Penczak (1986) Fish production in rivers: a review. *Polskie Archiwum Hydrobiologii*, **33**, 233–247.

Margalef, R. (1975) Human impact on transportation and diversity in ecosystems. How far is extrapolation valid? In: *Proceedings of the First International Congress of Ecology*. Centre for Agricultural Publishing and Documentation, Wageningen, Netherlands, pp. 237–241.

Martin, K.L.M., M.C. Lawson & H. Engebretson (1996) Adverse effects of hyposalinity from stormwater runoff on the aggregating anemone, *Anthopleura elegantissima*, in the marine intertidal zone. *Bulletin of the Southern California Academy of Science*, **95**, 46–51.

Mathieson, S. (1993) *Mercury accumulation by the eelpout Zoarces viviparus in the Forth estuary*. PhD Dissertation, University of Stirling, UK.

Mathieson, S., S.G. George & D.S. McLusky (1996) Temporal variation of total mercury concentrations and burdens in the liver of eelpout *Zoarces viviparus* from the Forth estuary, Scotland: implications for mercury biomonitoring. *Marine Ecology Progress Series*, **138**, 41–49.

McDowall, R.M. (1988) *Diadromy in Fishes*. Croom-Helm, London.

McLusky, D.S., V. Bryant & R. Campbell (1986) The effects of temperature and salinity on the toxicity of heavy metals to marine and estuarine invertebrates. *Oceanography and Marine Biology: An Annual Review*, **24**, 481–520.

McLusky, D.S., D.M. Bryant & M. Elliott (1992) The impact of land claim on macrobenthos, fish and shorebirds on the Forth estuary, Eastern Scotland. *Aquatic Conservation, Marine and Freshwater Ecosystems*, **2** (3), 211–222.

McLusky, D.S., S.C. Hull & M. Elliott (1993) Variations in the intertidal and subtidal macrofauna and sediments along a salinity gradient in the Upper Forth estuary. *Netherlands Journal of Aquatic Ecology*, **27**, 101–109.

Metayer, C., J.C. Amiard, C. Amiard-Triquet & J. Marchand (1980) Etude du transfert de quelques oligo-éléments dans les chaînes trophiques néritiques et estuariennes: accumulation biologique chez les poissons omnivores et super-carnivores. *Helgoländer Meeresuntersuchungen*, **34**, 179–191.

Methven, D.A. & C. Bajdik (1994) Temporal variation in size and abundance of juvenile Atlantic cod (*Gadus morhua*) at an inshore site off eastern Newfoundland. *Canadian Journal of Fisheries and Aquatic Sciences*, **51**, 78–90.

Miller, B.S. (1999) Mussels as biomonitors of point and diffuse sources of trace metals in the Clyde Sea area, Scotland. *Water Science and Technology*, **39**, 233–240.

Monosson, E. (1997) Reproductive and developmental effects of contaminants in fish populations: establishing cause and effect. In: (eds Rolland, R.M., M. Gilbertson & R.E. Peterson) *Chemically Induced Alterations in Functional Development and Reproduction of Fishes*. Proceedings from a session at the Wingspread Conference Center, 21–23 July 1995, Racine WI. Published by the Society of Environmental Toxicology and Chemistry (SETAC), Pensacola, Florida, USA, pp. 177–194.

Moss, R., A. Watson & J. Ollason (1982) *Animal Population Dynamics*. (Series eds Dunnet, G.M. & C.H. Gimingham). Chapman & Hall, London, New York, 75 pp.

Moyle, P.B. & J.J. Cech Jr. (1988) *Fishes: An introduction to ichthyology*, 2nd ed. Princeton Hall, Englewood Cliffs, New York, 559 pp.

Munkittrick, K.R. & D.G. Dixon (1989) A holistic approach to ecosystem health assessment using fish population characteristics. *Hydrobiologia*, **188/189**, 123–135.

Munkittrick, K.R. & J.F. Leatherland (1984) Abnormal pituitary-gonad function in feral populations of goldfish suffering epizooids of an ulcerative disease. *Journal of Fish Diseases*, **7**, 433–447.

Nacci, D.E., M. Kohan, M. Pelletier & E. George (2002) Effects of benzo[a]pyrene exposure on a fish population resistant to the toxic effects of dioxin-like compounds. *Aquatic Toxicology*, **57** (4), 203–215.

National Rivers Authority (1993) *The Quality of the Humber Estuary 1980–1990*. National Rivers Authority (Water Quality Series No. 12), Bristol, 108 pp.

Neff, J.M. (2002) Bioaccumulation in Marine Organisms. Effects of Contaminants from Oil Well Produced Water. Elsevier Science Ltd, Oxford, 452 pp.

Nicholson, A.J. (1954) An outline of the dynamics of animal populations. *Australian Journal of Zoology*, **2**, 9–65.

Odum, H.T. (1967) Biological circuits and the marine systems of Texas. In (eds Olson, T.A. & F.J. Burgess) *Pollution and Marine Ecology*. Wiley-Interscience, New York.

Odum, E.P. (1985) Trends expected in stressed ecosystems. *Bioscience*, **35** (7), 419–422.

Olla, B.L., W.H. Pearson & A.L. Studholme (1980) Applicability of behavioral measures in environmental stress assessment. *Rapports et Procés-verbaux des Réunions du Conseil International pour l'Exploration de la Mer*, **179**, 162–173.

Parrett, A. (1998) *Pollution impacts on North Sea fish stocks*. European Commission Directorate General XIV (Fisheries), Brussels, Report No. 96-083, 122 pp.

Pearson, T. & R. Rosenberg (1978) Macrobenthic succession in relation to organic enrichment and pollution of the marine environment. *Oceanography and Marine Biology Annual Review*, **16**, 229–311.

Phillips, D.J.H. & P.S. Rainbow (1994) *Biomonitoring of Trace Aquatic Contaminants*. Chapman & Hall, London, 371 pp.

Pohl, C. (1990) Skeletal deformities and trace metal contents of European smelt, *Osmerus eperlanus*, in the Elbe estuary. *Meeresforch*, **33**, 76–89.

Prigogine, I., G. Nicolis & A. Babloyantz (1972) Thermodynamics and evolution. *Physics Today*, **25**, 22–38.

Rapport, D.J. (1989) Symptoms of pathology in the Gulf of Bothnia (Baltic Sea): ecosystem response to stress from human activity. *Biological Journal of the Linnean Society*, **37** (1–2), 33–49.

Rolland, R.M., M. Gilbertson & R.E. Peterson (eds) (1997) *Chemically induced alterations in functional development and reproduction of fishes*. SETAC Technical Publications Series, Society of Environmental Toxicology and Chemistry, Pensacola, Florida, 224 pp.

Rose, K.A., E.D. Cowan, E. Houde & C.C. Coutant (1993) Individual based modelling of environmental quality effects on early life stages of fishes: a case study using striped bass. In: (ed. Fuiman, L.A.) *Water quality and the early life history stages of fishes* (American Fisheries Society Symposium). American Fisheries Society, Bethesda, Maryland, pp. 125–145.

Rothschild, B.J. (1986) *Dynamics of Marine Fish Populations*. Harvard University Press, Cambridge, Mass.

Sandström, O., P. Karås & E. Neuman (1991) Pulp mill effluent effects on species distributions and recruitment in Baltic coastal fish. *Finnish Fisheries Research*, **12**, 101–110.

Sauriau, P.G., J.F. Guillaud & B. Thouvenin (1994) *Qualité des eaux de l'estuaire de la Loire*, Vol. 2. Rapport CSEEL, France, 104 pp.

Sayer, M.D.J. & J. Davenport (1996) Hypometabolism in torpid goldsinny wrasse subjected to rapid reductions in seawater temperature. *Journal of Fish Biology*, **49**, 64–75.

Sayer, M.D.J. & J.P. Reader (1996) Exposure of goldsinny, rock cook and corkwing wrasse to low temperature and low salinity: survival, blood physiology and seasonal variation. *Journal of Fish Biology*, **49**, 41–63.

Sayer, M.D.J., R.N. Gibson & R.J.A. Atkinson (1995) Growth, diet and condition of goldsinny on the west coast of Scotland. *Journal of Fish Biology*, **46**, 317–340.

Schaaf, W.E., D.S. Peters, D.S. Vaughan, L. Coston-Clements & C.W. Krouse (1987) Fish population responses to chronic and acute pollution: the influence of life history stages. *Estuaries*, **10** (3), 267–275.

Sleiderink, H.M. & J.P. Boon (1995) Cytochrome-P450 1A response in North sea dab, *Limanda limanda*, from offshore and coastal sites. *Marine Pollution Bulletin*, **30** (10), 660–666.

Stagg, R.M., J. Rusin & F. Brown (1992) Na+, K+-ATPase activity in the gills of the flounder (*Platichthys flesus*) in relation to mercury contamination in the Firth of Forth. *Marine Environmental Research*, **33**, 255–266.

Stark, J.S. (1998) Effect of copper on macrobenthic assemblages in soft sediments: a laboratory experimental study. *Ecotoxicology*, **7**, 161–173.

Sulaiman, N., S. George & M.D. Burke (1991) Assessment of sub-lethal pollutant impact on flounders in an industrialized estuary using hepatic biochemical indexes. *Marine Ecology Progress Series*, **68** (3), 207–212.

Svelle, M., H. Aarefjord, H.T. Heir & S. Øverland (eds) (1997) *Assessment Report on Fisheries and Fisheries related Species and Habitat Issues*. Intermediate Ministerial Meeting on the Integration of Fisheries and Environmental Issues, 13–14 March 1997, Bergen, Norway. Ministry of Environment, Norway, 127 pp.

Thiel, R., H. Winkler & L. Urho (1996) Zur Veränderung der Fischfauna. In: (eds Lozán, J.L., R. Lampe, W. Matthäus, E. Rachor, H. Rumor & von H. Westernhagen) *Warnsignale aus der Ostsee*. Parey, Berlin, pp. 365–404.

Turnpenny, A.W.H., C. Demspey, M.H. Davis & J.M. Fleming (1988) Factors limiting fish populations in the Loch Fleet system, and acidic drainage system in Southwest Scotland. *Journal of Fish Biology*, **32** (1), 101–118.

Usher, M.B. (1972) Developments in the Leslie matrix model. In: (ed. Jeffers, J.N.R.) *Mathematical models in ecology*. Blackwell Scientific Publications, Oxford, UK.

van Egmond, R.A. (1993) *The effects of organic pollution on fish detoxification mechanisms and reproduction*. PhD Thesis, Napier University, Edinburgh.

Walker, C.H., S.P. Hopkin, R.M. Sibly & D.B. Peakall (1996) *Principals of Ecotoxicology*. Taylor & Francis Ltd., London, 321 pp.

Wells, P.G. (1999) Biomonitoring the health of coastal marine ecosystems: the roles and challenges of microscale toxicity tests. *Marine Pollution Bulletin*, **39**, 39–47.

West-Eberhard, M.J. (1989) Phenotypic plasticity and the origin of diversity. *Annual Reviews of Ecology and Systematics*, **20**, 249–278.

Whitfield, A.K. & M. Elliott (2002) Fishes as indicators of environmental and ecological changes within estuaries – a review of progress and some suggestions for the future. *Journal of Fish Biology*, **61** (Suppl. A) 229–250.

Williamson, M. (1972) *The Analysis of Biological Populations*. Edward Arnold, London.

Wingfield, J.C., D.L. Maney, C.W. Breuner, J.D. Jacobs, D. Lynn, M. Ramenofsky & R.D. Richardson (1998) Ecological bases of hormone-behavior interactions: The 'emergency life history stage'. *American Zoologist*, **38**, 191–206.

Wootton, R.J. (1990) *Ecology of Teleost Fishes*. (Fish and Fisheries Series 1.) Chapman & Hall, London.

Wu, R.S.S. (1999) Eutrophication, water borne pathogens and xenobiotic compounds: environmental risks and challenges. *Marine Pollution Bulletin*, **39**, 11–22.

Chapter 7
Molecular/Cellular Processes and the Population Genetics of a Species

L. Hauser, K.L. Hemingway, J. Wedderburn and A.J. Lawrence

7.1 Introduction

Although the effects of xenobiotics described in the previous chapters are very serious and range from molecular damage and cancer development to disease and death of individuals and reduction in population abundance, they are usually short-lived and quickly disappear once exposure has stopped. However, effects on the population genetic level may impact populations for many generations to come. For example, once genetic diversity is lost from a population, it will in the absence of immigration only be replenished on evolutionary time scales. Similarly, certain deleterious mutations accumulating in the germline may take many generations to be eliminated by selection. Considerable concern has developed over such potential chronic and transgenerational effects of contamination (Bickham *et al.*, 2000), and their consequences for population productivity and persistence. While much of the theory of mutation accumulation, selection and loss of genetic diversity is well developed, there is still a great need to:

(1) Establish firm links between damage at the molecular level, and long-term genetic changes at the population level
(2) Investigate the speed, extent and consequences of genetic adaptation to pollution
(3) Study the consequences of xenobiotics on the genetic diversity of exposed populations.

Such investigations are not only valuable to predict the long-term population effects of xenobiotic exposure, but may also provide a model system integrating variation at the molecular (DNA) level with genetic variation of quantitative traits. The relationship between molecular variation (estimated with a wide range of molecular markers) and genetic variation at quantitative traits (usually estimated with extensive breeding designs) has remained a fundamental problem in evolutionary biology and conservation genetics (Lande, 1996; Lynch, 1996). 'Phylogeographic and evolutionary ecotoxicology' (Staton *et al.*, 2001) may provide a model system to investigate this relationship, as many of the genetic and molecular processes in xenobiotic tolerance and resistance are well known, there are usually good time records of pollution events and experiments to assay pollutant susceptibility are well developed in a wide range of indicator organisms. The integration between population genetics and ecotoxicology therefore has not only applied, but also considerable fundamental scientific significance.

The aim of this chapter is therefore to introduce some of the central concepts in population genetics and discuss their relevance to ecotoxicology; to review empirical evidence for impacts of xenobiotics and their consequences; and to discuss the implications of concepts and evidence for the evolution of tolerance.

7.2 Evolutionary processes and concepts

7.2.1 Mutations

Mutation is the ultimate source of genetic variation, and is thus an essential process in evolution (Hartl, 1994). The nature of the mutation process in natural populations, however, remains largely unknown because individual events generally occur too infrequently for direct observation. Consequently, much of our interpretation of patterns of genetic variability in wild populations is based on unverified assumptions on rates and patterns of mutations in DNA sequences. Nevertheless, mutations have general characteristics which shape our views of their role in the evolution of populations and species. Many of these characteristics have already been discussed in Chapter 2; the population aspects of mutations will be considered here.

Although mutations may be biased, they can for most purposes be viewed as a random process and such random mutations are unlikely to be beneficial to the organism (Maynard-Smith, 1998). Indeed, most newly arising mutations are either silent (i.e. not affecting the amino acid sequences encoded for by the DNA), selectively neutral (i.e. do not affect the functionality of the gene product) or harmful to the organism. This is also true for mutations caused by mutagenic pollutants, which on average appear to be largely detrimental (Mukai *et al.*, 1972). The low mutation rates commonly observed at coding genes are probably a compromise between the genetic damage suffered due to deleterious mutations, and the beneficial effect of favourable mutations allowing adaptation and evolution (Hartl, 1994). Due to the polygenic inheritance of many adaptive traits, the introduction of novel genetic variation at quantitative traits by mutation can be quite rapid. As many traits are encoded by hundreds to thousands of loci, common estimates of mutation rates and the resulting traits are about 10^{-1} (Lynch & Walsh, 1998). For example, long-term selection lines of *Drosophila melanogaster*, initially with homozygous individuals, show that the accumulating effects of mutations can lead to appreciable response to selection equivalent to 7 to 8 phenotypic standard deviations in bristle number per 125 generations (Mackay *et al.*, 1994).

Mutations as a consequence of exposure to mutagenic chemicals have been extensively discussed in Chapter 2. However, many such mutations are somatic, not heritable, and thus will not be maintained in the gene pool. In contrast, germline mutations causing heritable genetic effects have been difficult to demonstrate empirically, mainly because they are very rare and individuals carrying deleterious mutations are often quickly eliminated from wild populations. Notable exceptions stem from the nuclear accident in Chernobyl, where increased mutation rates at minisatellite loci in humans, and at microsatellites in barn swallows, have been demonstrated (Dubrova *et al.*, 1996; Ellegren *et al.*, 1997). Similarly, there was an elevated minisatellite mutation rate in herring gulls nesting on contaminated urban sites (Yauk & Quinn, 1996).

In general, repeat regions such as microsatellites and minisatellites appear to be sensitive molecular markers for heritable genetic damage (Yauk, 1998), although mutations may be induced on genes encoding for DNA repair enzymes rather than the repeat regions themselves (Bickham *et al.*, 2000). Increased genetic diversity at allozyme, mtDNA and RAPD (random amplified polymorphic DNA) markers in mosquitofish and vole populations exposed to radionuclides provided additional evidence for increased mutation rates (Theodorakis & Shugart, 1997; Baker *et al.*, 2001).

In addition to such molecular evidence, there are indications for phenotypic effects of increased mutation rates. For example, the barn swallows in Chernobyl not only had higher mutation rates at microsatellites, but also showed increased albinism (white spots), some of which appeared to be heritable (Ellegren *et al.*, 1997). Additionally, reductions in fitness due to the accumulation of mutations have been shown in caged fly populations (Mukai *et al.*, 1972; Houle *et al.*, 1992, 1994). Such immediate expression of mutations is often the effect of acute exposure and may be quickly removed by selection (Cronin & Bickham, 1998). Large and stable populations will return to the pre-exposure mutation-selection equilibrium, where the occurrence of new mutations is balanced by their elimination by selection (Lynch & Walsh, 1998). However, recessive deleterious mutations (mutations whose effect is masked by the alternate unmutated allele) are 'invisible' to selection and may therefore accumulate in the population as 'mutational load' (Lynch & Walsh, 1998). Remarkably, at equilibrium, the mutational load at a locus is independent of the magnitude of the deleterious effects of mutations as the equilibrium frequency of a mutant allele depends on its deleterious effect on heterozygotes (Maynard-Smith, 1998). As such, the mutational load depends entirely on the mutation rate, a fact that is particularly relevant in populations exposed to mutagenic chemicals.

Although the dangers of xenobiotic exposure increasing the mutational load was identified as early as 1979 (Berry, 1980), there is still no empirical evidence from wild populations. The methodology of estimating mutational load by deliberate inbreeding is well developed (Lynch & Walsh, 1998), and could easily be carried out in model species with short generation time. Results would not only test the notion of an accumulation of pollutant-derived mutations in a gene-pool, but would also provide valuable information on the long-term effects of pollution, an issue of particular relevance for fish populations in recently cleaned European rivers and seas.

7.2.2 Gene flow

Genetic variability can also be introduced to a population by gene-flow, which is mediated by individuals which immigrate and successfully interbreed with the local population. In the short term, the increase in a population's genetic variation due to gene flow is often far greater than due to mutation (Futuyma, 1998).

Gene flow is a homogenising force: conspecific populations differ in allele frequencies only if gene flow is sufficiently low to allow genetic differentiation by selection or random genetic drift. Thus, gene flow may prevent adaptation to local environmental conditions by natural selection. In some cases, excessive, sometimes anthropogenic, gene flow has been shown to be detrimental for a population's fitness in its environment: in many salmonids, survival, disease resistance and homing accuracy was shown to be higher in native fish than in hybrids between native and introduced fish (Hindar *et al.*, 1991). This homogenising effect of gene flow may be more important in populations which rely on immigration from other populations for sufficient recruitment (source-sink populations), and where local adaptation may be impossible even under strong selection pressures (e.g. chironomid larvae (Groenendijk *et al.*, 2002), blue tit *Parus caeruleus* (Dias *et al.*, 1996)).

Molecular markers such as allozyme, mitochondrial DNA and microsatellites, are extremely powerful in identifying reproductively isolated populations with little gene flow to other populations (Carvalho & Hauser, 1994; Hauser & Ward, 1998). Clearly, if a population lives entirely in a polluted environment, the potential effects on the gene pool are much greater than if only a small part of the distribution area is affected. However, excessive gene flow may prevent local adaptation to pollution. The distribution of a population in relation to the area affected by pollution, and the extent of gene flow with other populations, are therefore important factors in the assessment of xenobiotic responses.

7.2.3 Selection

Ever since Darwin (1859), the concept of genetic change by natural selection has been central to our understanding of evolution (Maynard-Smith, 1998). On the intraspecific level, selection is a major factor in the development of local adaptation, and thus may contribute to the genetic differentiation among populations (Carvalho, 1993). Despite its importance in micro and macroevolution, selection is still an area with much argument and confusion (Endler, 1986). Until more recently, empirical studies in wild populations have concentrated on demonstrating its existence rather than attempting its quantification or understanding its mechanisms (Brodie *et al.*, 1995). Especially in molecular genetics with its supposedly neutral (that is, not selected) markers, selection is often seen as a 'nuisance factor', disturbing straightforward predictions from neutral models (e.g. Gauldie, 1991). One of the reasons for the difficulties associated with identifying selection may be that both quantitative and molecular methods cannot simultaneously determine all three agents for natural selection to operate (Endler, 1986):

(1) Variation among individuals in a trait
(2) A consistent relationship between that trait and fitness
(3) Heritability of that trait.

Relationships between quantitative traits and fitness can usually be inferred, and molecular markers are undoubtedly heritable; however, establishing the fitness value of a molecular marker may be as difficult as proving the heritability of a phenotypic trait in a wild population.

An additional complication in the investigation of selection in wild fish populations is the high phenotypic variability in fish. In particular, large differences in growth rate and body size can be observed between, as well as within, populations. For example, the size range between populations of Arctic char is over 4000%, which compares to 250% between species of Darwin's ground finches (Allendorf *et al*., 1987). Within populations, coefficients of variation of phenotypic characters are usually far greater than 10% in fish, whereas they rarely exceed 5% in other vertebrates. In contrast, heritabilities are generally lower in fish, at least in morphological characters (Purdom, 1993). Such high phenotypic flexibility was thought to reduce the selective differential between genotypes and thus slow down the rate of evolutionary change (Wright, 1931), but there is now evidence that the plasticity of a trait evolves independently from the trait itself (Via *et al*., 1995) and is thus an integral part in the evolutionary response to environmental change (Thompson, 1991).

It is a common misconception that selection may change heritable characteristics within a population only over thousands of generations. Experiments with caged flies show that the resistance to ethyl alcohol vapours increased from a 'knock-down' time of 5 minutes to 28 minutes in only 60 generations (Weber & Diggins, 1990). In fish, even faster reactions to selection pressures were observed in introduced guppy (*Poecilia reticulata*) populations in Trinidad. Male guppies from a site with high predation introduced to a habitat without predators evolved new heritable life history parameters within only 4 years (approximately seven generations), while female traits changed after 11 years (18 generations) (Reznick *et al*., 1990). This fast adaptation to new environmental conditions illustrates the importance of adaptation in the evaluation of population responses to xenobiotics.

It is worth considering how selection could be demonstrated by the molecular markers commonly employed by population geneticists. In some species, a higher sensitivity of certain allozyme genotypes to pollutants could be shown, for example, in mosquitofish (*Gambusia holbrooki*), stonerollers (*Campostoma anomalum*) and fathead minnows (*Pimephales promelas*) (Diamond *et al*., 1989; Newman *et al*., 1989; Guttman, 1994). Often the mechanism of selection remained unknown, though in a few cases selection acting on the protein product of an allozyme locus itself has been demonstrated conclusively by establishing the superiority of certain allelic products (*Gambusia holbrooki* (Kramer & Newman, 1994), *Campostoma anomalum* (Guttman, 1994)). More usually, however, the observed differentiation is caused by selection acting on loci linked to the allozyme locus, rather than the marker itself (Maynard-Smith & Haigh, 1974). This is particularly true for most DNA markers (RAPD, microsatellites), which are presumed to be largely neutral to selection. Nevertheless, neutral DNA markers linked to genes involved in pollution tolerance have been used to identify tolerant individuals (Theodorakis *et al*., 1999).

In addition to single-locus differentiation and changes in genetic variability, selection may be detectable by testing data against expectations under a neutrality model, i.e. no selection (Lewontin & Krakauer, 1973). If the observed data deviates significantly from expectations, the assumptions of the model, one of which is the absence of selection, are violated (Endler, 1986). Whilst it is difficult to discard alternative explanations, the test may still be used to indicate the presence of selection.

Genetic differentiation due to adaptation to pollutants may also be detectable by investigating temporal changes in genotype frequencies: if differentiation is due to selection, allele frequencies in the exposed populations would be expected to be similar to the 'starting trait distribution' (Endler, 1986), and only in the course of generations to respond to selection. Clearly, such starting traits could be genotype frequencies estimated with molecular markers as well as non-molecular characters, such as tolerance measures, though with the latter measures, the genetic basis of the observed phenotypic variation has to demonstrated.

Unfortunately, starting trait distributions are not usually available in ecotoxicological investigations, although it may be possible to monitor larvae or juveniles immigrating into a polluted area.

7.2.4 Random genetic drift

One of the most significant short-term effects of pollutant exposure is the reduction in population size due to increased mortality and reduced fecundity. Such a 'population bottleneck' causes important changes in the gene pool, which may further increase the population's vulnerability to extinction by demographic and environmental stochasticity. These processes in small populations are well known from population genetic models, and have been demonstrated empirically in wild populations. Subsequently, the three main concepts in the area – inbreeding, genetic drift and the effective population size – will be introduced and their relevance to the ecotoxicology of fish populations discussed.

As a simple hypothetical experiment, a large population may be divided into many subpopulations which are completely isolated from each other (Falconer, 1989). It is assumed that there is no mutation and no selection, and that each subpopulation has the same number of breeders. Considering a locus with two alleles, each subpopulation is a sample of the original population, with the mean allele frequency of all these samples being the allele frequency of the original population (p_0). The allele frequencies of the lines are distributed around this mean with the binomial variance for sample means, and the continuous sampling of alleles in each generation causes an irregular change in allele frequencies in each subpopulation (Fig. 7.1), only predictable in amount but not in direction. This process is called 'random genetic drift'.

Due to the random fluctuations in allele frequencies, some alleles will be lost from the population just by chance. The likelihood of loss depends on both initial allele frequency and effective population size, and is largest for rare alleles in small populations (Fig. 7.2). Although these expectations were formulated for selectively neutral loci, the exposure to new xenobiotics may rapidly increase the adaptive significance of previously neutral alleles. For example, genes of detoxification enzymes such as the P450 system, which are normally not expressed and thus presumably relatively neutral, are a crucial component of survival in polluted environments. It is therefore dangerous to presume that models based on neutral loci have no relevance for the adaptability and evolutionary persistence of wild populations.

Probably more significant in the short term is the effect of genetic drift on the effectiveness of selection. In small populations, even strong selection cannot compensate for the large erratic fluctuation caused by random genetic drift, and the population is likely to become monomorphic (Fig. 7.3). Due to the random nature of these allele frequency fluctuations in small populations, deleterious genes have roughly the same probability of fixation as beneficial ones (Frankel & Soulé, 1981). Clearly, in populations exposed to pollution, such a relaxation of selection may prevent adaptation to the pollutant, and may thus increase the effects of exposure.

Because genetic drift overrides selection in small populations, the loss of particular genotypes is independent of their mutational load, and so the fittest genotypes have an equal chance to become extinct as the more affected genotypes. Therefore, a random increase in

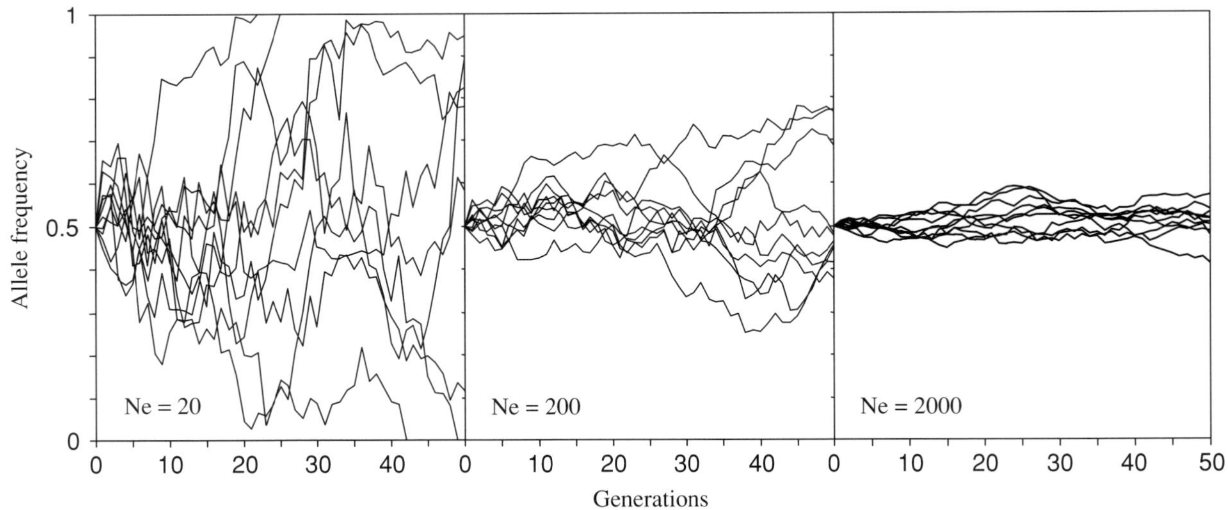

Fig. 7.1 Computer simulations of allele frequencies under the influence of random genetic drift. Temporal fluctuations in allele frequencies are stronger in smaller populations (left) than in larger (right). In the very small population (left) and in one run, one of the alleles was lost after only 20 generations, although it started at a frequency of 0.5. Calculations according to Falconer & Mackay, 1996.

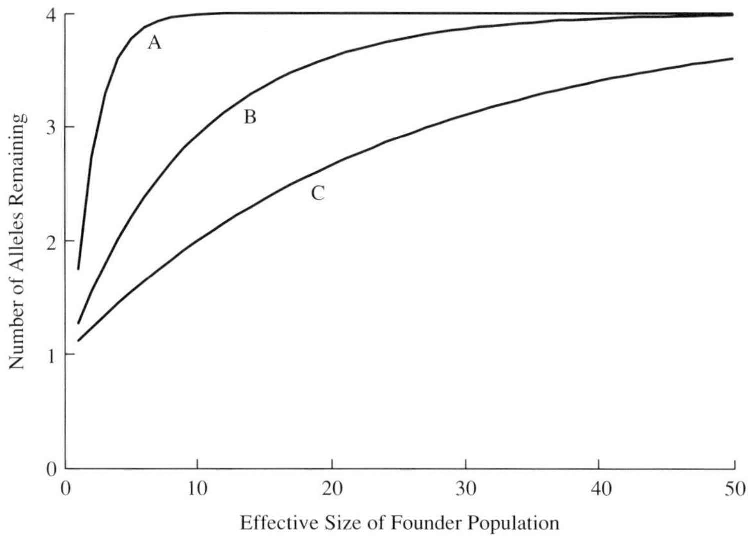

Fig. 7.2 The loss of rare alleles per generation in a small population. The number of alleles remaining at a locus with initially four alleles is shown. Initial allele frequencies are:
A: 4 alleles with equal frequency ($p_1 = 0.25$, $p_2 = 0.25$, $p_3 = 0.25$, $p_4 = 0.25$)
B: 3 moderately rare alleles ($p_1 = 0.85$, $p_2 = 0.05$, $p_3 = 0.05$, $p_4 = 0.05$)
C: 3 rare alleles ($p_1 = 0.94$, $p_2 = 0.02$, $p_3 = 0.02$, $p_4 = 0.02$).
There is no selection and mutation. See Frankel & Soulé (1981) for calculations.

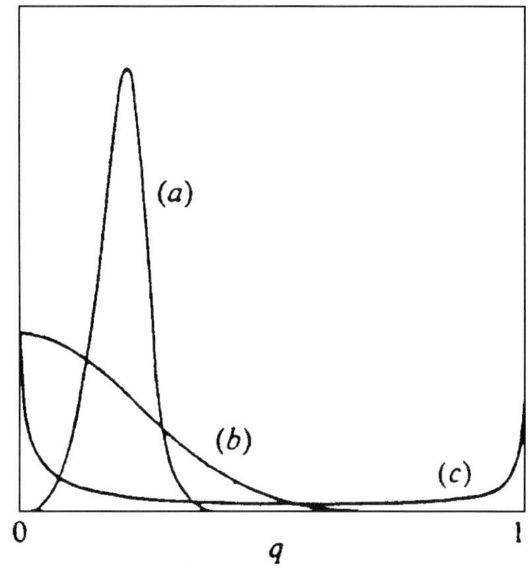

Fig. 7.3 Theoretical distribution of allele frequencies among subpopulations when the dispersion is balanced by mutation and selection. The graphs refer to a recessive allele with a mutation rate of 10^{-5} a selection coefficient (proportionate reduction in fitness of the less favoured genotype) of $2 * 10^{-4}$ and an N_e of (a) 250 000, (b) 25 000, (c) 2500. At small population sizes selection cannot maintain polymorphism against genetic drift. © D.S. Falconer 1975, 1989, reprinted by permission of Pearson Education Limited.

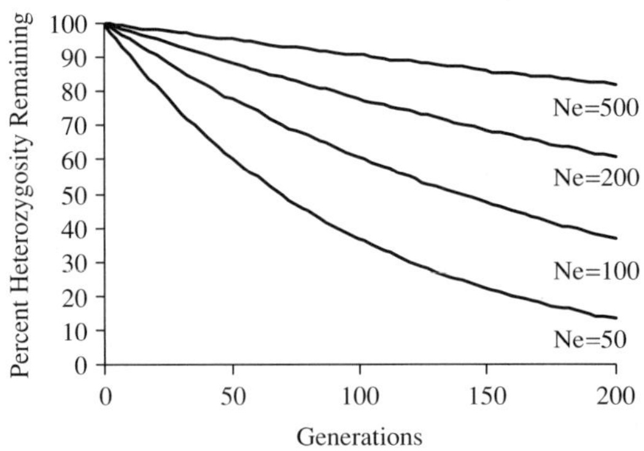

Fig. 7.4 Loss of heterozygosity in population bottlenecks of varying size. The percentage of initial heterozygosity remaining is shown. Calculated from equations in Crow & Kimura, 1970.

less fit genotypes will reduce the average fitness in the population, and may thus cause a reduction in population size, leading to a further increase in genetic drift. This positive feedback cycle is called 'mutational meltdown' and has been well established from computer models (Gabriel *et al.*, 1993). In particular, in populations where mutation rates are increased (Chapter 2) and abundance is reduced (Chapters 3 and 5) due to mutagenic pollutants, the extinction risk due to mutational meltdown may be quite considerable (Lynch *et al.*, 1995).

7.2.5 Inbreeding

The second, closely-related approach for describing processes during population bottlenecks is to consider the amount of inbreeding in finite populations. Inbreeding is the mating between related individuals which often carry the same alleles. Therefore, inbreeding in population genetic terms is 'the probability of an individual having both alleles at a locus derived from the same ancestral allele' (Crow & Kimura, 1970). As a consequence, the heterozygosity, that is the proportion of heterozygous individuals, decreases over time (Fig. 7.4).

In addition to reducing heterozygosity, inbreeding may result in the expression of recessive deleterious mutations in inbred and thus homozygous individuals. This expression of the mutational load of a population is called 'inbreeding depression', the reduction of fitness in inbred individuals. Inbreeding depression is well known from many domestic animals and plants (Thornhill, 1993), and there is also increasing evidence from wild populations (Crnokrak & Roff, 1999). Inbreeding can affect most fitness characters, such as fecundity, longevity, development and stress and disease resistance (Falconer, 1989; Frankham, 1998).

Although such individual inbreeding effects are well known, their consequences on the population level are less obvious. Compensatory ecological and genetic effects may reduce the population effects; for example, in great tits a higher egg mortality of inbred matings was compensated by higher recruitment of hatched offspring (van Noordwijk & Scharloo, 1981). Furthermore, because inbreeding causes the expression of recessive deleterious mutations, selection against such mutations will become more effective and they may be eliminated from the population. This process of 'purging' reduces the genetic load and may increase mean fitness in the population (Lande & Barrowclough, 1987), although its effectiveness depends on the deleterious effect of mutation, the speed of inbreeding and the genetic mechanisms of inbreeding depression (Bijlsma *et al.*, 2000). In contrast, if a decrease in average individual fitness causes a reduction in population size, inbreeding depression may further intensify (Lynch *et al.*, 1993, 1995; Lande, 1996). Additionally, there is evidence for a direct link between inbreeding and population extinction from captive *Drosophila* (Bijlsma *et al.*, 2000) and *Clarkia pulchella* (Newman & Pilson, 1997), together with wild butterfly metapopulations (Saccheri *et al.*, 1998). In several taxa, the reduction of inbreeding by introductions of conspecifics from other populations has been shown to increase population viability and abundance (Vrijenhoek, 1996; Madsen *et al.*, 1999; Ebert *et al.*, 2002). As such, there is strong evidence that the expression of deleterious mutations during inbreeding may significantly decrease population abundance and may cause population extinction.

Perhaps not surprisingly, the effects of deleterious mutations during inbreeding also depend on environmental conditions. In stressful environments, inbreeding has much stronger effects than under benign conditions (Bijlsma *et al.*, 2000). Moreover, there is evidence that inbreeding increases the sensitivity to stress (Dahlgaard *et al.*, 1995) as the expression of deleterious mutations is environment specific, and purging of such deleterious mutation during inbreeding depends on environmental conditions (Dahlgaard & Hoffmann, 2000). Deleterious mutations that are not expressed may increase in frequency due to genetic drift and may cause further inbreeding depression once environmental conditions change (Bijlsma *et al.*, 2000). Notably, this may also happen after the population has increased to reasonable levels and so a 'history of inbreeding' may cause population extinction of even moderately sized populations (Bijlsma *et al.*, 2000). In the current context, such effects are particularly important as xenobiotic exposure changes rapidly with changing polluters and so even apparently recovered fish populations may be at risk from extinction due to genetic and environmental stochasticity.

7.2.6 Effective population size (N_e)

An important consideration for the assessment of the population genetic effects of pollution is the effective population size, a concept that standardises the effects of mating systems, demographic stability and other factors across populations. Genetic models usually assume ideal populations, where the sex ratio is 1:1, the family size is randomly distributed among breeders, and each generation has the same number of individuals (Gall, 1987). Clearly, these criteria are seldom met in wild populations. Thus, in order to obtain valid predictions, the actual population size has to be converted to the 'effective population size' (N_e) according to the deviations from idealised conditions (Falconer, 1989). In simple terms, N_e can be seen as the size of an ideal population undergoing the same changes in genetic variability as the population observed.

In practice, N_e is very difficult to estimate for wild populations (Ryman *et al.*, 1981; Nelson & Soulé, 1987); however, there are several main factors affecting N_e. Firstly, all juveniles and non-breeding adults do not contribute to N_e, which is especially significant in highly fecund species with density-dependent recruitment. Many fish may in fact be the progeny of very few adults (Cushing, 1973; Nelson & Soulé, 1987). N_e is also reduced by an unequal sex ratio, which can be seen intuitively by considering a male spawning with several females (Frankel & Soulé, 1981). All the offspring of these females will be half-sibs or full-sibs and much more closely related and genetically more homogeneous than had the females each mated with a different male. When a large proportion of the offspring are sired by only a few males, some of the original genetic variability will be lost.

N_e is also inversely related to the variance of the lifetime family size (Nelson & Soulé, 1987). A large variance in lifetime family size has a similar effect as skewed sex ratios; large families contribute disproportionally to the next generation. The lifetime family size is especially important in iteroparous fish (Nelson & Soulé, 1987); firstly, as a result of the large fecundity of many fish and associated density-independent mortality, resulting in orders-of-magnitude variation in cohort strength (Cushing, 1973), and secondly, individuals surviving longer have more offspring than those dying shortly after maturity. In fish, this effect is enhanced by the fact that the fecundity increases with body size (Bagenal, 1973) and therefore with age. As such, a fish surviving twice as long after maturity has more than double the number of offspring.

If the size of the population varies largely between generations, the overall N_e is more affected by the smaller-sized generations than the large ones. The breeding survivors of a population 'crash' contain only a sample of the original genetic variation; all future generations will have a corresponding deficit in genetic variability (Frankel & Soulé, 1981).

Combination of the above factors means that except in heavily managed hatchery populations, N_e is often an order of magnitude smaller than the census number of individuals (Nelson & Soulé, 1987). In a review of terrestrial and freshwater animals, Frankham (1995) estimated N_e/N, that is the effective population size as a proportion of census numbers, to vary between 0.05 and 0.8, with a mean of 0.11. However, there is evidence that this ratio may be considerably lower in marine species, where reproductive success may be very biased among individual spawners (Hedgecock, 1994) and estimates suggest N_e/N ratios of 0.003 in red drum (*Sciaenops ocellatus*) (Turner *et al.*, 1999) and of 10^{-5} in New Zealand snapper (*Pagrus auratus*) (Hauser *et al.*, 2002). In fish populations exposed to pollutants, these estimates may be even lower, as many adults may not be reproducing because of poor health, endocrine disruption or other factors, and some year classes may effectively fail because of catastrophic pollution events. It may therefore be misleading to conclude from large census population sizes that populations are not in danger of losing genetic diversity, and indeed, in New Zealand snapper a decline in genetic diversity was found, despite an estimated census population size of 3 million fish (Hauser *et al.*, 2002).

7.2.7 The importance of genetic diversity

Reductions in population size (population bottlenecks), including those caused by pollution, may cause considerable loss of genetic diversity, and consequently reductions in viability, adaptability and evolutionary potential of affected populations. There are two main aspects of genetic diversity: the qualitative aspect, that is the allelic diversity, the number of different variants of a gene available in the population; and the quantitative aspect, or heterozygosity, which is the proportion of heterozygous individuals, and thus concerns both the number and the frequency of alleles. The distinction between these two aspects is important, as rare alleles do not contribute much to heterozygosity but may nevertheless be vital for the adaptation to new environmental conditions. Indeed, it is a dictum of evolutionary genetics that genetic diversity is necessary for evolutionary change (Carvalho, 1993). Genetic variability within a species may therefore be crucial to its evolutionary persistence (Frankham, 1995), and its ability to evolve and speciate, although on an evolutionary timescale, lost genetic variability may be regained by mutation (Nei *et al.*, 1975) and gene-flow.

If environments are affected by pollution, genetic variability may be a more medium-term or even short-term concern, as it is a necessity for adaptation to environmental changes (Lynch, 1996). Future environmental conditions, including pollutants, are usually unpredictable, and there is little understanding of the adaptive value of specific alleles or allele combinations (Ryman, 1991). The conservation of as much allelic diversity as possible is thus of utmost importance for the maintenance of the adaptability of a species (Frankel & Soulé, 1981).

However, the quantitative aspect of variability, heterozygosity, may be a crucial factor for the viability and short-term survival of a population (Frankham, 1995). Indeed, a large body of evidence has been collected demonstrating heterozygote advantage in fitness traits including survival, growth, reproductive success, disease resistance, oxygen consumption and other traits (Mitton & Koehn, 1975; Beardmore & Ward, 1977; Altukhov & Varnavskaya, 1983; Handford, 1983; Danzmann *et al.*, 1988; Gentili & Beaumont, 1988; Ferguson & Drahushchak, 1990; Vrijenhoek *et al.*, 1992; McAlpine, 1993). There is also some evidence for a heterozygote advantage in pollution tolerance. In mosquitofish, populations exposed to radionuclides showed higher allozyme heterozygosity than unexposed populations, possibly indicating selection for heterozygotes (Theodorakis & Shugart, 1997). In acute laboratory tests with low pH and high aluminium concentration, heterozygous mudminnows (*Umbra limi*) were more tolerant than individuals with less genetic diversity (Kopp *et al.*, 1992). In other studies, the relationship between heterozygosity and pollutant tolerance was less clear, often because of the superiority of one of the alleles with intermediate heterozygotes (Nevo *et al.*, 1981; Lavie & Nevo, 1982; Guttman, 1994; Kopp *et al.*, 1994).

The mechanisms underlying correlations between heterozygosity in supposedly neutral molecular markers, usually allozymes, and fitness parameters are not well understood (Vrijenhoek, 1996). Possible explanations include a heterozygote advantage at the loci under study (Zouros & Pogson, 1994), at effector loci closely linked to the molecular marker (Maynard-Smith & Haigh, 1974), or simply that more homozygous individuals are more likely to be inbred (Ledig *et al.*, 1983). It has also been suggested that allozyme heterozygotes may have a lower maintenance metabolic rate, and so an energetic advantage over homozygotes in stressful environments (Hummel & Patarnello, 1994; Bayne & Hawkins, 1997). Non-monomeric enzymes of heterozygotes may also be protected from direct chemical interactions with pollutants by heteroduplex chains which neither of the homozygotes possess (Hummel & Patarnello, 1994).

Despite many uncertainties, and the fact that many studies could not detect heterozygote advantages or even found evidence for superior performance in homozygotes (e.g. Christiansen *et al.*, 1977), it is widely accepted that there is a positive relationship between heterozygosity and fitness (Allendorf & Leary, 1986). Therefore, a population losing heterozygosity may be less able to cope with environmental stresses such as pollution.

In contrast to individual heterozygosity, the importance of genetic diversity for population survival is less well demonstrated. Although molecular markers have been used extensively to assess the level of genetic variation at adaptive traits, and thus the adaptive potential of the population, the correlation between molecular and quantitative variation is uncertain (Carvalho *et al.*, 2003). Since most adaptive characters are typically under polygenic control, and the expression of such genes is subject to genotype-environment interactions, the dynamics of quantitative traits is highly complex, and often independent of molecular measures (Butlin & Tregenza, 1998; Reed & Frankham, 2001). Moreover, the genetic architecture of complex traits, including dominance and epistasis, although dependent on population-level processes (Barton & Turelli, 1989), may respond to selection and evolutionary history in fundamentally different ways than molecular variation (Lynch, 1996; Hedrick, 2001). Although some studies demonstrated an increase or a reduction in molecular variation (e.g. Guttman, 1994), such differences in variability are likely to reflect past changes in effective population size and should not be taken as a predictor for adaptive genetic variation or indeed for ecological persistence or evolutionary potential.

However, populations exposed to anthropogenic pollutants offer the opportunity to investigate selective changes at specific genes involved in xenobiotic metabolism and carcinogenesis. The function of many of these genes, for example, those encoding for enzymes of the P450 system or oncogenes, in relation to specific pollutants is well known (Wirgin & Waldman, 1998), thus allowing the investigation of the temporal and spatial dynamics of specific gene loci under selection in wild populations. Furthermore, the level of expression of such genes can be assayed by modern molecular methods, and so the regulatory as well as the structural component of genotypic variation can be assayed. Hopeful beginnings in this direction have been made from research in insecticide resistance genes (Wilson, 2001). Research on such functional genes may not only provide insights into adaptation to xenobiotics, but may also help to bridge the gap between molecular diversity, adaptive genetic variation and fitness in wild populations.

7.3 Impacts and their consequences

7.3.1 Sublethal molecular and cellular response and the potential for selection

Despite living in contaminated environments, some marine organisms can successfully reproduce, develop and grow. This ability to tolerate a cocktail of contaminants is due to the possession of a variety of defence mechanisms. Much work has been focused on characterising and evaluating many of these mechanisms, or 'biomarkers', for their use in determining pollutant exposure and its effect. The presence of these biomarkers may confer a selective advantage to the organism by allowing, for example, the detoxification of pollutants and the faster removal of contaminants from the organism. Moore and Willows (1998) proposed that lysosome rich animals with high intralysosomal ROS (reactive oxygen species) generation, as well as high MDR (multi drug resistance), would be more tolerant of pollutant stress, and that differing lysosomal protective capacities underly species-specific sensitivity or resistance to pollutants. Holland-Toomey and Epel (1993) showed that sea urchin embryos from a pristine environment did not possess the MDR phenotype and were thus sensitive to hydrophobic toxins. Fossi *et al.* (1989) showed higher MFO (mixed function oxidase) activity in *Gobius niger* from a polluted port and this was suggested as possibly being a genetic adaptation.

Pollutants may exert selective pressures which are reflected in changes in the genotypic make-up of populations, potentially resulting in the greater expression of these mechanisms. This is illustrated by the evolution of genetically resistant populations at chronically polluted sites (Depledge, 1996). Several authors have shown a shift in genotype frequencies as a possible indicator of pollutant exposure (Mortimer & Hughes, 1991; Patarnello *et al.*, 1991; Hummel *et al.*, 1995; Snyder & Hendricks, 1997; Newman & Jagoe, 1998), but generally there is little indication of the phenotypic consequences of this selected genotype. The link between measurable change at the genotype, for example, changes in allele frequency, has yet to be fully linked with the greater expression of detoxification or protective biomarkers, such as CYP450.

Although organisms from a polluted site may express a higher rate of activity in, for example, the MFO system, it is difficult to ascertain whether this is part of a genetic response or physical acclimation, where highly conserved mechanisms of protection are activated.

A small number of authors have attempted to address this problem, and have examined the physiological changes that occur as a result of the genotype. Hoffman (1995) showed changes in glutathione S-transferase (GSH), MFO, and acetylcholinesterase (AChe) activity in *Chironomus riparius* after selection to DDT and malathion. Jerneloev (1988) showed increased tolerance to acid waters in the fish *Tribolodon hakoniensis* after long-term exposure (15 generations) to humic, acidified water. It was suggested that the mechanism for this is via chloride excretion cells which are part of the osmoregulatory system and which can provide acid tolerance through hydrochloric acid excretion. Selected fish were also shown to have higher numbers of red blood cells.

Hilbish and Koehn (1985) showed a greater ability for salinity adaptation in populations of *Mytilus edulis* that possessed a specific genotype, with an excretion rate of ammonia and amino acids nearly twice that of alternate genotypes. Lavie and Nevo (1982) showed that zinc resistance in marine gastropods probably depended on a decrease in permeability and an increase in excretory ability. Copper resistance was suggested to be conferred by a detoxifying mechanism allowing copper concentration in resistant animals. Lavie and Nevo (1982) suggested that allozyme variation, in their ability to form metal complexes, could explain the differential survivorship of allozyme genotypes.

Nagel and Voigt (1989) derived a cadmium tolerant population of *Chlamydomus reinhardtii* from a cadmium-sensitive cell wall-deficient strain by long-term selection. Cadmium tolerance was shown to be derived from alterations in the metabolic pathways associated with the chloraplast and not by increased efficiency of a particular detoxification system. Cytochrome P450 (CYP450) has been shown to be part of a selective response in insects by Zhang *et al.* (1997), who found that CYP450, part of the monoxygenase system, plays an important role in housefly resistance to pyriproxyfen, a juvenile hormone analogue.

Although there has been a large amount of work done on genetic adaptation to polluted environments, much of this has concentrated on gross differences in genotype with the aim of using genetic structure and frequency changes as a biomarker. Relatively little investigation has been made into the physiological consequences of selection or, conversely, the way in which activation of highly conserved adaptations, such as the MDR and MFO systems, may increase the potential for selection in an impacted environment.

Such investigations are made difficult by the highly subtle and discrete nature of many of these sublethal responses. The relatively long life span of many of the organisms under test makes experimental examination of selection a difficult undertaking. At present, there is insufficient information to fully understand causal mechanisms and satisfactorily link selective pressure to sublethal responses that may give rise to changes in populations, communities and ecosystem structure. Genetic variation should be examined in the laboratory in association with sublethal toxicity testing in an attempt to further establish causal mechanisms. However, the identification of the genes, gene complexes and modulators of gene expression which actually determine the characteristics of enzymes, metabolic processes, detoxification mechanisms and excretory systems is an extremely large task (Depledge, 1996).

7.3.2 Differential mortality and fitness effects

The selection of certain responses to pollutant-induced selection pressure will lead to the survival of certain genotypes at the expense of others. Critical laboratory experiments using marine organisms show that inorganic and organic pollutants cause fast differential mortality, indicating that allozyme genotypes are selected by the environment (Nevo, 1998). The sensitivity of allozymes to environmental stress through differential mortality reflects the adaptive nature of the surviving individuals (Moraga & Tanguy, 1997). Nevo *et al.* (1986) demonstrated that marine species with a higher genetic diversity were more resistant to pollutants. Hawkins *et al.* (1989) claimed a positive relationship between growth rates of individual mussels and heterozygosity measured at polymorphic enzyme loci. These results suggested that fitness is positively correlated with heterozygosity.

Theodorakis *et al.* (1999) demonstrated an increase in fecundity and a reduction in DNA strand breakage in fish (*Gambusia affinis*) from a contaminated site, that possessed a specific genotype, compared to those which did not.

At the sublethal level, the viability of an organism (i.e. its ability to survive) can be assessed by direct physiological measurement (e.g. heart rate activity or behaviour) or from extrapolation from sublethal biomarkers such as increased subcellular perturbation. Although there is a wealth of information on how genotype selection can effect individual survivorship, there is a dearth of experimental evidence linking genotype selection and phenotype expression. Furthermore, the consequences of sublethal responses in terms of whole animal mortality are poorly understood, which further increases the problem of establishing causality.

7.4 The evolution of tolerance

A significant proportion of literature looking at the potential effects of toxicant exposure on genetic variability of natural populations has focused primarily on two important issues: that toxicant exposure may result in a loss of adaptability and subsequently, the possibility of population extinction; and that it may select for resistant genotypes which may have reduced fitness (in relation to non-resistant genotypes) in the absence of exposure. If either of these issues is justified then this may have serious implications in terms of ecology and evolution (Forbes, 1999).

7.4.1 Intrapopulation diversity

Individuals of a particular species are generally not uniformly distributed in space, but exist in 'clusters' or local populations. This is, however, sometimes difficult to define as the boundaries between local populations are often blurred, with individuals tending to migrate from one local population to another.

In the study of evolution, the concept of a gene pool is useful (i.e. the aggregate of the genotypes of all the individuals in a population), and the existence of genetic variation is vital for evolution to occur. For example, if at a certain gene locus all individuals of a given population are homozygous for exactly the same allele, then evolution cannot take place at that locus because the allelic frequencies are not able to change from generation to generation. In contrast, in a different population where there are two alleles at a particular locus, evolutionary change can take place in this population, and one allele may increase in frequency at the expense of the other (Ayala & Kiger, 1984).

Individuals which have advantageous variations are more likely to survive and reproduce, and as a result, useful variations will become more prevalent through the generations, whilst harmful or less useful ones will be eliminated. Individuals may differ in phenotype (i.e. a morphological, physiological, biochemical or behavioural characteristic of an individual organism or group of similar individuals) as a result of genetic and/or environmental differences. The major sources of variation in phenotype as described by Futuyma (1998) include:

Differences in genotype (i.e. in the DNA sequence at one or more loci): Although most genetic variations can be transmitted through either eggs or sperm, some are strictly maternally or paternally inherited. Different genotypes often differ in phenotype.

Differences in environment: Features such as physiological and behavioural traits may be affected by immediate or very recent environmental conditions, and may change repeatedly throughout life. Additionally, differences which persist through part or all of an individual's lifetime may be caused by environmental differences experienced very early in development, or even in the egg.

Maternal effects: These refer to characteristics of the offspring which are not due to the genes they inherit from their mother, but rather to non-genetic influences, such as the amount or composition of yolk in her eggs, the type of maternal care she provides, or her physiological condition whilst carrying eggs or embryos. Differences among mothers may be due to either their genotypes, nutrition or other environmental factors.

Differences among individuals may therefore result from non-genetic maternal effects or by environmental factors which act on an embryo before birth/hatching, rather than genetic differences.

With respect to pollution, this may either kill an organism or effect a variety of processes such as increasing mortality or reducing somatic growth rate. It may elicit the expression of genes which would not otherwise have been expressed, i.e. the induction of enzymes, and consequently variation which was of little importance in the unpolluted environment may now distinguish survivors from non-survivors (Walker *et al.*, 1996). Therefore, survival may depend on having alleles which increase detoxication enzymes (Terriere, 1984; Oppenoorth, 1985).

Although pollutants act on individual organisms, at a higher level this may also have an effect on the population/community, and therefore both individual and population aspects may be integrated by looking at the effects on the gene pool. If a population is affected by a pollutant, and the appropriate genetic variation exists within that population, then the gene pool will change. In contrast, if pollutants affect individuals (e.g. cause death), but this in turn does not have an effect on the population, then there will be no effect on the gene pool (Moriarty, 1983).

With respect to changes in gene pools, Moriarty (1983) outlined the following points. Firstly, if the appropriate genetic variation does not exist within a population, then the gene pool will be unaffected by a pollutant even if the population is affected. Secondly, species differ in their susceptibility to pollutants and if the gene pool of one species is affected, it does not automatically follow that the numbers of all other species in that habitat which lack the appropriate genetic variability will have been reduced. Finally, an increase in the proportion of a particular genotype as the degree of exposure to a pollutant increases, does not by itself prove that the pollutant has caused the change in the gene pool.

Cuvinaralar and Aralar (1995) looked at resistance to a heavy metal mixture in *Oreochromis niloticus* (Cichlidae) progenies from parents chronically exposed to the same metals. Adult *Oreochromis niloticus* and their progenies (F-1) were exposed to a mixture of mercury, cadmium and zinc for a two month period. Survivors were grown to sexual maturity in a natural environment, spawned, and the progenies (F-2) of the exposed F-1 (EF(1)) exposed to another mixture of zinc, cadmium and mercury, both in a static and static-renewal system. For control purposes, another group of F-2 from unexposed F-1 (UF1) received the same treatment, and results showed that in both exposure systems, survival of the F-2 of EF(1) was significantly higher than those from UF1. Cuvinaralar and Aralar (1995) conclude that exposure of the parental stock resulted in the culling of individuals that were more susceptible to the heavy metals. The more resistant members of the population which have the ability to adapt to the toxicants were able to pass on the resistance to their offspring. Cuvinaralar and Aralar noted that the results were supported by other studies in the field which demonstrate high resistance in populations of organisms living in contaminated sites.

7.4.2 Interpopulation differentiation

Differences among populations arise by the transformation of genetic variation *within* populations into variation *among* populations, due to changes in allele frequencies that transpire differently from one population to another (Futuyma, 1998). Differences among different geographic populations of the same species have been studied in depth, and as a result these studies of geographic variation have provided an insight into the mechanisms of evolution.

If distinct forms or populations have overlapping geographic distributions, such that they occupy the same area and can frequently encounter each other, they are *sympatric*.

Populations with adjacent but non-overlapping geographic ranges that come into contact are *parapatric*, and populations with separated distributions are *allopatric*. The genetic and phenotypic differences between populations of a species vary from slight to very pronounced, and may occur over short or long distances. Often, allele frequencies (the proportion of gene copies in the population that are of that allelic type) differ among populations at some loci, but not others, and the same is true of phenotypic characters (Futuyma, 1998).

As already mentioned above, differences among populations arise by the transformation of genetic variation within populations into variation among populations. Similarly, with respect to variations between populations, Futuyma (1998) makes the following three points. Firstly, a species is not genetically uniform over its geographic range. Populations differ in allele and genotype frequencies, often considerably and in many different characteristics. Secondly, differences among populations range from slight to great, and it may be difficult or even arbitrary to determine if populations belong to one or more than one species. Conspecific populations vary in the degree of reproductive isolation, and even in characters that ordinarily distinguish higher taxa. Finally, variation among conspecific populations forms a continuum with variation among species.

With respect to interpopulation differentiation as a result of pollution, three 'natural system' studies were initiated at Miami University, Oxford, Ohio, in order to demonstrate whether a relationship existed between changes in population genetic structure and exposure to contaminants in natural systems. These studies additionally investigated whether a cause-and-effect relationship existed between contaminants and sensitivity to toxicity of individuals with different genotypes.

The first of these studies conducted by Gillespie and Guttman (1989) sampled fish and selected benthic macroinvertebrates from a stream flowing through a uranium reprocessing facility in south-western Ohio. Sites were sampled both above and below the production area as the stream had received a significant amount of radionuclide contamination from processing activities, and the level of contamination had remained high relative to upstream sites in recent years. Patterns for PGM (phosphoglucomutase) in stoneroller minnows (*Campostoma anomalum*) were found to indicate a marked shift in allozyme frequency between the upstream and downstream sites, with the majority of change occurring over a distance of less than 500 metres. Examination of genotypic proportions indicated the same pattern, with the PGM-BB genotype increasing significantly from upstream to downstream. Similarly, under laboratory conditions, toxicity testing of copper in fish collected from upstream of the facility, together with a local stream, suggested that stonerollers with the PGM-AA and PGM-AB genotypes were more sensitive to copper toxicity than fishes with the PGM-BB genotype. Therefore, *C. anomalum* with certain allozyme genotypes may be more sensitive to the toxic effects of specific contaminants and complex effluents than individuals with other genotypes (Guttman, 1994).

The second study undertaken by Kopp *et al.* (1992) used horizontal starch-gel electrophoresis to characterise the genetic structure of central mudminnow (*Umbra limi*) populations from acid-stressed (low pH/high aluminium) and non-acid stressed sites in the north branch of the Moose River (NBMR), New York. The aim of this study was to determine if environmental shifts were accompanied by detectable shifts in genetic structure. Central mudminnow populations in the NBMR were characterised by significantly lower heterozygosity levels at stressed sites than reference sites. Additionally, several genetic loci demonstrated consistent or moderately consistent genotypic shifts in comparisons between populations at reference and low pH/high aluminium sites. Populations at acid-stressed sites were also characterised by higher frequencies of one particular allozyme, suggesting that environmental conditions were acting as a selective force. In acute laboratory toxicity tests, the most tolerant fish were significantly more genetically variable, indicating that genetic diversity may be beneficial to the survival of individual fish experiencing physiological stress (Kopp *et al.*, 1992).

Similarly, in the third study, Benton *et al.* (1994) compared the genetic structure of mosquitofish (*Gambusia holbrooki*) and freshwater snail (*Helisoma trivolvis*) populations from two sites at the Department of Energy Savannah River site (South Carolina). Samples taken from one site were contaminated with high levels of arsenic, cadmium, chromium, copper and zinc in both the water column and sediments as a result of effluent from the nearby power plant, together with samples from a second, control site from a nearby reservoir (Par Pond). Distinct genetic patterns within the contaminated habitat, combined with data from other published work, suggested that selection for tolerant genotypes may have occurred in both species. In mosquitofish, a particular genotype associated with small body size appears to have been favoured in the contaminated environment (Benton *et al.*, 1994).

Weis *et al.* (1999) looked at the teratogenetic responses and degree of tolerance of *Fundulus heteroclitus* (mummichog) populations to methylmercury from both clean and polluted environments. They found that embryos of *F. heteroclitus* taken from unpolluted areas of eastern Long Island, New York, showed a wide variety of embryological malformations when exposed to methylmercury at concentrations greater than 50 $\mu g\ l^{-1}$. However, considerable variation in the severity of response was found in eggs from different females. Some females produced embryos that were extremely resistant to the exposure, whilst other females produced embryos that were very susceptible or of intermediate susceptibility. This degree of tolerance was thought to be linked with the number of dorsal fin rays of the female, and consequently this high variability in tolerance allows the population to be better able to withstand an influx of mercury contamination.

Weis *et al.* (1999) subsequently studied a mummichog population living in a contaminated estuary in New Jersey which has elevated levels of heavy metals and other contaminants within the sediment. Very few females in this population were found to produce eggs susceptible to methylmercury, and most embryos were tolerant with respect to the production of embryological malformations, although this tolerance is not seen in larvae after hatching, nor in adults. Phenotypic variability seen within the clean population was not seen in populations from the polluted estuary, and this may reflect reduced population genetic diversity. Adults from the polluted population were additionally found not to grow as well or live as long as the reference population, but become reproductive sooner, an evolutionary strategy for perpetuating a population in a stressful environment (Weis *et al.*, 1999).

Keklak *et al.* (1994) identified genetic differences between populations of mosquitofish from a uranium-contaminated stream using starch gel electrophoresis. Fish collected from the uncontaminated mainstream of Upper Three Runs Creek (South Carolina, USA) were found to exhibit greater genetic variability than those which were collected from the contaminated Tims Branch. A toxicity assay was performed to determine if these genetically distinct mosquitofish also displayed enhanced uranium tolerance, and times to death were compared for fish from an uncontaminated site and offspring of fish taken from the uranium-contaminated Tims Branch. Keklak *et al.* (1994) found that after seven days of exposure to 2.57 mg l^{-1} of uranium as uranyl nitrate, 98% and 96% of the population from the uncontaminated site had died in the replicate tanks. In contrast, the final mortality of the offspring from the population previously exposed to uranium was 25% and 57% in the replicate tanks. They concluded that fish derived from the uranium-contaminated site were more tolerant than those from the uncontaminated site, and that because these were second generation fish, this tolerance probably had a genetic basis.

However, Klerks and Lentz (1998) investigated the occurrence of adaptation to lead and zinc in the western mosquitofish *Gambusia affinis* inhabiting the industrially contaminated Bayou Trepagnier (Louisiana, USA). Levels of lead and zinc in water and sediment were found to be considerably higher in Bayou Trepagnier than in a nearby control stream, and tissue metal levels of mosquitofish were highly elevated for lead and to a lesser extent, zinc. Fish collected from Bayou Trepagnier and exposed to zinc in a 96 hour laboratory bioassay were found not to differ in their sensitivity to zinc from conspecifics collected from a control site. In contrast, Bayou Trepagnier fish were found to exhibit an increased resistance to lead, although this difference to lead between Bayou Trepagnier fish and control fish was no longer evident when both groups of fish were kept for 34 days under identical clean water conditions in the laboratory. Klerks and Lentz (1998) concluded that although a genetic basis of the difference in resistance between the two populations cannot be fully excluded, it appears that the elevated lead resistance in Bayou Trepagnier mosquitofish is due to acclimation (physiological, individually-based) rather than adaptation at the population level.

Similarly, Klerks *et al.* (1997) examined physiological acclimation, genetic adaptation and genetic differentiation in darter gobies (*Gobionellus boleosoma*) inhabiting a coastal marsh with a long history of PAH contamination. No acclimation was detected and a two week pre-exposure at the polluted site resulted in a decreased rather than an increased resistance in a subsequent laboratory exposure to polluted sediment. Additionally, fish collected from sites with elevated sediment PAH levels did not exhibit an increased resistance in bioassays with polluted sediment, confirming the lack of acclimation and indicating a lack of adaptation to the pollutants. Klerks *et al.* (1997) also detected no differences in frequencies of allozyme genotypes when comparing gobies from the polluted area to those from a nearby control site, and overall levels of heterozygosity were found to be similar in the two populations. They concluded that whilst lack of UV-induced toxicity in laboratory exposures could have played a role, at least four other factors may explain the apparent lack of adaptive responses and genetic differences. Firstly, bioavailability of the contaminants to the darter goby may have been low. Secondly, the contaminated marsh contained a large number of different chemicals, and acclimation, adaptation and genetic differentiation are expected to be less likely when more contaminants are involved. Thirdly, the hydrocarbon distribution at the contaminated marsh was very patchy so fish may avoid exposure to the highly-contaminated sediment. Fourthly, gene flow may be sufficiently high in this mobile species to prevent local adaptation.

With respect to invertebrates, Grant *et al.* (1989) investigated mapping the ecological impact of heavy metals on the estuarine polychaete *Nereis diversicolor* from Restronguet Creek, Cornwall, UK, using inherited metal tolerance. They collected animals from a variety of sites in Restronguet Creek (which has a history of metal contamination), the adjacent Mylor Creek and the uncontaminated river Avon, and these were subsequently bred in the laboratory. The offspring were grown until they were of sufficient size for toxicity testing, and simultaneous tests were carried out on field derived animals of the same size. Results showed that although the most contaminated site in Restronguet Creek was adjacent to the metal-laden Carnon River, animals from the site further upstream showed the greatest tolerance to both metals. Tolerance to copper declined steadily, moving from the head to the mouth of Restronguet Creek, and the mean survival time for worms from the mouth of the estuary was only slightly elevated from those of the Avon worms. In the adjacent Mylor Creek, the LT_{50} at a given copper concentration was found to be comparable to that for worms from the uncontaminated Avon estuary. In contrast, only animals from the most contaminated site and that further upstream showed tolerance to zinc. Animals from the rest of Restronguet and Mylor Creeks showed no evidence of being more tolerant to zinc than those from the uncontaminated Avon.

Given that it was the offspring of worms that were found to be substantially more tolerant to both copper and zinc, these overall observations indicated that copper and zinc tolerance in *Nereis* has a large heritable component. More recently, Millward and Grant (2000) have also described copper tolerance in nematode communities from the same estuarine system. Grant *et al.* (1989) suggest that these studies provide strong evidence for an ecological impact by both metals in Restronguet Creek, although the different spatial distributions of the two metal tolerances in *Nereis diversicolor* indicate that the two phenomena are not closely linked either genetically or physiologically. Additionally, as the degree of metal tolerance reduces rapidly as the level of contamination decreases, the authors suggest that metal tolerant individuals are competitively inferior (see section 7.4.4) to normal individuals in clean environments as found in other non-marine species (Luoma, 1977). The study by Grant *et al.* (1989) would seem to support the suggestion made by Luoma (1977) that mapping of tolerant ecotypes can be used to demonstrate real ecological impact of pollution. In the study with nematodes, Millward and Grant (2000) suggest that pollution-induced community tolerance (PICT) can be used rather than species specific studies, the advantage being that it is faster, requiring less taxonomic expertise.

In invertebrates, altered life history is seen as part of the syndrome with tolerant populations generally showing a shorter life cycle and higher reproductive effort (Posthuma & Van Straalen, 1993). Tolerance involves the modification of existing physiological mechanisms for pollutant detoxification. In *Drosophila melanogaster* it involves duplication of the metallothionein gene whilst in *Orchesella cincta* it may be associated with altered frequency of Got-alleles involved in altered energy use as suggested for other Krebs cycle enzymes in mosquitofish *Gambusia holbrooki* (Kramer *et al.*, 1992). In contrast, in arthropods from contaminated sites, tolerance was found to be associated with low 70 kDa stress protein levels. In this case it appears that selection of insensitive phenotypes in long-term polluted soil had occurred and was associated with increased selectivity in food choice (Kohler *et al.*, 2000). Long-term selection for high HSP70 levels was not observed, indicating a trade-off between this strategy and other fitness consequences.

7.4.3 The speed of adaptation

A direct correlation between the amount of genetic variation in a population and the rate of evolutionary change by natural selection was demonstrated mathematically with respect to fitness, by Sir Ronald A. Fisher in his Fundamental Theorem of Natural Selection (1930). Fisher states that: 'The rate of increase in fitness of a population at any time is equal to its genetic variance in fitness at that time' (fitness being used here as a measure of relative reproduction rate).

This Fundamental Theorem applies only under particular environmental conditions and strictly to allelic variation at a single gene locus. However, the greater the number of variable loci and the more alleles there are at each variable locus, the greater the possibility for change in the frequency of some alleles at the expense of others. This requires selection favouring the change of some trait(s) and that the variation be relevant for the trait(s) being selected (Ayala & Kiger, 1984).

Whether a population will be eliminated or adapt to an introduced stress depends on the rapidity of onset, severity of the stress, and the capacity of the population to adapt to it (Weis *et al.*, 1999). With respect to the mummichog (*Fundulus heteroclitus*), Mitton & Koehn (1975) suggest that the polygynous mating system of this species allows for rapid gene frequency changes and, therefore, rapid evolutionary responses to a variable environment. In invertebrates, genotypic adaptation can be achieved in a few generations (Posthuma & Van Straalen, 1993).

Weis *et al.* (1999) discussed the speed at which a shift of embryonic tolerance could occur in *F. heteroclitus* from an unpolluted site. They note that in 1982 an unusually large amount of rainfall and associated run-off of pesticides near one of their study sites resulted in over 40% of the females producing non-viable eggs. In addition, those fish which produced viable eggs generally produced tolerant ones with respect to abnormalities such as cardiac and skeletal defects (Weis & Weis, 1984). Accompanying the shift in methylmercury tolerance was a trend toward increased dorsal fin rays of the females that produced viable eggs (increased dorsal fin rays in female *F. heteroclitus* already being associated with increased embryonic tolerance to methylmercury (Weis *et al.*, 1999)). Weis *et al.* (1999) hypothesise that this influx of contaminants, including chlorinated hydrocarbon pesticides, resulted in the marked changes in reproductive success. They note that the normal variability in the population allowed some fish to produce viable eggs, and these were the ones whose eggs were more resistant to methylmercury, indicating that when a population is variable to begin with, a change in overall tolerance can happen very quickly. The following summer, the tolerance to methylmercury returned to its normal heterogeneous state (Weis *et al.*, 1999).

7.4.4 The costs of adaptation

Evolutionary response to pollution and contaminants, including xenobiotics, is known as 'resistance', and this resistance has a genetic basis. Pollution often results in unfavourable environmental change, and resistance usually defends organisms against the deleterious consequences of pollution. This defence may reduce an organism's mortality rate, although generally at the expense of another function, i.e. using energy and/or nutrients which could otherwise have been used for reproduction or somatic growth. Defence may therefore involve a trade-off between production and survival, i.e. increased survival may only be obtained at a cost of reduced growth or reproduction (Walker *et al.*, 1996).

The fitness of a particular allele depends on the environment in which the carrier lives and the allele may be termed 'resistant' if it increases the fitness of its carriers in a polluted environment. Similarly, in unpolluted environments, if resistant alleles are outperformed by 'susceptible' (i.e. non-resistant alleles), then the resistant alleles are said to have a 'fitness cost'. As previously mentioned, if alleles exist which affect production rate but not mortality rate, selection acts to maximise production rate. Similarly, if alleles affect mortality rate but not production rate, selection acts to minimise mortality rate. However, it is not always possible to alter one life history trait without affecting the other if mortality and production are involved in a trade-off (Walker *et al.*, 1996). It is likely that a decrease in mortality rate can only be achieved at the cost of a decrease in somatic growth rate, as resources allocated to protective mechanisms (defences) which decrease mortality rate are therefore not available for somatic growth. This trade-off does however depend on the organism's environment. An effective defence in one environment may have no impact in another (i.e. defences against pollution are of no value in unpolluted environments) and therefore trade-offs are environment-dependent (Walker *et al.*, 1996).

If genes are expressed in polluted environments that were previously absent or silent, then evolutionary outcomes may differ between polluted and unpolluted environments, and their populations may be genetically distinct. Such differentiation can be studied by a 'transplant experiment' which involves transplanting individuals from a number of environments to a common environment in which their life history components are then measured. Previous studies have indicated that, as predicted, resistant strains are fitter in polluted environments, and susceptible strains fitter in unpolluted environments. This results in a 'fitness cost of resistance', i.e. the resistant alleles which are fitter in the polluted environment are less fit than susceptibles in the unpolluted environment (Walker *et al.*, 1996).

Weis *et al.* (1999) looked at the eggs of *Fundulus heteroclitus* (mummichog) populations from two sites: Piles Creek, a contaminated estuary of the Newark Bay system (USA), and an uncontaminated area of eastern Long Island, New York (USA). They found that eggs of fish from the polluted site would only fertilise successfully at reduced salinities rather than full strength seawater (Bush & Weis, 1983). Eggs from the unpolluted populations were however found to fertilise successfully over a larger range of salinities from 10–30‰. If the eggs from the polluted site, stripped into 30‰ seawater, were subsequently transferred to 15‰ within one minute, successful fertilisation could occur. Weis *et al.* (1999) suggest that this population may have adapted so narrowly to the specific conditions of its habitat (salinity approximately 15–20‰) that it has lost some of its euryplasticity present in the unpolluted population. Consequently, the development of tolerance to methylmercury at the cost of reduced genetic variability in the population, may reduce their ability to deal with natural stresses or other types of pollution.

Genetic variation in relation to water quality was demonstrated by Heithaus and Laushman (1997) who used three freshwater fish species to investigate the effects of ecology, life history, and water quality on genetic variation. *Etheostoma caeruleum* (Rainbow darter), *E. blennioides* (Greenside darter) and *Campostoma anomalum* (Central stoneroller) were sampled from six streams of varying water quality. Using allozyme electrophoresis, the most ecologically specialised species, *E. caeruleum*, was found to be the least variable, *E. blennioides* intermediate in specialisation and variation, and the least specialised species, *C. anomalum*, the most variation. Populations in the river with the worst water quality (Huron River) were found to have the lowest within-population variation, and therefore genetic variation may be a useful indicator of water quality. Heithaus and Laushman (1997) note that genetic variation may result from selection associated with specific loci (e.g. PGM-2 in stoneroller minnows); however, indirect effects on population size probably contributed to the erosion of genetic variation. They concluded that ecology, life history and pollution tolerance data combine as predictors of species' risk of genetic erosion.

In laboratory-based selection experiments using *Drosophila melanogaster*, cadmium-resistant lines were found to pay a fitness cost in unpolluted environments. This was demonstrated with fecundity being reduced by 44% and emergence weight reduced by 4% in females (Shirley & Sibly, 1999). Back crosses indicated that resistance was due to a single sex linked gene. Furthermore, the life history traits affected were produced by a single gene and were dependent on the same metabolic pathway which also appeared associated with metallothionein production. In this species metallothionein production is known to be linked to genes on the X-chromosome. The study illustrates how single or closely linked genes can have large antagonistic pleiotropic effects on life histories (Shirley & Sibly, 1999).

Under a metal requirement hypothesis, it has been suggested that metal tolerant organisms, by virtue of their tolerance mechanism, are less efficient at the uptake and use of metals. This suggests that with metal tolerance comes an evolved dependency for high metal concentrations (Posthuma & Van Straalen, 1993). A consequence of this is that micronutrient deficiency might occur in metal tolerant organisms maintained in clean environments and that this might explain the reduced fitness of tolerant individuals in these situations. If this mechanism does explain the 'cost of tolerance', evidence to date suggests that it is very species specific. For example, in the midge *Chironomus riparius*, cost of cadmium tolerance was identified with animals showing high control mortality and increased larval development time (Postma *et al.*, 1995a). High control mortality has also been observed in metal tolerant *Nereis diversicolor* (Burlinson & Lawrence, in prep.). In the case of *Chironomus*, it appears that reduced growth and reproduction were a direct consequence of a zinc shortage. Mortality, however, was not reduced in the presence of zinc (Postma *et al.*, 1995b). This increased need for zinc may therefore provide evidence for the metal requirement hypothesis, despite the fact that zinc was not the metal for which the animals had acquired tolerance. However, in the plant *Mimmulus gutatus*, no evidence has been found to support the theory in relation to vegetative growth or reproduction (Harper *et al.*, 1997, 1998). Similarly, in *Nereis diversicolor*, no evidence has yet been found to support the theory in relation to survival, growth or reproduction (Burlinson & Lawrence, in prep). It is, therefore, likely that a cost to tolerance that involves an evolved dependency for high metal concentrations might only occur in specific cases. It seems intuitive to suggest that this might be linked to cases where the evolution of tolerance has involved the selection of insensitive phenotypes (section 7.4.2; Kohler *et al.*, 2000).

7.4.5 The identification of tolerance genes

Individuals which are homozygous resistant are known as 'resistant individuals'. The relative resistance of heterozygotes measures the degree of dominance of the resistant allele. This degree of dominance affects the speed with which an allele spreads, with advantageous dominant alleles spreading faster initially than recessive alleles. To establish the number of loci involved, breeding experiments over several generations are usually conducted using homozygous strains. A homozygous resistant strain is crossed with a homozygous susceptible strain, with the offspring being heterozygous at all loci. These offspring are then back crossed with the parental strains, and half the offspring of these back crosses are expected to be heterozygous if only one locus is involved. However, if more than one locus is involved, less than half of the offspring show resistance (Walker *et al.*, 1996). Statistical techniques can be used to estimate the number of genes involved, but the validity of these assumptions needs to be checked. When discrimination between genotypes is difficult, as is often the case in resistance studies, further experiments may be required and these may include repeated back crossing or the use of genetic markers which map the positions of resistant genes on the chromosomes. Using studies of this type, major genes (i.e. genes with large effects) are found in most cases of resistance (Walker *et al.*, 1996).

A variety of laboratory studies (Gillespie & Guttman, 1993; Guttman, 1994) have shown that organisms with different allozyme genotypes vary in their tolerance to the toxicity of various toxicants. As a result, specific allozyme genotypes may be selected for as they contribute to resistance with respect to the toxic effects of pollutants. Similarly, allozyme analysis could still be used as a marker of genetic susceptibility or as an estimate of genetic variation in populations, even if allozyme variation is not directly adaptive or contributes in a minor way to genetically-based resistance (Gillespie & Guttman, 1999). Genetic variation is an ecologically important variable, and allozyme variation may be a useful tool in monitoring populations of organisms that are susceptible to environmental chemical pollutants (Ben-Shlomo & Nevo, 1988; Gillespie & Guttman, 1993; Kopp *et al.*, 1994; Gillespie, 1996; Evenden & Depledge, 1997). Consequently, it may subsequently prove valuable for risk analysis in ecotoxicology (Guttman, 1994; Evenden & Depledge, 1997).

However, future research involving population genetic structure and ecotoxicology needs to be undertaken. According to Guttman (1994) this research should focus on determining the mechanism of sensitivity, documenting multigenerational effects of chronic laboratory exposure on population genetic composition, investigating whether previously stressed and genetically impacted populations are more susceptible to further natural and/or anthropogenic stressors, and establishing the utility of population genetic structure as a sensitive monitor of impacts in aquatic systems and their subsequent remediation.

7.5 References

Allendorf, F.W. & R.F. Leary (1986) Heterozygosity and fitness in natural populations of animals. In: (ed. Soulé, M.E.) *Conservation Biology: The Science of Scarcity and Diversity*. Sinauer Associates, Sunderland, Massachusetts, USA, pp. 57–76.

Allendorf, F.W., N. Ryman & F. Utter (1987) Genetics and fishery management: past, present and future. In: (eds Ryman, N. & F. Utter) *Population Genetics and Fisheries Management*. University of Washington Press, Seattle & London.

Altukhov, Y.P. & N.V. Varnavskaya (1983) Adaptive genetic structure and its relationship to intrapopulation, sex, age and growth differentiation in sockeye, *Oncorhynchus nerka* (Walb.). *Genetika*, **19**, 796–806.

Ayala, F.J. & J.A.J. Kiger (1984) *Modern Genetics*, 2nd ed. The Benjamin/Cummings Publishing Company, USA, 923 pp. + appendices.

Bagenal, T.B. (1973) Fish fecundity and its relation with stock and recruitment. *Rapports et Procès-verbaux de Réunions du Conseil International pour l'Exploration de la Mer*, **148**, 186–198.

Baker, R.J., J.A. Dewoody, A.J. Wright & R.K. Chesser (2001) On the utility of heteroplasmy in genotoxic studies: an example from Chernobyl. *Ecotoxicology*, **8**, 301–309.

Barton, N.H. & M. Turelli (1989) Evolutionary quantitative genetics – how little do we know? *Annual Review of Genetics*, **23**, 337–370.

Bayne, B.L. & A.J.S. Hawkins (1997) Protein metabolism, the costs of growth and genomic heterozygosity: experiments with the mussel *Mytilus galloprovincialis* L. *Physiological Zoology*, **70**, 391–402.

Beardmore, J.A. & R.D. Ward (1977) Polymorphism, selection and multilocus heterozygosity in the plaice, *Pleuronectes platessa* L. In: (eds Christiansen, F.B. & T.M. Fenchel) *Measuring Selection in Natural Populations*. Springer Verlag, Berlin, pp. 207–222.

Ben-Shlomo, R. & E. Nevo (1988) Isozyme polymorphism as monitoring of marine environments: the interactive effect of cadmium and mercury pollution on the shrimp, *Palaemon elegans*. *Marine Pollution Bulletin*, **19**, 314–317.

Benton, M.J., S.A. Diamond & S.I. Guttman (1994) A genetic and morphometric comparison of *Helisoma trivolvis* and *Gambusia holbrooki* from clean and contaminated habitats. *Ecotoxicology and Environmental Safety*, **29** (1), 20–37.

Berry, R.J. (1980) Genes, pollution and monitoring. In: (eds McIntyre, A.D. & J.B. Pearce) *Biological Effects of Marine Pollution and the Problems of Biomonitoring*. Imprimerie Bianco Luno A/S, Copenhagen, pp. 253–257.

Bickham, J.W., S. Sandhu, P.D.N. Hebert, L. Chikhi & R. Athwal (2000) Effects of chemical contaminants on genetic diversity in natural populations: implications for biomonitoring and ecotoxicology. *Mutation Research*, **463**, 33–51.

Bijlsma, R., J. Bundgaard & A.C. Boerema (2000) Does inbreeding affect the extinction risk of small populations? Predictions from *Drosophila*. *Journal of Evolutionary Biology*, **13**, 502–514.

Brodie, E.D., A.J. Moore & F.J. Janzen (1995) Visualising and quantifying natural selection. *Trends in Ecology and Evolution*, **10**, 313–318.

Burlinson, F.C. & A.J. Lawrence (in prep.) Heavy metal tolerance and its fitness consequences in *Nereis diversicolor*.

Bush, C.P. & J.S. Weis (1983) Effects of salinity on fertilization success in two populations of *Fundulus heteroclitus*. *Biological Bulletin*, **164**, 406–417.

Butlin, R.K. & T. Tregenza (1998) Levels of genetic polymorphism: marker loci versus quantitative traits. *Philosophical Transactions of the Royal Society of London Series B – Biological Sciences*, **353** (1366), 187–198.

Carvalho, G.R. (1993) Evolutionary aspects of fish distributions: genetic variability and adaptation. *Journal of Fish Biology*, **43** (A), 53–73.

Carvalho, G.R. & L. Hauser (1994) Molecular genetics and the stock concept in fisheries. *Reviews in Fish Biology*, **4**, 326–350.

Carvalho, G.R., C. van Oosterhout, L. Hauser & A.E. Magurran (2003) Measuring genetic variation in wild populations: from molecular markers to adaptive traits. In: (eds Hails, R., H.C.J. Godfray & J. Beringer) *Genes in the Environment*. Blackwell Publishing, Oxford, in press.

Christiansen, F.B., O. Frydenberg & V. Simonsen (1977) Genetics of Zoarces populations. X. Selection component analysis of the EstIII polymorphism using samples of successive cohorts. *Hereditas*, **87**, 129–150.

Crnokrak, P. & D.A. Roff (1999) Inbreeding depression in the wild. *Heredity*, **83**, 260–270.

Cronin, M.A. & J.W. Bickham (1998) A population genetic analysis of the potential for a crude oil spill to induce heritable mutations and impact natural populations. *Ecotoxicology*, **7**, 259–278.

Crow, J.F. & M. Kimura (1970) *An Introduction to Population Genetics Theory*. Harper & Row, New York.

Cushing, D.H. (1973) Dependence of recruitment on parent stock. *Journal of the Fisheries Research Board of Canada*, **30**, 1965–1976.

Cuvinaralar, M.L.A. & E.V. Aralar (1995) Resistance to a heavy-metal mixture in *Oreochromis niloticus* progenies from parents chronically exposed to the same metals. *Chemosphere*, **30** (5), 953–963.

Dahlgaard, J. & A.A. Hoffmann (2000) Stress resistance and environmental dependency of inbreeding depression in *Drosophila melanogaster*. *Conservation Biology*, **14**, 1187–1192.

Dahlgaard, J., R.A. Krebs & V. Loeschke (1995) Heat shock tolerance and inbreeding in *Drosophila buzzatii*. *Heredity*, **74**, 157–163.

Danzmann, R.G., M.M. Ferguson & F.W. Allendorf (1988) Heterozygosity and components of fitness in a strain of rainbow trout. *Biological Journal of the Linnean Society*, **33**, 285–304.

Darwin, C. (1859) *On the Origin of Species by Means of Natural Selection or the Preservation of Favored Races in the Struggle for Life*. Murray, London.

Depledge, M.H. (1996) Genetic ecotoxicology: an overview. *Journal of Experimental Marine Biology and Ecology*, **200** (1–2), 57–66.

Diamond, S.A., S.C. Newman, M. Mulvey, P.M. Dixon & P. Martinson (1989) Allozyme genotypes and time to death of Mosquitofish *Gambusia affinis* (Bairs & Girard), during acute exposure to inorganic mercury. *Environmental Toxicology and Chemistry*, **8**, 613–622.

Dias, P.C., G.R. Verheyen & M. Raymond (1996) Source-sink populations in Mediterranean Blue tits: evidence using single-locus minisatellite probes. *Journal of Evolutionary Biology*, **9**, 965–978.

Dubrova, Y.E., V.N. Nesterov, N.G. Krouchinsky, V.A. Ostapenko, R. Neumann, D.L. Neil & A.J. Jeffreys (1996) Human minisatellite mutation rate after Chernobyl accident. *Nature*, **380**, 683–686.

Ebert, D., C. Haag, M. Kirkpatrick, M. Riek, J. Hottinger & I. Pajunen (2002) A selective advantage to immigrant genes in a *Daphnia* metapopulation. *Science*, **295**, 485–488.

Ellegren, H., G. Lindgren, C.R. Primmer & A.P. Moller (1997) Fitness loss and mutations in barn swallows breeding in Chernobyl. *Nature*, **389**, 593–596.

Endler, J.A. (1986) *Natural Selection in the Wild*. Princeton University Press, Princeton, New Jersey, 337 pp.

Evenden, A.J. & M.H. Depledge (1997) Genetic susceptibility in ecosystems: The challenge for ecotoxicology. *Environmental Health Perspectives*, **105** (Suppl. 4), 849–854.

Falconer, D.S. (1989) *Introduction to Quantitative Genetics*. Longman Scientific & Technical, Harlow, UK.

Falconer, D.S. & T.F.C. Mackay (1996) *Introduction to Quantitative Genetics*. Longman Group Ltd, Essex, UK, 464 pp.

Ferguson, M.M. & L.R. Drahushchak (1990) Disease resistance and heterozygosity in rainbow trout. *Heredity*, **64**, 413–417.

Forbes, V.E. (1999) Genetics and Ecotoxicology – Insights from the Interface. In: (ed. Forbes, V.E.) *Genetics and Ecotoxicology*. Taylor & Francis Ltd, London, pp. 1–8.

Fossi, C., C. Leonzio & S. Focardi (1989) Detoxification mechanisms and adaptation phenomena in marine organisms. *Oebalia*, **15**, 885–891.

Frankel, O.H. & M.E. Soulé (1981) *Conservation and Evolution*. Cambridge University Press, Cambridge.

Frankham, R. (1995) Conservation genetics. *Annual Reviews in Genetics*, **29**, 305–327.

Frankham, R. (1998) Inbreeding and extinction: island populations. *Conservation Biology*, **12**, 665–675.

Futuyma, D.J. (1998) *Evolutionary Biology*, 3rd ed. Sinauer Associates Inc., Sunderland, Massachusetts.

Gabriel, W., M. Lynch & R. Bürger (1993) Muller's ratchet and mutational meldowns. *Evolution*, **47**, 1744–1757.

Gall, G.A.E. (1987) Inbreeding. In: (eds Ryman, N. & F. Utter) *Population Genetics and Fisheries Management*. University of Washington Press, Seattle & London, pp. 47–88.

Gauldie, R.W. (1991) Taking stock of genetic concepts in fisheries management. *Canadian Journal of Fisheries and Aquatic Sciences*, **48**, 722–731.

Gentili, M.R. & A.R. Beaumont (1988) Environmental stress, heterozygosity and growth rate in *Mytilus edulis*. *Journal of Experimental Marine Biology and Ecology*, **120**, 145–153.

Gillespie, R.B. (1996) Allozyme frequency variation as an indicator of contaminant-induced impacts in aquatic populations. In: (ed. Ostrander, G.K.) *Techniques in Aquatic Toxicology*. Lewis Publishers, Boca Raton, FL, pp. 247–275.

Gillespie, R.B. & S.I. Guttman (1989) Effects of contaminants on the frequencies of allozymes in populations of the central stoneroller. *Environmental Toxicology and Chemistry*, **8**, 309–317.

Gillespie, R.B. & S.I. Guttman (1993) Allozyme frequency analysis of aquatic populations as an indicator of contaminant-induced impacts. In: (eds Gorsuch, J., F.J. Dwyer, C.G. Ingersoll & T.W. La Pointe) *Environmental Toxicology and Risk Assessment*. ASTM STP 1173. American Society for Testing and Materials, Philadelphia, pp. 134–145.

Gillespie, R.B. & S.I. Guttman (1999) Chemical-induced changes in the genetic structure of populations: effects on allozymes. In: (ed. Forbes, V.E.) *Genetics and Ecotoxicology*. Taylor & Francis, London, pp. 55–77.

Grant, A., J.G. Hateley & N.V. Jones (1989) Mapping the ecological impact of heavy metals on the estuarine polychaete *Nereis diversicolor* using inherited metal tolerance. *Marine Pollution Bulletin*, **29** (5), 235–238.

Groenendijk, D., S.M.G. Lucker, M. Plans, M.H.S. Kraak & W. Admiraal (2002) Dynamics of metal adaptation in riverine chironomids. *Environmental Pollution*, **117**, 101–109.

Guttman, S.I. (1994) Population genetic structure and ecotoxicology. *Environmental Health Perspectives*, **102** (12), 97–100.

Handford, P. (1983) Age related allozymic variation in the cyprinid fish *Alburnus alburnus*. *Canadian Journal of Zoology*, **61**, 2844–2848.

Harper, F.A., S.E. Smith & M.R. Macnair (1997) Can an increase in copper requirement in copper-tolerant *Mimulus guttatus* explain the cost of tolerance? II. Reproductive phase. *New Phytology*, **140**, 637–654.

Harper, F.A., S.E. Smith & M.R. Macnair (1998) Can an increase in copper requirement in copper-tolerant *Mimulus guttatus* explain the cost of tolerance? I. Vegetative growth. *New Phytology*, **136**, 455–467.

Hartl, D.L. (1994) *Genetics: a study guide*. Jones and Bartlett Publishers International, UK.

Hauser, L. & R.D. Ward (1998) Population identification in pelagic fish: the limits of molecular markers. In: (ed. Carvalho, G.R.) *Advances in Molecular Ecology*. IOS Press, Amsterdam, pp. 191–224.

Hauser, L., G.J. Adcock, P.J. Smith, J.H. Bernal Ramírez & G.R. Carvalho (2002) Loss of microsatellite diversity and low effective population size in an overexploited population of New Zealand snapper (*Pagrus auratus*). PNAS **99** (18), 11742–11747.

Hawkins, A.J.S., J. Rusin, B.L. Bayne & A.J. Day (1989) The metabolic/physiological basis of genotype-dependent mortality during copper exposure in *Mytilus edulis*. *Marine Environmental Research*, **28** (1–4), 253–257.

Hedgecock D. (1994) Does variance in reproductive success limit effective population size of marine organisms? In: *Genetic and Evolution of Aquatic Organisms* (ed. Beaumont A.R.). Chapman & Hall, UK, pp. 122–134.

Hedrick, P.W. (2001) Conservation Genetics: where are we now? *Trends in Ecology and Evolution*, **16**, 629–636.

Heithaus, M.R. & R.H. Laushman (1997) Genetic variation and conservation of stream fishes: influence of ecology, life history, and water quality. *Canadian Journal of Fisheries and Aquatic Sciences*, **54** (8), 1822–1836.

Hilbish, T.J. & R.K. Koehn (1985) Genetic variation in nitrogen metabolism in *Mytilus edulis*: Contribution of the Lap locus. In: (ed. Gibbs, P.E.) *Proceedings of the nineteenth European marine biology symposium*. Plymouth, Devon, UK, pp. 497–504.

Hindar, K., N. Ryman & F. Utter (1991) Genetic effects of cultured fish on natural fish populations. *Canadian Journal of Fisheries and Aquatic Sciences*, **48**, 945–957.

Hoffman, E.R. (1995) *Biochemical, fitness, and genetic effects of DDT and malathion on two populations of Chironomus riparius: Population and insecticide specific response to selection for resistance*. PhD Dissertation, Institution unknown.

Holland-Toomey, B.H. & D. Epel (1993) Multixenobiotic resistance in *Urechis caup* embryos: Protection from environmental toxins. *Biological Bulletin*, **185**, 355–364.

Houle, D., D.K. Hoffmaster, S. Assimacopoulos & B. Charlesworth (1992) The genomic mutation rate for fitness in *Drosophila. Nature*, **359**, 58–60.

Houle, D., K.A. Hughes, D.K. Hoffmeister, J. Ihara, S. Assimacopoulos, D. Canada & B. Charlesworth (1994) The effects of spontaneous mutations on quantitative traits. I, Variances and co-variances of life history traits. *Genetics*, **138**, 773–785.

Hummel, H. & T. Patarnello (1994) Genetic effects of pollutants on marine and estuarine invertebrates. In: (ed. Beaumont, A.R.) *Genetics and Evolution of Aquatic Organisms*. Chapman & Hall, London, pp. 425–433.

Hummel, H., R.H. Bogaards, C. Amiard-Triguet, G. Bachelet, M. Desprez, J. Marchand, H. Rybarczyk, B. Sylvand, Y. Dewit & L. Dewolf (1995) Uniform variation in genetic traits of a marine bivalve related to starvation, pollution, and geographic clines. *Journal of Experimental Marine Biology and Ecology*, **191**, 150.

Jerneloev, A. (1988) *Physiolgical mechanisms for acid tolerance in fish*. Swedish Environmental Research Institute, Stockholm.

Keklak, M.M., M.C. Newman & M. Mulvey (1994) Enhanced uranium tolerance of an exposed population of the eastern mosquitofish (*Gambusia holbrooki* girard 1859). *Archives of Environmental Contamination and Toxicology*, **27** (1), 20–24.

Klerks, P.L. & S.A. Lentz (1998) Resistance to lead and zinc in the western mosquitofish *Gambusia affinis* inhabiting contaminated Bayou Trepagnier. *Ecotoxicology*, **7** (1), 11–17.

Klerks, P.L., P.L. Leberg, R.F. Lance, D.J. McMillin & J.C. Means (1997) Lack of development of pollutant-resistance or genetic differentiation in darter gobies (*Gobionellus boleosoma*) inhabiting a produced-water discharge site. *Marine Environmental Research*, **44** (4), 377–395.

Kohler, H.R., M. Zanger, H. Eckwert & I. Einfeldt (2000) Selection favours low HSP70 levels in chronically metal stressed soil arthropods. *Journal of Evolutionary Biology*, **13**, 569–582.

Kopp, R., S.I. Guttman & T.E. Wissing (1992) Genetic indicators of environmental stress in central mudminnow (*Umbra limi*) populations exposed to acid deposition in the Adirondack Mountains. *Environmental Toxicology and Chemistry*, **11**, 665–676.

Kopp, R., T.E. Wissing & S.I. Guttman (1994) Genetic indicators of environmental tolerance among fish populations exposed to acid deposition. *Biochemical Systematics and Ecology*, **22**, 459–475.

Kramer, V.J. & M.C. Newman (1994) Inhibition of glucosephosphate isomerase allozymes of the mosquitofish, *Gambusia holbrooki*, by mercury. *Environmental Toxicology and Chemistry*, **13** (1), 9–14.

Kramer, V.J., M.C. Newman, M. Mulvey & R. Ultsch (1992) Glycolysis and Krebs cycle metabolites in mosquitofish *Gambusia holbrooki*, Girard 1859, exposed to mercury chloride: alozyme genotype effects. *Environmental Toxicology and Chemistry*, **11**, 357–364.

Lande, R. (1996) The meaning of quantitative genetic variation in evolution and conservation. In: (eds Szaro, R.C. & D.W. Johnston) *Biodiversity in Managed Landscapes: Theory and Practice*. Oxford University Press, Oxford, pp. 27–40.

Lande, R. & G.F. Barrowclough (1987) Effective population size, genetic variation, and their use in population management. In: (ed. Soulé, M.E.) *Viable Populations for Conservation*, Cambridge University Press, Cambridge, pp. 87–124.

Lavie, E. & E. Nevo (1982) Mercury selection of phosphoglucose isomerase allozymes in marine gastropods. *Marine Biology*, **71**, 17–22.

Ledig, F.T., R.P. Guries & B.A. Bonefiled (1983) The relation of growth to heterozygosity in pitch pine. *Evolution*, **37**, 1227–1238.

Lewontin, R.C. & J. Krakauer (1973) Distribution of gene frequency as a test of the theory of the selective neutrality of polymorphisms. *Genetics*, **74**, 175–195.

Luoma, S.N. (1977) Detection of trace contaminant effects in aquatic ecosystems. *Journal of the Fisheries Research Board of Canada*, **34**, 436–439.

Lynch, M. (1996) A quantitative-genetic perspective on conservation issues. In: (ed Hamrick & J. Avise) *Conservation Genetics: Case Histories from Nature*. Chapman & Hall, London, pp. 471–501.

Lynch, M. & B. Walsh (1998) *Genetic and Analysis of Quantitative Traits*. Sinauer Associates, Sunderland, Massachusetts, 980 pp.

Lynch, M., R. Bürger, D. Butcher & W. Gabriel (1993) The mutational meltdown in asexual populations. *Journal of Heredity*, **84**, 339–344.

Lynch, M., J. Conery & R. Burger (1995) Mutation accumulation and the extinction of small populations. *American Naturalist*, **146**, 489–518.

Mackay, T.F.C., J.D. Fry, R.F. Lyman & S.V. Nuzhdin (1994) Polygenic mutation in *Drosophila melanogaster*: estimates from response to selection of inbred strains. *Genetics*, **136**, 937–951.

Madsen, T., R. Shine, M. Olsson & H. Wittzell (1999) Restoration of an inbred adder population. *Nature*, **402**, 34–35.

Maynard-Smith, J. (1998) *Evolutionary Genetics*. Oxford University Press, Oxford, 330 pp.

Maynard-Smith, J. & J. Haigh (1974) The hitch-hiking effect of favourable genes. *Genetic Research*, **23**, 23–35.

McAlpine, S. (1993) Genetic heterozygosity and reproductive success in the green treefrog, *Hyla cinerea*. *Heredity*, **70**, 553–558.

Millward, R.N. & A. Grant (2000) Pollution-induced tolerance to copper of nematode communities in the severely contaminated Restronguet Creek and adjacent estuaries, Cornwall, United Kingdom. *Environmental Toxicology and Chemistry*, **19**, 454–461.

Mitton, J.B. & K. Koehn (1975) Morphological adaptation to thermal stress in a marine fish, *Fundulus heteroclitus*. *Biological Bulletin*, **151**, 548–559.

Moore, M.N. & R.I. Willows (1998) A model for cellular uptake and intracellular behaviour of particulate-bound micropollutants. *Marine Environmental Research*, **46** (1–5), 509–514.

Moraga, D. & A. Tanguy (1997) Effects of anthropogenic factors on genetic diversity in the marine bivalve *Crassostrea gigas*: search for genetic markers. In: (eds Feral, J.P. & G. Boucher) *Biodiversity in dispersive environments, Congress of the French marine network*. Vol. 4. Banyuls sur Mer France Laboratoire Arago, France, pp. 355–365.

Moriarty, F. (1983) *The Study of Pollutants in Ecosystems*, 2nd ed. Academic Press, London, 289 pp.

Mortimer, M.R. & J.M. Hughes (1991) Effects of organophosphate pollution on genetic structure in 2 species of estuarine crabs. *Marine Pollution Bulletin*, **22**, 352–359.

Mukai, T., S.I. Chigusa, L.E. Mettler & J.F. Crow (1972) Mutation rate and dominance of genes affecting viability in *Drosophila melanogaster*. *Genetics*, **72**, 335–355.

Nagel, E. & J. Voigt (1989) *In vitro* evolution and preliminary characterization of a cadmium-resistant population of *Chlamydomonas reinhardtii*. *Applied Environmental Microbiology*, **55**, 526–528.

Nei, M., T. Maruyama & R. Chakraborty (1975) The bottleneck effect and genetic variability in populations. *Evolution*, **29**, 1–10.

Nelson, K. & M. Soulé (1987) Genetical Conservation of Exploited Fishes. In: (eds Ryman, N. & F. Utter) *Population Genetics and Fisheries Management*. University of Washington Press, Seattle & London, pp. 345–368.

Nevo, E. (1998) Molecular evolution and ecological stress at global, regional and local scales: The Israeli perspective. *Journal of Experimental Zoology*, **282**, 95–119.

Nevo, E., T. Perl, A. Beiles & D. Wood (1981) Mercury selection of allozyme genotypes in shrimps. *Experientia.*, **37**, 1152–1154.

Nevo, E., R. Noy, B. Lavie, A. Beiles & S. Muchtar (1986) Genetic diversity and resistance to marine pollution. *Biological Journal of the Linnean Society*, **2**, 139–144.

Newman, D. & D. Pilson (1997) Increased probability of extinction due to decreased genetic effective population size: experimental populations of *Clarkia pulchella*. *Evolution*, **51**, 354–362.

Newman, M.C. & R.H. Jagoe (1998) Allozymes reflect the population-level effect of mercury: simulations of the mosquitofish (*Gambusia holbrooki* Girard) GPI-2 response. *Ecotoxicology*, **7** (3), 141–150.

Newman, M.C., S.A. Diamond, M. Mulvey & P. Dixon (1989) Allozyme genotype and time to death of mosquitofish, *Gambusia affinis* (Baird & Girard) during acute toxicant exposure: a comparison of arsenate and inorganic mercury. *Aquatic Toxicology*, **15** (2), 141–156.

Oppenoorth, F.J. (1985) Biochemistry and genetics of insecticide resistance. In: (eds Kerkut, G.A. & L.I. Gilbert) *Comprehensive Insect Physiology Biochemistry and Pharmacology*. Vol. 12. Pergamon Press, Oxford.

Patarnello, T., R. Guinez & B. Battaglia (1991) Effects of pollution on heterozygosity in the barnacle *Balanus amphitrite* (Cirripedia, Thoracica). *Marine Ecology Progress Series*, **70**, 237–243.

Posthuma, L. & N.M. Van Straalen (1993) Heavy metal adaptation in terrestrial invertebrates: a review of occurrence, genetics, physiology and ecological consequences. *Comparative Biochemistry and Physiology*, **106C**, 11–38.

Postma, J.F., A. Van Kleunen & W. Admiraal (1995a) Alterations in life-history traits of Chironomus riparius (Diptera) obtained from metal contaminated rivers. *Archives of Environmental Contamination and Toxicology*, **29**, 469–475.

Postma, J.F., S. Mol, H. Larsen & W. Admiraal (1995b) Life-cycle changes and zinc shortage in cadmium-tolerant midges, *Chironomus riparius* (Diptera), reared in the absence of cadmium. *Environmental Toxicology and Chemistry*, **14**, 117–122.

Purdom, C.R. (1993) *Genetics and Fish Breeding*. Chapman and Hall, London.

Reed, D.H. & R. Frankham (2001) How closely correlated are molecular and quantitative measures of genetic variation? A meta-analysis. *Evolution*, **55** (6), 1095–1103.

Reznick, D.A., H. Bryga & J.A. Endler (1990) Experimentally induced life-history evolution in a natural population. *Nature*, **346**, 357–359.

Ryman, N. (1991) Conservation genetics considerations in fishery management. *Journal of Fish Biology*, **39** (A), 211–224.

Ryman, N., R. Baccus, C. Reuterwall & M.H. Smith (1981) Effective population size, generation interval, and potential loss of genetic variability in game species under different hunting regimes. *Oikos*, **36**, 257–266.

Saccheri, I.J., P.M. Brakefield & R.A. Nichols (1998) Severe fitness depression and rapid fitness rebound in the butterfly *Bicyclus anyana* (Satyridae). *Evolution*, **50**, 2000–2013.

Shirley, M.D.F. & R.M. Sibly (1999) Genetic basis of a between-environment trade-off involving resistance to cadmium in *Drosophila melanogaster*. *Evolution*, **53**, 826–836.

Snyder, C.D. & A.C. Hendricks (1997) Genetic responses of *Isonychia bicolor* (Ephemeroptera: Isonychiidae) to chronic mercury pollution. *Journal of the North American Benthological Society*, **16**, 651–663.

Staton J.L., N.V. Schizas, G.T. Chandler, B.C. Coull & J.M. Quattro (2001) Ecotoxicology and population genetics: the emergence of 'phylogeographic and evolutionary ecotoxicology'. *Ecotoxicology*, **10**, 217–222.

Terriere, L.C. (1984) Induction of detoxification enzymes in insects. *Annual Review of Entomology*, **29**, 71–88.

Theodorakis, C.W. & L.R. Shugart (1997) Genetic ecotoxicology. II. Population genetic structure in mosquitofish exposed *in situ* to radionuclides. *Ecotoxicology*, **6**, 335–354.

Theodorakis, C.W., T. Elbl & L.R. Shugart (1999) Genetic ecotoxicology IV: survival and DNA strand breakage is dependent on genotype in radionuclide-exposed mosquitofish. *Aquatic Toxicology*, **45**, 279–291.

Thompson, J.D. (1991) Phenotypic plasticity as a component of evolutionary change. *Trends in Ecology and Evolution*, **6**, 246–249.

Thornhill, N.W. (ed.) (1993) *The Natural History of Inbreeding and Outbreeding*. University of Chicago Press, Chicago.

Turner, T.F., L.R. Richardson & J.R. Gold (1999) Temporal genetic variation and the female effective population size of red drum (*Sciaenops ocellatus*) in the northern Gulf of Mexico. *Molecular Ecology*, **8**, 1223–1229.

Van Noordwijk, A.J. & W. Scharloo (1981) Inbreeding in an island population of the great tit. *Evolution*, **35**, 674–688.

Via S., R. Gomulkiewicz, G. de Jong, S.M. Scheiner, C.D. Schlichting & P.H. van Tienderen (1995) Adaptive phenotypic plasticity: consensus and controversy. *Trends in Ecology and Evolution*, **10**, 212–217.

Vrijenhoek, R.C. (1996) Conservation genetics of North American desert fishes. In: (eds Hamrick, J.L. & J.C. Avise) *Conservation Genetics: Case Histories from Nature*. Chapman & Hall, London.

Vrijenhoek, R.C., E. Pfeiler & J. Wetherington (1992) Balancing selection in a desert stream dwelling fish, *Poeciliopsis monacha*. *Evolution*, **46**, 1642–1657.

Walker, C.H., S.P. Hopkin, R.M. Sibly & D.B. Peakall (1996) *Principals of Ecotoxicology*. Taylor & Francis Ltd., London, 321 pp.

Weber, K.E. & L.T. Diggins (1990) Increased selection response in larger populations. II. Selection for ethanol vapour resistance in *Drosophilia melanogaster* at two population sizes. *Genetics*, **125**, 585–597.

Weis, J.S. & P. Weis (1984) A rapid change in methylmercury tolerance in a population of killifish, *Fundulus heteroclitus*, from a golf course pond. *Marine Environmental Research*, **13**, 231–245.

Weis, J.S., N. Mugue & P. Weis (1999) Mercury tolerance, population effects, and population genetics in the mummichog, *Fundulus heteroclitus*. In: (ed. Forbes, V.E.) *Genetics and Ecotoxicology*. Taylor & Francis Ltd, London, pp. 31–54.

Wilson, T.G. (2001) Resistance of *Drosophila* to toxins. *Annual Reviews in Entomology*, **46**, 545–571.

Wirgin, I. & J.R. Waldman (1998) Altered gene expression and genetic damage in North American fish populations. *Mutation Research – Fundamental and Molecular Mechanisms of Mutagenesis*, **399** (2), 193–219.

Wright, S. (1931) Evolution in Mendelian populations. *Genetics*, **16**, 97–159.

Yauk, C. (1998) Monitoring for induced heritable mutations in natural populations: application of minisatellite DNA screening. *Mutation Research*, **411** (1), 1–10.

Yauk, C.L. & J.S. Quinn (1996) Multilocus DNA fingerprinting reveals high rate of heritable genetic mutation in herring gulls nesting in an industrialized urban site. *Proceedings of the National Academy of Sciences of the United States of America*, **93** (22), 12137–12141.

Zhang, L., K.J. Harada & T. Shono (1997) Genetic analysis of pyriproxfen resistance in the housefly, *Musca domestica* L. *Journal of Applied Entomology and Zoology*, **32**, 217–226.

Zouros, E. & G.H. Pogson (1994) The present status of the relationship between heterozygosity and fitness. In: (ed. Beaumont, A.R.) *Genetics and Evolution of Aquatic Organisms*. Chapman & Hall, London, pp. 135–145.

Chapter 8

From Population Ecology to Socio-Economic and Human Health Issues

K. Crean and C. Lacambra

8.1 Introduction

8.1.1 Aims and objectives

Earlier chapters in this book have developed the scientific themes associated with the impact of xenobiotics on fish and fish populations. The purpose of this chapter is to link pollution impacts on fish with the wider world of social, economic and human health aspects. This is achieved through the following specific objectives:

(1) To explain through the bio-economic model how the biological, social and economic spheres of activity interact in terms of fish exploitation systems
(2) To briefly describe the structure of the EU fisheries sector, including the nature of the production system, policy environment and market for fish
(3) To review what is meant by quality of the fish and how quality can be influenced by environmental and anthropogenic activities
(4) To examine the impact of anthropogenic activities on the quality of fish populations
(5) To examine case studies which show the social, economic and health impacts of xenobiotic influences in relation to fish (and shellfish) populations
(6) To discuss the implications for the health of human populations that consume fish.

This is a prodigious task as each of the aforementioned fields has its own theoretical basis, research 'hot spots' and component sub disciplines. Therefore, it would be reasonable only to lay down the broad framework that governs the interactions of xenobiotic impacts in the wider domain of human activities. The bulk of the chapter will therefore be devoted to the socio-economic aspects with only a passing mention of human health consequences of xenobiotic influences. The chapter will establish some of the principles of interaction between the biological, social and economic parameters in the context of the bio-economic model and its variants. It will also draw attention to the main legal instruments that are set down to control xenobiotics influences in fish. These sections will be used as a basis to develop the discussion of the impact of xenobiotic influences on the 'quality' of fish and fish populations in the European Union (EU) but, where appropriate, bringing in other case study examples.

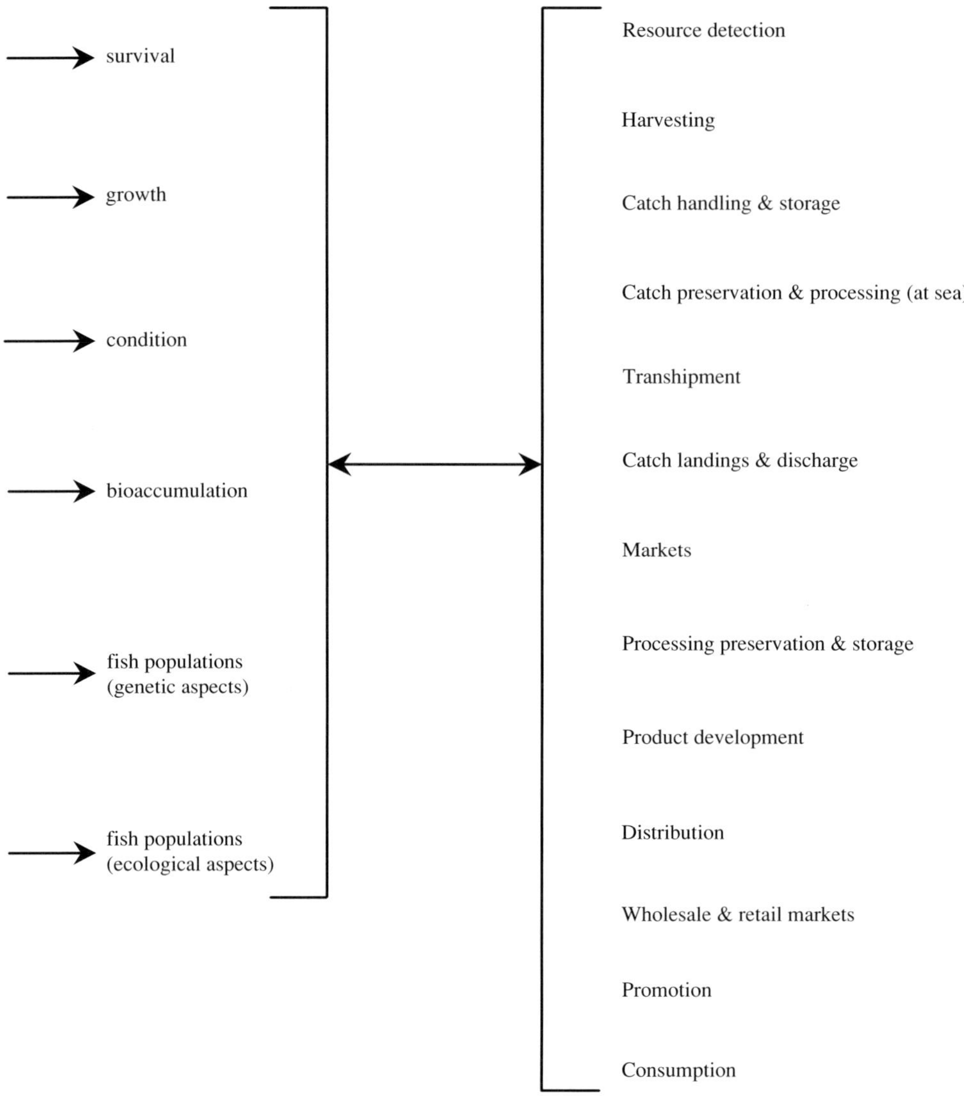

Fig. 8.1 Socio-economic aspects interfacing with xenobiotic factors.

Figure 8.1 shows in summary how xenobiotic influences on fish and fish populations (as shown in terms of survival, growth, condition and levels of bioaccumulation) might equate with activities in the production chain for fish products. This chain embraces the main locations where social, economic and health aspects of human populations, dependent on fish, come to the fore.

8.1.2 The bio-socio-economic model

Whilst the domain of the biology and ecology of fish populations is separate from socio-economic parameters, the disciplines are brought together in the context of the bio-economic

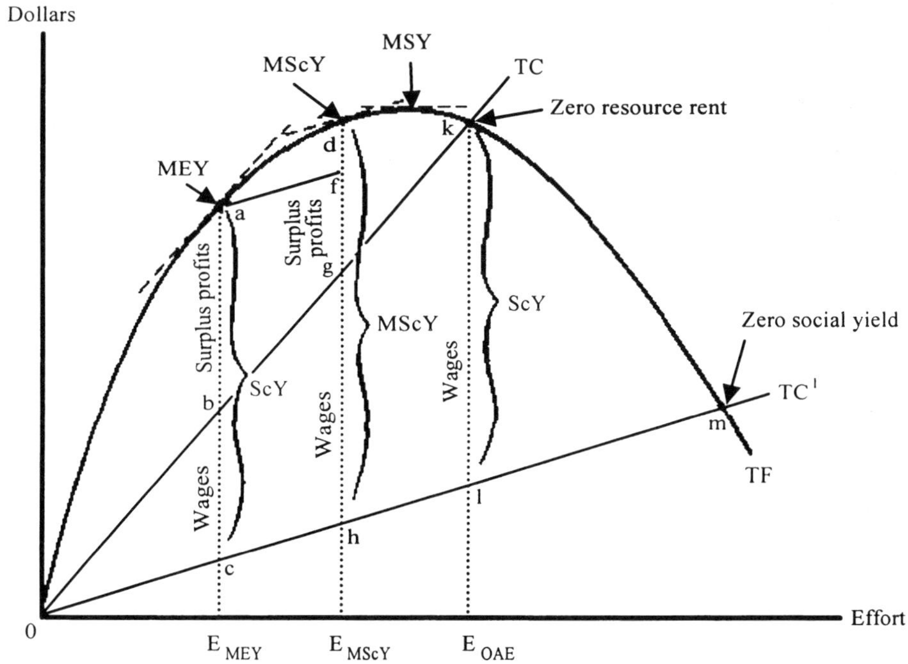

Fig. 8.2 Maximum social yield (MScY) in the absence of alternative employment opportunities (social yield (ScY) = wages + profits) (Panayotou, 1982). Reprinted with permission from the Food and Agriculture Organization of the United Nations, Rome, Italy.

model (Gulland, 1977; Panayotou, 1982; Hannesson, 1993). The history of models of this type dates to the mid-1950s (Schaefer, 1954; Beverton & Holt, 1957) and were created in an attempt to synthesise variables that would improve the management of production in marine fisheries. Essentially, the model consists of a stock yield curve (see TF in Fig. 8.2) which shows the effects of a series of inputs of fishing effort in terms of the yield (that can be measured as revenue, tonnage, etc. but in this case US dollars). This is therefore the meeting place where it is possible to assess the effects of xenobiotic influences on fish stocks, as those influences – largely independent of fishing effort – will exert their own effects on individual fish and ultimately therefore, fish populations.

The Schaefer curve is probably the most famous illustration for promoting the concept of maximum sustainable yield (MSY), which has been a starting place for many of the arguments relating to resource sustainability (Schaefer, 1954). The curve models the interaction of fishing effort on a single stock, and the zenith of the curve (MSY) is the point where to input further effort leads to a reduction of yield per unit effort (Fig. 8.2). Ideally, it would be best to locate the level of fishing effort at, or just below, MSY; however, in practice this is difficult to do without risking exceeding the zenith and incurring the ensuing biological and economic penalties. More preferable for fisheries managers and resource users is to locate fishing effort at a point that gives the optimum economic benefit without threatening the stock: this point is termed maximum economic yield (MEY). Thus in Fig. 8.2, for E (MEY) the return on the level of effort invested in the fishery yields substantial profits once the costs of fishing have been accounted for.

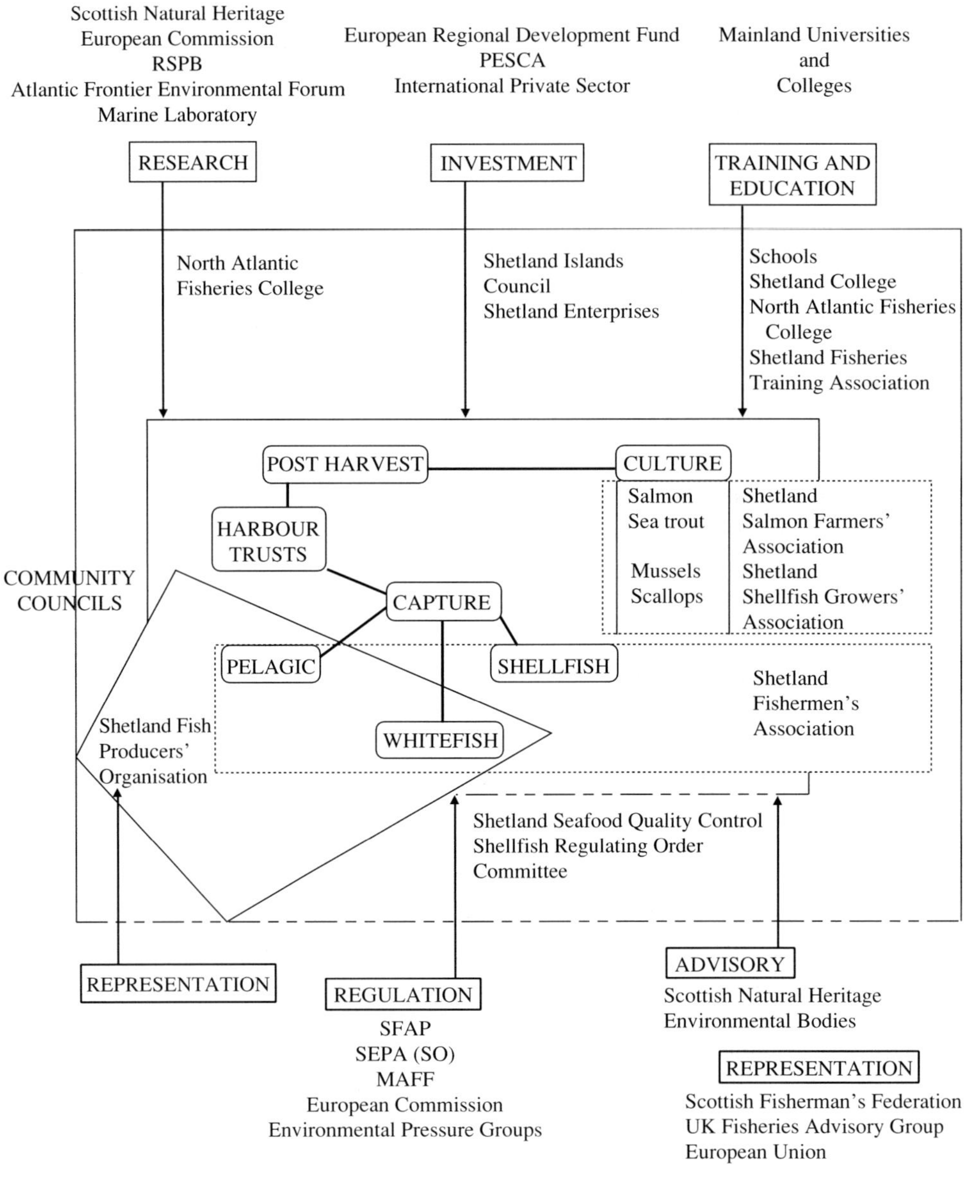

Fig. 8.3 The Shetland fish sector.

Intermediate between MEY and MSY is a further theoretical point where overall effort has increased, yet there is a better dispersion of costs between profit and the earnings of individual fishermen involved in the fishery. This point in Panayotou's model (Panayotou, 1982) is known as maximum social yield (MScY). Clearly from Fig. 8.2, it can be seen that the generation of surplus at either MScY or MEY levels of effort is determined by the position of the total costs of fishing (line TC). This line emanates from the origin and bisects the stock curve (TF) at point K where resource rent derived from the fishery is zero (see zero resource rent in Fig. 8.3). The resource rent may be defined as the difference between the value of output (i.e. fish catch) and the opportunity cost of labour and capital needed to produce it, and represents the income that has been earned by the owner of the natural resource (Collins *et al.*, 1998).

It is possible in practice for the TC line to move towards either the x or y-axis. These movements are likely be triggered by changes in the costs associated with fishing, e.g. divergences in fuel costs, labour, insurance and other financial variables. Should the level of fishing effort be at the point E_{OAE} then it could be seen that the yield of stock has now fallen (per unit effort) because the level of effort has built up beyond MSY. Thus this has the effect of reducing the spawning stock biomass and eliminating the 'profit' surplus. The line TC1 exaggerates this effect and shows a point on the curve where the stock has collapsed and the economic and social benefits have ceased to flow.

The bio-economic model has a utility, at least conceptually, in demonstrating the link between the biological, social and economic domains. Nevertheless, the model also attracts criticism in that in practice, it has proven very difficult to determine the points of MSY, MScY and MEY. Furthermore, the model is based on the exploitation of a single stock of fish and this seldom reflects the reality of most commercial fisheries where a multispecies model would be more appropriate. Nevertheless, it is clear that lethal xenobiotic influences would have a marked effect on the yield curve (TF) in that the standing stock of fish would be reduced and thus there would be repercussions on the economic and social yields from the fishery. Similarly, revenues would also be depressed in the yield curve if sublethal xenobiotic effects affected the *quality* of individual fish in the catch.

Calculating the economic and social dislocation caused by xenobiotic influences needs to take account of not only the loss of resource rent but also the change in fiscal values overtime. This would involve calculations of net present value to arrive at a determination of a more realistic assessment of the impact (Lipsey, 1963). Economic damages measured this way are equivalent to a capital sum that has been lost as a consequence of a once-only event (Collins *et al.*, 1998). This approach is important in that there are instances where catastrophic xenobiotic effects have resulted in the full or partial closure of a fishery and have thus adversely affected future income streams (Cohen, 1995; Goodlad, 1996).

Figure 8.3 shows a diagrammatic representation of a fish sector, in this case using Shetland as an example. This is in effect a qualitative but detailed description of the interrelationships between the 23 000 inhabitants of Shetland and the fish resources upon which the local economy is dependent (Crean, 1999). The diagram shows how the membrane of the Community Councils embraces the key economic, social and organisational characteristics of the Islands' activities.

8.2 The fish sector of the European Union

8.2.1 Introduction

The real world interactions of the biological, social and economic variable of the Panayotou's bio-economic model can be observed in the fish sector of the European Union (EU). The term fish sector denotes the existence of an industrial base that depends on the commercial production of large quantities of captured and cultured fish. The production centres are located in the 15 countries that make up the EU, but by far the greatest activity takes place in those countries that have a maritime boundary (Saltz, 1991). There is also a north-south separation where the activities of the so-called 'Common Pond' of the northern EU states account for a far greater production than those states with a Mediterranean boundary.

The EU fleet captured 6.777 million tonnes of fish in 1994; however, almost as much fish again as was caught in this period was imported to EU countries from other fish producing nations (MAFF, 1995). The highest catches of fish and shellfish in the EU were made by vessels from Denmark, Spain, the UK and France. 80% of the catch of EU fish came from within the Common Pond, which is bounded by the 200-mile Exclusive Economic Zone. The catch was valued at over 5 billion ECU. The EU fleet at that time consisted of 96 133 vessels which had a gross registered tonnage of 1.954 million. The engine power of this fleet was 7.923 million kW. Nearly 300 000 persons were involved full-time in catching fish and up to a further 1 million involved in the ancillary industries of shipbuilding and repair, fish processing and marketing. In addition, the EU produces large quantities of cultured fish and this is a rapidly expanding business. In the UK, for example, the 1998 production of cultured fish totalled 130 825 tons, with an estimated value of 449 949 ECU (Federation of European Aquaculture Producers, 1999).

Clearly, the fish sector is a developing component of the federal economy of the EU and assumes great importance in the relative microcosm of fisheries-dependent areas, e.g. Brittany, Galicia, Humberside, Jutland and Shetland (European Commission, 1992).

8.2.2 The Common Fisheries Policy (CFP)

The fundamental policy that governs the interaction of the social and economic variable in the fish sector of the EU is the Common Fisheries Policy (CFP). In the context of the relationship between the xenobiotic and socio-economic variable it is important to briefly establish the nature of this policy. The following paragraphs describe the origins, structure and function of the policy as it affects the fish sector of the EU.

The CFP has its origins in the Treaty of Rome (1957) and this is the foundation upon which the early aims and objectives were established. At that time the fish sector was regarded very much as a minor part of a common market dedicated to trade in agricultural products. In practice the aims and objectives of the CFP have been developed incrementally (European Commission, 1991; Symes *et al.*, 1994). The following statement gives some idea of this situation:

> 'to provide for rational and responsible exploitation of living aquatic resources and of aquaculture, while recognising the interest of the fisheries sector in its long term development and in its economic and social conditions and the interests of consumers taking into account the biological constraints, with due respect for the marine ecosystem.'

The CPF's adherence to the Treaty of Rome has led to the establishment of the principle of 'equal access' for Community fishermen with respect to the disposition of fish resources. Whilst this approach, up to a point, satisfies the social equity aspect of the policy it has not proved a suitable basis for management of the community's fish stocks (Symes, 1996).

Indeed, implementation of this principle has been constrained by two countervailing forces: 'relative stability' and 'rights of establishment', both of which support the status quo in terms of access to particular fisheries. Several derogations have been obtained to prevent full expression of 'equal access', ostensibly to provide protection to inshore fishermen. The most notable include the reservation of a 12 mile limit for the coastal states' own fishing interests.

Historically, the CFP has comprised three distinctly separate policy features relating to structures, markets and conservation. The building of the CFP was initiated in 1976 and was largely disposed to making provision for access to fish resources, the guarantee of supplies of fish to the consumers and the maintenance of market conditions. The development of a conservation policy came later with the passing of two regulations: the first identified appropriate technical measures, and the second setting out the details of total allowable catches (TACs) and national quotas for fish stocks. It was intended that these regulations would operate for a 20-year period. A mid-term review was undertaken in the early 1990s, the revised CFP resulted in the establishment of a Community system for fisheries and aquaculture production to run until 2002 (European Commission, 1991).

Currently, the CFP is driven by a legislative programme that requires action by the European Commission: regular revision, as with the annual review of TACs; periodic revision of the Multi-Annual Guidance Programme (MAGP) now in its third series (1993–96); occasional action in response to specific, often short-run, problems such as introduction of temporary measures to protect the EU's coastal markets from the effects of surges in imported fish supplies; and the continuing development of CFP, which is likely to take on increasing significance with the approaching expiry date for existing legislation. The need for joint actions, integrating fisheries and related policy areas, is a growing feature. One example is the evolution of a joint strategy for coping with the socio-economic impacts of the structural and conservation measures in conjunction with DGs V (Employment, Industrial Relations and Social Affairs) and XVL (Regional Policies), including the incorporation of fisheries measures in the revised system for the European Structural Funds and the development of new measures for financial aid to the fishing industry (Symes *et al.*, 1994).

8.2.3 The crisis in EU fisheries: interdependence or independence in relation to xenobiotic influences?

Despite the existence of the CFP and the attempts to bolster its flexibility to respond to challenges in the social, economic and environmental domains, the policy has not been able to tackle problems that have unerringly drawn the fish sector into crisis. The Community's policy lacks a formal structure (indeed is apparently divorced from any identifiable planning methodology) and has been changed ad hoc in response to the problems of the industry, with a strong emphasis always on maintaining the status of the commercially exploited fish stocks. Unfortunately, the major fish stocks upon which the Community depends have, despite intense regulation, slipped beyond the brink of serious overfishing to the point where they are no longer sustainable. The failure to slow the rate of stock decline, let alone develop sustainable fisheries, has led the fishing industries of the Community into a crisis which has deepened over the last 20 years (European Commission, 1994).

The conditions of the north-east Atlantic exemplify the problem. This prolific fishery was probably the first in the world to exhibit signs of overfishing in the modern era. The region has recorded one of the longest periods of sustained decline, production falling from 13.2 million tonnes in 1976 to 11.1 million tonnes in 1992, a reduction of 16% (Symes, 1996). The catastrophic erosion of fisheries productivity in this region is disturbing given that fishery managers have not been able to sustain the stocks despite the extensive expertise and advanced scientific methodologies that have been made available (Pope, 1982; Sparre, 1994). One of the overwhelming difficulties has been that the managers have found that in practice that it has proved virtually impossible to balance catches with catching power, this despite implementing a capacity reduction process – the Multi-annual Guidance Programme (MAGP). Unfortunately, the modest achievements in reducing capacity have tended to be offset by continuing efficiencies in the development of new technology. Thus, it has proved necessary to continually re-appraise MAGP targets. Despite the operation of three MAGP phases, fleet cuts by as much as 40% were still being mooted by the Commission in 1996 (*Fishing News*, 1996). The Commission criticises the ineffective implementation of MAGP in that a balance has not been reached between catches and vessels, but there have been substantial losses sustained by the coastal communities of the Community. The fishing ports of Hull and Grimsby, UK, for example, once boasted one of the largest distant water fishing fleets in the world, numbering 180 vessels. However, during a profound period of restructuring initiated in the late 1970s and early 1980s, this fleet was reduced to five vessels. The decline in the number of fishing vessels has not been confined to East Yorkshire, nor the distant water subsector of the fleet, and overall in England and Wales during the period 1983 to 1989 there was a capacity reduction of 38% in gross registered tonnage. With respect to the actual vessel numbers of 1983 this represents a loss of a quarter of the fleet (European Commission, 1992).

The profound changes in the structure of the fleet have impacted on the regions dependent on fishing. It has proved difficult to assess the importance of fishing to the coastal communities of England and Wales in absolute terms (numbers unemployed, loss of earnings, etc.). However, a regional socio-economic study of the UK fisheries sector undertaken in the early 1990s (European Commission, 1992) commented that:

> 'difficult adjustment problems will occur in most of the 23 local areas examined. They will be most serious in Grimsby, Newlyn, Whitby, Lowestoft, Hull, Brixham and Fleetwood where concentrations of fishing activities are greatest.'

There was obviously a substantial loss of jobs in fishing at this time, but this is not apparent from the figures that were available.

Inevitably, this adverse socio-economic situation has given rise to conflict at all levels in the fisheries of the Community, whether at sea between different fleets or around the negotiating table between different nations. In the majority of cases the problem is that there are simply insufficient resources available to sustain the coastal communities that depend on fishing for their livelihoods. As a consequence, in an effort to preserve their businesses, large numbers of fishermen have opted to become part of the illegal trade in fish (Symes, 1996). The real size of commercial catches is further obscured as a result of the vast tonnage caught in the 'Common Pond' of the Community that whilst brought on board fishing vessels is not landed in port and for a variety of reasons is thrown away at sea. The following statement taken from the European Commission (1991) gives some idea of the scale of this problem:

> 'in the North Sea discards of haddock may exceed what is retained from a single trawl. The global estimate for 1985 was 460 million discarded individuals, whereas landings amounted to 500 million . . .'

One of the most acute symptoms of the crisis in the EU fish sector is the phenomenon of discards (Crean & Symes, 1994). This paper identified the principle cause as the failure of management regulations generated by centralist authorities operating an access regime where there is effectively a 'free for all' approach to exploitation strategy. However, when access arrangements are controlled by communal institutions, discarding tends not to occur, as there are intrinsic controls in place that will limit catching efficiency; and supply is closely tied in to local demand. Crean and Symes (1994) argued that discards are a symptom of the incoherence and discord of existing policy, and therefore systemic changes to the regulatory system are necessary and should be accompanied by a radical alteration of the decision-making structure. They compared the discrete, selective and balanced approach of exploiting resources held in the trust of the community and the wasteful practice of discarding that is made legitimate by the CFP. As such, the fishermen, the food industries that use fish and the maritime communities dependent on fishing as a source of income and employment have all suffered from the effects of uncontrolled exploitation. This has been demonstrated by a sharp decline in the numbers of full-time fishermen, shortages of certain prime fish species and the depression or even disappearance of fishing as an occupation in the coastal communities of the EU (Collet, 1996; Delbos & Prémel, 1996).

The European Commission has tried to respond to these difficulties, and would argue that it has made, and continues to make, serious efforts to address the problems of the fisheries sector. A variety of regulatory schemes have been tried to alleviate the crisis (European Commission, 1991). Unfortunately, despite determined approaches to counteract overexploitation, the introduction of new regulatory measures has failed to stem the erosion of the fisheries and as a result, the European Commission now feels that it must go even further in its attempts to protect the fish stocks. As such, they have approved a 30% cut in fish catches to be brought in over a five year period.

The failure to alleviate the crisis demonstrates that there are problems which are deeply rooted and not easily remedied, and most believe that even the newly proposed cuts in catches will not contribute to solving the Community's fisheries problems. It has been stated that it is the design, content and inadequacy of the CFP itself that is the main source of the problems that afflict the fisheries sector (Holden, 1994; Symes, 1995). The fundamental assumptions and basis of the CFP are in question, and the UK industry is not alone in calling for political reform whilst some are calling for withdrawal from the CFP (Ashworth, 1996).

On examination of the CFP in more detail, it is apparent that the policy relies heavily on the assumption that it is possible to manage the fisheries by applying strategies based on bio-economic theory. However, there is a growing recognition that the principles of the theory are insufficient by themselves to enable the development of a workable management policy for fisheries. Bio-economic theory fails to model in an appropriate way the true, unpredictable behaviour of fisheries in an uncontrollable (some believe chaotic) ecological system (Wilson & Kleban, 1992). Instead, it has substituted a false notion of predictability for stock behaviour (and economic returns) upon which the basic working principles of the CFP, TACs and quotas are established. It is argued that a more realistic exploitation regime would need to be based on a multispecies model. Unfortunately, these are by definition more complex and would likely be more difficult to implement. This dilemma has been neatly summarised by Holm (1995):

> '. . . where the simplifications of the single species model destroy its predictive capabilities, the realism of multi-species models create unmanageable complexities . . .'

In an effort to countermand the inherent problems in the CFP, the maritime governments of the Community have deployed an intense and excessive regulatory strategy. This in turn has led to the development of an operational split in sectoral activities: the fishermen and their organisations have become the production component whilst the government has adopted responsibility for the regulatory function. This dichotomy has led to 'fall out' being generated by the adverse interaction of the fishermen and the resource regulators. The fishermen argue that they have been disenfranchised by the policy generating and implementation mechanisms (Symes, 1996). They have responded by circumventing the regulatory system to maintain a competitive edge at the expense of the resource.

The effects of the crisis in the fisheries of the Community have been felt acutely in the UK. Whilst it would be simplistic to blame the Community's role in the management of fisheries for all that has happened since the UK joined the EEC, the British fishermen would point to the loss of fishing opportunities that followed as the single most devastating change in that period. The very terms of the UK's entry with respect to fisheries have left a permanent scar in some quarters of the country and this has always been a source of discontent to be fuelled by the various crises that have hit the Community's fisheries. The UK's fishing industry suffered considerably at the time of the declaration of Exclusive Economic Zones (EEZs), and its virtual exclusion from distant water grounds was hampered at that time by membership of the EEC. As a consequence, UK landings from distant water grounds fell from 246 000 tonnes in 1975 to 10 000 tonnes in 1986 (Symes, 1992). Unfortunately, to the chagrin of the British fishing industry, the shortfall in distant water landings could not be made up from increased catches in the seas under the control of the Community, including the coastal waters of the UK. A particularly inflammatory statistic at the time of UK accession to the Community was that approximately 70% of the stocks available to Community fishing vessels lay within the notional EEZ of the UK, but the UK had access to 37% of these resources (Ashworth, 1996). This sense of lost 'ownership' of fish stocks has never been satisfactorily dealt with by any of the measures emanating from Brussels and to this day, some 25 years on, is still an issue (Ashworth, 1996).

Clearly, the situation with respect to problems inherent in the CFP and the general decline in the base of fish stocks could be exacerbated by xenobiotic effects of anthropogenic origin. The key factors acting in concert would have the capacity to further reduce the viability of the EU's fish stocks and the industries dependent on them. The social and economic repercussions would be profound, as resource rents, already parlous with respect to non-viability, would become negative (Collins *et al.*, 1998).

8.3 The quality of individual fish (intrinsic and extrinsic characteristics), scarcity and its effects on consumer health and behaviour

The preceding section has outlined the broad political, social and economic factors which affect fish production in the EU. These factors impact not only on the viability of fish populations but also on the composition and quality of specimens and individual fish species. Therefore, within section 8.3, the principles of how fish quality is defined and how changes are determined will be discussed. Fish quality is a complex area of objective and subjective interactions, and the quality in the suitability of fish as food is influenced by both intrinsic and extrinsic factors.

8.3.1 Intrinsic quality in fish

Intrinsic quality is the sum of those attributes which are inherent in the harvested raw material (Connell, 1990; ASMI, 1998). It is determined by the physico-chemical condition of the fish at the time of harvest, and by the individual's species, sex, stage of maturity and other physiological characteristics. Intrinsic quality can change with the fish's condition, which may be regarded as an index of fatness (i.e. the net effect of all physiological processes). The only way in which man can intervene with the intrinsic quality of the fish is by fish cultivation (aquaculture). Indirectly, man influences the intrinsic quality of the fish by preserving or damaging the environment, and the environment condition is therefore reflected in the physical and chemical condition of the organism present in the environment. Some intrinsic characters such as levels of pollutants in the tissue are clearly anthropogenic.

Generally, the larger size individuals can command a higher price for two main reasons. Firstly, there is more flesh on the larger specimens and they are easier to handle and process. However, size is not necessarily related to the quality of flavour, although it has been reported that when other variables are equal, large specimens of individual fish are of higher quality than smaller fish of the same species (Burgess *et al.*, 1965). Secondly, Connell (1990) indicated that small fish of the same species tend to have a higher post-rigor pH than larger species, which leads to greater microbiological activity. The rate of deterioration varies with size, although larger fish tend to deteriorate at a slower rate than smaller fish (both of the same and differing species). Between species, the rate of deterioration additionally varies with habitat and season, possibly due to the composition of the inner biological flora and the environment in which they live.

In general, the sex of a fish is relatively unimportant in influencing market prices. However, for some species such as the capelin (*Mallotus villosus*), there is a segregation between sexes in the shoals and this will influence the value of the catch as the roe carries a price premium (Connell, 1990).

Physiological changes attributable to seasons occur in all fish species (Connell, 1990) but tend to be less noticeable in shellfish and more noticeable in pelagic species such as sardines, sprat, herring, mackerel and anchoveta. Poor condition (i.e. emaciation) usually relates to post-spawning periods and winter conditions and this is reflected in the flesh composition which when eaten, is generally soft and gelatinous. The quality (condition) is also affected after intensive feeding periods. For example, in herring, fat is stored as a liquid rather than a solid and as a result, the fish are susceptible to a damaging condition called 'belly-burst'. Within the same species, the poor condition season may vary between different fishing grounds and can extend through several months. Differences may even be found between individuals caught at the same ground.

8.3.2 Extrinsic quality in fish

By contrast, extrinsic quality is determined by the methods employed in the harvesting, handling, processing and storing of fish catches. Correct handling procedures will result in fish of high extrinsic quality, and incorrect handling methods will result in loss of extrinsic quality. Deterioration in the extrinsic quality is caused by the action of microorganisms and other biochemical changes which take place in the fish tissues after death. The most important factors which influence the maintenance of fish quality post harvest are the storage and handling conditions up to the point of consumption (Connell, 1990), and this includes exposure of the fish to stress during capture and landing aboard the fishing vessel. Of particular importance is the duration of the soaking time which may be defined as the 'time that the fish (in the fishing gear) is in the water prior to being landed'. Throughout the catching, handling, landing and marketing process, intrinsic quality remains constant whereas extrinsic quality will change.

The quality of fish in markets is determined by size, freshness, species, fat content, age and sex. Depending on the market, the individual characteristics of fish quality will show a differential. Overall, the markets require good quality products that will not affect the consumer's health and will satisfy their needs.

Raw material for a processor may be the finished product for a retailer, and thus each of them values quality according to their needs and takes different measures to achieve good quality. It is expected that prices reveal the effort on quality, and often in markets the measure of quality, and thus the price, is only related to the species and the size of the individuals. Even though size is very important, other aspects should, and in some cases are, also taken into account by many consumers (Amhed & Anderson, 1994; Nielsen *et al.*, 1997).

Rigor mortis (or *rigor*) is the term used to describe the stiffening of muscle tissue that occurs shortly after an organism has died. It occurs in all vertebrates but length of time before onset of *rigor* and the duration of *rigor* are influenced by a number of intrinsic and extrinsic factors. In fish, the species, size of individuals, condition and method of capture all influence the rapidity of the onset of *rigor* (Burgess *et al.*, 1965; Hultin, 1984). Once *rigor* has ended, tissue decomposition is accelerated.

The manipulation of the conditions which affect *rigor* in fish has a pronounced effect on fish quality. In most commercial activities, operators seek to delay the onset of *rigor* and slow the rate at which the fish catch passes through *rigor*.

Primary deterioration signs are easier to detect on the external surfaces of gills and/or organs, which is where freshness and spoilage tests are done. Quality deterioration in fresh raw fish has two different causes – microbiological and non-microbiological (Connell, 1990). The first is attributable to microorganism activity on the external surfaces and the gut of fish. Prior to *rigor*, the high pH of the fish flesh inhibits microbial activity; however, once *rigor* has passed, microorganisms are free to invade the fish. The subsequent degeneration of the fish flesh is characterised by a well-defined sequence of changes in biochemical and organoleptic properties of the fish (Connell, 1990; Seafish, 1998). There are marked changes in odour and flavour of the fish flesh and there are more wide-scale changes to the overall morphology of the fish. These changes include an intensification of bacterial slime on skin and gills, increasing cloudiness of the fish's eyes, loss of gloss to the fish skin, loss of scales, and the peritoneum becomes dull and is easier to separate from the internal wall (Connell, 1990).

Non-microbiological deterioration is mainly caused by the action of enzymes naturally present in the flesh; in life they are involved in tissue building but after death they become involved in degradative reactions (Connell, 1990). Thus, the development of rancidity is catalysed by enzymatic actions after mortality of those fish species that contain oils.

pH can be utilised to determine the quality of the fish. After death, glycogen in the fish is converted into lactic acid, lowering the flesh pH. Bacteria, which cause spoilage, are more active in flesh which has a high pH (Connell, 1990). This particular factor has been suggested as causing poor quality characteristics of some species such as the 'gaping' in filleted cod (Burgess *et al.*, 1965), 'chalkiness' in halibut (Geiger & Borgstrom, 1962) and 'honeycombing' in tuna.

8.3.3 The fish trade and quality

The fish sector and its associated markets have evolved rapidly since the early 1970s with the emphasis shifting from quantity to quality. World-wide fish stocks are diminishing and although the aquaculture sector is growing, the marine harvest is still the main contributor to the market. Concepts such as sustainable development, health and quality programs and conservation of species are having their impact. Quality is an attribute that enhances the value of the product, not just because of health factors but also because consumers are willing to pay more for a fresh fish.

According to the Fisheries Commission, the Community market in fisheries products is now characterised by six traits:

(1) Major changes in patterns of consumption, leading to two separate markets: fresh and processed or manufactured products
(2) Dominant share of the retail share in fisheries products
(3) A large and increasing gap between demand and supply
(4) Globalisation and dovetailing with the international markets
(5) Increased, integration in the industry, leading to a reduction in time and costs of transport, given by the evolving, technology in communication and information systems
(6) Consumers are more aware and demanding of the nutritional, health and hygiene requirements of products.

Objectives that are driving the new market include the following: responsible fishing and trading; greater involvement of operators in the management of the market; encouraging competitiveness among the Community producers; promoting market transparency and integration through co-operation between the various parts of the industry; keeping the market open to international trade in a context of a fair competition; and promoting quality in the field of fishery products. To accomplish these objectives it is suggested that quality remains an aim through all levels of the industry and that supply should closely match consumers' requirements in terms of price, quality, availability and regularity.

Quality and health standards should be developed (i.e. labels) for imported and exported fish products in accordance with international commitments and signed instruments. This action would benefit both the consumers and the industry. The former are increasingly more aware and interested in the quality and nutritional value of their meals. Such characteristics are considered higher in fresh, packed products that are highly perishable. This would also benefit the industry since a better quality would add value to the product.

Market organisations should encourage responsible fishing, promote dynamic and professional organisations, and implement improved mechanisms for achieving better quality; it is the Commission's duty to make proposals along these lines. As a result, the market would be ruled by marketing standards (minimal capture sizes), production structure, intervention arrangements – adapting these to the specific characteristics of the market and its segmentations (fresh or processed) and market transparency by giving information of the product to the consumer (name, production methods, origin, freshness category), and by certifying responsible fishing. In both cases demand should act as the arbiter penalising conduct against stock conservation and/or environmental protection.

These factors outline issues affecting the quantity and quality of fish populations, and have set the scene for a more detailed consideration of the social and economic input on xenobiotic influence.

8.4 Xenobiotic influences on fish quality

When a fish is subject to xenobiotic influences in polluted waters, it will begin to accumulate toxins in its body. The degree to which they affect the fish and its different organs will vary with the substance, its concentration, the species affected and the organ involved. The most affected organs are kidney, liver, skin, gills, heart and gonads, and activity can vary depending on the substance. This can be detected by observing the changes in enzyme activities (i.e. peroxidase activity), which are often used as bio-indicators of the fish and environmental conditions (Bhatnagar *et al.*, 1992; Lichtenfels *et al.*, 1996; Vethaak *et al.*, 1996; Chitra & Kumar, 1997).

Fish will additionally become stressed by xenobiotic influences and this in turn will affect the immune system, making individuals more susceptible to diseases and parasites (Grinwis *et al.*, 1998). The relationship between pollutants, stress and parasites is sometimes used as a bio-indicator of the ecological condition of the water body (Zharikova *et al.*, 1995).

Other, more subtle effects of pollutants in fish have also been reported and among them are the possible changes in hormones by endocrine-disrupting chemicals that can affect fish reproduction. Pollutants have additionally been linked to the development of skin diseases as a consequence of bacterial infection, although it is difficult in some situations to separate these effects from those attributable to other factors, i.e. the catching and handling process can also result in skin damage.

Clearly, all these interactions will affect the consumers' perceptions of the fish. Any increase in the condition of the fish will make the fish more desirable for consumers, whereas any increase in the bioaccumulation of pollutants will decrease the value. The consumers' reactions will be reflected in their willingness to purchase fish and therefore, there may be impacts on fish price. In some instances, consumers will exhibit resistance to the purchase of fish that are obviously marked or otherwise damaged. To some extent these superficial marketing problems can potentially be avoided by eliminating damaged fish or creating products that do not show the superficial damage. Nonetheless, these unwarranted changes will give rise to adverse socio-economic effects.

Curiously, in some fish production networks the presence of parasites does not deter the consumption of fish that are affected. In the Netherlands, the consumption of raw herring fillets is common despite the presence of parasites that are transmitted to man. In Russia, the cestode parasite *Ligula* sp. is eaten as a delicacy. This example serves to reinforce the message that fish quality and the preferences shown by human consumers are very subjective phenomena.

There are methods of assessing basic fish quality that can be applied objectively to fish flesh and these are valuable tools in establishing background quality standards against which xenobiotic influences might be measured. As such, it is possible to measure spoilage in fish using physical and chemical methods.

In the EU fish markets where the catch is first landed, inspectors grade the quality of landings based on a series of criteria that are used to describe the superficial aspects of the fish. The fish are graded by size of the individuals, the odour and freshness of the gills, whether the eye and skin are opaque, the elasticity of the fish flesh, and the cleanliness of the body cavity. These properties have been formalised into organoleptic quality scales that can be applied to both fresh and cooked fish flesh (Burgess *et al.*, 1965). There has also been the development of a portable meter that can determine changes in the electrical conductivity of the fish skin. The meter equates the stages of deterioration of fish quality with changes in the dielectric properties occurring when the muscle is degraded (Ferri *et al.*, 1995). More advanced techniques are also used under laboratory conditions such as chromatographic and spectroscopic methods for determining enzyme and tissue degradation. Molecular techniques are used to determine damage to DNA and the production of tumours and cancers. Chemical methods may additionally be used to determine the rate of spoilage of fish tissues and these involve measuring the denaturating of trimethylamine oxide, levels of total nitrogenous bases, and the build-up of rancidity (Connell, 1990).

The interactions between xenobiotic influences on fish quality and marketability are summarised in Fig. 8.4. Lipton and Strand (1997) summarised the xenobiotic effects on the economics of the fish sector stating that the impact would depend on factors such as the response of the fish stock to pollutants, the responsiveness of the consumers to price changes, and the magnitude of the perceived change in quality of fish products.

These changes would influence users who might be grouped as follows:

(1) Commercial users (and their dependent consumers)
(2) Recreational user groups
(3) Non-users.

The effects on commercial users and consumers would be very much related to the effects on the supply of and quality of commercially exploited fish species (prices of fish which are affected by the quality of the fish and therefore by the responses of fish to pollutants). It

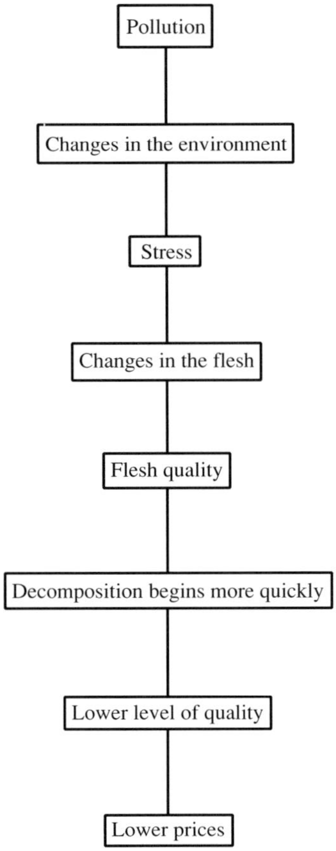

Fig. 8.4 Schematic representation of the impact of xenobiotics on fish quality and marketability.

would also depend on the capacity of consumers to recognise the difference and to value quality of the fish products. Unlike the commercial users, the losses to the recreational user groups may not be manifested in the market economy. Nevertheless, the interests of recreational fish users would be affected by pollution incidents. Finally, there would be those groups of users who are willing to pay to conserve an environment free of pollution, even if they do not consume the fish resource or practise any sport or recreational activity in relation to the fish population.

With these categories in mind, it is now appropriate to turn to a series of case studies that exemplify the impact of xenobiotic processes and incidents on the social, economic and health dimensions of life in the EU.

8.4.1 Ciguatoxin and red tides

Certain bio-toxins present in the marine environment, e.g. para-haemalytic shellfish toxin, ciguatoxin, clupeotoxin, are believed to originate from the influence of xenobiotic events. Their socio-economic effects are quite marked in that they restrict fishing activity (and the marketing and consumption of fish products) in the areas where they are prevalent. For example, in parts of the Caribbean and Pacific Ocean, fish of all species are not consumed due to the high incidence of ciguatoxin poisoning. Whilst this can have a beneficial effect on the stock base through the reduction in fishing effort (Caddy, 1984), the toxin is believed to have a xenobiotic origin. In the Solomon Islands, it is widely believed that the modification of the coral reef environment as a consequence of the Second World War and subsequent nuclear testing has resulted in the presence of ciguatoxin in the coral reef food chains (Ruff, 1989). The dumping of waste metallic compounds, industrial wastes, and shipwrecks can act as a trigger on toxins affecting algae, invertebrates and fish (Idler, 1972; Ruivo, 1972). Similarly, the excess heat generated by power plants seems to favour the proliferation of organisms linked with ciguatera toxin in tropics waters (Nitta, 1972). Nitta (1972) also notes that red tides occur in Tokyo Bay, even in winter, as a consequence of eutrophication.

8.4.2 Organochlorine pesticides

Organochlorine pesticides are some of the most widely distributed chemicals on the planet and influence all living organisms. They may play a role in the sublethal effects on fish and this problem will reveal itself in a deterioration of the condition of individual fish. Unfortunately, as with many sublethal xenobiotic effects, the socio-economic consequences of pesticides in fish are not readily understood in the majority of cases. However, in Uganda the practice of killing fish using pesticides (Thiodan is the compound implicated) has resulted in the deaths of at least 12 people, the closure of the fisheries for tilapia and the suspension of export of Nile perch fillets to the European Union. The loss to the country's economy since April 2000 is believed to be in the region of US $5 million (Odhiambo, 1999). The impact on the Ugandan local and national economy has been devastating. Worse has come about with the suspension of the fish trade between the EU and the other riparian states of Lake Victoria (Odhiambo, 1999). The fish export industries of Uganda, Kenya and Tanzania have all been closed down. Kenya's loss is much greater than the other states as the EU imports about 56% of the Kenyan Nile perch catch valued in export terms at US $66 million.

The overall socio-economic fallout for the Lake people is inevitably difficult to assess but it is likely that somewhere in the region of one million fish catchers and processors are affected, whereas those of hinterland populations dependent on this industry maybe multiplied by a factor of ten.

In 1991 the fisheries along the coast of California, Washington and Oregon had to be closed due to the high levels of pesticides found in razor clams, mussels and Dungeness crabs. Pelicans were killed by the consumption of contaminated anchovies and this led to the closure of the fishery (Villac *et al.*, 1993) and resultant socio-economic dislocations.

8.4.3 Heavy metals

Among the metals, mercury and lead are considered to be the most dangerous pollutants. All other metals, even though toxic in high concentrations, seem to primarily influence the flavour and colouring of the animals. Metals appear to have a stronger influence compared to other substances in colouring fish tissues. Copper ions are believed to discolour oyster tissue, and iron is associated with green discoloration of cod fillets (Ruivo, 1972).

One of the most famous cases involving the xenobiotic effects of heavy metals in fish and shellfish took place on the island of Kyushu in Japan in the late 1950s. Here the contamination of the coastal waters by traces of heavy metals was believed to have caused the outbreak of poisoning in which over 2000 people were affected (Borgstrom, 1962). The symptoms included numbness and speech deformities. Seriously affected patients showed an increased amount of amino acids in the urine and of gamma globulin in the blood. The mortality rate was 33%. Heavy metal pollution as a result of mercury contamination was implicated in the disease outbreak and measures were taken to close the fishery. To date, there have been 2252 sufferers diagnosed with Minimata disease in the past 36 years. Among them 1043 have died (Harada, 1995).

8.4.4 The effects of hydrocarbons

Worldwide, the general trend to reduce oil spills may be improving or, in US waters, at least holding steady. Data from the Oil Spill Intelligence Unit, which tracks oil spills worldwide, showed a decline in the total volume of oil spilled from 1978 to 1980 compared with 1981 to 1990, with occasional megaspills inflating the total. The number of megaspills (greater than 10 million gallons) has also decreased, with 15 reported during the first 3 years that data was kept, and only 11 reported during the remaining 10 years (US Congress, 1990). However, there are many sources of oil pollution in the ocean other than tanker and oil rig accidents, which collectively account for approximately 12.5% and 1.5% respectively of all spilled oil worldwide, according to a 1985 report from the National Academy of Sciences. Human activity on the land and natural seepage are significant sources of ocean oil pollution; municipal and industrial waste discharge and run-off account for about 37% and natural sources almost 8% of the total of oil spilled into the ocean (Baker *et al.*, 1990).

Petrochemicals in the aquatic environment have given rise to some of the most catastrophic xenobiotic incidents killing large numbers of birds, fish, whales, seals, otters and fish. However, it appears that schools of fish may in some instances be able to swim away from the contaminants, although fish in ocean hatcheries, floating fish eggs, and larvae can be destroyed (Hap Pritchard & Charles, 1991). Oil spills affect the odour and flavour of the fish directly or may be ingested into the fish through feeding on contaminated marine organisms. This is important in terms of the impact it may have on fish quality for as Ruivo (1972) showed, hydrocarbons can be retained in the fish and might also be concentrated, becoming bonded to proteins. Thus the incorporated oils or products destroy the food value.

8.5 Case studies

8.5.1 Oil spills

8.5.1.1 The Exxon Valdez *oil spill*

Prior to the 1989 *Exxon Valdez* spill, a rich community of 24 taxa represented the Bay's flora and fauna (Goldberg, 1991), and this diversity was seriously reduced after the spill. By 1991 when the hydrocarbons' concentration was reduced, the re-colonisation process had begun. However, in 1993 a reduction of the diversity occurred again when the hydrocarbon concentrations were low. However, according to Jewett *et al.* (1993), this particular reduction was caused by a natural hypoxia-anoxia process and not by the spill.

Stekoll *et al.* (1993) made a comparison between systems that were affected by the *Exxon Valdez* oil spill and similar areas not affected. He found statistically significant differences relating to algae and invertebrate biomass, abundance and percentage cover. These effects were observed until 1991 and were found to vary with tidal elevation, habitat and region. The fauna found to be most affected included *Fucus* sp., limpets, mussels, barnacles, littorinids and oligochaetes with intertidal communities suffering the most severe consequences of the spill and clean-up operations. A wide number of taxa were affected and included brown algae (*Fucus gardieri*), the limpet (*Tectura persona*), three species of barnacle (*Chthamalus dalli, Semibalanus balanoides*, and *Balanus galndula*), the mussel *Mytilus trossulus*, two littorinid species, and oligochaetes. Most were affected negatively either directly or indirectly and some were affected by direct contact with the oil spill or via the loss of habitat. However, the abundance of *Chthamalus dalli* and certain oligochaetes were found to increase (Highsmith *et al.*, 1993; Hooten & Highsmith, 1993).

With respect to fish communities, the impacts of the spill were analysed by Brown *et al.* (1993). The pacific herring in Prince William Sound were heavily impacted as almost 50% of their eggs were deposited in the trajectory of the oil spill. The effects shown in the larvae included premature hatching, low weights, reduced growth and increased morphological and genetic abnormalities.

Similarly, Geiger *et al.* (1993) analysed the effects of the spill in the pink salmon population of the site. They found that approximately 1.9 million adult wild pink salmon failed to return to the area in 1990 as a result of the effects of the spill on the critical nearshore life-stage, with almost 23% of the potential wild-stock production being lost. In a comparison of juvenile pink salmon from contaminated and non-contaminated waters in 1989, the results showed that salmon were significantly smaller and had reduced growth rates in oiled water when compared to non-oiled locations. This was thought to have affected their survival to adult stage and the migratory return to Prince William Sound in 1990 (Wertheimer *et al.*, 1993; Willette, 1993). Analysis of the visceral tissues of the juvenile pink salmon and chum salmon revealed that ingestion of oil was an important source of flesh contamination (Carls *et al.*, 1993). Other taxa such as harpacticoid copepods and epibenthic crustaceans (prey resources for the juvenile salmon) were not affected by the spill. In some cases, their abundance increased in the areas that were polluted (Celewycz & Wertheimer, 1996).

In addition, toxicity tests were undertaken for finfish and shellfish in the area of the *Exxon Valdez* oil spill. The finfish from all study sites were found to be safe to eat, whilst high concentrations of aromatic contaminants were reported in the shellfish (Fall & Field, 1993).

Cohen (1995) employed a market model to evaluate the economic losses of the 1989 *Exxon Valdez* spill on Alaska's fisheries. The upper limit of the accident's first-year social costs on these resources was estimated at $108 million, approximately 27% of the ex-vessel value. Second year effects may have been as high as $47 million.

8.5.1.2 The Braer *oil spill*

The *Braer* spill involved the spillage of oil off Shetland, a group of islands located off the northern coast of the United Kingdom. On 5 January 1993, a Liberian-registered oil tanker, the *Braer*, ran aground off the southern tip of the Shetlands when the engines could not be restarted after they became flooded with seawater. On 12 of January 1993, the tanker subsequently broke into three sections after it was continually thrown against the rocks of the island. The entire cargo of 85 000 tons (620 000 barrels) of Norwegian Gulfaks crude oil destined for export to Canada spilled into the sea around the southern end of Shetland. None of the oil could be recovered from the tanker whilst it was aground due to the high winds (up to 100 mph) and the rough seas. However, fortunately for Shetland, the Gulfaks crude oil which the *Braer* was carrying is not a typical North Sea oil. It is lighter and more easily biodegradable than other North Sea crude oils, and this, in combination with some of the worst storms seen in Shetland (naturally dispersing the oil by wave action and evaporation), prevented the event becoming an even bigger disaster by preventing an oil slick from forming on the surface, and breaking the spill up quickly (http://www.wildlife.shetland.co.uk/braer/, accessed August 2002).

According to one report, 'around 30 percent of the oil has been deposited in the sediments of two basins . . . this oil will slowly break down . . .' (*New Scientist*, 1993). The dispersion of the spill was aided by chemical dispersants dropped from the air. The dispersants broke the oil into globules that sank below the surface, and so helped to moderate the effects on sea birds in immediate danger of oiling (*The Economist*, 1993).

The influence of the spill progressed beyond the range originally predicted and fishermen found oil in their catch in areas where it had not been expected, with tidal action spreading the oil underwater around Shetland's 900 miles of coastline. According to the fishermen, by 1996 the fishing grounds were still depleted and the shellfish extracted still affected by the oil. The spill threatened the populations of seabirds, salmon, sea trout, grey seals and other species on and around the islands. In addition, a fine mist of oil particles drifted over the islands a few days after the spill occurred and left an oily residue on the island's sheep. However, it was determined that 'the level of air pollution after the spill was extremely low despite the oily mist' (*New Scientist*, 1993).

Initially, it was thought that the damage caused by the *Braer* oil spill would be similar to that of the *Exxon Valdez* in 1989. However, the actual impacts of the *Braer* were less severe even though 'it was the twelfth largest spill in history' (*Discover*, 1994). The damage to the wildlife was as follows:

> 'The official death tolls – the number of carcasses recovered – included 1542 seabirds, several thousand pounds of commercially farmed salmon, 10 grey seals, and 4 otters. Two of the otters were run over by a camera crew covering the spill, however, and the other two probably died of old age' (*Discover*, 1994).

Following the spill, and in pursuit of its statutory functions, the Fisheries Research Services (FRS) Marine Laboratory in Aberdeen gave priority to the monitoring of commercial species, which included herring, mackerel and demersal fish such as cod and haddock, in addition to various wild shellfish. The monitoring also extended to farmed salmon and shellfish. On 8 January 1993, the former Department of Agriculture and Fisheries (DAFS) declared a Fisheries Exclusion Zone, under the Food and Environmental Protection Act 1985 (FEPA) around the southern part of Shetland in order to protect the public from exposure to potentially contaminated fish and shellfish. This Fisheries Exclusion Zone prohibited the taking of farmed or wild fish and shellfish from defined areas and on 27 January, after further surveillance, this Zone was extended 5 miles to the west (Fisheries Research Services, 2000).

Wild fish were found to be contaminated, but at lower levels than those seen in farmed salmon, presumably as a result of being able to swim away from the areas of greatest pollution. Round fish such as cod and haddock were generally less contaminated than flatfish such as lemon sole and dab, as a result of not being in direct contact with the sediment. Concentrations of polycyclic aromatic hydrocarbons (PAHs) in wild fish fell rapidly (being at reference concentrations by the middle of March 1993), and the Exclusion Zone for wild fish species was lifted on 23 April 1993 (Fisheries Research Services, 2000). However, salmon farms on the south-west coast of Shetland were particularly affected by the oil spill, and caged fish in the most affected areas which could not swim away were severely tainted and contaminated with PAHs. Rapid uptake of the contaminants was followed by a rapid loss, but not back to reference conditions which took approximately 150 days. The Scottish Office finally lifted the FEPA order for this species on 8 December 1993 (Fisheries Research Services, 2000).

With respect to shellfish, the degree of contamination varied with habit, e.g. crustacean shellfish such as *Nephrops* showed severe contamination due to living in impacted sediments. Shellfish were slower to return to reference conditions, and the FEPA Exclusion Order was lifted for crustaceans, with the exception of the Norway lobster (*Nephrops norvegicus*) on 30 September 1994. The filter feeding bivalve molluscs such as mussels, queens and scallops accumulated the greatest concentrations of PAHs of all fish and shellfish, and the FEPA Order for scallops and queens was eventually lifted on 9 February 1995. Restrictions continued for mussels and *Nephrops* and the last FEPA Order was finally lifted in March 2000 (Fisheries Research Services, 2000).

With respect to tourism, it was estimated that the Shetland Islands would lose £18.2 million of tourist revenue by the year 2000 as a result of the oil spill (*The Scotsman*, 1994) and people from the islands noted that tourism was down, citing the oil spill as the cause. Lack of job alternatives on these small islands means that the tourism trade is vital to their economic well-being. The physical damage to the environment and resources around Shetland was much lower than anticipated and the rough seas helped to break up the light crude oil fairly rapidly. However, the major impact of the spill has been in terms of the damage done to the reputation of the Shetland as a pristine environment, with many islanders being rightly worried about the resulting fall in tourism.

The damage to the Shetland seafood market was however more serious with major buyers no longer purchasing Shetland seafood products after the disaster. Marks & Spencer together with Japanese buyers pulled out of the trade in Shetland salmon (Goodlad, 1996), and it has taken considerable time to rebuild confidence in the Shetland industry. The Islands' fisheries were however compensated for their losses. Compensation given for the destruction of the 1991 and 1992 salmon stocks was £7.176 million and £12.118 million respectively. The effects of the Exclusion Zone on the fishing fleets attracted compensation of £1.363 million for the demersal fleet and £4.57 million for the shellfish fleet. Compensation for market losses in reduced prices for Shetland seafood products was worth a further £3.781 million. However, these payments do not really reflect the economic losses (of reputation) that will stretch into the future.

8.5.1.3 The Sea Empress *oil spill*

The grounding of the *Sea Empress* in February 1996 caused one of the largest and most environmentally damaging oil spills in European history (SEEEC, 1998a). The *Sea Empress* released approximately 72 000 tonnes of crude oil and approximately 360 tonnes of heavy fuel oil between 15 and 21 February 1996 when it grounded off Milford Haven in south-west Wales. Further oil leaked out as the tanker was moved out of Milford Haven on the way to Belfast, Northern Ireland. The *Sea Empress* spill is the third largest tanker spill in UK waters, and may have been the most destructive.

The rich and productive waters off the coast of south-west Wales support thriving populations of ecologically and economically important shellfish and fish. Rocky shores are colonised by the edible periwinkle *Littorina littorea* and the edible blue mussel *Mytilus edulis*. Subtidal kelp forests on rocky coasts support large populations of the edible crab *Cancer pagurus* and the lobster *Homarus vulgaris*. The crawfish *Palinurus vulgaris* is also harvested. Beds of cockles and oysters support significant fisheries within Milford Haven and the large estuaries of Carmarthen Bay. An important offshore fishery has developed for whelks which are exported to the Far East. With respect to fish, herring are usually caught within Milford Haven at this time of year. There are also substantial offshore fisheries for rays, flatfish and sea bass. Migratory sea fish such as salmon and sea trout pass through the area on their way to the rivers of Carmarthen Bay, UK.

Of the official quantity spilled, only 3000 tonnes are estimated to have been recovered during the clean-up operation, with more than 100 km of coastline of outstanding beauty seriously polluted by oil. The coastline within Milford Haven was heavily oiled, and outside the Haven much of the oil moved south and then eastwards parallel to the south coast of Pembrokeshire, affecting this coastline as far as Pending Sands in Carmarthen Bay. Ecosystems of conservation, fishery and recreational importance were affected. Some oil reached Skomer Island north-west of the spill, but no oil was observed north of St David's Head. Lundy Island in the Bristol Channel received light oiling, and some pellets of oil reached the Irish coast; however, no oiling of the coast of mainland Devon and Cornwall was reported.

A massive clean-up operation was launched, both at sea and onshore. Some oil was recovered mechanically at sea, and between 17 and 25 February 1996, 445 tonnes of chemical dispersants were used to break up the oil into dispersed droplets in order to reduce the risk to the coastline and to birds at sea. A policy was adopted of not using dispersants within 1 km of the shoreline so as to avoid shallow water, where the dispersed oil would be less readily diluted, thereby increasing the risk to marine life (SEEEC, 1996).

The effects of the spill extended from the marine environment to the tourist economy of the region. With respect to the impacts on wildlife, birds at sea were severely impacted during the early weeks of the spill and by 1 June 1996, more than 6900 oiled birds of at least 28 species had been recovered dead or alive. In contrast, the grey seal (*Halichoerus grypus*) population around Skomer Island appeared little affected and impacts to subtidal wildlife were limited. However, significant damage was caused to shorelines as a result of bulk oil. Limpet mortality (*Patella* spp.) was found to vary between 5% and 50% (although up to 90% at one site), and there were significant impacts on periwinkles (*Littorina* spp.), barnacles (*Cirripedia*) and sea slaters (*Ligea* spp.). There were mass strandings of lower shore bivalves, urchins and starfish, and significant patches of vegetation were oiled, with some damage to the rare narrow-leaved eelgrass *Zostera angustifolia* and *Zostera noltii* (SEEEC, 1996).

The *Sea Empress* oil spill has presented considerable difficulties for the fishing industry within the region. As a precautionary measure, a ban on all fishing and the collection of edible plants and seaweeds was imposed from St. David's Head to Port Eynon on the Gower shortly after the spill. Fishing for salmon and sea trout was also banned in the rivers flowing into this area. Samples of various species of fish, shellfish and crustacea were analysed for oil content over subsequent months, and the closure orders were lifted in stages as analysis showed that oil concentrations in different classes of organism had returned to background levels. Salmon and sea trout did not appear to have been affected and the ban on these was lifted on 3 May 1996, whilst that for other fish was lifted on 21 May 1996. The ban for crabs and lobsters outside the Milford Haven waterway and for whelks throughout the whole area was lifted on 29 August 1996, and over the following year, the ban was progressively lifted in stages until all remaining restrictions were removed on 12 September 1997 (SEEEC, 1998b).

The question has been raised as to who should pay the costs related to the accident. Knowing the inherent dangers of its business, the oil industry has set up a system which enables individual companies to remain apparently blameless for spills of their own toxic cargo. The *Sea Empress* was built in Spain, owned and skippered by a Norwegian, registered in Cyprus, managed from Glasgow, chartered by the French, crewed by Russians and was flying a Liberian 'flag of convenience'.

However, Texaco gave some financial help to clean up the coastline as a goodwill gesture. Liability for the environmental damage was covered up to a limit of $15 million by the Scandinavian insurer Skuld. Additional costs including compensation claims were covered by the International Oil Pollution Compensation Fund, to which most oil companies contribute. The compensation claim period is 3 years. After this, further compensatable impacts are not covered. This raises the concern that the impacts of pollution on fish reproduction, fecundity and larval survival may only become apparent after this three-year period when compensation to fishermen is no longer available.

8.5.2 Claims and compensations

Until now, the costs of pollution to the fishing industry have been considered in terms of what it is losing. However, it is also important to mention that despite the economic cost that pollution causes directly by diminishing the production capacity of an ecosystem, there are other economic effects that should be considered. Oil spills represent the best examples, involving costs of damages to the environment and economical effort invested to recover the damage and/or compensate the people affected by those incidents. This section details some information regarding the amount of money spent on claims and compensation in oil spills (Table 8.1).

Compensation is limited by the Civil Liability Convention (1969) and by the Fund Convention (1992). In general, the costs of an oil spill are divided between the bodies that undertake response actions and the polluter. International regimes have established a two-tier system in which both the individual tanker owner and the cargo receivers take the costs.

Table 8.1 Maximum amount of compensation available ($ million) for various sizes of tankers (White, 1999).

Gross Tonnage	1969 Liability Convention	1971 FUND	1992 Liability Convention	1992 Fund Protocol
5000	0.8	81.0	4.0	182.0
25 000	4.0	81.0	15.3	182.0
50 000	8.1	81.0	29.5	182.0
100 000	16.2	81.0	57.8	182.0
140 000	18.9	81.0	80.6	182.0

The amount of compensation in each case also depends on the country's membership to any Convention. The value of the limits for compensation in a national currency will therefore depend on the exchange at a particular time, 1 SDR = $1.35, according to 1969 and 1992 Conventions. The limits of liability are expressed in Special Drawing Rights (SDR).

Pollution damage is defined slightly differently depending on each Convention's guidelines; however, compensation claims generally tend to involve actions for prevention measures, property damage, economic loss and/or restoration or reinstatement of impaired environments.

Fisheries are affected by property damage and economic loss, firstly by costs of cleaning contaminated fishing gear and mariculture installations, and secondly by fishing activities being prevented and tourism reduced. Damage might appear some time later; claims should be presented as soon as signs of damage appear, and court action can take place up to 6 years after the incident.

8.6 Conclusions

The early sections of this chapter set the scene with respect to the European fish sector which, despite difficulties in obtaining supplies of fish, is buoyant and is a meaningful contributor to the GDP of the EU. There are substantial problems in the context of management of the fish stocks and many of the prime commercial species are in decline as the result of adverse fisheries-dependent factors, notably excessive fishing effort. As has been argued earlier, the EC through its CFP has battled to come to terms with problems in the industry but has not been entirely successful in either limiting fishing capacity or eliminating some of the most wasteful practices such as discarding. Thus, xenobiotic effects come into play in a situation where the resource base is under threat from fisheries-dependent variables. Now there are justifiable concerns that efforts to preserve fish stocks may be undermined by deleterious changes in the environment. Sublethal and catastrophic incidents of xenobiotic origin now pose a fisheries-independent threat to stocks.

The preceding chapters in this book have set down markers in an array of studies that have been directed at the problem of establishing xenobiotic influences on fish. This has established that there is a dichotomy of scale of interactions between the biological/physiological impacts and socio-economic outcomes. It is helpful therefore to divide the biological/physiological impacts in terms of their influence on the macro socio-economic scale of fish populations and the micro socio-economic scale of interaction on individual fish specimens. Where this is most relevant is in the expression of the sublethal effects of xenobiotics on fish. It is in this domain that the complex interactions that take place and the reactions they engender are least understood. However, where there are catastrophic impacts, the interactions are more clear-cut and it is less meaningful to separate influences impacting on individual fish/fish populations.

The catastrophic influences are well documented in terms of their socio-economic impacts and there is a substantial body of information that relates the ecology of catastrophic xenobiotic events with the socio-economy. A number of indicative case studies have been used to show how the cost of xenobiotic disasters can be measured. In addition, through political negotiation, scales of compensation can be agreed which give a definitive link between xenobiotics, fish populations, adverse physiological effects and the socio-economy of areas dependent on fishing for their livelihoods.

One of the most important gaps in information related to this is the impact of future discounted loss of revenues as a consequence of xenobiotic influences. A good example of this would be Shetland and those users who were affected by the *Braer* spill. In this situation, a scale of compensation payments was agreed for those who were affected by the disaster. However, markets for Shetland seafood used to pay a price premium for Shetland seafood because it was produced in a 'pristine environment'. Whilst the validity of this statement in an ever-changing ecosystem could be argued, in a marketing context the phrase has a different value and validity. If it is assumed, for example, that the mark-up in price for farmed salmon or locally harvested scallops was 10% on the average value before the oil spill, because the specimens were deemed to be produced in a 'pristine environment', the cost to producers of losing the market advantage as a result of the pollution would be significant. This is because the market is unlikely to restore the price premium for some time, if ever. The economic loss to the Shetland seafood industry therefore becomes the value of discounting this loss into the future for a finite, or maybe even infinite, time. This figure could potentially dwarf the already agreed financial packages.

The linkages and scale of eco-physiological impacts of xenobiotics are least understood at the sublethal level. The sublethal impacts clearly have an impact on the 'quality' of fish populations, but there is also an important microdimension associated with the impacts of individual fish on marketing channels and consumers. Food quality standards have never been more important and most societies, in developed countries at least, are being shaped to seek out and respond to quality signatures. Indeed, supported by a rapid growth in international food standards and codices of handling, preservation, processing and product marketing, those who set and police food standards are reaching ever further down the food production chain in an effort to prescribe quality. Thus there is a very real meeting point, even if it is not always adequately understood where the sciences of measuring and interpreting xenobiotic influences interact with the sciences of food hygiene, quality assurance, health promotion and consumer protection.

One of the most powerful and all-encompassing developments in the world's food industries during the 1990s was the adoption of the so-called HACCP system, as applied to food production and safety assurance for consumers. HACCP is the acronym for Hazard Analysis Critical Control Pathway, a standard methodology applied to minimising health risks in all sectors but none more important than the food industry. In essence, the HACCP system is used to review and investigate each stage of production in the chain of events that takes raw food material from the production or harvesting stage to the consumer. Whilst every stage of the chain is important, increasingly the HACCP users are looking for guarantees as to the safety and reliability of the *raw material* that is the starting place for the development of consumer foodstuffs.

The quality of fish and its impacts on human populations (as consumers) is possibly the most important meeting place between the biological impacts of pollution on fish and its socio-economic relevance. Thus, studies on the physiological changes associated with xenobiotic impacts on fish tissues can be related to changes in the quality (or perception of quality) of fish as food. For example, it is known that sublethal xenobiotics can reduce the capacity of cells to store glycogen in fish tissues. This in turn will influence the rapidity with which fish pass into *rigor* after death and therefore will reduce the shelf-life duration of the fresh fish and products derived from it. Likewise, the capacity of fish lipids to absorb xenobiotic compounds, and the levels at which they are stored sublethally in adipose tissue, are the sort of information that may soon be required to satisfy HACCP requirements and therefore initial acceptance criteria for raw materials. The importance and powerful socio-economic consequences of not being able to guarantee this sort of information were vividly revealed in East Africa in 1999. Here in the fishery for Nile perch of Lake Victoria there were instances of 'fishermen' harvesting fish, using commercially used pesticides as a method of capture. There were cases of local people being poisoned and this led to banning by the EU of imports of Nile perch fillets from some of the riparian countries. This has had serious repercussions on the economies of Tanzania and Uganda with many fishermen, processors and communities being disadvantaged as a result.

Looking to the future, it is clear that the importance of knowledge of the processes and impacts of xenobiotics will only become more critical. Whilst the current understanding of the interactions is incomplete, the information that is available is being used. The food industry researchers are joining hands with the fish physiologists, geneticists and ecologists in the pursuit of more definitive knowledge of the responses of individual fish and fish populations to xenobiotic change. In instances where it is necessary to measure the impact of catastrophic events involving the effects of xenobiotics on fish/fish populations, the natural scientists have a role to play in determining socio-economic impacts. This has been used to determine levels of financial and economic compensation for individuals and communities. Whilst information on these processes is only beginning to be gathered, and for the most part the interactions of the key variables are little understood, the field of co-operation is set to expand and hopefully become more effective.

8.7 References

Amhed, F. & R. Anderson (1994) Fishery resources, consumption and import trends, and biotechnology developments in the USA. *Fisheries Research*, **19**, 1–15.

Ashworth, J. (1996) *Save Britain's Fish – The Common Fisheries Policy*. Paper presented at the Colloquium on the Politics of Fishing, Department of Politics, University of Newcastle, 17–18 September 1996.

ASMI (1998) Recommended whitefish quality guidelines for fishing and processing operations. http://www.commerce.state.ak.us.

Baker, J.M., R.B. Clark, P.F. Kingston & R.H. Jenkins (1990) *Natural recovery of cool water marine environments after an oil spill*. Presented at the Thirteenth Annual Arctic and Marine Oil Spill Program Technical Seminar.

Beverton, R.J.H. & S.J. Holt (1957) On the dynamics of exploited fish populations. UK Ministry. *Agriculture and Fisheries Investigations*, **2**, 533.

Bhatnagar, M.C., A.K. Bana & Tyagi (1992) Respiratory distress to *Clarias batrachus* (Linn) exposed to endosulfan – a histological approach. *Journal of Environmental Biology*, **13**, 227–231.

Borgstrom, G, (ed.) (1962) *Fish as Food: Production, Biochemistry and Microbiology*. Academic Press, London, 725 pp.

Brown, E.D., T.T. Baker, J.E. Hose, R.M. Kocan, G.D. Marty, M.D. McGurk, B.L. Norcross & J. Short (1993) Injury to the early life history stages of Pacific herring in Prince William Sound after the *Exxon Valdez* oil spill. In: (eds Rice, S.D., R.B. Spies, D.A. Wolfe & B.A. Wright) *Proceedings of the Exxon Valdez Oil Spill Symposium*. American Fisheries Society Symposium, No. 18, USA, pp. 440–447.

Burgess, G.H.O., C.L. Cutting, J.A. Lovern & J.J. Waterman (1965) *Fish Handling and Processing*. Tory Research Station, HMSO, Edinburgh.

Caddy, J.F. (1984) *Indirect approaches to the regulation of fishing effort*. Paper presented at the Expert Consultation on the Regulation of Fishing Effort (fishing mortality), 17–26 January. FAO, Rome.

Carls, M.G., L. Holland, M. Larsen, J.L. Lum, D.G. Mortenson, S.Y. Wang & A.C. Wertheimer (1993) Growth, feeding and survival of pink salmon fry exposed to food contaminated with crude oil. In: (eds Rice, S.D., R.B. Spies, D.A. Wolfe & B.A. Wright) *Proceedings of the Exxon Valdez Oil Spill Symposium*. American Fisheries Society Symposium, No. 18, USA, pp. 608–618.

Celewycz, A.G. & A.C. Wertheimer (1996) Prey availability to juvenile salmon after *the Exxon Valdez* oil spill. In: (eds Rice, S.D., R.B. Spies, D.A. Wolfe & B.A. Wright) *Proceedings of the Exxon Valdez Oil Spill Symposium*. American Fisheries Society Symposium, No. 18, USA, pp. 564–577.

Chitra, K.Y. & N.S.R. Kumar (1997) Effect of water pollution on peroxidase activity in fish, *Channa gachua*. *Journal of Environmental Biology*, **18**, 191–194.

Civil Liability Convention (1969) *International Convention on Civil Liability for Oil Pollution Damage*. International Maritime Organisation (IMO). Came into force 19 June 1975.

Cohen, M.J. (1995) Technological disasters and natural resource damage assessment: an evaluation of the Exxon Valdez oil spill. *Land Economics*, **71** (1), 65–82.

Collet, S. (1996) Leviathan management or customary administration: the search for new institutional arrangements. In: (eds Crean, K. & D. Symes) *Fisheries Management in Crisis: A Social Science Perspective*. Blackwell Science, Oxford, pp. 99–109.

Collins, A., M. Stapleton & D. Whitmarsh (1998) Fishery-pollution interactions: A modelling approach to explore the nature and incidence of economic damages. *Marine Pollution Bulletin*, **36** (3), 211–221.

Connell, J.J. (1990) *Control of Fish Quality*, 3rd ed. Fishing News Books, Oxford UK.

Crean, K. (1999) Centralised and community-based fisheries management strategies: case studies from two fisheries dependent archipelagos. *Marine Policy*, **23** (3), 243–257.

Crean, K. & D. Symes (1994) The discards problem: towards a European solution? *Marine Policy*, **18** (5), 422–434.

Delbos, G. & G. Prémel, 1996. The Breton fishing crisis in the 1990s: local society in the throes of enforced change. In: (eds Crean, K. & D. Symes) *Fisheries Management in Crisis: A Social Science Perspective*. Blackwell Science, Oxford, pp. 129–140.

Discover (1994) *A Disaster that Wasn't*. Discover, January 1994, p. 69.

European Commission (1991) *Report 1991 from the Commission to the Council and the European Parliament on the Common fisheries Policy*. European Commission, Report No. SEC(91)2288, Brussels.

European Commission (1992) *Regional, Socio-Economic Study in the Fisheries Sector*. Directorate-M General for Fisheries. XIV/373/92, Brussels.
European Commission (1994) *Communication from the Commission to the Council and the European Parliament on the Crisis in the Community's Fishing Industry*. European Commission, Report No. COM(94)335 final, Brussels.
Fall, J.A. & L.J. Field (1993) Subsistence uses of fish and wildlife before and after the *Exxon Valdez* oil spill. In: (eds Rice, S.D., R.B. Spies, D.A. Wolfe & B.A. Wright) *Proceedings of the Exxon Valdez Oil Spill Symposium*. American Fisheries Society Symposium, No. 18, USA, pp. 819–836.
Federation of European Aquaculture Producers (1999) United Kingdom aquaculture production and price data. http://www.fishlink.co.uk/feap/uking.html.
Ferri, M., P. Mattei, T. Civera & S. Gili (1995) Considerations on the employment of electrical conductivity in freshness assessment of nine Mediterranean fish species. *Industrie Alimentari*, **34**, 1277–1282.
Fisheries Research Services (2000) *Case Study: The Grounding of the MV Braer*. Aquatic Environment Information Leaflet, Fisheries Research Services (FRS) Marine Laboratory, Aberdeen, 2 pp.
Fishing News (1996) Baldry's 10 point plan!: fishermen must be involved in management. *Fishing News*, 4323, 1.
Fund Convention (1992) *International Convention on the Establishment of an International Fund for Compensation for Oil Pollution Damage*. International Maritime Organisation (IMO). Adoption 27 November 1992; came into force 30th May 1996 (this convention updates the international convention on the establishment of an international fund for compensation for oil pollution damage 1971 which was originally adopted on 18 December 1971 and came into force on 16 October 1978).
Geiger, E. & G. Borgstrom (1962) Fish protein – nutritive aspects. In: (ed. Borgstrom, G.) *Fish as Food Volume II Nutrition, Sanitation and Utilization*. Academic Press, New York, pp. 29–91.
Geiger, H.J., B.G. Bue, S. Sharr, A.C. Werheimer & T.M. Willette (1993) A life history approach to estimating damage to Prince William Sound pink salmon caused by the *Exxon Valdez* oil spill. In: (eds Rice, S.D., R.B. Spies, D.A. Wolfe & B.A. Wright) *Proceedings of the Exxon Valdez Oil Spill Symposium*. American Fisheries Society Symposium, No. 18, USA, pp. 487–498.
Goldberg, R. (1991) *Marine Oil Spills*. Case study: *Exxon Valdez* and others. Academy of Natural Sciences, May 1991.
Goodlad, J. (1996) Effects of the *Braer* oil spill on the Shetland seafood industry. *Science of the Total Environment*, **186**, 127–133.
Grinwis, G.C.M., A. Boonstra, E.J. VandenBranhof, J.A.M.A. Dormans, M. Engelsma, R.V. Kuiper, H. vanLoveren, P.W. Wester, M.A. Vaal, A.D. Vethaak & J.G. Vos (1998) Short-term toxicity of bis(tri-n-butyltin)oxide in flounder (*Platichthys flesus*): Pathology and immune function. *Aquatic Toxicology*, **42**, 15–36.
Gulland, J.A. (1977) *Fish Population Dynamics*. J.Wiley, London.
Hannesson, R. (1993) *Bioeconomic Analysis of Fisheries*. Blackwell Science, Oxford.
Hap Pritchard, P. & F. Charles (1991) EPA's Alaska oil spill bio-remediation project. *Coastal Environment Science and Technology*, **25** (3), 372–379.
Harada, M. (1995) Minimata disease – methylmercury poisoning in Japan caused by environmental pollution. *Critical Reviews in Toxicology*, **25**, 1–24.
Highsmith, R.C., T.L. Rucker, M.S. Stekoll, S.M. Saupe, M.R. Lindeberg, R.N. Jenne & W.P. Erickson (1993) Impact of the *Exxon Valdez* oil spill on intertidal biota. In: (eds Rice, S.D., R.B. Spies, D.A. Wolfe & B.A. Wright) *Proceedings of the Exxon Valdez Oil Spill Symposium*. American Fisheries Society Symposium, No. 18, USA, pp. 212–237.
Holden, M. (1994) *The Common Fisheries Policy*. Blackwell Science, Oxford.

Holm, P. (1995) *Fisheries management and the domestication of nature*. Paper presented to the Fifth Annual Common Property Conference: Reinventing the Commons. Bodí, Norway, 24–28 May, 1995.

Hooten, A.J. & R.C. Highsmith (1993) Impacts on selected intertidal invertebrates in Herring Bay, Prince William Sound, after the *Exxon Valdez* oil spill. In: (eds Rice, S.D., R.B. Spies, D.A. Wolfe & B.A. Wright) *Proceedings of the Exxon Valdez Oil Spill Symposium*. American Fisheries Society Symposium, No. 18, USA, pp. 249–270.

Hultin, H.O. (1984) Postmortem biochemistry of meat and fish. *Journal of Chemical Education*, **64** (4), 289–298.

Idler, D.R. (1972) Effects of pollutants on quality of marine products and effects on fishing. In: *Marine Pollution and Sea Life*. Fishing News Books, Blackwell Science Oxford, pp. 535–540.

Jewett, S.C., T.A. Dean & D.R. Laur (1993) Effects of the *Exxon Valdez* oil spill on benthic invertebrates in an oxygen-deficient embayment in Prince William Sound, Alaska. In: (eds Rice, S.D., R.B. Spies, D.A. Wolfe & B.A. Wright) *Proceedings of the Exxon Valdez Oil Spill Symposium*. American Fisheries Society Symposium, No. 18, USA, pp. 440–447.

Lichtenfels, A.J., G. Lorenzi-Filho, E.T. Guimaraes, M. Macchione & P.H. Saldiva (1996) Effects of water pollution on the gill apparatus of fish. *Journal of Comparative Pathology*, **115** (1), 47–60.

Lipsey, R.C. (1963) *An Introduction to Positive Economics*, 4th ed. Weindenfeld & Nicolson, London.

Lipton, D.W. & I.E. Strand (1997) Economic effects of pollution in fish habitats. *Transactions of the American Fisheries Society*, **126**, 514–518.

MAFF (1995) *Sea Fisheries Statistics – United Kingdom 1994*. Ministry of Agriculture, Fisheries and Food, The Stationery Office, London.

New Scientist (1993) Shetland oil spill did little harm. *New Scientist*, 26 June 1993, p. 8.

Nielsen, N., E. Sorensen & K. Grunert (1997) Consumer motives for buying fresh or frozen plaice – a means-end chain. In: (ed. Lutten, J.B.) *Approach Sea Food from Consumer to Producer, Integrated Approach to Quality*. Elsevier Science BV, Denmark.

Nitta, T. (1972) Marine pollution in Japan. In: *Marine Pollution and Sea Life*. Fishing News Books Oxford, pp. 77–81.

Odhiambo, N. (1999) Europe bans East African fish caught with poison. Dar es Salaam, Tanzania, 13 May 1999. Environmental News Service.

Panayotou, T. (1982) *Management concepts for small-scale fisheries: economic and social aspects*. FAO Fisheries Technical Paper, No. 228, 53 pp.

Pope, J. (1982) *Background to scientific advice on fisheries management*. Vol. 54. MAFF Directorate of Fisheries Resources, Laboratory Leaflet, Lowestoft.

Ruff, T.A. (1989) Ciguatera in the Pacific: a link with military activities. *Lancet*, **8631** (1), 201–204.

Ruivo, M. (1972) *Marine Pollution and Sea Life*. Fishing News Books, Oxford.

Saltz, P. (1991) *The European Atlantic Fisheries: Structure, Economic Performance and Policy*. Agricultural Economics Research Institute, The Hague.

Schaefer, M.B. (1954) Some aspects of the dynamics of populations important to the management of commercial marine fisheries. *Inter-American Tropical Tuna Commission Bulletin*, **1** (2), 27–56.

Seafish (1998) *Quality Assurance, Fish Purchase Specifications*. Sea Fish Industry Authority, Edinburgh.

SEEEC (1996) *Initial Report, July 1996*. *Sea Empress* Environmental Evaluation Committee, Rivers House, Cardiff.

SEEEC (1998a) *Sea Empress* Environmental Evaluation Committee. The *Sea Empress* oil spill. In: (eds Edwards, R. & H. Sime) *Proceedings of the Conference held in Cardiff on 11–13 February 1998*. Chartered Institute of Water and Environmental Management, UK.

SEEEC (1998b) *The Environmental Impact of the Sea Empress Oil Spill. Sea Empress* Environmental Evaluation Committee, Rivers House, Cardiff.

Sparre, P.J. (1994) The data requirements for basic fisheries bio-economics. Presented at ICES Workshop *Improving the link between fisheries science and management: biological, social and economic considerations*. St. Johns, Newfoundland.

Stekoll, S.S., L. Deysher, R.C. Highsmith, S.M. Saupe, Z. Guo, W.P. Erickson, L. McDonald & D. Strickland (1993) Coastal habitat injury assessment: intertidal communities and the *Exxon Valdez* oil spill. In: (eds Rice, S.D., R.B. Spies, D.A. Wolfe & B.A. Wright) *Proceedings of the Exxon Valdez Oil Spill Symposium*. American Fisheries Society Symposium, Number 18, USA, pp. 177–192.

Symes, D. (1992) The common fisheries policy and UK quota management. *Ocean Coastal Management*, **18**, 319–338.

Symes, D. (1995) The European pond: who actually manages the fisheries? *Ocean Coastal Management*, **27** (1–2), 29–46.

Symes, D. (1996) Fishing in troubled waters. In: (eds Crean, K. & D. Symes) *Fisheries Management in Crisis: A Social Science Perspective*. Blackwell Science, Oxford, pp. 3–16.

Symes, D., K. Crean, M. Mohan & J. Phillipson (1994) *The UK fishing industry: policy formulation, implementation and monitoring*. Working paper 1 compiled for the European Commission. Devolved and Regional Management Fisheries System for Fisheries. EU funded research project No. AIR-2CT93-1392: DG XIV SSMA. University of Hull, Hull, UK.

The Economist (1993) The wreck of the *Braer*. 9 January 1993, p. 50.

The Scotsman (1994) The tainted isles come clean. 5 January 1994, p. 8.

US Congress (1990) *Coping with an Oiled Sea: An Analysis of Oil Spill Response Technologies*. Office of Technology Assessment, Washington, DC.

Vethaak, A.D., J.G. Jol, A. Meijboom, M.L. Eggens, T. Aprheinallt, P.W. Wester, T. VandeZande, A. Bergman, N. Dankers, F. Ariese, R.A. Baan, J.M. Everts, A. Opperhuizen & J.M. Marquenie (1996) Skin and liver diseases induced in flounder (*Platichthys flesus*) after long-term exposure to contaminated sediments in large-scale mesocosms. *Environmental Health Perspectives*, **104**, 1218–1229.

Villac, M.C., D.L. Roelke, E.P. Cahvez, L.A. Cifuentes & G.A. Fryxell (1993) *Pseudonitzschia austriallis frenquelli* and related species from the west coast of the USA – Occurrence and domoic acid production. *Journal of Shellfish Research*, **12**, 457–465.

Wertheimer, A.C., M.J. Bax, A.G. Celewycz, M.G. Carls & J.H. Landingham (1993) Harpacticoid copepod abundance and population structure in Prince William Sound, one year after the *Exxon Valdez* oil spill. In: (eds Rice, S.D., R.B. Spies, D.A. Wolfe & B.A. Wright) *Proceedings of the Exxon Valdez Oil Spill Symposium*. American Fisheries Society Symposium, Number 18, USA, pp. 551–563.

White, I.C. (1999) MAFF, London.

Willette, M. (1993) Impacts of the *Exxon Valdez* oil spill on the migration, growth and survival of juvenile pink salmon in Prince William Sound. In: (eds Rice, S.D., R.B. Spies, D.A. Wolfe & B.A. Wright) *Proceedings of the Exxon Valdez Oil Spill Symposium*. American Fisheries Society Symposium Number 18, USA, pp. 533–550.

Wilson, J.A. & P. Kleban (1992) Practical implications of chaos in fisheries: ecologically adapted management. *Maritime Anthropological Studies*, **5** (1), 67–75.

Zharikova, T.I., N.A. Izyumova & M.A. Stepanova (1995) On features of bream infection by ectoparasites in a number of areas of Gorki-reservoir. *Russian Journal of Ecology*, **26**, 117–120.

Chapter 9
The Role of Modelling in Fish and Fishery Ecotoxicology

A.J. Lawrence

9.1 Introduction

There are many substantial problems in terms of management of European fish stocks, and many of the prime commercial species are in decline as a result of adverse fisheries-dependent factors, most notably excessive fishing effort. The EU has been unsuccessful in limiting fishing capacity and eliminating some of the most wasteful practices such as discarding. Thus, pollution effects impact on a system where the resource base is under threat from fisheries-dependent variables. As a result, there is now a justifiable concern that efforts to preserve fish stocks may be undermined by deleterious changes in the environment. Therefore, sublethal incidents of xenobiotic origin now pose a real fisheries-independent threat to stocks. However, the paradox here is that fisheries-induced stress may be so severe that it masks any other stressors such as pollution.

The previous chapters of this book have identified several underlying tenets of the effects of pollution on fish. Firstly, that the response to pollution at lower levels of biological organisation (cell, individual) are more cause specific than the higher levels. Secondly, that there is a quicker response to pollution at lower levels of organisation. Thirdly, that each level has an inherent ability to withstand or absorb the effects of the stressor such that a progression of impact through each biological level cannot be assumed.

The problem lies in determining the impact of pollution on fish and fisheries at the higher levels of organisation, based on the clear responses seen at the lower levels. No empirical study has yet been performed which demonstrates this clear linkage. Furthermore, because of the destructive nature of some of the techniques employed in ecotoxicology studies, it is currently impossible to show this linkage in any one individual or population. Consequently, some other approach must be adopted and it is here that mathematical modelling has an important role to play in the overall assessment of effects of pollution exposure.

The aim of this chapter is, therefore, to briefly review the types and potential role of modelling in ecotoxicological studies. Before this, a summary of the findings of the earlier chapters of the book will be presented with regard to observed linkages in pollution impacts, and in relation to the conceptual model presented in Chapter 1. Finally, gaps in the scientific literature will be outlined both in relation to the overall assessment of pollution impacts on the hierarchic levels of organisation in fish, and with regard to the use of modelling to resolve these issues.

9.2 Summary of the effects of pollution on fish

The earlier chapters of this book have highlighted the evidence which can be drawn from the current literature to support the mechanistic pathways presented in the conceptual model shown in Chapter 1. These pathways can be used to link pollution impacts at the molecular and subcellular level to those at the population and community level, together with their socio-economic ramifications. Using biomarkers, pollution damage to biological membranes through cell injury can be linked with increased protein breakdown, tissue and organ atrophy, energetic loading, reduced performance and reproductive output. It also seems intuitive to extrapolate from this, unequivocal data on functionally impaired health to a negative impact on the population in terms of lack of recruitment and mortality as a result of impaired reproductive success. However, whilst it can be shown that the mechanisms do occur and that responses can be detected, it is mostly not possible to determine the magnitude of the mechanism/response nor, except in a few cases, the transmission of a response at one level to a response at another.

9.2.1 Cellular and molecular responses

The underlying basis of all stress-induced pathological and physiological change is damage or other change to life processes at the molecular and subcellular levels of organisation. Knowledge of sublethal distress signals has grown rapidly over the last 20 years, often drawing on the reservoir of understanding of these processes in humans and other vertebrates. The main advantage to be gained from the use of biomarkers is that early biochemical and subcellular changes in cells can hopefully be linked to pathological end-points through an integrated multitiered approach. This would not only detect early warning distress signals but would also provide prognostic capability for predicting the likely consequences for the health of individuals in a population. In addition, such molecular and cellular alterations will frequently reflect exposure to particular types of causative agents, thereby offering advantages over non-specific responses at higher biological levels.

Potential biomarkers include alterations in intracellular membranes (endoplasmic reticulum, lysosomes, endosomes, transport vesicles), genotoxicity (DNA adducts, hydrophobic adducts, micronuclei), specific proteins or enzymes (metal binding proteins, stress proteins, oncoproteins, cytochrome P450, multidrug resistance protein) and inhibition of cholinesterase by neurotoxins. Some of these markers are indicative of cell injury and potential damage to health, whilst others are indicative of exposure to certain classes of xenobiotic.

9.2.2 Damage to DNA

Damage to DNA may occur by oxygen radicals, adduct formation or mutagenic chemicals. Adducts potentially lead to mutations and tumour formation although the empirical demonstration of a link between these has been difficult. DNA repair mechanisms can reverse some damage although the efficiency of the process will be affected by physiological factors and life history stage. Direct chemical effects on chromosomes including sister chromatid exchange, and micronucleae production has also been observed. However, there is still limited evidence of quantitative links between damage at the genetic and molecular level and individual health, fecundity and population productivity. This would be most likely through damage to the germ cell line rather than somatic cells.

9.2.3 Physiological responses

Links can be demonstrated between induction of enzymes and proteins employed in detoxification, and protection and increased protein degradation, protein turnover and cellular energetics within an individual. The recognition that reduced protein turnover consistently underlies lower energy expenditure, with benefits including longer survival following pollution impact, may be central to understanding how species survive stressed environments. The link between this trait and multilocus heterozygosity within populations may also be crucial in providing a mechanistic link between physiology and population fitness and survival. However, these studies have primarily taken place using invertebrate models and are yet to be demonstrated in fish.

9.2.4 Immune system responses

A compromised immune system will clearly be detrimental to the individual. However, there are currently no clear links between this and effects on populations. Currently, most information and evidence is available on xenobiotic exposure, cellular processes and pathological effects in the liver. However, evidence for demonstrable links between cellular responses and pathological impacts in other important tissues and organs is surprisingly lacking. Many of these organs, such as the eyes and olfactory system, may be essential in, for example, ensuring reproductive success through the detection of visual or chemical cues. It is only through the detection of these cues that appropriate behavioural responses, leading to reproduction, may ensue.

9.2.5 Reproductive system responses

One of the central processes on which pollutants can impact with consequences for population and yield is the reproductive process. Indeed, some studies have shown that the most important pollution impact on fish stock and fisheries is the loss of reproduction. Pollutants may impact on this system in a number of ways. Through cellular and whole animal energetics, egg size and fecundity may be reduced. DNA damage may impair fertilisation or cause embryo damage. Egg quality and fecundity might also be reduced by the action of endocrine disrupters, and pollutants may affect sperm motility. Fecundity can additionally be reduced by impaired gonad development, reduced spawning ability, reduced egg number and weight. However, it can also be reduced by high mortality of early life history stages such as eggs, larvae and juveniles.

In addition to direct mortality, xenobiotic impacts on larvae are possibly one of the most important life history stages in which to determine pathological effects, through the development of larval aberrations. Whilst various forms of larval abnormality have been described and the proportion of these within the ichthyoplankton can be high, it is currently not possible to present a clear link between larval aberrations and population effects.

9.2.6 Population responses

Despite the number of mechanisms by which reproduction and fecundity can be impacted, and the evidence for reduced individual fecundity as well as the conceptual logic of prediction on the population level, causal links between xenobiotic damage to individuals and population recruitment are difficult to demonstrate. This is one of the most pressing problems in environmental toxicology. Knowledge of the mechanisms of toxicity and the process of cellular injury leading to pathology, disease and reproductive impairment advanced greatly during the 1990s. In contrast, population ecology has not advanced at the same rate in its understanding of environmental influences on population fluctuations.

The problem lies in the effect of unknown population density-dependent factors that may compensate for the loss of early life history stages. Indeed it has been suggested that density-dependent effects may sustain a constant population growth rate despite up to 60% acute mortality, although this has not been demonstrated in fish. Consequently, investigation of density-dependent factors in recruitment dynamics is another of the most important tasks needed to predict population level effects of pollutants as a consequence of a detectable subcellular response.

Whilst evidence of a clear causal link between toxicological effects on individuals and population responses is lacking, several field studies have demonstrated reduced abundance in wild populations most probably caused by pollution-related reductions in recruitment. Consequently, it can be argued that it is not the effect of pollution on individual fish or the consequence to populations and communities that need to be investigated, but the links between the two, possibly leading to biomarkers of imminent population collapse.

One of the ways in which attempts have been made to quantify this link is through the use of modelling. Where these models have been applied, they indicate an impact of pollution on the population level that may be comparable in magnitude to the effect of fishing pressure. However, most of these models are very sensitive to changes in survival estimates of eggs and larvae, and these are the very stages that are difficult to estimate due to spatial and temporal dynamics of the ichthyoplankton.

The importance of density-dependent factors for population responses to pollution also points to the need to consider the whole ecosystem and community together with their ecological relationships. Several ecologists have argued that the single species approach to ecotoxicology is no longer adequate. Species interactions may produce surprising outcomes when several species are exposed to pollutant. For example, there are cases in the literature of fish populations expanding under pollution stress due to the loss of a more sensitive competitor for a specific food resource.

The importance of larval and juvenile survival to population stability and yield has also been highlighted in ecological studies. Juveniles and the youngest age classes are a very important component of the production level in many populations, this being a reflection of the high growth rates in the early life history stages which are characterised by higher $P/\bar{B}$ ratios than larger older fish. Consequently, it has been demonstrated that an increase in fishing mortality or equivalent decrease in early life-stage survival caused by toxic effects of a pollutant would cause stock declines as older and weaker populations exhibit a decrease in reproducing individuals. Furthermore, it has been shown that if pollution-induced mortality occurs after a period of high density-dependent mortality, the decline in yield would be more severe than that caused by an equivalent increase in fishing mortality.

9.2.7 *Population genetic responses*

Homeostatic processes may operate at different levels of biological organisation. Classically, homeostasis is used at the physiological level to denote adjustment to perturbation without a resultant reduction in fitness to survive. However, it can also be used at the cell, population and community level, because each level has the ability to absorb change although, at the population level, there may be a cost in terms of reduced fitness to survive.

Density-dependent mechanisms compensating for reduction in recruitment may provide enough time for a population to adapt to pollutants. Such adaptations have been demonstrated in a number of vertebrate and invertebrate aquatic species. This genotypic adaptation is generally manifested through the development of tolerance to particular pollutants. However, with adaptation comes a cost to tolerance which may impact the future fitness of the population. Additionally, there may be significant survival consequences for populations exhibiting reduced heterozygosity, such as those populations showing tolerance, if impacted by additional environmental stressors.

9.2.8 *Socio-economic response*

At the moment, it is the catastrophic impacts of pollution that are best documented in terms of their socio-economic impact. From these, it has been shown how the cost of pollution can be measured. Furthermore, through negotiation it can be shown how scales of compensation can be agreed, giving links between adverse physiology/fish population effects and the socio-economy of areas dependent on fishing.

The biological effects of pollution can be demonstrated especially at the lower levels of organisation where the effects of pollution, as a reduction in biological fitness or integrity, are apparent. However, at the socio-economic level, it is the perception of the quality as well as, and/or rather than, the defined reduction in quality or quantity which is important. As such, quantifying the effects of pollution (real or perceived) on the fishery is extremely difficult.

9.3 The role of modelling of pollution impacts on fish and fisheries

The conceptual model presented in Chapter 1 provides a simplistic schematic diagram of the principal links between molecular responses to pollution and higher order impacts on physiology, health, reproduction, population and socio-economics. However, the evidence provided to illustrate or support the links in the model has often come from a variety of disparate studies, some on fish but others on invertebrate, human or rodent systems. There are no comprehensive studies which detail all of the aspects and links within the aquatic environment.

Within the conceptual model, the arrows linking each part of the model can be regarded as processes which require quantification. As demonstrated here, it is not yet possible to do this. Consequently, some caution has to be taken when viewing the model in direct relation to fish. Currently, no study has demonstrated all of the impacts and links between the hierarchic levels of response in any one species. However, if it were possible to quantify each of the processes noted in the model then it would be possible to create iterative algorithms which would provide a clear predictive capacity to pollution-based studies. For example, it would be possible to estimate the reduction in a fish stock, or its economic value, based on the induction or increase in expression of an enzyme or detoxification process.

Whilst this is not yet possible, it does highlight the potential role and value of modelling in this field. Indeed, the possible use of models in toxicity studies and ecotoxicology has already been recognised and some progress has been made in this field. A number of models have been developed which examine aspects of pollutant impact at various levels of organisation. These fall mainly into individual-based models, population-based models and ecosystem-based models. Furthermore, models have been developed which try to evaluate the socio-economic impact to a fishery caused by chronic pollution events.

9.3.1 Individual-based models

Individual-based models have been available for some time (Barnthouse, 1992). Using these, toxic impacts on physiology and metabolism of individual organisms have been developed. These have been developed further using fish bioenergetics and growth models (Nisbet *et al.*, 1997; Gamito, 1998). Models have been developed to examine the uptake and accumulation of pollutants. For example, the accumulation and body burden of PCBs in trout (*Salvelinus namaycush*) has been examined using a bioenergetics-based pollutant accumulation model which highlighted the importance of the food chain route of exposure in this fish (Luk & Brockway, 1997). Muller and Nisbet (1996) have taken this further by linking an energy budget model of individual growth and reproduction to a toxicant bioaccumulation model and a model of toxicant action on the individual.

One modelling approach that is proving instructive in risk assessment is the physiologically-based pharmacokinetic models (PBPK). The PBPK model for a chemical comprises a mechanistically-based description of its uptake, distribution and clearance (Haddad *et al.*, 1998). These models are generally used to perform analysis on an individual basis although they can take account of variability in populations. They can include tissue metabolism and protein binding and are, therefore, useful in toxicological risk assessment. They allow examination of the impact of biochemical processes on the pharmacokinetics of a particular chemical. In addition, they can be used to describe the impact of pharmacokinetic factors on the variability of individual risk. For example, they can be used to determine the impact of differences in key metabolism enzymes within a general population (Clewell & Andersen, 1996). These include differences resulting from multiple genotypic expression including cytochrome P450 polymorphisms as well as normal variation in enzyme activities.

Whilst primarily developed to evaluate human risk in relation to drug development, PBPK models are now also being applied to environmental risk assessment. This, however, can and should be developed much further and potentially applied, for example, to fish-based studies. In the meantime, the further development of the models continues to make the adoption of this approach even more relevant. For example, Kim *et al.* (2001) used a PBPK model to examine chemical pharmacokinetics, and hence toxicological risk, when a chemical is present as a low dose over time. This clearly has important ramifications and potential value in ecotoxicology studies of fish that are likely to be exposed to low doses of toxicants over a broad area during their lifetime.

9.3.2 Population-based models

Population-based models have been available for some time and since the mid 1980s these have been linked with toxicity data to assess the impact on populations of sublethal pollution (Barnthouse *et al.*, 1990). In some cases, a reproductive potential index has been used to quantify the response of the modelled population to pollution exposure (Barnthouse *et al.*, 1987). However, as highlighted in Chapter 6, more commonly the population approach has employed age-structured models derived from the Leslie matrix (Barnthouse *et al.*, 1990; Landahl *et al.*, 1997; Johnson *et al.*, 1998). Where these models have been employed, they indicate that sublethal pollution impacts on individual fish could substantially reduce the size of fish populations in contaminated areas if density-dependent effects are weak. With the concern over endocrine disruption in fish and wildlife, a matrix model has recently been used to determine the effects of endocrine disrupting chemicals on populations of fathead minnow (*Pimephales promelas*) and two species of bird (Gleason & Nacci, 2001). This study highlighted the importance of the life history strategy of the species in relation to the population consequence of induction of markers of endocrine effects at the individual level.

Leslie matrix type models were initially employed to assess the likelihood of population survival in a conservation context. Very popular now in the field of conservation biology is the use of population viability analysis (PVA). PVA are simulation type models. These are deterministic-based models built on a matrix type base that also build in the impact of environmental and demographic stochastic events. They are used to estimate the extinction probabilities of endangered species over a period of time determined by the specific definition of a minimum viable population used for the species.

Importantly, PVA can be used to evaluate the impact of current and future threats that endangered species face (Brook *et al.*, 1997). For example, PVA has been used to assess the likelihood of survival of the Florida manatee under various environmental scenarios (Marmontel *et al.*, 1997). It has been linked with power analysis to assess the outcome of various management options on the survival of a population of bottlenose dolphin in Scotland (Thompson *et al.*, 2000), and has been used to assess the potential survival of salmon populations on the west coast of the USA (Ratner *et al.*, 1997).

Hansen and Johnson (1999) argued that ecotoxicology data should be built into PVA. This highlights the need for the integration of the two disciplines at this level. Conservationists could then build ecotoxicology data into PVA to better assess the viability of endangered species, and ecotoxicologists could use models developed for conservation in order to determine the consequence of ecotoxicology data on the species or population of concern.

The use of PVA models could be particularly important for studies involving fish because as with many endangered species, pollution represents an additional threat imposed on a population already under threat from a number of other demographic and environmental stochastic processes, including habitat loss and overexploitation. In one study, pollution data has been incorporated into a PVA. The southern sea otter recovery team developed a PVA which included the probability of the population size of sea otters off the coast of California being affected by large-scale oils spills (Ralls *et al.*, 1996).

Despite their potential, a number of concerns and limitations have been raised regarding the use of matrix-based population models and PVA (Ellner *et al.*, 2002; Kendall & Fox, 2002; Reed *et al.*, 2002). These can be categorised into issues concerning the type of end-point used in the models, the lack of incorporation of genetic-based information into models, the problem of density dependence, the problem of double dipping and the problem of incorporating early life-stage information into the models.

In relation to the type of end-point used in these matrix type models, most are generally based on abundance. However, Power and Power (1995) argue that this is inappropriate for assessing population level risk posed by sublethal pollution. Comparison of an individual-based modelling framework for brook trout (*Salvelinus fontinalis*) with a Leslie matrix approach resulted in critical differences between the two. Power and Power (1995) argued that these occurred because of a lack of population size structure information in the matrix bases approach.

A significant problem with matrix-based models is that they treat all individuals within the population as the same. They take no account of the genetic differences within the population despite the importance of population heterozygosity and genetic stochastic event to population survival. However, the potential importance of genetic differences and their effects on fish population dynamics are now being explored using modelling. For example, a simulation model (FITPOP) has been developed that indicates that subtle differences in genetic and population parameters can have significant implications on population size, fitness and sensitivity (McKenna, 2000).

Genetic issues are also relevant in PVA if, such as inbreeding depression, they affect population persistence. This is particularly important for small and isolated populations or those that have low growth rates (Allendorf & Ryman, 2002). It is possible to account for the loss of vital rates as a consequence of inbreeding depression and loss of heterozygosity. This has been quantified in several studies in relation to juvenile and adult survival and reproduction (Ralls *et al.*, 1988; Jimenez *et al.*, 1994; Balou, 1997; Kendall & Fox, 2002).

The problem of density dependence and its effects on population responses to sublethal pollution impact has been highlighted as a critical factor, making prediction of population level impacts difficult (Chapter 6). In addition, whilst most models can incorporate density-dependent factors, in practice they have not. This may limit their use in assessing the impact of reduced recruitment on populations but they are still useful in identifying the data needed for improved analysis. However, Leslie matrix-based approaches and PVA have the capacity to build in density-dependent effects. The study by Landahl *et al.* (1997), for example, builds in density-dependent effects on English sole (*Pleuronectes vetulus*) populations. However, the model results suggest that density-dependent regulation does not completely compensate for declines in population growth rate.

The issues surrounding density-dependent compensation mechanisms and their consequence to populations, are also being debated in relation to the value PVA models (Reed *et al.*, 2002). However, density dependence and spatial heterogeneity issues can be built into

PVA and where this has been attempted, it has been found that they may considerably influence the estimation of extinction risk (Wiegand *et al.*, 2002).

The issue of double dipping has been raised by Brook (2000) in relation to PVA. Double dipping occurs when basic population demographic parameters are taken from a population and built into a model before adding the impacts of demographic or environmental stochastic processes such as pollution impact. The problem here is that the basic demographic information is often taken from populations already impacted by the environmental process. Consequently, the stochastic impact is built into the basic model and then its impact is assessed on this population. Ultimately, this can lead to overly pessimistic projections of population change (Brook, 2000).

Earlier chapters of this book have highlighted the importance of larval and juvenile survival to population stability. In addition, it has been shown that there can be a high proportion of pollution-induced larval abnormalities seen in the plankton. However, the consequence of this to the future population size and structure is not known. To add to this, Leslie matrix type models are particularly sensitive to changes in survival estimates of larvae. This highlights the importance of good age specific growth and mortality data, especially for early life-stages, on overall population dynamics and model outcomes. The problem here is that it is this data that is most difficult to collect.

However, models are again being developed to examine and describe growth and survival of these vital early stages (Arino *et al.*, 1998). For example, a model was developed to examine the density-dependent or compensation consequences on fish populations that had suffered high human-induced mortality on immature stages (Nisbet *et al.*, 1996). This model indicated that most, but not all, forms of compensation failed to prevent a decline in adult stock as a consequence of enhanced larval mortality. Furthermore, the authors stated that whilst there is evidence for density-dependent adult fecundity in marine fish, there is no evidence that immature stages can compensate for reduced larval density whether through increased growth or survival. Consequently, Nisbet *et al.* (1996) argue that optimistic outcomes in marine fish require density-dependent mechanisms that have not been proven in any marine fish anywhere.

9.3.3 Ecosystem-based models

As already noted, there is growing argument among ecologists to consider the whole ecosystem and community in pollution-related studies. By doing this, it is argued that density-dependent factors and other species-related interactions can be better assessed. Within the field of modelling, ecosystem-based approaches are already being used as an alternative to age structured population-based models. Ecosystem models attempt to quantify material flow and interspecies interactions and are valuable because they show the totality of pollution impact at an appropriate scale (Barnthouse, 1992). In this context, the use of fish has been highlighted as of value in monitoring ecosystem health, at the appropriate scale, because fish are often wide ranging and migratory and consequently the life history stages are exposed to all anthropogenic impacts (Attrill & Depledge, 1997).

Ecosystem models are dynamic and incorporate numerical data obtained from the field. Results are represented as spatial-temporal variations of the basic components of an ecosystem affected by external factors including anthropogenic impacts. Components of the model may be linked hierarchically from processes involving individual components, through interactions between the components to the whole system, incorporating hydrobiological, hydrochemical and hydrodynamic components. Ecosystem models have been integrated with ecoscreening models to give an estimation of the state of an ecosystem (Solovjova, 1999). As a result, the possibility of estimating annual risk for populations of organisms, based on variations in phytoplankton and zooplankton biomass, were demonstrated (Solovjova, 1999).

One of the problems for determining the ecosystem responses to pollution load is related to second level interactions. These are the interactions occurring between species such as competition and predator prey relationships. As noted in Chapter 3, sublethal pollution in an environment does not necessarily lead to a reduction in the population of the species of interest. By releasing the species from competitive or predator prey interactions, it is possible that pollution impacts can result in an increase in population of the species. However, Preston and Snell (2001) have suggested that assessing the toxic effect on population growth of individual species may provide sufficient data to predict toxic effects on species interactions (competition, predator/prey) as well. In their study, models accounting for behavioural toxicity in combination with reproductive toxicity yielded similar quantitative results to those using reproductive toxicity alone. Consequently, it appeared that use of the intrinsic rate of population growth as an end-point is a good predictor of both direct and indirect toxic effects. This is likely to be due to the density dependence of competition and predator prey interactions (Preston & Snell, 2001).

Whilst this only represents evidence from a single study at this stage, the outcome of the work by Preston and Snell (2001) may argue against the use of ecosystem-based models in ecotoxicology. This would also have the additional benefit of reducing the level of data required for good population risk assessments of the species of concern.

9.3.4 New bioeconomic models incorporating sublethal pollution impacts

Clearly, mathematical modelling has an important role to play in an overall assessment of pollution impact on fish and fisheries. As already highlighted, these models are principally separated into individual, population and ecosystem-based approaches, each of which have their own advantages and disadvantages.

However, the scope of this volume goes further to consider the socio-economic consequences of pollution impacts on fish and fisheries. In Chapter 8, the standard bioeconomic model was described and discussed in relation to pollution impacts on fish quality. This was followed by examples in which catastrophic pollution events and their impacts on local fisheries were described. These included economic assessments, which due to the nature of the impact, were relatively easy to quantify in terms of monetary loss.

Whilst catastrophic events are useful in illustrating the socio-economic impact of pollution on fisheries, they do not describe the consequences of sublethal, low dose impacts. Ultimately, this type of impact and socio-economic cost is of more interest because it is both more insidious and more common. However, it is also much more difficult to determine.

Consequently, a fourth modelling approach is beginning to develop in ecotoxicology studies on fish, which examines the economic impact on fisheries of sublethal pollution. For example, Collins *et al.* (1998) have developed a simple bio-economic model of fishery-pollution interaction to assess the impacts of xenobiotics on a hypothetical fishery. The model examined the response of a fishery to pollution-induced reductions in catch rates and the impact of this on the size and composition of the fish population. These were then examined under acute and chronic scenarios.

Under the chronic scenario, a reduction of biomass in all age classes was noted, followed by a reduction of fishing effort in the fishery. This reduction of effort was related to reduced catch per unit effort and profit per vessel. However, as boats withdrew from the fishery, the stock partially recovered. Under the acute pollution scenario the pollution impact caused the fish biomass to fall sharply. However, recovery took place quickly due to both flushing of the contaminants and reduced fish mortality. In this case, profitability recovered and after a few years exceeded the break-even levels, although these gains were only short-lived.

The economic cost was measured as the lost value of the commercial harvest. This was calculated as the cumulative discount loss in resource rent (fish catch and the capital needed to produce it) following the discharge. For a time after a contamination event, the resource rent became negative. This was compounded because over time the value of money changes. Consequently, a discount factor was built into the model. The result of this was that the loss of resource rent after the impact carried greater weight than any temporary surplus earned in the recovery years. As such, there was a loss which was related to the magnitude of the pollution event and the degree to which this affected mortality and fecundity.

An additional important issue raised by this study was that the level of economic damage to a fishery also depended on the way fishing effort reacted. Economic damage was inversely related to response rate. The slower the fishermen switched to an alternative fishery, the longer and more extensive the unprofitability and the greater the social cost. Collins *et al.* (1998) argued that the speed of adjustment depended on the ability to switch to other harvestable stocks. This meant that economic damage to the fishery depended as much on this as it did on the ecotoxicology of the incident.

Whilst a number of critiques are made in relation to the model and some of its assumptions (see Collins *et al.*, 1998), development of approaches like this are likely to be relevant to policy issues and therefore policy makers.

9.3.5 The validity of modelling

Clearly, at the heart of every model are a set of assumptions which ideally should be validated with appropriate testing. Where they are not, they should be highlighted as constraints or issues when considering the outputs of the model.

One of the biggest problems with models is the lack of basic scientific data for specific organisms or systems being tested. As a consequence, data or information is often extrapolated across species or phyla. Indeed, in the earlier chapters of this book where information on mechanistic links between various processes has been absent in the fish literature, it has been drawn from information on invertebrate and mammal systems. However, there is a significant risk in this. For example, Barnthouse *et al.* (1990) note that uncertainties accumulated in extrapolating from toxicity test data to population level effects differ substantially depending on the type of test, and on the taxonomic distance between the tested species and the species of interest for risk assessment.

This concern about extrapolating data from different species or making empirical links using data from different species and groups has been highlighted in one study (Delistraty, 1999). Examining the relationship between cancer slope factor (CSF) and acute toxicity data in rats and salmonid fish, it was shown that whilst a significant correlation existed between CSF and rat LD50 data, no correlation was found in the salmonid data. Furthermore, rat and fish acute toxicity data was not significantly correlated. Consequently, the prediction of carcinogenic potency, based on CSF, was only shown in rats. This may have significant ramifications for fish toxicology and the aim of linking biomolecular responses of pollution with higher order effects. As noted in Chapter 4, whilst neoplasms and lesions in fish are well described, the linkage of these to ROS production or genotoxicity has not been proven and consequently, basic assumption of mechanistic links are yet to be confirmed.

This does not invalidate the value of models or indeed the interpretation of information from them. However, it does highlight the need to remember that, until confirmed empirically via experimentation, the outcomes of models should be considered as predictive. Unfortunately, model results are often accepted as real, without an appropriate level of caution.

In the field of conservation this has lead to the argument that PVA should only be used to evaluate alternative management options (Ellner *et al.*, 2002; Reed *et al.*, 2002). To counter this, Brook *et al.* (2002) argue that conservation biology is a crisis discipline which requires decisions. Consequently, it has to use the best information to hand and the best tools to support the decision-making process. This may include poor data or data from other species where necessary.

Fisheries science, like conservation biology, may also be considered a crisis discipline. However, in relation to ecotoxicology assessment, the potential shortcomings of using poor data, or data from unrelated species, has already been highlighted. This paradox between data and outcome was highlighted eloquently by Schaffer (1994) who noted that 'we live in a model rich but data poor world'. However, one of the additional benefits of using modelling is that the models can often highlight areas where additional data is required. Ultimately, in relation to fish, the problem lies in the fact that despite the vast number of studies which have taken place to examine the impacts of pollution, there are still substantial gaps in understanding the processes or linkages between effects in even the common commercial species.

9.4 Gaps in current understanding

There are many gaps in the current understanding of the processes and links between each of the hierarchic levels of impact of pollution on fish. These gaps occur in the current evidence to support mechanistic links in the hierarchy of response and are also species specific and pollutant specific.

9.4.1 Molecular and cellular response and genotoxicity

In relation to the molecular and subcellular response to pollution and genotoxicity, there is a lack of clear-cut evidence linking pollution-related overproduction of reactive oxygen species (ROS) with direct damage to DNA in fish, and the transferability of results obtained from mammalian cells is uncertain. In addition, the implications of ROS-induced genetic damage for the health of the individual and population of fish are unknown. In particular, it is not known whether oxidative damage to DNA caused by peroxisome proliferators or other xenobiotics can induce hepatocarcinomas in fish. Related to this, there are poorly defined effects of neoplasms on fish physiology, growth and reproduction. Furthermore, there is little evidence to show that increased levels of antioxidant enzymes, ROS scavengers, MT or HSP confer protection against DNA damage in fish tissues or cells.

9.4.2 Molecular and cellular response and physiological processes

In relation to physiological processes, there is some evidence for mechanistic links between subcellular effects and a number of processes. However, one of the important steps in the response hierarchy would appear to be through organism energetics and resource partitioning. Most studies in this field relate to invertebrates and repeat studies need to be conducted on scope for growth, adenylate energy charge and cellular energy allocation in fish. Specifically, reduced scope for growth (SfG) or cellular energy allocation (CEA) needs to be related to impacts on reproduction through reduced fecundity. CEA, whilst only recently developed, would appear to offer an opportunity to develop this work in fish although in this case the procedure needs to be verified.

In addition, many stress responses in fish and their physiological consequence are effected through the hypothalmic-pituitary-interrenal axis through the release of cortisol. This hormone controls many stress response functions from energy mobilisation to immune response and behaviour. The axis itself is controlled by a number of neuroendocrines including seratonin. However, the specific interrelationships between all of the endocrine systems, their control of stress responses and the consequence of overstimulation or suppression of cortisol, need far greater understanding in specific species of concern.

9.4.3 Molecular and cellular response and immune effects

The immune response and fish pathology is still not well understood regarding disease progression. Little is known about the production of tumours as a result of pollution detoxification mechanisms in a quantitative cause-effect link. This is needed in order to use the response in a predictive manner and to allow extrapolation between species and from laboratory to the field. Little is known about how immune effects link into decreased individual growth, reproduction and survival or how natural factors and xenobiotics perturb the metabolism of essential nutrients such as vitamins and trace elements. Except for a few specific cases, the mechanistic links from cellular response to organ dysfunction is limited and general organ pathology is poorly understood. Importantly, there is little or no knowledge of the reversibility of health and pathological effects. Furthermore, baseline information on normal processes such as larval development is often lacking in all but a few species.

9.4.4 Molecular and cellular response and reproduction

Impacts on the reproductive process are central to the hierarchic response to pollution exposure. However, to understand this, basic knowledge of reproductive and endocrine systems in relevant fish species, especially regarding signal transduction and hormone receptor systems, is required. This must extend to information on the effects of disrupting chemicals on neuroendocrine differentiation, function and behaviour as well as the hypothalamic-pituitary-gonadal axis. In addition, knowledge of the period during larval development when specific species are sensitive to sex reversal by oestrogens is required, as is knowledge of whether intersex males suffer any reproductive dysfunction. The deleterious consequence of unscheduled vitellogenesis and zona radiata protein synthesis needs to be examined both in terms of impact on fecundity and on fish physiological energetics. Finally, there are clear gaps in relation to the impact on individual fecundity and overall reproductive output and population stability.

In the majority of studies into larval development, embryo and larval stages are exposed to contaminants for short periods. There are few studies that consider sequential effects of contaminants on gametogenesis, egg release, fertilisation, embryo and larval development and recruitment. In addition, there is an urgent need to develop new techniques to measure biochemical, physiological, histological and morphological effects in embryos.

9.4.5 Population responses

Linking impacts of xenobiotics on the reproductive process or fecundity of an individual to population level effects is crucial and is currently the most difficult to prove experimentally. At this level, there is a great deal of noise in the system and density-dependent processes may act to completely compensate for any pollution impact at the individual level. Unfortunately, what factors constitute the density-dependent mechanism in most situations is poorly understood or defined. A far greater understanding of the processes causing population fluctuation of particular species under natural conditions is required. To overcome the combined problem of pollution impact at the individual level and density-dependent compensation at the community level would require multispecies, multigenerational studies. These have not currently been undertaken, presumably because of the facilities that would be required and the time that would be taken to perform such studies. One possible way of overcoming this would be to use tropical species with short generation times as surrogates for temperate species of concern. However, this would again generate difficulties associated with extrapolation of data between species.

9.4.6 Population genetic responses

Multigenerational studies are also required to examine the effects of chronic contamination on population genetic composition. This would help to determine whether previously stressed and genetically impacted populations are more susceptible to further natural or anthropogenic stressors and establish the utility of population genetic structure as a monitor of impacts in aquatic systems.

There is limited information available on interindividual and interpopulation genetic differences in the response to xenobiotics. This information would be vital for the prediction of population level effects of pollution, and also for the assessment of the scope for adaptation increasing tolerance. As a result, further studies are needed to locate genes related to resistance traits as well as to document mutation rates in identifiable genes under conditions of stress. These studies should relate mutational damage to changes in allele frequencies and the structure of the gene pools to assess the fitness consequences of genotoxin-induced mutation in terms of individual and population level effects.

9.4.7 Socio-economic impact

In relation to bio-economic impacts of pollution on fisheries, this investigation has shown that there is a dichotomy of scale of interaction between the biological/physiological impact and socio-economic outcomes. Biological impacts can be divided in terms of their influence at the macro and microeconomic scale. Impacts of catastrophic events at the macro scale can be established but at the microscale, with sublethal impacts on individual fish this is more complex and the interaction between the two is poorly understood. Food quality standards are now very important and society is responding to quality signatures. Consequently, there is a real meeting point, not currently well understood, where the science of measuring and interpreting pollution impact interacts with the science of food hygiene, quality assurance, health promotion and consumer protection.

9.5 Summary

There are currently no comprehensive studies that detail all aspects and links of pollution impact within the aquatic environment. Within the conceptual model, presented in Chapter 1, the arrows linking each part of the model can be regarded as processes which require quantifying. However, it is not yet possible to quantify these processes or links in any one species.

There are clearly two crucial stages in the hierarchy of pollution responses which require far greater study. The first is the link between the impact on the reproductive process (however this may have occurred), reduced individual fecundity, impacts on larval and juvenile survival and pathology and the consequence of these impacts at a population level. The second is the effect of population density-dependent factors on the population response. This highlights the need for studies at the multispecies ecosystem level over several generations. The problem of linkage at this stage is crucial because it has resulted in a credibility gap for those seeking to make policy decisions based on probable impact on populations and yield. Evidence can be provided to demonstrate pollution impact at each level within an individual but this cannot be extrapolated to population level responses. It is the population response that is of concern to policy makers, together with clear evidence of causation.

As noted earlier, the conceptual model draws on information from fish, invertebrates and other vertebrate systems to develop the mechanistic links between the hierarchic levels of response to pollution. Whilst this is of value in allowing interpretation of processes and allows gaps to be filled for any one group or species, it may also lead to misinterpretation for any one specific group or species. Despite this, some have argued that extrapolation of toxicological and pathological processes across species and phyla is a crucial requirement to link impacts of pollution from molecular to ecosystem using a mechanistic approach to a holistic problem.

The difficulty with this extrapolation is that the life history traits for each species, at the crucial stages identified above, i.e. reproductive processes, recruitment, population demography and interactions, are unique. Extrapolation between species of fish at this level will most likely, therefore, be misleading, extrapolation between fish and other groups such as invertebrates and mammals even more so. Yet it is the impact at this level which is of primary interest to policy makers.

In addition, whilst there is good information on several of the aspects of pollution impact on fish, most of this information is on freshwater species. There is far less information available on marine species and less still on marine species of commercial importance. Consequently, scientists and policy makers will have to decide on how to tackle this problem. They can accept and rely on extrapolation of data from other species, with all of the intrinsic problems associated with this, through the use of mathematical models. Alternatively, funding will have to be directed to baseline studies of fundamental underlying biology as well as pollution studies in particular species of interest.

Management issues often work on the basis of the precautionary principle, i.e. that an effect of pollution will be manifested at some level of biological organisation and that this will reduce the population's fitness for survival. There is the additional assumption with the precautionary principle that a change at one level will be transmitted to another level. However, another important issue highlighted by this report and which must be considered in the context of pollution impacts on population and fishery yield is the capacity for the system to absorb change. Whilst the conceptual model sets out the likely mechanistic links between cellular response and population impact, what it does not highlight are the number of points within the system in which homeostatic feedback processes can work to reduce or compensate for any impact seen at the cellular level.

This book has shown for example the presence of DNA repair mechanisms and the fact that many cellular responses are natural detoxification and repair processes. It is only when these systems are overloaded or damage expressed beyond a certain level that impact at the next hierarchic level might be anticipated. The higher up the hierarchy, the more noise there is in the system and consequently, the more difficult it is to determine a pollution signal. It has been argued, for example, that mortality of fish larvae is so high under natural conditions that it will not matter if a few more die as a result of pollution. However, it could be argued that an increased toll of just a few percent would have dramatic consequences for the relevant stock.

Many of these problems may be overcome with the use of mathematical modelling. However, these models will only be as good as the information and numbers that go into them. This once again highlights the need for far better understanding of the basic population biology of the species, especially for larval and juvenile stages and over the first one or two years. Consequently, whilst it can currently be stated that the conceptual model works at the level of highlighting potential mechanistic links between the hierarchic levels of response, it currently has no predictive capacity at all. This predictive capacity will only be developed when numbers can be attached to some of the arrows linking the separate compartments of the conceptual model.

9.6 References

Allendorf, F.W. & N. Ryman (2002) The role of genetics in population viability analysis. In: (eds Beissinger, S.R. & D.R. McCullough) *Population Viability Analysis*. University of Chicago Press, Chicago, 496 pp.

Arino, O., M.L. Hbid & R.B. de la Parra (1998) A mathematical model of growth of population of fish in the larval stage: density-dependent effects. *Mathematical Biosciences*, **150**, 1–20.

Attrill, M.J. & M.H. Depledge (1997) Community and population indicators of ecosystem health: Targeting links between levels of biological organisation. *Aquatic Toxicology*, **38** (1–3), 183–197.

Ballou, J.D. (1997) Ancestral inbreeding only minimally affects inbreeding depression in mammalian populations. *Journal of Heredity*, **88** (3), 169–178.

Barnthouse, L.W. (1992) The role of models in ecological risk assessment: A 1990s perspective. *Environmental Toxicology and Chemistry*, **11**, 1751–1760.

Barnthouse, L.W., G.W. Suter II, A.E. Rosen & J.J. Beauchamp (1987) Estimating responses of fish populations to toxic contaminants. *Environmental Toxicology and Chemistry*, **6**, 811–824.

Barnthouse, L.W., G.W. Suter II & A.E. Rosen (1990) Risks of toxic contaminants to exploited fish populations: influence of life history, data uncertainty and exploitation intensity. *Environmental Toxicology and Chemistry*, **9**, 297–311.

Brook, B.W. (2000) Pessimistic and optimistic bias in population viability analysis. *Conservation Biology*, **14**, 564–566.

Brook, B.W., L. Lim, R. Harden & R. Frankham (1997) Does population viability analysis software predict the behaviour of real populations? A retrospective study on the Lord Howe Island woodhen *Tricholimnas sylvestris* (Sclater). *Biological Conservation*, **82**, 119–128.

Brook, B.W., M.A. Burgman, H.R. Akcakaya, J.J. O'Grady & R. Frankham (2002) Critiques of PVA ask the wrong questions: throwing the heuristic baby out with the numerical bath water. *Conservation Biology*, **16**, 262–263.

Clewell III, H.J. & M.E. Andersen (1996) Use of physiologically based pharmacokinetic modelling to investigate individual versus population risk. *Toxicology*, **111**, 315–329.

Collins, A., M. Stapleton & D. Whitmarsh (1998) Fishery-pollution interactions: A modelling approach to explore the nature and incidence of economic damages. *Marine Pollution Bulletin*, **36** (3), 211–221.

Delistraty, D. (1999) Relationship between cancer slope factor and acute toxicity in rats and fish. *Human and Ecological Risk Assessment*, **5**, 415–426.

Ellner, S.P., J. Fieberg, D. Ludwig & C. Wilcox (2002) Precision of population viability analysis. *Conservation Biology*, **16**, 258–261.

Gamito, S. (1998) Growth models and their use in ecological modelling: an application to a fish population. *Ecological Modelling*, **113**, 83–94.

Gleason, T.R. & D.E. Nacci (2001) Risk of endocrine-disrupting compounds to wildlife: Extrapolation from effects on individuals to population responses. *Human and Ecological Risk Assessment*, **7**, 1027–1042.

Haddad, S., J. Withey, S. Lapare, F. Law & K. Krishnan (1998) Physiologically-based pharmacokinetic modelling of pyrene in the rat. *Environmental Toxicology and Pharmacology*, **5**, 245–255.

Hansen, L.J. & M.L. Johnson (1999) Conservation and toxicology: Integrating the disciplines. *Conservation Biology*, **13**, 1225–1227.

Jimenez, J.A., K.A. Hughes, G. Alaks, L. Graham & R.C. Lacy (1994) An experimental study of inbreeding depression in a natural habitat. *Science*, **266**, 271–273.

Johnson, L.L., J.T. Landahl, L.A. Kubin, B.H. Horness, M.S. Myers, T.K. Collier & J.E. Stein (1998) Assessing the effects of anthropogenic stressors on Puget Sound flatfish populations. *Journal of Sea Research*, **39** (1–2), 125–137.

Kendall, B.E. & G.A. Fox (2002) Variation among individuals and reduced demographic stochasticity. *Conservation Biology*, **16**, 109–116.

Kim, C.S., J.A. Sandberg, W. Slikker Jr., Z. Binienda, P.M. Schlosser & T.A. Patterson (2001) Quantitative exposure assessment: application of physiologically-based pharmacokinetic (PBPK) modelling of low-dose, long-term exposure of organic acid toxicant in the brain. *Environmental Toxicology and Pharmacology*, **9**, 153–160.

Landahl, J.T., L.L. Johnson, J.E. Stein, T.K. Collier & U. Varanasi (1997) Approaches for determining effects of pollution on fish populations of Puget Sound. *Transactions of the American Fisheries Society*, **126** (3), 519–535.

Luk, G.K. & F. Brockway (1997) Application of a polychlorinated biphenyl bioaccumulation model to Lake Ontario lake trout. *Ecological Modelling*, **101**, 97–111.

Marmontel, M., S.R. Humphrey & T.J. O'Shea (1997) Population viability analysis of the Florida manatee (*Trichechus manatus latirostris*) 1976–1991. *Conservation Biology*, **11**, 467–481.

McKenna Jr., J.E. (2000) FITPOP, a heuristic simulation model of population dynamics and genetics with special reference to fisheries. *Ecological Modelling*, **127**, 81–95.

Muller, E.B. & R.M. Nisbet (1996) Modelling the effects of toxicants on the parameters of dynamic energy budget models. In: (eds Dwyer, J., T. Doane & M. Hinman) *6th Symposium on Environmental Toxicology and Risk Assessment: Modelling and Risk Assessment*. ASTM STP 622. American Society for Testing Materials, Philadelphia, pp. 71–81.

Nisbet, R.M., W.W. Murdock & A. Stewart-Oaten (1996) Consequences for adult fish stocks of human-induced mortality on immatures. In: (eds Schnitt, R.J. & C.W. Osenberg) *Ecological Impacts: Concepts and Application in Coastal Habitats*. Academic Press, UK.

Nisbet, R.M., E.B. Muller, A.J. Brooks & P. Hosseini (1997) Models relating individual and population responses to contaminants. *Environmental Modelling and Assessment*, **2**, 7–12.

Power, M. & G. Power (1995) A modelling framework for analysing anthropogenic stresses on brook trout (*Salvelinus fontinalis*) populations. *Ecological Modelling*, **80**, 171–185.

Preston, B.L. & T.W. Snell (2001) Direct and indirect effects of sublethal toxicant exposure on population dynamics of freshwater rotifers: a modelling approach. *Aquatic Toxicology*, **52**, 87–99.

Ralls, K., J.D. Ballou & A. Templeton (1988) Estimates of lethal equivalents and the cost of inbreeding in mammals. *Conservation Biology*, **2**, 185–193.

Ralls, K., D.P. Demaster & J.A. Estes (1996) Developing a criterion for delisting the southern sea otter under the US Endangered Species Act. *Conservation Biology*, **10**, 1528–1537.

Ratner, S., R. Lande & B.B. Roper (1997) Population viability analysis of spring chinook salmon in the South Umpqua River, Oregon. *Conservation Biology*, **11** (4), 879–889.

Reed, J.M., L.S. Mills, J.B. Dunning Jr., E.S. Menges, K.S. McKelvey, R. Frye, S.R. Beissinger, M.-C. Anstett & P. Miller (2002) Emerging issues in population viability analysis. *Conservation Biology*, **16**, 7–19.

Schaffer, M.L. (1994) Population viability analysis: determining natures share. In: (eds Meffe, G.K. & C.R. Carrol) *Principles of Conservation Biology*. Sinaauer Associates Inc., Sutherland, USA, pp. 195–196.

Solovjova, N.V. (1999) Synthesis of ecosystemic and ecoscreening modelling in solving problems of ecological safety. *Ecological Modelling*, **124**, 1–10.

Thompson, P.M., B. Wilson, K. Grellier & P.S. Hammond (2000) Combining power analysis and population viability analysis to compare traditional and precautionary approaches to conservation of coastal cetaceans. *Conservation Biology*, **14**, 1253–1263.

Weigand, K., K. Henle & S.D. Sarre (2002) Extinction and spatial structure in simulation models. *Conservation Biology*, **16**, 117–128.

Index

Note to index: diagrams and tables have page numbers in **bold**.